Primate Behavior and Human Origins

This comprehensive introduction demonstrates the theoretical perspectives and concepts that are applied to primate behavior, and explores the relevance of non-human primates to understanding human behavior. Using a streamlined and student-friendly taxonomic framework, King provides a thorough overview of the primate order. The chapters cover common features and diversity, and touch on ecology, sociality, life history, and cognition. Text boxes are included throughout the discussion featuring additional topics and more sophisticated taxonomy. The book contains a wealth of illustrations, and further resources to support teaching and learning are available via a companion website. Written in an engaging and approachable style, this is an invaluable resource for students of primate behavior as well as human evolution.

Glenn E. King is Professor Emeritus of the Department of History and Anthropology at Monmouth University, USA.

Primate Behavior and Human Origins

Glenn E. King

Routledge
Taylor & Francis Group
LONDON AND NEW YORK

First published 2016
by Routledge
2 Park Square, Milton Park, Abingdon, Oxon OX14 4RN

and by Routledge
711 Third Avenue, New York, NY 10017

Routledge is an imprint of the Taylor & Francis Group, an informa business

British Library Cataloguing-in-Publication Data
A catalogue record for this book is available from the British Library

Library of Congress Cataloging-in-Publication Data
King, Glenn E.
Primate behavior and human origins / Glenn E. King.
pages cm
1. Primates--Behavior. I. Title.
QL737.P9K535 2016
599.815--dc23
2015017537

ISBN: 978-1-138-85316-4 (hbk)
ISBN: 978-1-138-85317-1 (pbk)
ISBN: 978-1-315-72300-6 (ebk)

Typeset in Goudy
by Saxon Graphics Ltd, Derby

To the people who have meant the most to me –

In my professional life
Sherwood L. Washburn
William A. Lessa
William P. Mitchell

and in my personal life
Barbara Barrett
Cynthia King
David King
Kimberly Spurgeon
Edith King

Contents

Figures

Preface

In graduate school primate behavior lured me away from other fascinating specialties and actual experiences did nothing to change that feeling. The first was a sweltering summer in New Mexico, observing a colony of captive chimpanzees that impressed me with their intelligence, strong personalities, and complex relationships. Next, I studied three kinds of baboons (a loose term for certain large monkeys) in the Los Angeles Zoo. The fog of rising at dawn began to lift when Billy, the adult male mandrill, gave me the seemingly fearsome snarl that is really a greeting in his species (Figure 0.1).

Figure 0.1 A male mandrill "grins"—a friendly expression. © abzerit via iStock

In Africa I observed baboons in their natural setting. Expected behavior patterns were spiced by unusual occurrences, such as a male baboon lifting an infant over his head in a movement reminiscent of a human father's play. Ordinary events sometimes assumed unexpected dimensions—the simultaneous crunching of crisp fruits by 30 or 40 baboons became hilarious. And there were close encounters like those at the Serengeti Lodge, where a vervet strolled into the restaurant and a baboon tried to become my roommate.

Primate studies also bring you into contact with interesting people, many of them nice. I was swindled in Nairobi, but invited to a dinner of roast goat by a chance acquaintance. I was bawled out by a pompous minor official in a Tanzanian park, but invited to share meat ("Karibu chakula") with a hospitable game warden. I was frustrated by elusive baboons in South Africa, but comforted by one of the best cooks and one of the kindest couples I have ever met. And I learned that two people can communicate about important things despite rudimentary knowledge of one another's languages.

Acknowledgements

William A. Lessa provided a floundering new graduate student at UCLA with support and direction, nurturing a continuing interest in cultural anthropology. Sherwood L. Washburn opened a new world of primate behavior in relation to human evolution. He guided me to my doctorate at Berkeley, along with Phyllis Dolhinow. William P. Mitchell escorted me through the intricacies of professorship at Monmouth University and became a good friend along the way.

Dr. John Cavallo and Dr. Rob Blumenschine were essential to my first field trip to Africa. I'm also grateful to the good people of Kenya and Tanzania, who facilitated my work and made my stay pleasant, especially my driver Daniel. When problems arose in my work in South Africa, my troubled soul was soothed by the kindness of Jesse Felthuis, manager and chef extraordinaire at the Makhasa Game Reserve, and by the very generous help and hospitality of Xander and Susan Combrink at the Mkhuze Reserve. Meeting and talking with Zulu people at Makhasa was a great pleasure.

Many people encouraged me in the long trek to this book. Foremost is my fabulous wife Barbara Barrett, who was always there with whatever I needed most, whether it was techinal advice, a loving word or a snarky remark. I'm also grateful to Cynthia King, Al King, Jack Bufton, Mary Carol Stunkel, and many others at Monmouth University in New Jersey and St. Mark's Episcopal Church in Highland, Maryland. Special thanks go to the Department of History and Anthropology at Monmouth and to Dr. Kenneth Stunkel, who again provided both moral and immoral support.

Brian Barrett made valuable comments on parts of the manuscript. Lin Gaylord midwifed the project. Katherine Ong and Lola Harre of Routledge guided it to maturity with cheerful skill and great patience. It has been a great pleasure working with them.

An undertaking like this one requires ready access to an enormous and varied professional literature. I am deeply grateful to the following individuals for making the resources of the Monmouth University Library available to me: Dr. Thomas Pearson, Provost; Dr. Ravindra Sharma, Dean of the Library; and Grey Dimenna, Esq., University Attorney.

I'm very thankful to all the colleagues who have made their work available via mailed reprints in the past and online in more recent times. The personal and laboratory websites of colleagues have been enormously helpful and ResearchGate and Academia.edu. have also been of great value. For images in this book I want to acknowledge the generosity of Yvonne de Jong and Tom Butynski; Rhett Butler and Mongabay, and with special thanks, Stacey Tecot, Eric Matthews, Chimpanzee Sanctuary Northwest, and Curt Busse.

Introduction

A Book about Humans and Other Primates

This book will introduce the reader to our closest living relatives. Monkeys and apes are the most familiar members of a large group that includes lemurs, lorises, and tarsiers. Primates are fascinating animals that are worthy of study in their own right. Nevertheless, due to shared ancestry, we have many physical and behavioral features in common with other primates and this is also a legitimate focus of scientific inquiry. Information about other primates is widely used to shed light on human evolution and modern human characteristics.

This text is unusual in the extent to which it combines the study of nonhuman primate behavior with applications of that study to understanding our ancestors and ourselves. Probably every introduction to primate behavior gives a nod to the subject's anthropological relevance, but this book makes it an important theme. All major groups of primates are covered, but there is an overall emphasis on the ones that are more relevant to understanding humans and their ancestry.

Organization of the Book

The first chapter introduces the primates in a general way. The second surveys methods, concepts, and general theories used in the study of primate behavior. Chapter 3 describes behavior patterns that occur in many or most of the species and introduces additional theoretical perspectives. The rest of the book covers important features of major subgroups with emphasis on relatively distinctive traits. Each chapter discusses theoretical perspectives that are helpful for understanding the described behavior.

This survey of nonhuman primates leads up to a review of how the data are used to better understand recent humans and their prehistoric ancestors. Three chapters deal with the last common ancestor shared by humans and nonhuman primates, the further evolution of our early ancestors, and the continuing influence of this evolutionary heritage on the behavior of recent and contemporary humans. These chapters give critical consideration to relevant primate models. The text follows the convention of using the word *primate* alone to stand for nonhuman primates. Where humans are included in a statement, this is indicated where necessary by phrases such as *humans and other primates* or *primates, including humans*.

The final chapter attempts an overview of primate conservation, including common problems and some distinctive issues arising from primate diversity. I hope that concern for the conservation of primates will have been heightened by contact with their fascinating behavior and their relevance to understanding ourselves.

Reading the Text

The descriptions of primate behavior are arranged under four headings: Ecology, Sociality, Life Cycle, and Cognition (Table 0.1 summarizes the definitions for easy reference). This will facilitate comparison of the various primates described in the book. *Ecology* encompasses all the relationships between a kind of animal and everything in its environment. *Sociality* refers to interactions among members of a group and the group structure that this behavior represents. The *Life Cycle* is the typical path of an individual's development from infancy to death. Each stage has its distinctive behaviors and relationships. *Cognition* is used broadly here to include all mental capabilities and processes, ranging from intelligence to emotion. Social traditions are based on mental capabilities, so they are considered under this heading.

Table 0.1 Categories of Behavior

Ecology	Relationships between a kind of animal and relevant features of its environment.
Sociality	Interactions among group members and the group structure that this behavior represents.
Life Cycle	Developmental stages of the individual from birth to death.
Cognition	Mental capabilities and processes, including transmission of traditions.

In comparative studies it is important to have some idea of how widespread behavioral patterns are. Several non-technical terms, as defined for this book, will be used to provide a rough idea of trait distributions (Table 0.2 summarizes the definitions for easy reference). This is intended to provide a feeling for the frequency of the behavior patterns without using misleadingly precise formulations. *Ubiquitous* behavior patterns are defined here as being found in virtually all well-studied species of a particular taxon. This term is used instead of *universal*, because it is impossible to establish that there are no exceptions to a widespread pattern. *Typical* behavior patterns are widely distributed throughout a taxon, but with a significant number of exceptions. A *recurrent* behavior pattern occurs several times in a taxon without being frequent enough to be called typical. Because it is often difficult to distinguish typical from recurrent behavior patterns with available evidence, the word *common* is used for a behavior pattern that is at least recurrent and possibly typical.

Table 0.2 Terminology for Frequency of Behavior Patterns

Ubiquitous	Found in virtually all well-studied species of a taxon.
Typical	Widely distributed in a taxon, but with numerous exceptions.
Recurrent	Appearing several times in a taxon, but not typical.
Common	Recurrent and possibly typical (data insufficient to distinguish).

Textbooks often present research data in their original precise form, e.g. "the average infant weighed 453.5 grams". This gives a misleading impression of exactness because the average from one study will not be exactly the same as the average in another. Most numbers in this book are rounded off. The infants cited above would be described as weighing "about 450 grams".

The most important terms in the text are emphasized in two ways. Those that are unfamiliar or that have a technical meaning are given in boldface the first time that they appear in the text (e.g. **catarrhine**). These are listed in the Glossary. The importance of some well-known terms is emphasized with the conventional use of italics (e.g. *only*). These do not appear in the Glossary.

Saving Space

This section describes several measures that have been taken to allow enough space for relatively comprehensive coverage of the subject. First, there are bibliographic liberties. The sources for each chapter have been divided into **Basic Bibliography** and **Additional References**. The Basic Bibliography consists of general works that provide fundamental information and concepts that are agreed upon by most experts. These are rarely given specific reference within the text. Additional References (mostly journal articles) are indicated in the text in parentheses (e.g. Jones 2013). These serve several purposes. Some augment the Basic Bibliography by providing additional information or updates. Others represent divergent opinions on controversial issues. Still others provide interesting sidelights, such as unique observations.

Many individual references have been shortened. Most of the journal titles have been abbreviated and Table 0.3 provides a key to those abbreviations. Book publishers have also been abbreviated and Table 0.4 provides a key. In any instance where a book has more than two editors or a paper has more than two authors, the additional individuals are indicated by 'et al.'. All sources can be found in their complete form at the Routledge companion website for this book [link here].

Table 0.3 Abbreviations for Journal Titles

AA	American Anthropologist
AJPA	American Journal of Physical Anthropology
AJP	American Journal of Primatology
AR	Anatomical Record
AB	Animal Behaviour
AC	Animal Cognition
AABS	Applied Animal Behaviour Science
BE	Behavioural Ecology
BE&S	Behavioral Ecology and Sociobiology
BP	Behavioural Processes
BL	Biology Letters
BMCEB	BMC Evolutionary Biology
CB	Current Biology
ESR	Endangered Species Research
EHB	Evolution and Human Behavior
EA	Evolutionary Anthropology
FP	Folia Primatologica
H&B	Hormones and Behavior
ICB	Integrative and Comparative Biology
IJCP	International Journal of Comparative Psychology
IJP	International Journal of Primatology
JAAWS	Journal of Applied Animal Welfare Science
JCP	Journal of Comparative Psychology
JEPG	Journal of Experimental Psychology: General
JZL	Journal of Zoology London
PTRSB	Philosophical Transactions of the Royal Society B, Biological Sciences
PRSB	Proceedings of the Royal Society B, Biological Sciences
YPA	Yearbook of Physical Anthropology

Table 0.4 Abbreviations for Book Publishers

California: UCP	University of California Press
Chicago: UCP	University of Chicago Press
CUP	Cambridge University Press
HUP	Harvard University Press
OUP	Oxford University Press

Beyond this Introduction, all Tables have been placed on the website to allow ample space for comparative data and the opportunity to update them as more information becomes available in the rapidly growing and changing field of primatology.

Companion Website

Additional resources to accompany the book are available at: **www.routledge.com/cw/king**

The website also provides the following lists as study aids for each chapter:

1. Study questions to guide reading.
2. Technical terms that were introduced in the chapter.
3. The most important hypotheses that have been introduced in the chapter. This includes all theoretical constructs, whether they have been designated as hypotheses, theories, models, or something else.
4. Additional readings with annotations.
5. Links to other websites that relate to the content of the chapter.
6. DVDs and online videos that relate to the chapter.

DVDs and online videos are briefly described on the website. Some are the direct product of researchers, but others are commercialized to varying degrees. Some of the latter are childish or exploitive, but many provide sound scientific information. There is one pervasive problem with the commercialized productions (many taken directly from TV shows), which is that even the best may be heavily edited to produce melodrama. One staple is the death of an infant or juvenile, which led one of my students to say in exasperation, "Why don't you just show us *Old Yeller*?" Of course, deaths do occur and this must be faced in order to achieve an understanding of the subjects. However, a balanced perspective is necessary.

Zoos are another obvious resource for the study of primate behavior. The animals can be seen alive and unedited and students can take notes on their behavior. However, zoo primates should be considered with caution because living conditions may drastically alter their behavior, varying from minimally modified natural behavior to severe pathologies. Fortunately for the animals and us, disturbed behavior has become less frequent and less intense as zoo conditions have improved. Some zoos with good primate collections are listed on the website.

1 The Primates

Meet Your Relatives

What's that thing? A monkey? An ape? It looks like my grandpa.

There are several hundred species of primates in the world today. Together they constitute what zoologists call an **order** of animals, which is a group with a common ancestor dating back tens of millions of years. This chapter surveys the primates in terms of their biological features, evolutionary origins, and main subgroups. The figures in this chapter and the next will illustrate the diversity of the primates.

Primate Biology and Behavior

Anatomical traits facilitate some behaviors and limit others. Wings make flying possible for most birds, but the reduced wings of ostriches show that flight is no longer possible for them. Physiology is also crucial: hormonal secretions (e.g. testosterone) and neurological processes shape behavior and vice versa. Biological traits raise issues of **function** and behavior, i.e. how the traits work and what the results are. For instance, birds must flap their wings in order to fly. Complex neurophysiological processes underlie the motivation to eat that we call hunger.

Common Primate Traits

Primate characteristics are illuminated by knowledge of the order's mammalian heritage. Some of these characteristics are shared, such as internal temperature regulation (warm bloodedness), live birth, and milk secretion. Other primate features contrast with those of most other mammals (Box 1.1). Some of these have been retained while other mammals changed. For instance, the five digits on each hand and foot contrast with specializations such as paws and hooves. Some distinctive primate traits are unusual innovations, such as the location of the eyes.

Compared to most other mammals, primate eyes are located closer together and toward the front of the head. This produces **binocular vision**, the ability to fix both eyes on the same thing simultaneously (Figure 1.1). The result is depth perception: getting two views of the same object allows the brain to calculate the distance to that object. Primates as a group have the largest binocular visual fields among mammals (Heesy 2004). Large eyeballs also enhance primate vision, and the visual centers in the brain are correspondingly large and complex.

Box 1.1 Distinctive Primate Traits

This table summarizes important features that are common among primates and relatively rare among other mammals. The broadly functional categories emphasize connections with behavior.

Grasping

- mobile digits, especially first digits of hand and foot
- flattened nails on some or all digits
- friction surfaces on tips of digits
- sensitive nerve endings on the surfaces of appendages.

Locomotion

- long, muscular legs
- hindlimb dominance (for both braking and acceleration).

Senses

- orbital convergence (for binocular vision)
- postorbital bone structure (protection of the eye).

Dentition

- four types of teeth: incisors, canines, premolars, molars.

Figure 1.1 A greater galago focuses its binocular vision. © Y.A. de Jong and T.M. Butynski, www.wildsolutions.nl

The **orbits** are the bony sockets that hold the eyes, and orbital convergence is a term for the relative closeness of primate eyes. Primate orbits are reinforced by extra bone at the rear that replaces the tough fiber found in most other mammals. This orbital closure creates a complete ring of bone around each eye and functions as protection against external dangers and/or unusual chewing stresses. Eye damage and loss are more critical for primates than for many other animals.

Primates are largely **arboreal**, meaning that they conduct most or all of their activities in trees. Their typical climbing pattern is distinctive. Rather than digging in with claws, they grasp with flexible extremities that can be called hands and feet rather than paws (Figure 1.2). Flat nails in place of claws support the fingers and toes. The five digits on each hand and foot (called **pentadactyly**) can be spread to facilitate gripping. Grip is further enhanced by friction surfaces with curved ridges on the digits and on the palms and soles. The term **dermatoglyphics** can be used for both the skin patterns and the study of them. Nerve endings in these surfaces provide feedback on how the appendages are interacting with the environment.

The **hallux** (the same digit as the big toe in humans) is widely divergent in virtually all primates and is capable of a pincer-like grasp more powerful than that of the hands (Figure 1.3). In many primates the thumb is opposable to some extent: it diverges from the other digits and can rotate to make some degree of contact with them. Primates use their flexible hands and feet to grasp all sorts of objects, from branches to food items.

Primate limbs are also flexible. Paired bones in the forearms and lower legs allow a higher degree of rotation than in other mammals. The collarbone (absent from many other mammals) is a strut that allows the arm to rotate into a variety of positions. Climbing creates a tendency toward vertical posture, which also distinguishes primates from the majority of mammals.

Figure 1.2 A red-bellied lemur demonstrates its grasping ability. © Stacey Tecot

Figure 1.3 A chimpanzee shows its opposable hallux, the primate equivalent of the human big toe. © Curt Busse

Primates appear to resemble other mammals in being **quadrupedal**, broadly defined as walking on four limbs on a relatively level surface (Figure 1.4). However, many components of primate quadrupedalism are especially suited to arboreal movement (Schmitt 2010). During each step the animal bends its arms and legs. This **flexion** reduces stress on the limbs (Larney and Larson 2004). It results in a crouching posture when walking on branches, which lowers the center of mass and enhances balance. Features of the primate hindlimbs and ankles facilitate leaping.

The primate forelimb is used for foraging and other manipulations, as well as diversified **locomotion** (any movement from one place to another). This requires mobility at the expense of stability, which makes the forelimb more vulnerable to stresses. Limb flexion alleviates some stress, but **hindlimb dominance** is also a factor: while most mammals place the greater part of their weight on the forelimbs, the distribution in most primates is roughly equal or is greater on the hindlimbs (Hanna et al. 2006). During arboreal movement the rear legs provide propulsion and a resting foundation, while the forelimbs provide guidance (Schmitt 2010).

Primates can process diverse foods because they have retained four kinds of teeth from their mammalian ancestors. The front teeth perform a variety of grasping functions. Incisors, relatively thin and flat, are used for nipping. The pointed canines can be used for fighting as well as food processing actions such as stripping bark. Premolars and molars (cheek teeth) are essential for chewing. Each has protruding structures, typically three or four, called cusps. Cheek teeth can pierce food with the cusps, shear with crests that link the cusps, and crush or grind between the occlusal surfaces. The pattern of molar cusps and their low profile in primates enhance crushing at the expense of other actions. This allows primates to obtain greater nutritional benefits from hard or tough foods such as nuts, seeds, roots, and insects with hard exoskeletons.

Figure 1.4 A captive monkey on a net displays the flexed limbs and crouching posture of arboreal locomotion. © Jaydenwong via iStock

The number of teeth in each category varies among primates. The typical array for a particular group is expressed by a **dental formula** that represents one side of the mouth, upper and lower. For example, the human dental formula is 2.1.2.3 / 2.1.2.3. Each quadrant of the jaw contains two incisors, one canine, two premolars, and three molars. Although some human individuals never grow third molars, this simply exemplifies the variation that exists in virtually every biological trait.

Primate Evolution and Classification

Evolution, as Darwin wrote, is descent with modification: the biological features of a group of organisms undergo hereditary changes across generations. There are several causes of evolutionary change but we will be concerned with **natural selection**, a process in which traits that favor successful reproduction become more common.

Evolution and Natural Selection

In modern terms evolution is a process of change in the characteristics of a breeding population, a group of organisms capable of reproductive interaction. Visible changes are the result of shifts in the **gene pool**, which is the totality of genes that are circulating among the members of the population, from one generation to the next. Evolution affects anatomy, which shapes and limits behavior, and affects behavior more directly through changes in the brain and the endocrine system.

Natural selection is an evolutionary process in which some genes become more or less common over generations because the traits that they influence affect reproductive success. Individuals who have more offspring than others in the same population because of genetic differences will transmit more genes to future generations, with the proviso that their offspring are **viable** (i.e. they can also reproduce). As a result of this process, the genes in question become more common and may ultimately be found in all members of the population.

Natural selection takes place in varied environments, so it usually favors traits that result in survival and reproduction in a certain kind of animal in a certain setting. An angry rabbit that fights a bobcat will probably do this only once. Rabbits that become frightened and run away are more likely to survive and reproduce. Those that start sooner and run faster are the ones most favored by natural selection. They transmit the genes underlying their anatomy and behavior to the next generation. Contrary to a widespread misunderstanding, it is the members of the same population (rabbits in this case) who compete for reproductive success. Predators (e.g. bobcats) are not competitors—they are part of the environment.

Evidence for evolution includes **fossils**, the physical remains of living things. Most of them are bones or teeth preserved by mineralization. However, fossils can also be other signs of the past, such as footprints. The fossil record has many gaps, so comparison of related living species has been used to infer evolutionary events. Since our main concern is the behavior of living primates, we will make little use of the fossil record throughout most of this book. Fossils will become more important when we look at reconstructions of our ancestors' behavior in Chapters 27 and 28.

Molecular primatology (Di Fiore et al. 2011), the comparative study of biomolecules such as DNA, makes vital contributions to our understanding of relationships between individuals, populations, and species. This approach to genetics illuminates mating systems, social structure, kinship, and emigration. It is essential to the reconstruction of population histories and phylogenetic relationships. Many consider molecular evidence to be superior to morphology for this purpose, because alternate forms are more clearly defined. One of the most important aspects of molecular studies is that they can provide dates for evolutionary events based on (assumed) constant rates of mutational change in the biomolecules of organisms.

In addition to DNA, chromosomes have long provided a molecular basis for inferences about evolution and behavior, and new techniques are making them more valuable. Increasing use is being made of RNA, enzymes, and proteins in cell membranes. **Genomics** goes beyond older molecular approaches, being the study of the complete DNA sequence of an organism. It promises more complete and sophisticated comparisons.

Classification

Biological **classification** is the ordering of living things into groups on the basis of their relatedness (Groves 2001). **Taxonomy** is the study and practice of classification. Groups that are formally classified according to these rules are called **taxa** (singular **taxon**). The definition of relatedness among taxa is the subject of some controversy, but most primatologists (and many other scientists) are interested in taxa based on **phylogeny**—that is, ancestor–descendant relationships. More concisely, a taxon is a group of organisms thought to have a common ancestor. This criterion is called **monophyly**, and a monophyletic group is called a **clade**.

Species in a clade share many traits due to common origins. If such traits have undergone little evolutionary change, they are described as **ancestral** or **primitive**. Such traits may have originated before the clade itself appeared (for example, the hairy integument that primates share with other mammalian orders). **Shared derived characters** identify a clade. This means that the common traits have changed from the condition of earlier ancestors, such as the differentiation of primate binocular vision from the visual apparatus of earlier mammals. Much taxonomic work involves distinguishing derived from ancestral features.

Classification is generally considered to be hierarchical in that each taxon can be subdivided into smaller ones (or conversely, nested within larger ones). The Order Primata, for example, can be divided into *suborders* and/or *infraorders*. Each of these can be divided into *superfamilies*, and so on. This raises the question of the level at which a given taxon should be placed. Levels are often identified by distinctive endings to taxonomic names (e.g. *-oidea* for a superfamily). If taxonomists change the level, the name changes (e.g. *-oidea* becomes *-idae* for the family level).

One criterion, influentially advocated by Groves (2001), is that the level of classification should reflect the age of the clade. However, this means that the name may fluctuate with new evidence and interpretations regarding the age of the common ancestor. It also means that those who reject the age criterion may use different terminology. This kind of problem applies at every level of classification. Another taxonomic issue is the identification and naming of new taxa within a recognized one, e.g. creating several new species out of a single one.

Technical controversies about the grouping and naming of taxa can be very confusing to non-specialists. According to Groves (2001), this is how it must be. However, non-taxonomists hope for restraint and compromise (Asher and Helgen 2010; Bruner 2013). Untrammeled taxonomic change is a barrier to communication. At the foundation of this book is a simplified classification of living primates, which is intended to facilitate comparative learning about primate behavior (Box 1.2). It uses informal terms wherever such terms are available and unambiguous. Informal terms are likely to stay the same while formal terminology changes. For example, "Old World monkeys" (the monkeys of Asia and Africa) are the same group of species no matter what taxonomic level they are placed at or name they are given. However, since some contact with technical classification is unavoidable, particular issues and problems will be discussed in boxes throughout the book.

Box 1.2 A Simplified Classification of Extant Primates

This classification is based on Fleagle (2013), but is simplified to facilitate communication about behavior. It is intended to provide a framework for studying, understanding, and remembering broad patterns of primate behavior. The words in parentheses are informal terms that are used in the text. This classification differs from Fleagle in what is omitted (see website for a complete taxonomy). The omitted levels of classification will be discussed in the appropriate chapters later in the book.

SUBORDER STREPSIRRHINI (strepsirrhines)
- Infraorder *Lorisiformes* (lorisiforms)
- Infraorder *Lemuriformes* (lemuriforms)

SUBORDER TARSIIFORMES (tarsiers)
- Infraorder *Tarsiiformes* (tarsiers)

SUBORDER ANTHROPOIDEA (anthropoids)
Infraorder *Platyrrhini* (platyrrhines, i.e. New World monkeys)
Infraorder *Catarrhini* (catarrhines, i.e. Old World monkeys and apes)

The underlined taxa represent the three major clades introduced in this chapter. They have been formally recognized by experts other than Fleagle, but placed at different taxonomic levels with different labels, including "semiorder" (e.g. Masters et al. 2013).

Fleagle's suborders and infraorders provide the basic framework for this book, as will be seen in the arrangement of chapters. Lower level taxa are also used where they facilitate understanding of important variation in primate behavior. The relationships of this scheme to formal taxonomies (and the controversies they entail) will be discussed in boxes throughout the book.

Major Primate Taxa

Living primates can be divided into three major clades (or lineages) that emerged near the origin of the primate order and have continued to the present day (Beard 2013). In our informal terminology these are the strepsirrhines, tarsiers, and anthropoids. Tarsiers are sometimes treated as a **grade**, i.e. a notional intermediate stage between strepsirrhines and anthropoids. However, the fossil, molecular, and behavioral evidence all indicate that living tarsiers and their ancestors form a clade that reaches back almost to the origin of the primates. That is not changed by the fact that they form a larger clade with the anthropoids (called the haplorrhines) (Beard 2013). We are most interested in the distinctive behavioral outcomes of tarsier evolution. In addition to primitive features shared with strepsirrhines and derived traits shared with anthropoids, "tarsiers have many distinctive features all their own" (Fleagle 2013).

Strepsirrhines

Strepsirrhines (Figures 1.1 and 1.2) differ from other primates in several traits that they share with non-primate mammals. The majority of species are nocturnal, which means they are active mainly at night. The concomitant sensory equipment makes the strepsirrhine face resemble those of mammals such as dogs. The short snout ends in a **rhinarium**, a hairless and moist area around the nostrils that takes scent molecules from the air. Other information comes from prominent ears and from **vibrissae** ("whiskers") at the sides of the face. A bone called the postorbital bar protects each eye. This strut, probably an ancestral primate trait, bridges the gap in the rear of the orbit but does not form a complete socket like that of a human.

Strepsirrhines also display highly distinctive derived traits. The **toothcomb** is a dental structure in the lower jaw composed of incisors and canines that protrude at an angle. A strepsirrhine sometimes grooms itself by running the toothcomb through its fur. There is a hardened structure under the tongue (called the sublingula) that is used to remove the resulting debris from the toothcomb. The so-called **grooming claw** on the second toe of each foot is not really a claw like those of other mammals, because it is not sharp or bilaterally compressed. Rather, the strepsirrhine structures are elongated and curved nails that stand at an angle to the dorsal surfaces of the toes rather than flat on top of them.

Strepsirrhine eyes glow brightly when light strikes them in the darkness because the **tapetum lucidum**, a layer of tissue at the rear of the retina, reflects light back toward the pupil. This recycling effect allows the animal to see more in the dark. Other nocturnal mammals also have such structures, but the primate tapetum lucidum functions with different chemicals. This indicates a separate evolutionary origin.

Tarsiers

Living tarsiers are nocturnal, but lack the sensory equipment of strepsirrhines (Figure 1.5). Tarsier eyes glow in the dark, but with a dull orange color. Instead of a tapetum lucidum the eye has a **fovea**, a specialized section of the retina in which light-sensitive cells are closely packed. This intensifies images and gives tarsiers the most acute low-light vision among primates (Jablonski 2003). Each of the tarsier's eyes is larger than the animal's brain, which results in cranial proportions that are unique among the primates. The eye is protected by a postorbital flange, which covers more space than the strepsirrhine bar.

Figure 1.5 A tarsier in vertical clinging position. © Elecstasy via iStock

Tarsiers are less dependent on smell than strepsirrhines and lack the naked rhinarium and the philtrum. Instead, hair encroaches closely around the nostrils. Vibrissae are absent. Tarsier hands are proportionally longer than those of any other living primate, suitable for grasping prey and thin vertical substrates. There are large grooming claws on the second and third toes of each foot, contrasting with the single claw on each foot among strepsirrhines.

Anthropoids

Anthropoids are the most numerous of extant primates. Monkeys, apes, and humans are the living members. Monkeys are quadrupedal primates with (in almost all species) tails, found in the Americas and the Old World (Figure 1.4); apes are long-armed and tailless inhabitants of Africa and Asia (Figure 1.6).

Figure 1.6 A male chimpanzee illustrates ape characteristics: broad chest, long arms, and short legs. © Eric C. Matthews

Most anthropoids are larger than other primates and have relatively shorter trunks. Other than some species in one genus, all living anthropoids are diurnal. Like tarsiers (whose ancestors may have been diurnal), anthropoids have a fovea that intensifies images. Their visual acuity exceeds that of all other mammals and is matched only by birds of prey. Some form of color vision is also typical of anthropoids. Each eye is supported and protected by a **postorbital septum** that forms a complete bony cup behind the eyeball. As in tarsiers, the tapetum lucidum is absent and facial features indicate a shift away from olfaction. The anthropoid nose is dry and often covered with hair.

Summary

The Order Primata is composed of strepsirrhines, tarsiers, and anthropoids. The last lineage contains monkeys, apes, and humans. Primates have diverged into many species, so classification is essential to understanding their evolutionary relationships and doing comparative studies of behavior. The mammalian heritage of the primates is important, but the order also displays distinctive and unusual physical and behavioral traits. Primates climb by grasping. Binocular vision gives them depth perception, and diverse teeth facilitate omnivory. The three primate lineages are ancient, and each displays important features that distinguish it from the others.

Sources

Basic Bibliography

Begun, D. (Ed.) (2013). *A Companion to paleoanthropology*. Malden, MA: Wiley-Blackwell. [Chapters 18, 19, 20]

Campbell, C.J., et al. (Eds.) (2011). *Primates in perspective* (2nd ed.). NY: OUP.

Fleagle, J.G. (2013). *Primate adaptation and evolution* (3rd ed.). NY: Academic Press.

Vaughn, T.A., et al. (2015). *Mammalogy* (6th ed.). Burlington, MA: Jones & Bartlett Learning.

Additional References

Asher, R.A., & Helgen, K.M. (2010). Nomenclature and placental mammal phylogeny. *BMCEB*, *10*, 102 (9pp.).

Beard, K.C. (2013). Anthropoid origins. In D. Begun (Ed.), *A Companion to paleoanthropology* (pp. 358–375). Malden, MA: Wiley-Blackwell.

Bruner, E. (2013). The species concept as a cognitive tool for biological anthropology. *AJP*, *75*, 10–15.

Di Fiore, A., et al. (2011). Molecular primatology. In C.J. Campbell et al. (Eds), *Primates in perspective* (2nd ed.) (pp. 390–416). NY: OUP.

Fleagle, J.G. (2013). *Primate adaptation and evolution* (3rd ed.). NY: Academic Press.

Groves, C. (2001). *Primate taxonomy*. Washington DC: Smithsonian Institution Press.

Hanna, J.B., et al. (2006). Forelimb and hindlimb forces in walking and galloping primates. *AJPA*, *130*, 529–535.

Heesy, C.P. (2004). On the relationship between orbit orientation and binocular visual field overlap in mammals. *AR Part A*, *281*, 1104–1110.

Jablonski, N.G. (2003). The evolution of the tarsiid niche. In P.C. Wright et al. (Eds), *Tarsiers* (pp. 35–49). New Brunswick, NJ: Rutgers University Press.

Larney, E., & Larson, S.G. (2004). Compliant walking in primates: elbow and knee yield in primates compared to other mammals. *AJPA*, *125*, 42–50.

Masters, J., et al. (2013). What's in a name? Higher level taxonomy of the prosimian primates. In J. Masters et al. (Eds.), *Leaping ahead, advances in prosimian biology* (pp. 3–9). NY: Springer.
Schmitt, D., et al. (2010). Forelimb compliance in arboreal and terrestrial opossums. *AJPA, 141*, 142–146.

2 The Study of Primate Behavior

It took a little while to get used to the bright blue testicles of the vervet males, often seen as they ran away. Now the monkeys are used to the primatologist and he can focus on features that distinguish them as individuals. Understanding their complex relationships with other troop members is far more satisfying than any colorful novelty.

This chapter introduces the study of primate behavior in terms of *research* (observing, recording, and experimenting) and *interpretation* (using theory to understand the results of research). Watching primates for scientific purposes can be a more complicated and arduous activity than some people might imagine. It presents a variety of problems that call for a variety of techniques. Studies of wild and captive animals each have their own distinctive advantages and difficulties. They also have much in common and their findings often complement one another. Understanding what has been seen requires well-defined analytical techniques and reference to a variety of theoretical frameworks.

Fieldwork

C.R. Carpenter performed the first scientific field research on primates in the 1930s. Field studies burgeoned after 1950, led by Sherwood Washburn (baboons in Africa), Kinji Imanishi (Japanese monkeys), and their many students. Chimpanzee research started by Jane Goodall in the 1960s brought primate fieldwork to the attention of the general public. Since its early days, primate field research has gained enormously in complexity and sophistication.

The environment for fieldwork can be anything from baking grasslands to leech-infested swamps. Primatologists have shown great fortitude and courage in their research. I write this with admiration, having worked in the highlands of Tanzania where dryness made temperatures of 90 degrees almost pleasant and the night air contained no mosquitoes because of the altitude.

The Goals of Fieldwork

The goal of field research may be **general study** or **hypothesis testing.** General study gathers data on a variety of observable behaviors. With sufficient information, later researchers can go into the field with particular questions in mind. These are best framed as testable hypotheses; that is, they are amenable to procedures used to decide whether a postulated relationship or explanation is probable beyond the level of chance.

Primate fieldwork shifted from general study to hypothesis testing in the 1970s and 1980s. However, there are arguments for renewed vigor in general studies. First, we still know very little. Many species remain unstudied, and in many cases where research has been done, we only know about one or a few populations or social groups. Second, many unstudied populations and taxa are rapidly disappearing due to human intrusions. Third, data recorded in general studies can be used to test hypotheses that have not yet been formulated. For example, the idea that primates have some kind of culture, now widely accepted, was originally based on data from prior general studies.

Making Contact

Making contact with the subjects can be a problem in fieldwork. There is some danger, but most primates are smaller than humans and even the most formidable (such as gorillas and baboons) will usually avoid humans rather than attack. Avoidance by the subjects is the greater problem for fieldworkers and requires **habituation**. This is the process in which the animals become accustomed to the presence of researchers so that they will allow sufficiently close observation (Figure 2.1).

Habituation demands patience, caution, and ethical concern for the subjects. The observer should be visible to the animals, but not close enough to disturb them. Eventually it may be possible to follow the subjects and stay near them or even walk among them. This process can take years, though it may be shorter for some species. The observer should not follow the subjects too persistently if they attempt avoidance or flight, because this causes stress and distorts natural behavior. Success comes from regular and frequent neutral contact with the same individuals. **Provisioning**, providing food on a regular basis, has been used

Figure 2.1 A fieldworker near habituated chimpanzees. © Curt Busse

in some field studies to bring the subjects close. However, provisioning is problematic because it creates dependence, alters normal behavior, and encourages dangerous interactions between primates and neighboring humans, such as crop raiding. Possible transmission of disease is an ethical problem because it is a danger to the subjects as well as the researcher.

Identification

Identification of individuals is necessary in many studies (in examinations of social relationships, for instance). This can come with familiarity, but photographs are helpful because they can be studied at leisure. Sometimes luck and insight combine to make things easier; for example, fieldworkers found that the nose of every gorilla forms a unique pattern (Fossey 1983; Schaller 1963). Marking of some kind may be necessary for species that are small or active, live in conditions of low visibility, are studied by different researchers, live in large groups or groups of changing composition, or do not have sufficiently distinctive individual characteristics. Marking methods include tattoos, ear tags, and collars or belts.

Marking raises questions about manipulating the animals, especially because it requires trapping. Traps should be humane for ethical reasons and also to avoid causing distorted behavior after release. A broad survey showed that humane trapping causes little or no change in primate behavior. The most common device is a cage baited with an attractive food. Primates present difficulties not found in other animals. Some can escape by raising the door with their hands. Some can learn to get the bait without entering the trap or without triggering automatic closure. On the other hand, some individuals become "trap-happy" and enter the cages more often than the researcher desires.

Anesthesia is needed for marking methods that are painful or animals that are hard to handle. It is usually needed to obtain information about physical characteristics, including measurements and collection of samples such as blood. Repeated capture increases risk of stress, injury, and death. Therefore, the maximum amount of information should be obtained whenever an animal is trapped and anesthetized for any reason.

Technology

Technology has become increasingly important and elaborated in fieldwork. Primatologists record behavior with photographs and video. They have long used recorded vocalizations and other sounds to study the reactions of their subjects. More recent is the use of **telemetry**, automatic measurement and transmission of data by wire, radio, or other means from remote sources to receiving stations. Telemetry provides identification and tracking of subjects collared with radio transmitters. Primate researchers now use the Global Positioning System (GPS) to demark locations on the ground more accurately and Geographic Information Systems (GIS) to create more accurate maps of primate habitats. Data recording by personal electronic devices saves an enormous amount of time and allows for protection of data by storage in more than one place.

Data Collection and Analysis

The researcher collects data by **sampling** the subjects—that is, by observing a number of individuals who accurately represent the total study population (Altmann 1974). This step may be simple (limited to adult males, for example) or complex (encompassing all age–sex categories), depending on the research question. One widely used technique is the **instantaneous scan**. The observer surveys the sample subjects at regular intervals and records what they are doing (or where they are) at each time. Continuous observation of particular individuals is required if the study question concerns the duration or rate of occurrence of a behavior. Instantaneous scan can cause brief but important behaviors to be missed. It is most useful in studying ordinary behavior in categories or groups such as *juveniles* or *cliques*.

Focal animal sampling records all the activities of a single individual over a predetermined period of time. When that time is up, the researcher follows another individual. This approach provides information about duration and frequency, but it requires that the behavior of other group members is ignored if they are not interacting with the focal animal. Another drawback is that results are invalidated if the researcher loses contact with the subject during the predetermined observation period.

Ad libitum sampling means the recording of behavior without a predetermined scheme. This may be part of a general study, but it is especially important with regard to rare and significant events such as the birth of an infant or an injurious fall from a tree. It is essential to break away from any scheduled procedure (or from resting or eating lunch) to record such events.

The collected data are quantified for statistical analysis. **Descriptive statistics** simply represent the behavioral patterns in the data, which are usually reported as frequencies. Percentages convey relative results, such as the amount of total daily active time taken up by a particular behavior. **Inferential statistics** explore the probability that a described relationship is *significant*, meaning that it is not due to chance.

Understanding behavior requires understanding the conditions that affect it. Fieldworkers record diverse aspects of the environment such as forest structure, daily light cycle, temperatures, and the availability and characteristics of plant foods and animal foods. Fieldworkers usually collect samples of plants and animals to be identified and/or analyzed in laboratories.

Dung samples are especially informative about behavior. They provide food remains with the assurance that the primates actually ingested that food (behavioral observations of feeding can be compromised by poor visibility). Feces can also be used to monitor hormonal changes in the subjects, which have implications for reproduction and for behavioral states like stress. Finally, the samples provide genetic material that can be used in studies ranging from phylogeny to kinship. The structure of scats can be used to assign them to a species, but association of a scat with a particular individual requires observation of the deed, or at least strong circumstantial evidence.

Captive Studies

Captive study refers to research on animals that are to some degree restrained by humans. This encompasses a wide variety of situations with a wide variety of effects on primates. The best case for the subjects is one in which they are maintained in a natural social group in an open-air enclosure in a favorable climate. This occurs at some research facilities and in some zoos. The main research advantages are that the animals can always be found, can be studied at any time, and in some cases can be seen from optimal locations. Some behavioral experiments can be performed, such as removing a particular animal to see how social relationships in the group are altered.

Manipulation of Captives

Researchers can subject caged primates to more extensive manipulation. This is considered vital because the scientists can then use numerous sophisticated techniques to record data (Stevens and Carlson 2008). Experiments may be relatively benign, such as a temporary set-up in which a female can choose between two potential sexual partners without being coerced by either. Most intelligence tests also fall into the benign category.

At the other extreme is the **deprivation experiment**, in which the opportunity to learn a behavior is taken away from the subjects. The usual goal is to explore which components of behavior are learned and which can develop without experience. In primates, for example, typical facial expressions are likely to appear despite deprivation.

In some experiments subjects have been deprived of all normal stimuli, resulting in severe trauma. Many primatologists oppose such experiments, as expressed in a letter signed by 56 students of primate behavior (King and LaFleur 2015). In a less drastic approach, humans care for the subjects in place of their mothers (e.g. Sugita 2008). Another method is **cross fostering**, in which learning deprivation is achieved by giving the subject a foster mother of a non-primate species (females of various mammal species readily accept such adoptions).

Physically intrusive studies involve deliberate damage to the animal, such as lesions in particular portions of the brain in order to observe the behavioral effects. Alternatively, electrodes may be permanently implanted in the brain so that particular sections can be stimulated when desired. There is skepticism about conclusions based on animals traumatized by severe deprivation or by physical damage. Some observed behavioral deficits (inability to copulate, for instance) might be due to general pathology rather than specific lesions or rearing conditions. Many scientists also have ethical reservations about injurious experiments.

Distorted behavior can arise from the captive situation itself, and this may compromise both experiments and naturalistic observations. Captive animals expend little or no time and energy in foraging and feeding; they are protected from predators; and they are to a large extent protected from and treated for disease and injuries (Stevens and Carlson 2008). In addition to such drastic departures from normal ecology, captivity may alter social behavior; for example, heightened aggression may occur. Some such effects occur in laboratory groups, but not in zoos (Hosey 2005). Mallapur and Choudhury (2003) showed that primates confiscated from touring zoos, circuses, and animal traders exhibited higher levels of abnormal behaviors than did animals in larger, recognized zoos.

Anecdotes and Anthropomorphism

An **anecdote** describes a unique or unusual event. Any study of primates is likely to provide anecdotes because their behavior is so complex and variable. Researchers also get anecdotes from local residents in the field, staff in the laboratory, and a variety of other people. Non-scientists sometimes portray these accounts as proof of something, but a single case or a few unusual ones can never support a general proposition. However, it is important to distinguish between anecdotes mistakenly intended as evidence and anecdotes used for other purposes.

Some anecdotes point the way to future research. Accounts of tool use by chimpanzees were anecdotal in the beginning (van Lawick-Goodall 1968), but turned into one of our most fertile areas of research (McGrew 2004). A second benefit of anecdotes is that they can suggest the behavioral potential of a species. In a species that frequently uses tools to get food, a few individuals may also use objects to pick their teeth or clean their noses. Bates and Byrne (2007) argued that creative behavior occurs at low frequencies in nonhuman animals and is typically missed by standard observation methods. They provide criteria for the reliability of such anecdotes. First, they should be recorded immediately after observation. Second, the observer should be someone familiar with the species and the individuals involved. Third, separate reports should be collated to demonstrate a pattern.

Finally, anecdotes can amuse or touch us and remind us that the study of primate behavior is fun and can be enlightening in ways other than scientific. An example is Kortlandt's (1962) account of a chimpanzee that sat and watched a sunset for perhaps 15 minutes. This single incident can inspire humanistic consideration of the nature of art or religion and the nature of animals. Smuts (2001) described a comparable experience. Late in the day the baboons she was following sat down at the edge of a small, still pool and gazed into the water in complete silence for half an hour.

The last two examples raise the issue of **anthropomorphism**, the interpretation of animal behavior as if the subjects were human. The epitome of this phenomenon is cartoons where animals speak and behave exactly as if they were human. With regard to real animals, a frequent manifestation is the idea that animals found in pairs are mated for life and have a romantic relationship. Many such relationships fall far short of lifetime tenure and display little or no evidence for affectionate feelings.

Objections to alleged anthropomorphism are often couched in terms of **Morgan's Canon**. This rule, formulated by an early comparative psychologist, is often cited as prohibiting the attribution of so-called higher mental processes to nonhuman animals. Morgan actually argued that such attributions should not be made *without a strong foundation* (Karin D'Arcy 2005). The temptation to engage in anthropomorphism is greater with primates than most other animals because the primates really are so much like us. They share a real biological heritage with us, which provides the "strong foundation" needed for legitimate comparison. The argument often revolves around the idea that apes are like us, but it makes just as much sense to say that we are like apes. Interviews with prominent primate fieldworkers led Rees (2007) to the conclusion that "careful anthropomorphism" is valuable for the insights that it can provide. Lloyd Morgan did very little research with primates (Maple 1979). Had he done more, he might have qualified his rule.

Alleged anthropomorphisms are increasingly testable. For instance, the same areas of the brain may be activated in human and monkey mothers in comparable circumstances. At a minimum this indicates that both species possess the same motivational system. One might go further and infer that the monkeys feel much the same emotion as humans (even if calling it *love* might be going too far). The key, as in any scientific research, is to formulate a precise hypothesis and find a way to test it. A variety of scientific disciplines have developed ways to test hypotheses about animal behavior.

Explaining Primate Behavior

The Science of Primate Behavior

Primatologists generally regard their discipline as a science. Scientific thinking means a number of different things, one of these being objectivity. Complete objectivity is impossible for an individual, but it can be approached in science because diverse participants can counterbalance one another's biases. Primatology has been an international enterprise (Ji and Jiang 2004; McNeilage and Robbins 2006; Terry 1983), from the first field studies done by Americans, Europeans, and Japanese to increasing contributions from Indian, Chinese, and African scientists. Equally important, a male bias in the early years was offset by an influx of female scientists (Fedigan 1994). Some bias in the social structure of the field seems to persist (Addessi et al. 2012), but we have reaped the benefit of data and insights from female primatologists. Primatology as a field of inquiry seems well situated for the pursuit of objectivity.

The purpose of scientific study is explanation, which requires hypothesis testing. **Competing hypotheses** (alternative explanations for the same phenomenon) are an important part of this process. Some people will dismiss a hypothesis by saying that they already have an explanation for something and "don't need" the alternative. For example, they reject biological explanations for human behavior because social science already explains some of the facts. However, science is not about needs. It is about finding the best explanation for the currently known facts.

In evaluating competing hypotheses it is important to consider whether or not they are **mutually exclusive**; that is, whether acceptance of one precludes the other. Some apparently competing hypotheses are actually complementary. For instance, some scientists attribute color vision to the consumption of fruit and others to the consumption of young leaves. These views are not necessarily contradictory, because most primates eat both fruit and leaves.

Tinbergen's Questions

Many different scientific disciplines have contributed to hypotheses about primate behavior and the means of testing them. The lines separating the primatological fields have blurred because, regardless of emphasis, most practitioners appreciate the overlaps among them (Maestripieri 2003). Whatever disciplinary names they use, students of primate behavior are concerned with four or five kinds of explanation (though particular theorists often emphasize one over the others). This framework is largely based on the work of ethologist Niko Tinbergen (1963), who postulated four questions that have to be answered in order to fully understand a pattern of behavior (Barrett et al. 2013; Bateson and Laland 2013). The following summary combines Tinbergen's formulations with some newer concepts and terminology.

Proximate mechanisms are the immediate causal antecedents of particular behaviors in individuals. They include stimuli from the environment (e.g. the sight of a predator) and internal signals (e.g. feeling hungry). The threatened individual may flee and the hungry one will probably eat. Generalized internal factors such as hunger or fear can be considered motivational systems. These are often specified by terms associated with emotions, which can be considered the animal's subjective perception of the activated system. Cognitive processes—such as perception, memory, and reasoning—are also internal causes of behavior. The physiological basis for these causes lies in hormonal systems, neurological processes, and the interactions between them.

Other endogenous factors are more specific. **Attentional mechanisms**, for example, make certain stimuli (snakes, for instance) more likely to be noticed than others. Other mechanisms cause the form of some behaviors to be relatively stereotypical for a species (e.g. facial expressions). A behavior with such a stereotypical tendency is called a **modal action pattern** (abbreviated to MAP).

Functions (see Chapter 1) are the effects or consequences of behavior in relation to the physical and/or social environment. For example, the function of hunger is to motivate food acquisition and processing. The function of facial expressions is to communicate with others. The term function refers only to recent relationships and not to evolutionary origins. The current function(s) of a trait may not be the same as those it had in the past. Functional interpretation does not necessarily imply a genetic basis for the trait. "Current utility" has been suggested as an alternative that clarifies the issue (Bateson and Laland 2013).

Ultimate causes are the conditions thought to account for the origins of particular behaviors. These include past environments and the action of natural selection. Hypotheses of ultimate causation often assume that the behaviors in question are under some degree of genetic influence. Ultimate causation can provide an explanation for the functional significance of a trait; however, as noted above, functions may change over time.

The widely used word **adaptation** has been subjected to so much argument and redefinition that it may cause more confusion than it alleviates. Broadly speaking, an adaptation is a trait that facilitates survival and/or reproduction. The word is also used for the process or processes by which adaptations come into existence. Some scientists would reserve the term adaptation for the origin of a trait by natural selection, which is often highly speculative. Confusion about the meaning of adaptation can be alleviated somewhat by using the concepts of function and ultimate cause wherever possible.

None of this presupposes awareness on the part of the primates (including humans) that their behavior is related to natural selection. For example, the presumption is that primates (including most humans) seek sex because they are motivated to engage in that interaction (loosely speaking, they like it). They are not consciously trying to transmit genes to the next generation, although this may be the functional/adaptive significance of the behavior.

Misunderstanding arises in part because behavioral scientists often press everyday words into use when their connotations seem appropriate. An important example is **strategy**, which is a short way of referring to certain functions. For instance, to a primatologist, "sexual strategy" encompasses behavioral patterns that lead to a successful copulation (including searching for mates, selecting one, courting, and defeating competitors). The primary hypothesis is that this complex pattern of behavior is based on proximate mechanisms that were probably favored by natural selection. Many complex behavioral systems are based on a series of relatively simple proximate mechanisms (Kummer 2002).

Ontogeny refers to the growth and change of the individual from conception to death. **Development** is a similar concept, but is more often used for the period before maturity. Ontogenetic explanations explore the complex processes that link a growing individual to the proximate and functional aspects of behavior. This approach investigates the ways in which endogenous mechanisms are established during the growth and maturation of the individual.

Some behaviors appear at or shortly after birth (such as the grasping reflex in primates, including humans). Others manifest themselves somewhat later (for example, walking and language in humans). Early-emerging behaviors are especially likely to be under a significant degree of genetic influence, a condition that may be described as **innate**. Mameli and Bateson (2011) pointed out that the term has taken on a number of meanings. However, several of them connect closely with genetic influence, e.g. "developed without learning" and "history of natural selection". It might clarify some issues if we use *innate* to denote genetic influences on components of behavioral patterns, rather than whole patterns.

The mechanisms underlying innate components may be fully formed at birth but they are likely to require some degree of **maturation**; that is, they must undergo some development before they are fully operative. Learning is probably involved in the ontogeny of virtually all primate behaviors and is predominant in many, perhaps most. Following the development of a behavior gives us more insight into its components and causes, such as the balance between genetic and experiential factors. Genetic influences may make certain kinds of behavior relatively easy to learn. In some cases a behavior is readily learned only during a particular period of life, called a **sensitive period**. Imprinting in birds is a well-known example. For instance, newly hatched ducklings quickly learn to follow the first thing that they see (even a human).

Beyond limited phenomena such as sensitive periods, ontogenetic studies make us aware of broader stages in life. Some researchers have applied psychological concepts (such as those of Piaget) to the cognitive development of primates. Others have made it clear that adaptive behavior varies with an individual's stage of life. Annoying primate infants can expect a great deal of tolerance, for example, but juveniles must be more careful not to provoke adults.

Phylogeny is the pattern of evolution within and between taxa. It is popularly depicted as the branching of a tree, though this is often an over-simplification. Phylogeny is the basis for taxonomic classification. It is also a way to understand behavior. Phylogenetic analysis places the behavior of living animals in the context of evolutionary history (Ossi and Kamilar 2006). It postulates that ancestry can explain the distribution of behaviors among related species.

The **comparative method** is the vehicle for phylogenetic analysis. Variation of a trait is mapped onto the known phylogenetic relationships of living species. In its simplest form this guides us to the identification and explanation of evolutionarily conservative traits, such as climbing in primates. Facial responses to sweet and bitter tastes provide another simple example. A variety of primates (including human infants) display similar reactions, suggesting that they are phylogenetically connected (Weiss and Santos 2006). A further interpretation is that the shared response is part of a motivational system that originated as an adaptation for choosing or avoiding potential foods. Evidence for ancestral behavior can also provide a baseline for understanding derived behavior. Comparison of varying forms in living species can suggest the sequence in which the forms evolved.

Summary

Primate behavior is studied in the field and in captivity. Sampling and statistical analysis are essential. Recording and interpretation should avoid anthropomorphism, but recognize that we are primates and share traits with our close relatives. Causes of behavior are proximal, functional, ontogenetic, and phylogenetic. Hypotheses are tested and competing hypotheses are compared. Many primatologists consider their work relevant to understanding human behavior.

Sources

Basic Bibliography

Campbell, C.J., et al. (Eds.) (2011). *Primates in perspective* (2nd ed.). NY: OUP. [Part Three and Chapter 46]

Setchell, J.M., & Curtis, D.J., Eds. (2011). *Field and laboratory methods in primatology* (2nd ed.). NY: CUP.

Additional References

Addessi, E., et al. (2012). Is primatology an equal-opportunity discipline? *PLONE*, *7*, e30458.

Altmann, J. (1974). Observational study of behavior: sampling methods. *Behaviour*, 49, 227–267.

Barrett, L., et al. (2013). Taking note of Tinbergen, or: the promise of a biology of behaviour. *PTRSB*, *368*, doi: 10.1098/rstb.2012.0352.

Bates, L.A., & Byrne, R.W. (2007). Creative or created: using anecdotes to investigate animal cognition. *Methods*, *42*, 12–21.

Bateson, P., & Laland, K.N. (2013). Tinbergen's four questions: an appreciation and an update. *TREE*, *28*, 712–718.

Fedigan, L.M. (1994). Science and the successful female: why there are so many women primatologists. *AA*, *96*, 529–540.

Fossey, D. (1983). *Gorillas in the mist*. New York: Houghton Mifflin Harcourt.

Hosey, G.R. (2005). How does the zoo environment affect the behaviour of captive primates? *AABS*, *90*, 107–129.

Ji, W., & Jiang, X. (2004). Primatology in China. *IJP*, *25*, 1077–1092.

Karin D'Arcy, M.R. (2005). The modern role of Morgan's Canon in comparative psychology. *IJCP*, *18*, 179–201.

King, B.J., & LaFleur, M. (2015). Letter to the President of the American Society of Primatologists.

Kortlandt, A. (1962). Chimpanzees in the wild. *Scientific American*, *206* (5).

Kummer, H. (2002). Topics gained and lost in primate social behavior. *EA*, *11*, 73–74.

Maestripieri, D. (2003). The past, present, and future of primate psychology. In D. Maestripieri (Ed.), *Primate psychology* (pp. 1–16). Cambridge MA: HUP.

Mallapur, A., & Choudhury, B.C. (2003). Behavioral abnormalities in captive nonhuman primates. *JAAWS*, *6*, 275–284.

Mameli, M., & Bateson, P. (2011). An evaluation of the concept of innateness. *PTRSB*, *366*, 436–443.

Maple, T. (1979). Great apes in captivity. In J. Erwin et al. (Eds.), *Captivity and behavior* (pp. 239–272). NY: Van Nostrand Reinhold.

McGrew, W. (2004). *The cultured chimpanzee*. NY: CUP.

McNeilage, A., & Robbins, M.M. (2007). Primatology comes to Africa. *AJE*, *45*, 1–3.

Napier, J.R., & Napier, P.H. (1985). *The natural history of the primates*. Cambridge, MA: MIT Press.

Ossi, K.M., & Kamilar, J.M. (2006). Environmental and phylogenetic correlates of *Eulemur* behavior and ecology (Primates: Lemuridae). *BE&S*, *61*, 53–64.

Rees, A. (2007). Reflections on the field – primatology, popular science and the politics of personhood. *Social Studies of Science*, *37*, 881–907.

Schaller, G.B. (1963). *The mountain gorilla*. Chicago: UCP.

Smuts, B. (2001). Encounters with animal minds. *Journal of Consciousness Studies*, *8*, 293–309.

Stevens, N.J., & Carlson, K.J. (2008). Bridging gaps between experimental and naturalistic approaches in the study of primate behavior. *IJP*, *29*, 1395–1399.

Sugita, Y. (2008). Face perception in monkeys reared with no exposure to faces. *PNAS*, *105*, 394–398.

Terry, M.W. (1983). Geographic origin of journal articles on nonhuman primates. *AJP*, *4*, 179–182.

Tinbergen, N. (1963). On aims and methods of ethology. *Zeitschrift fur Tierpsychologie*, *20*, 410–433.

van Lawick-Goodall, J. (1968). The behaviour of free-living chimpanzees in the Gombe Stream Reserve. *Animal Behaviour Monographs*, *1*, 161–311.

Weiss, D.J., & Santos, L.R. (2006) Why primates? The importance of nonhuman primates for understanding human infancy. *Infancy*, *9*, 133–146.

3 Primate Ecology and Behavior

Common Features

A young primate moves aimlessly through the trees, emitting high-pitched sounds, desperately seeking the social group that it has lost. Eventually, the mother approaches and the youngster quiets.

This chapter tries to describe behavioral patterns that are common across the primate order. That is, they are widely recurrent, typical, or even ubiquitous among primates as a whole. An important caveat is that many published papers about "primates" are really about Old World anthropoids or even narrower groupings. This presentation is based on descriptions and explanations that seem widely applicable.

Primate Ecology

Most primates inhabit tropical and subtropical rainforests and woodlands. **Woodlands** are areas with varying proportions of tree and grass cover. A tropical **rainforest** is a more complex ecological community, consisting of vegetational layers called **strata** (Corlett and Primack 2011; Woodward 2012). The lowest level is the **understory**, which is sparse at ground level due to limited light and moisture. Farther up, a **canopy** or two are comprised of continuous layers of leaves. **Emergents** are widely spaced trees rising above the rest. Vines and similar plants are intertwined with the trees. Each plant species has its own distinct flowering and/or fruiting season(s). Dry seasons are usually brief and a complete cessation of rain is rare.

Primates are often called omnivores, meaning that they have diversified diets that include both plants and animals as food. Though broadly correct, this obscures enormous variation in particular foods and the ways they are used. **Preferred foods** are those eaten when there is a choice. These are of *high quality*, i.e. rich in protein and/or energy and easily digestible. Fleshy tropical fruits are an energy source containing vitamins that facilitate the body's use of energy. Figs, whether they are preferred or not, are common in primate diets for several reasons (Shanahan et al. 2001). The trees (*Ficus*) occur in almost every primate habitat, the only genus for which this is true. Figs are available throughout the year and present no defenses (such as hard shells). They contain a variety of nutrients (though particular species vary in food value).

Primates typically prefer young leaves to mature ones because the former contain more protein and are lower in **secondary compounds**. These are chemicals that plants secrete to protect themselves from destruction by animals: toxins (poisons such as cyanide) and digestion inhibitors (e.g. **tannins** that interfere with protein uptake). Other factors also contribute to diet choices. For instance, protein concentrations in some leaves vary through the day. Some taxa are called **folivorous** because they specialize in leaf consumption to varying degrees.

Primates are unusual mammals in consuming **exudates**, substances secreted by trees and some other plants (Burrows and Nash, 2010). **Gums**, the exudates favored by primates, are found immediately under tree bark where their function is to repair damage. They rival fruit in sugar content, but are more difficult to digest. At least 69 primate species eat exudates, some routinely and some opportunistically, mostly during dry seasons. Where rainfall is irregular, gums may be a more reliable and rapidly renewing source of carbohydrates than fruit. Feeding sites are small, but a single tree may contain numerous sites.

Primates select certain plants for mineral content: leaves are richest, followed by bark and fruit. Primates may also obtain minerals through **geophagy**, eating soils and clays (Ferrari et al. 2008). Other explanations for geophagy include toxin adsorption, control of diarrhea, and pH adjustment of the gut. These hypotheses are not mutually exclusive, and it is doubtful that such a widespread phenomenon has a single explanation (Krishnamani and Mahaney 2000).

Animal foods provide primates with proteins and fats. Fat contains more than twice the energy of carbohydrates per unit of weight. Protein supplies amino acids for maintenance of the body. Most important to primates are nine amino acids that their bodies cannot synthesize. **Faunivory** denotes animal consumption by primates. The term **carnivory** is misleading due to the connotation of meat (vertebrate muscle tissue), which is limited or absent in most primates. Insects are the most important component of primate faunivory, but primates commonly eat other invertebrates, such as spiders and snails (McGrew 2014).

Primate prey are typically small and often cannot escape—for example, grubs, caterpillars, and insects that infest fruits. However, primates sometimes exert themselves to obtain animals like grasshoppers and rodents. Flexible hands facilitate capture, and the food value of larger animals can offset the energy costs of vigorous pursuit. If the prey can be controlled manually, the primate usually dismembers it or eats it whole. If it poses a danger (such as a bite) or is likely to escape, the primate will often kill or immobilize it quickly with a bite to the head and/or neck (King and Steklis 1984).

Some foods have a **patchy distribution**, meaning they exist in irregularly distributed clumps. Fruit distribution is patchy over a large area, but variation can occur within a single tree: warmer, drier upper crowns may produce bigger fruits at higher densities (Houle et al. 2007). Leaves may be everywhere, but the distribution of nutritious and easily digestible leaves can be patchy. Variation through time is also important. Seasonal cycles, rainy and dry for most primates, affect most fruit species. This is why the perpetual availability of figs stands out. Leaves are less affected, but they are not immune. Insect availability also varies, because insects depend on plant foods.

Fallback foods are eaten when preferred foods are scarce, usually during a dry season, and can be critical for survival. Fallback foods are inferior in terms of nutritional quality and/or the energy needed to process them. Proximately, these foods are less attractive due to taste and/or difficult processing. Fallback foods are sometimes conflated with *keystone resources*. The latter are theoretically critical for entire ecological communities. A fallback food may be important to just one species and it may not be essential to survival, especially if other fallbacks are available.

When primate species are **sympatric** (living in the same habitat), food competition is usually ameliorated by occupation of different **niches** (ways of relating to the environment). For instance, they may depend on different foods, forage at different times, or use different forest strata. Certain bats and squirrels also seek fruit, but birds are probably more important competitors.

Primates face a variety of dangers. Arboreal activity risks serious falls (Lovell 1991). Young individuals, less experienced and more likely to be careless, are particularly susceptible. Fighting and frights also result in accidents that cause death instantly, or through infection or susceptibility to predators. Predators may attack from any direction. Raptors can take primates from the trees (McGraw and Berger 2013). Cats and other terrestrial predators wait on the ground, and some can climb. Snakes eat a variety of primates (though none that weigh more than 13kg) and poisonous snakes may strike defensively (Headland and Greene 2011). An ongoing controversy addresses the possibility that orbital convergence and binocular vision are adaptations to snake detection (Wheeler et al. 2011). The degree of danger from snakes seems to vary by continent.

Fieldworkers found that their presence reduced attacks on primates by inhibiting the predators. This hampered assessment of the natural **predation rate**, the frequency of successful attacks on a group or population. **Predation risk** refers to the rate of unsuccessful attacks or other encounters with predators. Risk assessment includes primate responses to their perception of the danger (avoidance of a location, for example). If the predation *rate* for a given population or species really is low, it may be due to effective countermeasures by the potential victims rather than to low *risk*. However, even a low rate may be demographically significant because peaks in predator pressure can drastically reduce primate populations.

Vulnerability is the likelihood that a particular individual or kind of individual will be taken by a predator. Common sense may suggest that small-bodied species, and females and immatures in most species, are highly vulnerable. However, few field studies supported that idea and some found that large adult males were the most frequent victims. Relevant factors include greater visibility, more meat per kill, and/or availability because adult males sometimes confront predators.

Anti-predator behavior takes time from other activities, including foraging and social interaction, and can be costly in terms of energy expenditure. Primates in trees usually avoid raptors by moving lower and/or hiding in dense vegetation. Trees and cliffs provide refuge from terrestrial predators. **Alarm calls** are vocalizations that seem to warn conspecifics about predators. An alternative hypothesis holds that the calls notify the predator that the targets have been alerted. These functions are not mutually exclusive. Calling sometimes continues long after the predator has disappeared so that it no longer seems to be a warning; however, continued calls may convey that a predator has hidden rather than departed. A hypothesis of proximate cause is based on emotion: continuing alarm calls may express a high level of arousal that does not quickly subside.

Primates sometimes confront predators. In **mobbing**, the most common action, primates approach a potential predator and emit frequent loud calls while engaging in vigorous physical activity such as jumping (Crofoot 2012). An episode can last as long as 30 minutes. Mobbing is like an alarm call in that it notifies the predator that a sneak attack is impossible. It may confuse or even frighten some predators, especially younger individuals. Finally, it can teach young primates to recognize dangerous animals.

Vigilance, special visual attention to surroundings, seemed likely to function in predator detection. Doubts arose from the realization that vigilance could be directed internally, toward potential mates or competitors. This stimulated two kinds of problem-oriented research. Some fieldworkers looked for correlations between vigilance and natural predator encounters; others performed experiments simulating predator risk. These studies agreed that vigilance increases along with predator risk.

Van der Post et al. (2013) provided an additional perspective. This pertained to the issue of **free riders**, individuals who benefit from a behavior even though they never provide reciprocal benefit. If free riders benefit, the individual who performs a behavior has no selective advantage and the behavior should not evolve. The researchers postulated that predator detection is distance dependent, so that individuals closest to a given predator (and therefore more likely to become targets) would gain the greater survival benefit from vigilance. This mechanism seems compatible with others pertaining to predator defense.

Beyond predators, primates belong to larger ecological communities. Many primates contribute by dispersing the seeds in the fruits that they eat, propagating the trees and plants that they come from. The primates benefit, of course, but so do other frugivores and the plants themselves (Heymann and Hsia 2015). Less often noted is that there are some beneficial ways in which primate folivores alter the environment (Chapman et al. 2013).

Sleep ends a primate's day (Nunn et al. 2010). Most of its physiological functions seem to be characteristic of mammals generally, e.g. enhancement of the immune system and consolidation of memory. Two ecological hypotheses may have special relevance to primates. One holds that variation in sleep duration results from a trade-off between basic individual functions and ecological constraints on particular species (Capellini et al. 2008). The other hypothesis explains complete unconsciousness in sleep as a response to predation (Lima and Rattenborg 2007). Relative immobility reduces conspicuousness, and the period of helplessness is shorter because neural maintenance proceeds more quickly. Lighter sleep may have evolved to permit a quicker response to an immanent predator. Social contact during sleep may enhance relationships and provide thermoregulation, but it also increases the danger of disease transmission.

Primate Sociality

Primate groups differ from those of most other social mammals by integrating adults of both sexes on a long-term basis. Often the same individuals are together for years. Kappeler and van Schaik (2002) defined a **society** as a set of conspecific animals that regularly interact more with each other than with other conspecifics. **Social organization** denotes characteristics of a society, including size, sexual composition, and degree of cohesion in space and time. Other primatologists have included *relationships* (recurring patterns of interaction between individuals) and added that at least some interactions are affiliative (friendly or at least non-aggressive) (Mueller and Soligo 2005; Mueller and Thalman 2000).

Kappeler and van Schaik argued that the inclusion of mating patterns in social organization is neither necessary nor useful. Societies with similar patterns of day-to-day interaction may have different mating systems. In addition, mating with outsiders is a recurring pattern among primates, resulting from individuals seeking extra-group encounters or from temporary intrusions into a group.

Two main types of social organization recur throughout the primate order. One is based on a long-term relationship between a particular male and female, usually termed a **pair** (Figure 3.1). This kind of social unit is twice as common among primates as in other mammals. Pairs are usually accompanied by one or more offspring, and sometimes by one or a few other adults. I suggest the term **pair-group** to denote a social unit that is clearly centered on a single pair, but which may contain one or a few peripheral adults. The status of "extra" adults can be clarified by their interactions with others and by the demography of the population. There are also societies with multiple adults, ranging in size from a handful to several hundred. Such a society is commonly called a **troop** or **community** (Figure 3.2).

Figure 3.1 A pair of lesser galagos. © Y.A. de Jong and T.M. Butynski, www.wildsolutions.nl

Figure 3.2 Part of a baboon troop. © Glenn E. King

Benefits of social life seem to include protection from predators and access to resources. Multiple sets of eyes and ears enhance detection of predators and cooperative responses such as mobbing may deter them. Society also provides allies for competition with conspecifics for essential resources, mainly food.

Silk (2007) argued that most studies provide evidence for these benefits in the short run, but fail to demonstrate long-term effects on fitness. She called this the **phenotypic gambit** (following A. Grafen). In other words, most studies provide evidence for function but not adaptation. Ebensperger et al. (2012) tried to avoid this trap in a survey of group-living mammals and found "modest" positive effects on fitness that are greater in the tropics. The point about the tropics is especially relevant to primates.

Costs of social life include greater visibility to predators, increased chances of disease transmission, and heightened competition for resources within the group. **Contest competition** involves direct conflict between individuals and occurs when the resource can be monopolized. Patchy distribution of fruit, for instance, allows individual control of feeding sites (which may be as small as one tree or bush). More even distribution in the environment usually elicits **scramble competition**: individuals try to obtain as much of a limited resource as possible without engaging in direct conflict. Isbell and Young (2002) added a temporal factor that they called depletion time. If a site can be depleted quickly, an individual can use up the food before others arrive.

Dominance relationships, in which one participant habitually defers to the other, often mitigate conflict. This applies to relations between communities and between individuals within a community. To avoid the endless arguments about the definition and scope of dominance, we will begin with a simple approach and elaborate as necessary. Dominance is an **agonistic** relationship in which previous fights and/or threats have resulted in routine deference (such as leaving a contestable food source). Benefits and costs vary from one taxon to another.

Some primatologists think that agonistic behavior has been over-emphasized. A massive survey of the literature (which nevertheless excluded nocturnal strepsirrhines) found that 85 to 95 percent of primate social interactions are affiliative (Sussman et al. 2005). Furthermore, where different levels of agonistic behavior could be distinguished, most conflicts were mild. The researchers acknowledged that aggression and competition are significant, but argued that affiliation, cooperation, and tolerance are at the core of primate social life. This is because long-term benefits come from **mutualism**, actions that benefit both parties without active cooperation. However, it is also significant that a single fight causing serious injury or death can have an enormous effect on lifetime reproductive success.

Dispersion, the distribution of individuals in space, is an important social variable. Members of a *dispersed society* spend most or all of their time alone or in small subgroups. In a *cohesive society* most or all members are usually together. These terms are broadly descriptive and relativistic. Societies within each category vary as to the nature of internal social relationships.

Primate societies are shaped by emigration, the departure of individuals from the natal group or area. Males typically emigrate in about half of primate taxa, and both sexes do so in perhaps 30 to 40 percent of species (Field and Guatelli-Steinberg 2003). **Philopatry** denotes the tendency of one sex to remain in the natal place. Primates tend toward the mammalian pattern of female philopatry.

Philopatry tends to result in continuing associations between kin in the same group or area. Interacting with relatives may mitigate competition and/or facilitate cooperation, as predicted by the concept of **kin selection**: insofar as relevant behaviors have a genetic basis, natural selection should favor positive behavior or at least tolerance toward relatives. Relatives are likely to have some of the same genes, including those influencing behavior toward kin. Thus, favorable treatment of relatives results in such genes becoming more common in a population.

According to Silk (2009), kinship plays an important role in the structure of social systems and the development of social relationships among primates. Kinship contributes to the patterning of affiliative behaviors, including helping behavior such as cooperation and sharing. The length and intensity of the primate offspring–mother relationship creates the possibility (unusual among mammals) that it will continue into the adulthood of the offspring. This connection can bring an individual into close contact with siblings and other maternal relatives.

If genetically related animals do treat each other differently, this raises the question of a proximate basis for **kin recognition**, i.e. distinguishing relatives from other members of the social group. Childhood familiarity is probably the most important mechanism in primates as well as other mammals. However, phenotypic cues may also be involved, e.g. physical appearance, vocal qualities, or odors.

Male competition for mates can take the contest or scramble form. Dominant males typically get more matings (though a monopoly of all available females is rare). In many cases genetic evidence shows that dominant males actually have more offspring. **Sperm competition** is a kind of scramble: physiological mechanisms differentiate the probability of conception for two males who copulate with the same female. Several features recur where sperm competition is important: large testes, greater sperm motility, and higher ejaculation frequencies. In many primates the penis contains a bone called the **baculum**, which may facilitate sperm transport within the female reproductive tract (and perhaps the removal of rival sperm). Another possible function of the baculum is the stimulation of fertility-enhancing female responses.

Infanticide is a potential reproductive strategy for male mammals. The death of an infant returns the mother to a reproductively available state and the killer has an opportunity to copulate with her and produce an offspring of his own. It may also be significant that he eliminates a competitor's issue. This is called the Sexually Selected Infanticide Hypothesis because it relates to reproductive success and has a differential effect on males and females.

Some scientists doubted that primate infanticide existed, while others interpreted it as an aberration caused by abnormal conditions (especially habitat destruction by humans). Despite evidence from numerous species under a variety of conditions (van Schaik and Janson 2000), some still doubt the significance of infanticide. The current issue is the effect on the fitness of males (the phenotypic gambit again). In 25 of 49 cases that provided relevant data, infanticidal males definitely gained mating access to the mothers, and in 13 more cases access was probable or possible. However, the rarity of the event in many species casts lingering doubt on its importance as a reproductive strategy. The Sexual Selection Hypothesis illustrates the important concept of **sexual conflict** in reproductive strategies. Infanticide may benefit the male, but losing a healthy infant is a great cost to the female. According to the Hired Gun Hypothesis of sociality, females accept males into group membership as protectors against infanticide.

As in other mammals, the basic male strategy is copulation with multiple females to maximize the number of offspring. Even so, there are reasons for males to exercise **mate choice** (behavior that increases the likelihood of copulating with certain individuals rather than others). One reason is that the production of ejaculate may be energetically expensive and therefore it is not to be wasted. Another is that contests with other males may pose the risk of serious injury. A third is that females vary in reproductive and parenting capabilities.

Dominant females, for instance, may have shorter interbirth intervals and a higher survival rate for their offspring. Males often reject adolescent females, who are less likely to conceive and less likely to raise offspring successfully.

Despite coercion by some males (Muller and Wrangham 2009), female primates also exercise mate choice. Sometimes they choose co-resident dominant males, but in many species they accept or even solicit outsiders: peripheral males, new immigrants, or strangers. Ultimate causes for mating with unfamiliar males include the avoidance of inbreeding, the production of genetically diverse offspring, possible male investment in the young, and deterrence of infanticide. The latter two reasons are based on the possibility that the unfamiliar male will be the father of the female's next infant. In the context of the infanticide strategy, natural selection would obviously favor males who do not eliminate their own genes by killing their own offspring. Memory of copulation with the mother is a postulated proximate mechanism for male restraint.

A female may choose a partner because he seems likely to protect her and her offspring from other males and also from competitive females. An alternative hypothesis is that females choose males for the quality of their genes (Good Genes Hypothesis), as signaled by some external characteristic, but primate evidence is limited. In many species females choose multiple partners because any one male partner may be infertile or depleted; multiple partners are likely to provide genetically diverse offspring; and confusion of paternity may counter infanticide. Many complexities in female mating behavior seem to be related to this last function (Kappeler and van Schaik 2002). Unlike males, females cannot increase the number of their progeny by having many sex partners.

Copulation is usually in the dorsoventral position typical of mammals; that is, the male mounts the female from behind. However, taking advantage of their flexible anatomy, primate males often use their arms, hands, and/or feet to hold their partners. The most widespread pattern among primates is a single brief intromission (less than three minutes) and the male withdraws promptly after ejaculation.

The Socioecological Model encompasses various aspects of social organization. Originally, Wrangham (1980) identified and explained *female-bonded* societies, in which females have close social relationships with each other because they cooperate to defend high-quality food patches against other groups. The benefit to all group members outweighs the costs of competition among them. Elaboration of the model postulated that females arrange themselves in space to optimize access to essential foods (while minimizing risks) because food is the main limit on their reproductive success. Males must then organize their behavior to seek access to females, which predicts that the number of males in a group will be influenced by the number of females. Comparative analysis demonstrated this correlation and established the direction of causality by showing that changes in male numbers lag behind changes in female numbers (Lindenfors et al. 2004).

Another comparative study showed that the evolution of pairs in primates and other mammals is related to the distribution of females in small exclusive ranges (Komers and Brotherton 1997), which is consistent with the central role of female distribution in the Socioecological Model. This analysis examined and rejected two competing hypotheses for pairing: (1) need for paternal care; and (2) inability of males to monopolize more than one female.

Majolo et al. (2008) tested the relationship of group size with food competition and predator pressure. They tried to avoid the phenotypic gambit by connecting ecological factors with representations of individual fitness: (1) access to resources represented by fecundity; and (2) predator pressure represented by survival of juveniles (the most vulnerable segment of a primate population). Meta-analysis provided partial support for decreasing fecundity in larger groups, but no significant difference in juvenile survival. Thus the role of food competition in the Socioecological Model was supported, but the role of predation was not.

The Socioecological Model has been criticized because it does not explain important features of primate life such as prolonged non-sexual relationships (Janson 2000). Numerous factors have been added (summarized by Overdorff and Parga 2007), resulting in a "synthetic" model to which Thierry (2008) objected on the grounds that addition of so many factors made the model untestable. Identification of particular ecological influences on primate behavior is one thing, he argued, but combining them into a comprehensive model is another.

Socioecological theories now face a new challenge from the growing interest in **phylogenetic signals** in primate behavior (Schultz et al. 2011; Silk 2011). These are similarities between species that are hypothesized to arise from their common ancestry, possibly from shared derived sets of genes. However, ecology and phylogeny can be complementary explanations, as Tinbergen's questions clarify.

Primate Life Cycle

Pregnancy imposes substantial costs on any mammalian mother. Protein needs increase by as much as 30 percent during primate gestation and infant growth, which often elicits greater consumption of leaves and insects. Along with energy expenditure, the cost of maternity includes greater vulnerability to predation. Thus a mother makes an evolutionary investment of energy and risk in a particular offspring that might have gone to one or more others. Then, if she fails to nurse the newborn the investment is lost. However, lactation is a further investment of energy, which is lost if the mother does not act in other ways to preserve the infant. Therefore a variety of nurturing behaviors, such as protecting and warming the young, were favored by natural selection. The period of infant care and the following pregnancy add up to the **inter-birth interval** (IBI), the amount of time that elapses between successive births from a given female.

Primate reproduction conforms to a life history pattern that can be characterized as "lower and slower": relatively long pregnancies, large newborns, low reproductive rates, slow postnatal growth, late maturity, and a long lifespan. Gestation length is positively correlated with adult body weight in mammals, so it is significant that adult primates are smaller than many long-gestating mammals (such as elephants). Neurological development may be a major factor: the newborn primate's brain is about 12 percent of its body size, compared with 6 percent in most mammals (Whitten and Brockman 2001). Having a small number of offspring at one time facilitates long-term nursing and other care, which explains the primate tendency toward single birth. Vinicius and Mumby (2013) disagree with the conventional dichotomous view of primate life history and see significant variation along a continuum.

Allocare (infant care by any animal other than the mother) is a recurring trait among primates. An infant attracts attention from other group members, who may hold, groom, carry, and/or play with it. Young adult females that are **nulliparous** (have not yet given birth) usually show the most interest and perform the most allocare behaviors. The mother can benefit from these opportunities to rest or to forage more efficiently, while young females probably gain care-giving experience. The importance of experience is demonstrated by the fact that **primiparous** mothers (those with their first infants) may commonly make mistakes that can harm the infants (although there is some contrary evidence).

Some older females are solicitous toward a borrowed infant, but some treat it roughly and even keep it away from the mother until it dies of exposure, thirst, or starvation. At this point allocare overlaps with reproductive competition. In some cases females direct more overtly hostile behavior toward the offspring of others, ranging from mild aggression to infanticide.

Mammalian **neonates** (newborn offspring) can be classified as altricial (relatively helpless) or precocial (relatively advanced). Some primatologists consider primate young to be altricial because they depend on their mothers for a long time for feeding, transport, protection, and social stimulation. Mammalogists tend to view primate neonates as precocial because they have fur and functional eyes and ears, and most can cling to the mother within a few hours or days. Ventral clinging by very young infants during the mother's locomotion (Figure 3.3) is a typical precocial trait in primates. The divergence of opinion about primate neonates is largely a matter of terminology rather than fact. It demonstrates the importance of precise and explicit definition of terms.

Primate life history displays traits that are correlated with precociality in mammals: long ovarian cycle, prolonged gestation, single birth, large neonatal brain, and prolonged lactation. The low reproductive rate is compensated by a relatively high survival rate due to extended maternal care. Precocial species live in stable environments with relatively predictable food bases, such as the tropics (the typical primate habitat).

Infancy is the time between birth and weaning. A primate infant contributes to the relationship with its mother through behaviors such as seeking the nipple and, later, following. When separated, both youngster and mother display anxiety in the form of searching behavior and vocal signals (usually squeaks of some kind). The close mother–offspring relationship facilitates the transmission of learned behavior (food choices, for

Figure 3.3 Ventral carriage of an infant by a female macaque. © topten22photo via iStock

instance). Misadventures in trial-and-error learning of social behavior by infants sometimes annoy adults. Maternal protection may be delayed or inhibited by the dominance status of the other adult, but tolerance of infant mistakes is common among primates.

In more than half of species surveyed for the trait, infant hair color contrasted with that of adults (Figure 3.4). According to one hypothesis, the infant coat attracts allocare. However, comparative analysis demonstrated no correlation between coat color and care from others (Treves 1997). Furthermore, the natal coat is replaced long before the infant becomes independent. Natal pelage may function as a so-called *paternity cloak*, disguising cues that might tell an adult male whether or not the infant is his. Without such information an unrelated male may be less likely to attempt infanticide.

Weaning terminates infancy, typically around the time that a youngster reaches about one-third of adult body mass. Mothers usually initiate weaning and offspring resist. Data from every major taxon indicate that conflict is widespread, intense, often prolonged, and concerns resources (Maestripieri 2002). Some of it seems to be about activity scheduling rather than allocation of resources, but the two interpretations are not mutually exclusive. The ultimate explanation (Trivers 1974) is that mothers have evolved to maximize successful reproduction by distributing resources to the maximum number of offspring over the course of a lifetime. On the other hand, each offspring attempts to maximize its own survival and health by obtaining an increased share of resources. The mother's most adaptive strategy is to propel the infant toward independence as soon as its survival is probable, while the infant would benefit from prolonging exploitation of the mother at the expense of future siblings.

Mammals with complex societies (elephants, for instance) require a **juvenile** stage in which the individual is independent of the mother but infertile and short of full growth. The

Figure 3.4 Female baboons with black and brown infants. Black is the neonatal color phase. © Curt Busse

primate juvenile phase lasts for months or years. Physical growth is slow as the individual accumulates experience that refines behavior. Some of this experience is applied to social life immediately, because larger juveniles receive little tolerance. However, adult hostility is mitigated by the fact that the juvenile is not a social or sexual competitor.

Play can be loosely defined as behavior with no immediate practical result (although a precise definition has proved to be elusive). It may entail manipulation of objects, solitary body movements (jumping, for example), or engagement with a conspecific (most commonly mock fighting) (Figure 3.5). Play begins in infancy and peaks during the juvenile period of social mammals (Pellis and Iwaniuk 2000). It is the single most prominent behavior in juvenile primates (Kappeler et al. 2003).

These activities provide the young with data that can be applied to complex problems later in life. Play motivates an individual to become active, involved, and integrated into its world (Dolhinow 1999). Given the expressive behavior associated with play and the lack of immediate external rewards, it is not anthropocentric to postulate that play is fun, i.e. produces pleasurable sensations. This is the emotional dimension of the motivational system. **Exploratory behavior**, the investigation of novel objects and places, can be considered a form of play.

Juvenile mortality rates are typically high (Walters 1987). Play may be a contributing factor since serious injuries occur, and juveniles are probably more vulnerable to some predators because they lack size and experience. Nutritional stress is significant: protein needs increase by up to 30 percent, and consequently weaned young must often increase their consumption of leaves and insects. Need is exacerbated by small body size that makes the digestion of low-quality food more difficult. Juveniles are especially at risk during times of scarcity because they lack foraging experience and they receive no direct help in finding

Figure 3.5 Young baboons in social play. © Glenn E. King

food. Juvenile feeding problems are eased by deciduous dentition that allows adjustments to adult diet while growth continues.

Adolescence begins with the onset of puberty, but the end of the period is less clear. According to Wallen and Zehr (2004), it is marked by attainment of adult reproductive physiology. Walters (1987) used the criterion of "effective reproduction". Females go through a period of adolescent sterility and adult males may reject them. Males in many species must attain full growth in order to compete for females, which may not occur until years after sexual maturity. The term *subadult* usually refers to an individual that is beyond adolescence and can designate the period between physiological and social maturation. This is a distinctive life-stage because of individual changes and because subadults play distinctive roles in primate societies.

Dispersal (emigration) from the natal group usually takes place around the time of puberty, or at least before adulthood. Most emigrants end up in another society, but they may spend a period of time alone or with other recent emigrants. Kin relationships can survive dispersal: relatives may emigrate together or a lone emigrant may connect with previously dispersed relatives. *Dispersal* should not be confused with the *dispersed* condition of societies that are not cohesive.

Costs of dispersal include greater exposure to predators, the need to become familiar with a new home range, and hostility from residents of other groups. Dispersal can be explained by competition in the natal society for resources and/or mates. Another possibility is **inbreeding avoidance** (inhibition of reproduction with close relatives, which can have harmful genetic consequences). A survey by Field and Guatelli-Steinberg (2003) suggested that both causes are involved, varying from one genus to another, but that resource competition is more important than inbreeding avoidance. The cost of inbreeding may in some cases be outweighed by benefits from associating with relatives.

Explanations for dispersal raise questions about levels of causation. For example, competition can be interpreted in both functional and proximate terms simultaneously: the functional behavior of departure is stimulated by proximate factors such as aggression or deprivation. In contrast, inbreeding avoidance is strictly an ultimate explanation, since the animals are not aware of genetics. In the few cases where we have evidence, the proximate mechanism for inbreeding avoidance seems to be contact in early life. According to the Childhood Familiarity Hypothesis, which has been applied to humans and nonhuman primates, close acquaintance with another individual while immature will result in a lack of sexual interest in that individual after puberty. Females especially lack sexual interest in their prepubertal associates because maternal investment makes a defective offspring more costly.

Adulthood is marked by two major developments in primates, as in other mammals. One is full sexual maturity; thus, the size and age of males and females at first breeding are important life history variables from one species to another. The other major development is full growth to species-typical body size and proportions. Adult body size is correlated with many critical variables in reproduction and life history, including gestation length, number and sex of offspring, size and age at weaning, size and age at first breeding, interbirth interval (IBI), and lifespan. This is why comparisons among mammals and among primates must control for body size (Kappeler et al. 2003; Purvis et al. 2003). However, size is probably not the causal variable: the body size of a species seems to be derived from its life history strategy rather than vice versa, and that strategy is adapted to environmental conditions. Maturity comes late in primates compared to most other mammals. Combined with the tendency toward single birth, this means that the reproductive rate is low.

Continuation of social play into adulthood is a recurring pattern. The most common form is mock fighting. In most cases an adult plays with an immature individual after the young one initiates the interaction. However, adult primates also play with each other and these interactions may have important functions. According to the Social Assessment Hypothesis, an individual uses play to get information about the partner and perhaps to manipulate the partner. A survey of 71 primate species, representing all living families, supported the prediction that play fighting would be more prominent in situations of social uncertainty (Pellis and Iwaniuk 2000). Uncertainty occurs when social interactions are infrequent or erratic, so that the partners are not familiar with each other. Play can provide information and perhaps establish a dominance relationship without real fighting.

The length of the lifespan has a typical maximum for a given species (Zihlman and Bolter 2004). Primates are among the longest-lived mammals, even after correction for body size. This can be explained by low vulnerability to environmental causes of death, resulting from body size, arboreal life, and living in social groups (Austad and Fischer 1992). Long life is associated with social complexity and a reduction in serious fighting (Kappeler et al. 2003). A long adult life provides added time for reproduction, which may be important in relation to the low reproductive rate of primates. One or two additional offspring might be a significant percentage of the total produced by a female during her lifetime.

Primate Cognition

Some scientists argue that primate mental abilities are relatively simple. They consider attribution of mental complexity to nonhuman primates, such as an understanding of causality, to be anthropomorphic (Penn and Povinelli 2007). Along somewhat the same lines, Barrett and Rendall (2010) offered several evolutionary principles to guide the study of cognition:

1 Animals will know only what is needed for survival and reproduction, because natural selection will not favor the costly storage of unnecessary information.
2 Cognition is **embodied**—i.e. it emerges from the integration of brain, body, and environment—because it evolves to engage the world, not reflect on it.
3 Cognition is **distributed** in that the process emerges from interaction among conspecific individuals.

The third principle is illustrated by alternative interpretations of a female grooming a mother. Some would say that the groomer intends deception with the goal of handling the baby. A simpler account views the interaction as a series of behavioral adjustments: an interested female approaches a mother; the tense mother turns her back; the other female responds to the presented back by grooming; and the relaxing mother allows the other to handle the baby.

Two points can be made here. First, these theories pertain to elementary primate behaviors and do not necessarily encompass the complex actions of some primates (e.g. the sequential use of tools). Second, primates may have begun their existence with fairly complex cognitive skills. Mammals in general have greater mental capabilities than scientists have attributed to them in the past (Tomasello and Call 1997). All studied species have (albeit varied) skills of learning, memory, quantification (numbers and ordering), discrimination (distinguishing among stimuli), and categorization (conceptualizing types of stimuli). Many species have complex spatial memory and navigation skills, and search for objects when they disappear. Some species engage in object manipulation, tool use, and other forms of physical problem solving.

A major issue pits the idea of general intelligence against theories of **specific intelligence**. The latter view is that mental processes are divided into components responsible for particular areas of adaptive behavior. Such components are known by various terms, including **domain** and **module**. For example, a mental domain might encompass fear reactions, and a module of that domain might govern escape from predators. Against this, the general intelligence approach would attribute most or all fear responses to learning.

The hypothetical aggregate of specific mechanisms has been called **core knowledge**, a set of systems dedicated to learning various adaptive behaviors (Shutts et al. 2009). One version of core knowledge identifies four systems (some would probably say domains) representing objects, actions, number, and space, and perhaps a fifth that represents social partners. This interpretation is intermediate between two extreme ideas: (1) the mind as a completely flexible learning mechanism; and (2) the mind as a set of special-purpose mechanisms, each with a very specific function.

These interpretations of mental processes do not necessarily imply corresponding physical components in the brain. Neurological research shows that at least some behavior is based on circuits that connect structures in different parts of the brain. If such a system governs one kind of behavior, it can be considered a *mental* domain or module even though its *physical* basis is dispersed.

Deaner and van Schaik (2006) advocated the idea of general intelligence and offered a meta-analysis of cognitive tests given to primates (Box 3.1). Some taxa did better than others across a range of diverse tests, which the analysts considered evidence for domain-general cognitive ability. None of the taxa performed especially well in any particular areas of cognition, which fails to support the idea of domain-specific abilities. However, the researchers noted that they could not exclude the possibility of domain-specific abilities in cognitive areas not tested.

Box 3.1 A Sample of Cognitive Tests Administered to Nonhuman Primates

These are experimental tests of mental function commonly administered to nonhuman primates (Deaner and van Schaik 2006). Each test is described and then its supposed implications are stated in italics.

Patterned-string problems

The subject is shown an array of strings or wires, one of which is attached to a food reward. In many versions the strings are crossed or otherwise misleading. The subject is allowed to pull just one of the strings to try to obtain the reward. *Solution of the problem may represent learning in some cases, but not where the subject's first correct response is on the first trial. In these cases mental representation of spatial relations among objects can be inferred.*

Invisible displacement

The subject sees a reward placed in a container, which is then passed behind one or more of an array of barriers. The subject is then shown that the container is now empty and allowed to search behind the barriers. *If the subject only looks behind those barriers that hid the container, it is inferred that he/she can mentally represent the existence and spatial movement of unseen objects.*

Tool use

The subject is given an object that can be used to achieve an apparent purpose. *Success indicates that the subject understands how his/her actions affect the object and how the object affects another object or the substrate. This has implications for causal reasoning, spatial representation, and motor coordination.*

Object discrimination learning set

The subject is first presented with two objects and given a reward for selecting the one that has been arbitrarily chosen as "correct". Then the same problem is posed with two novel objects. *If the subject solves the second problem more quickly than the first, then a learning set has been formed. This indicates that the subject has formulated an abstract rule or hypothesis (e.g. "win-stay, lose-shift"), rather than just learned simple discrimination.*

Reversal learning

A subject learns to discriminate between two objects to get a reward. Then the reward status is changed from one object to the other. The speeds of learning in the two situations are compared. *If effective initial learning leads to poor reversal performance, the subject had achieved simple associative learning. If effective initial learning leads to effective reversal, the subject has formulated an abstract rule/hypothesis.*

Oddity learning

The subject is presented with three visual stimuli, two identical and one "odd". The subject is rewarded for choosing the differing stimulus. *Success indicates use of a relational or abstract concept.*

Sorting

The subject sorts objects according to categories based on apparent stimuli and is rewarded for success. *This tests for the ability to form abstract concepts and to use them for categorization.*

Delayed response

The subject sees a reward being hidden in one of two locations. After a delay he/she is allowed to choose one location to search. *A good performance has implications for memory and for the retention of a mental representation after the stimulus is gone.*

One long-standing theory of primate mental capabilities holds that they mainly resulted from ecological challenges, such as the need for mental mapping of resource locations (Milton 1980). Zuberbühler and Janmaat (2010) saw support for this thesis in the association of increased fruit size with the radiation of modern-type primates after 85 million years ago. They asserted that primates have been very successful in monopolizing arboreal space in most tropical forests because of their cognitive skills.

The alternative viewpoint attributes primate mental capabilities to complex social life and the developmental processes that prepare individuals for it. The length of the juvenile period in primates correlates with the size of the social group and with the relative size of the non-visual neocortex, the part of the brain that is responsible for learning and intelligence (Joffe 1997). Lewis (2000) compared play in a limited but diverse sample of captive primates and found that social play, but not solitary or object play, was correlated with the neocortex size. This theory has been expanded in numerous publications about the **social brain**.

Cunningham and Janson (2007) argued for balance between social and ecological explanations, because the processes are intertwined. Primates learn much about the environment, such as predator identification and tool use, from other members of their social group. Awareness of both social rank and resource availability are likely to influence an individual's decision to follow another animal to food or to avoid the other in order to prevent competition.

Summary

Primates live in societies containing adults of both sexes. They eat fruit for energy and leaves and/or animals for protein. Allogrooming controls parasites and facilitates relationships. Dispersal shapes group composition. Violence is occasional, but can have serious consequences. The Socioecological Model probes interrelationships between environmental and social patterns. Basic male reproductive strategy tries to maximize fertilizations, and infanticide may be one tactic. Lengthy gestation constrains mothers to nurse and care for offspring. Play and exploration facilitate learning. A long juvenile period allows extensive learning. Adolescence is a time of social and sexual transition. Long life makes early learning useful and partially offsets low reproductive output. Primates probably have greater cognitive abilities than most mammals.

Sources

Basic Bibliography

Campbell, C.J., et al. (Eds.) (2011). *Primates in perspective* (2nd ed.). NY: OUP. [Parts Four, Five, & Six]

Dixson, A.F. (2012). *Primate sexuality*. NY: OUP.

Ghazanfar, A., & Platt, M. (Eds.) (2010). *Primate neuroethology*. NY: OUP.

Kappeler, P.M., & Pereira, M.E. (Eds.) (2003). *Primate life histories and socioecology*. Chicago: UCP.

Mitani, J.C., et al. (Eds.) (2012). *The evolution of primate societies*. Chicago: UCP. [Parts 2, 3, 4, & 5]

Reed, K.E., & Bidner, L.R. (2004). Primate communities: past, present, and possible future. *YPA*, *47*, 2–39.

Additional References

Austad, S.N., & Fischer, K.E. (1992). Primate longevity: its place in the mammalian scheme. *AJP*, *28*, 259–261.

Barrett, L., & Rendall, D. (2010). Out of our minds: the neuroethology of primate strategic behaviour. In M.L. Platt & A.A. Ghazanfar (Eds.), *Primate neuroethology*. NY: OUP.

Burrows, A.M., & Nash, L.T. (Eds.) (2010). *The evolution of exudativory in primates*. NY: Springer.

Capellini, I., et al. (2008). Phylogenetic analysis of the ecology and evolution of mammalian sleep. *Evolution 62*, 1764–1776.

Chapman, C.A., et al. (2013). Primates as ecosystem engineers. *IJP*, *34*, 1–14.

Constantino, P.J., & Wright, B.W. (2009). The importance of fallback foods in primate ecology. *AJPA*, *140*, 599–602.

Corlett, R.T., & Primack, R.B. (2011). *Tropical rain forests*. Oxford: Wiley-Blackwell.

Crofoot, M.C. (2012). Why mob? Reassessing the costs and benefits of primate predator harassment. *FP*, *83*, 252–273.

Cunningham, E., & Janson, C. (2007). A socioecological perspective on primate cognition, past and present. *AC*, *10*, 273–281.

Deaner, R.O., and van Schaik, C.P. (2006). Do some taxa have better domain-general cognition than others? A meta-analysis of nonhuman primate studies. *Evolutionary Psychology*, *4*, 149–196 (human-nature.com/ep, last accessed 26 July 2015).

Dolhinow, P. (1999). Play: a critical process in the development system. In P. Dolhinow (Ed.), *The nonhuman primates*. Mountain View, CA: Mayfield Publishing Co.

Ebensperger, L.A., et al. (2012). Direct fitness of group living mammals varies with breeding strategy, climate and fitness estimates. *Journal of Animal Ecology*, *81*, 1013–1023.

Ferrari, S.F., et al. (2008). Geophagy in New World monkeys. *FP*, *79*, 402–415.

Field, M., & Guatelli-Steinberg, D. (2003). Dispersal and the inbreeding avoidance hypothesis. *Primate Report 67*, 7–60.

Headland, T.N., & Greene, H.W. (2011). Hunter-gatherers and other primates as prey, predators, and competitors of snakes. *PNAS*, *108*, 1470–1474.

Heymann, E.W., & Hsia, S.S. (2015). Unlike fellows – a review of primate-non-primate associations. *Biological Reviews*, *90*, 142–156.

Houle, A., et al. (2007). Intratree variation in fruit production and implications for primate foraging. *IJP*, *28*, 1197–1204.

Isbell, L.A., & Young, T.P. (2002). Ecological models of female social relationships in primates. *Behaviour 139*, 177–202.

Janson, C.H. (2000). Primate socio-ecology, the end of a golden age. *EA*, *9*, 73–86.

Joffe, T.H. (1997). Social pressures have selected for an extended juvenile period in primates. *Journal of Human Evolution*, *32*, 593–605.

Kappeler, P.M., & van Schaik, C.P. (2002). Evolution of primate social systems. *IJP*, *23*, 707–740.

Kappeler, P.M., et al. (2003). Primate life histories and socioecology. In P.M. Kappeler & M.E. Pereira (Eds.), *Primate life histories and socioecology*. Chicago: UCP.

King, G.E., & Steklis, H.D. (1984). New evidence for the craniocervical killing bite in primates. *JHE*, *13*, 469–481.

Komers, P.E., & Brotherton, P.N. (1997). Female space use is the best predictor of monogamy in mammals. *PRSLB*, *264*, 1261–1270.

Krishnamani, R., & Mahaney, W.C. (2000). Geophagy among primates. *AB*, *59*, 899–915.

Lesku, J.A., et al. (2008). Phylogenetics and the correlates of mammalian sleep: a reappraisal. *Sleep Medicine Reviews 12*, 229–244.

Lewis, K.P. (2000). A comparative study of primate play behaviour: implications for the study of cognition. *FP*, *71*, 417–421.

Lima, S.L., & Rattenborg, N.C. (2007). A behavioral shut-down can make sleeping safer. *AB*, *74*, 189–197.

Lindenfors, P.F., et al. (2004). Females drive primate social evolution. *BL*, *271*, S101–S103.

Lovell, N.C. (1991). An evolutionary framework for assessing illness and injury in nonhuman primates. *AJPA*, *34*, 117–155.

Maestripieri, D. (2002). Parent–offspring conflict in primates. *IJP*, *23*, 923–951.

Majolo, B., et al. (2008). Costs and benefits of group living in primates: group size effects on behaviour and demography. *AB*, *76*, 1235–1247.

Marshall, A.J., & Wrangham, R.W. (2007). Evolutionary consequences of fallback foods. *IJP*, *28*, 1219–1235.

McGraw, W.S., & Berger, L.R. (2013). Raptors and primate evolution. *EA*, *22*, 280–293.

McGrew, W.C. (2014). The other faunivory: the significance of insects and insect resources for nonhuman primates, modern humans, and extinct hominins. *JHE*, *71*, 4–11.

Milton, K. (1980). *The foraging strategy of howler monkeys: a study in primate economics*. NY: Columbia University Press.

Mueller, A.E., & Soligo, C. (2005). Primate sociality in evolutionary context. *AJPA*, *128*, 399–414.

Mueller, A.E., & Thalman, U. (2000). Origin and evolution of primate social organisation: a reconstruction. *Biological Reviews*, *75*, 405–435.

Muller, M.N., & Wrangham, R.W. (Eds.) (2009). *Sexual coercion in primates and humans*. Cambridge, MA: HUP.

Nunn, C.L., et al. (2010). Primate sleep in phylogenetic perspective. In P. McNamara & 2 co-editors, *Evolution of sleep* (pp. 123–144). NY: CUP.

Overdorff, D., & Parga, J. (2007). The new era of primate socioecology: ecology and intersexual conflict. In C.J. Campbell, et al. (Eds.), *Primates in perspective*. NY: OUP.

Pellis, S.M., & Iwaniuk, A.N. (2000). Adult-adult play in primates. *Ethology*, *106*, 1083–1104.

Penn, D.C., & Povinelli, D.J. (2007). Causal cognition in human and nonhuman animals: a comparative, critical review. *Annual Review of Psychology*, *58*, 97–118.

Preston, B.T., et al. (2009). Parasite resistance and the adaptive significance of sleep. *BMCEB*, *9*, 7.

Purvis, A., et al. (2003). Primate life histories and phylogeny. In P.M. Kappeler & M.E. Pereira (Eds.), *Primate life histories and socioecology*. Chicago: UCP.

Rees, A. (2009). *The infanticide controversy*. Chicago: UCP.

Schultz, S., et al. (2011). Stepwise evolution of stable sociality in primates. *Nature*, *479*, 219–222.

Shanahan, M., et al. (2001). Fig-eating by vertebrate frugivores: a global review. *Biological Reviews*, *76*, 529–572.

Shutts, K., et al. (2009). Core knowledge and its limits: the domain of food. *Cognition*, *112*, 120–140.

Silk, J.B. (2007). The adaptive value of sociality in mammalian groups. *PTRSB*, *362*, 539–559.

Silk, J.B. (2009). Nepotistic cooperation in nonhuman primate groups. *PTRSB*, *364*, 3243–3254.

Silk, J.B. (2011). Evolutionary biology: the path to sociality. *Nature*, *479*, 182–183.

Sussman, R.W., et al. (2005). How important are affiliation and cooperation in the evolution of primate sociality? *AJPA*, *128*, 84–97.

Thierry, B. (2008). Primate socioecology, the lost dream of ecological determinism. *EA*, *17*, 93–96.

Tomasello, M., & Call, J. (1997). *Primate cognition*. NY: OUP.

Treves, A. (1997). Primate natal coats: a preliminary analysis of distribution and function. *AJPA*, *104*, 47–70.

Trivers, R.L. (1974). Parent–offspring conflict. *American Zoologist*, *14*, 249–264.

van der Post, D.J., et al. (2013). A novel mechanism for a survival advantage of vigilant individuals in groups. *AN*, *182*, 682–688.

van Schaik, C.P., & Janson, C.H. (Eds.) (2000). *Infanticide by males*. NY: CUP.

Vinicius, L., & Mumby, H. (2013). Comparative analysis of animal growth: a primate continuum revealed by a new dimensionless growth rate coefficient. *Evolution*, *67*, 1485–1492.

Wallen, K., & Zehr, J.L. (2004). Hormones and history: the evolution and development of primate female sexuality. *Journal of Sex Research*, *41*, 101–112.

Walters, J.R. (1987). Transition to adulthood. In B.B. Smuts, et al. (Eds.), *Primate societies*. Chicago: UCP.

Wheeler, B.C., et al. (2011). Predictors of orbital convergence in primates: a test of the snake detection hypothesis of primate evolution. *JHE*, *61*, 233–242.

Whitten, P.L., & Brockman, D.K. (2001). Strepsirrhine reproductive ecology. In P.T. Ellison (Ed.), *Reproductive ecology and human evolution*. NY: Aldine de Gruyter.

Woodward, S.L. (2012). Biomes of the world. Department of Geospatial Science, Radford University. https://php.radford.edu/~swoodwar/biomes/ (last accessed 26 July 2015).

Wrangham, R. (1980). An ecological model of the evolution of female-bonded groups of primates. *Behaviour*, *75*, 262–300.

Zihlman, A.L., & Bolter, D.R. (2004). Mammalian and primate roots of human sociality. In R.W. Sussman & A.R. Chapman (Eds.), *The origins and nature of sociality*. Hawthorne, NY: Aldine de Gruyter.

Zuberbühler, K., & Janmaat, K. (2010). Foraging cognition in nonhuman primates. In M.L. Platt & A.A. Ghazanfar (Eds.), *Primate neuroethology*. NY: OUP.

4 The Strepsirrhine Suborder

A small creature busily searches for plant foods and insects in the trees in the dark. It seems to be alone. But then it calls, and another answers. At dawn they reunite and settle into their hidden sleeping place.

Extant strepsirrhines represent one of the three major primate lineages. They retain many ancestral mammalian traits, but also display some striking derived features. Chapter 1 introduced them, Box 4.1 provides reminders, and Figure 4.1 illustrates some of their traits.

Box 4.1 Strepsirrhine Traits

- majority of species nocturnal
- naked rhinarium facilitates olfactory sense
- prominent ears represent acute hearing
- vibrissae provide tactile information
- tapetum lucidum enhances night vision
- toothcomb and grooming "claws" are important derived traits.

Strepsirrhine olfaction is like that of most other mammals in anatomy and function; growth and maturational characteristics of the system are basically similar to that of rodents (Garrett et al. 2013). This includes the **vomeronasal organ**, located at the base of the nasal cavity, that augments the external olfactory apparatus. Olfaction plays a major role in both ecology and social behavior. The vibrissae are attached to distinctive musculature in the upper lip, so that they can be used actively in spatial recognition and navigation (Muchlinski et al. 2013). Large and membranous ears are made mobile by distinctive muscles.

A large-scale phylogenetic reconstruction concluded that ancestral primates were nocturnal, and that this has been retained in the majority of extant strepsirrhine species (Santini et al. 2015; supported by Joffe et al. 2014). Though it enhances night sight, the tapetum lucidum reduces visual acuity so that strepsirrhines probably do not see things very sharply by our standards. Typical strepsirrhine vision, as in most other mammals, is comparable to red–green color blindness in humans. They have just two kinds of *cones*, the color-identifying retinal cells that are sensitive to various wavelengths of light. This is the basis for the term **dichromatic vision**.

Figure 4.1 A lesser galago illustrates strepsirrhine traits: large ears, large eyes, and naked rhinarium. © Y.A. de Jong and T.M. Butynski, www.wildsolutions.nl

Box 4.2 Strepsirrhine Taxonomy

The obvious derived features of the strepsirrhines, the toothcomb and the grooming claws, have been described in the text. Chemicals used by the tapetum lucidum differ from those of other nocturnal mammals, which indicate that it is derived. The postorbital bar of the strepsirrhines differs from the flange of the tarsiers and the fully enclosed orbit of the anthropoids. However, it is probably an ancestral trait that originated with primates that preceded the differentiation of the three lineages represented today.

There is controversy about the phylogenetic significance of primate reproductive organs. The bicornuate uterus of the strepsirrhines is characterized by two projections or "horns". This, like the multiple pairs of nipples, seems to be associated with an ancestral pattern of multiple births. The outermost membrane of the embryo (the chorion) attaches to the internal epithelial lining of the uterus without penetrating it to any degree. This epitheliochorial placentation has been characterized as the simplest kind of connection between the embryo and the wall of the uterus (Martin 1990). However, it is now seen as a derived trait of the strepsirrhines (Carter and Enders 2013).

There is a consensus that the strepsirrhine taxon encompasses two clades, but debate about further classification continues. A common opinion is that these lineages should be separated at the infraorder level as Lorisiformes and Lemuriformes (Fleagle 2013; Kappeler 2012), placing them just below the Suborder Strepsirrhini.

Masters et al. (2013) arrived at a somewhat different conclusion by integrating molecular and fossil evidence and strictly applying taxonomic rules and the history of primate taxonomy. They concluded that two broader categories should occupy the infraorder level, and that Lorisiformes and Lemuriformes (keeping the same names) should be reduced to the next level, called "series". The separation date between lorisiforms and lemuriforms has been placed at 55 to 60 million years ago (Finstermeier et al. 2013).

There are two living subdivisions of strepsirrhines, which we will call lorisiforms and lemuriforms. The former are scattered across the forests of Africa and Asia, while the latter are found only on the island of Madagascar. This chapter reviews general features of the strepsirrhines. The subdivisions are covered in the following chapters.

Strepsirrhine Behavior

Strepsirrhine Ecology

Most strepsirrhines are small, like nocturnal mammals in general. With few exceptions they weigh between 0.1kg and 1kg. Their size has important effects on metabolism, subsistence, and susceptibility to predation. **Metabolism** refers to the extraction of energy from the environment, mostly through food, and its application to life processes. The **basal metabolic rate** (BMR) is an adult's minimal rate of energy expenditure when it is resting and not directly affected by factors such as digestion, temperature, or stress. Life processes (e.g. growth) and activities (e.g. running) require additional energy.

BMR is correlated with body mass in terms of the **Jarman/Bell Principle** (Gaulin 1979): metabolic rate increases in proportion to mass, but more slowly. Smaller species need more energy per unit of body mass to maintain body temperature because heat loss is increased by a high surface-to-volume ratio. Thus small species require high-energy foods. Insects qualify, but they come with costs. Each insect is a small package of nutrients, elusive in some cases, and the supply varies seasonally. Dependence on insectivory tends to constrain body size. According to **Kay's Threshold,** this adaptation is viable for animals weighing up to about 500g.

Larger primates have slower metabolisms, using less energy per unit of body mass, and can survive on more mixed diets. Many combine fruit with either insects or leaves. Folivores tend to be the largest in a taxon. They spend a lot of time gathering and digesting food that is relatively low in quality but abundant. Lower metabolic rates permit the slow physiological process of digesting foods such as leaves.

The strepsirrhine diet emphasizes fruit and insects, a high-energy diet appropriate to the small size of most strepsirrhines. Some exceptions are based on a recurrent pattern of consuming gums. The toothcomb, used as a scoop, facilitates feeding on gums and fruit pulp in some species (Eaglen 1980).

Thermoregulation is an important factor in metabolism that is related to piloerection, the involuntary raising of the hair that provides insulation for the body. It is controlled by muscles called MAP (musculi arrectores pilorum). The strepsirrhine body is poorly supplied with MAP (probably an ancestral condition), which has resulted in a variety of metabolic and behavioral patterns for the conservation of energy (Chaplin et al. 2014).

Strepsirrhine hands are more flexible and controlled than the forepaws of other mammals, even those that are highly arboreal. These primates can grasp objects (such as plant parts and small prey) with one hand, and they tend to contact supports (such as branches) with thumb and forefinger around them. However, they show little individual control of digits and tend to grip objects between fingers and palm. Strepsirrhines rarely groom fur with their fingers (Goosen 1987). They groom by licking, as do other mammals, and they use their specialized claws and teeth.

A positional system called **vertical clinging and leaping** (VCL) is recurrent (Figure 4.2). The strepsirrhine rests in a vertical position while clinging to a vertical substrate such as a small tree trunk or stalk of bamboo. It moves from one resting point to another by jumping with the aid of long, powerful hind legs. The animal pushes off with the legs and makes first contact with them. Gebo (2011) pointed out that leaping across gaps is a typical primate behavior, so the vertical clinging component is the distinctive feature. Leaping from that vertical posture is also unusual. Common specializations of ankle and knee facilitate the use of vertical supports in VCL. Nevertheless, the pattern seems to have evolved separately in distantly related strepsirrhine lineages.

Small size increases vulnerability to a wider range of predators, including owls at night, but it may also aid **crypsis** (i.e. avoiding attention). Foraging alone at night, most strepsirrhines

Figure 4.2 A Kenya coast galago demonstrates the clinging component of vertical clinging and leaping. © Y.A. de Jong and T.M. Butynski, www.wildsolutions.nl

are relatively inconspicuous. When day arrives they sleep in clumps of foliage, tree-holes, or constructed nests. The choice among these refuges varies across individuals, species, seasons, environments, and situations (for example, some species make nests only when infants are present). Despite these potential defenses, many strepsirrhines are subject to substantial rates of predation and doubts have been raised about the effectiveness of crypsis.

Nocturnal mammals can detect predators with their sense of smell. However, the primate visual system sets them apart, as shown by comparison to other nocturnal mammals in a meta-analysis by Prugh and Golden (2013). They found that moonlight tends to suppress nocturnal activity by various non-primate species. In contrast, all the strepsirrhines in their sample (ten species in five genera of both infraorders) increased their activity on moonlit nights (the same was true of one genus of anthropoids). These researchers offered one note of caution. All the primates in their sample are frugivores or omnivores, so no solid conclusion can be drawn about insectivores. This is important because the level of moonlight activity results from a balance between avoiding predators and finding food.

Strepsirrhine Sociality

Solitary species are common among strepsirrhines in the sense that adults are usually alone during the nightly period of activity (Kappeler 2012). However, there are often social connections that are not immediately apparent. These are maintained by vocal and olfactory communication. In some cases these connections become clear when members of a society sleep together. Longer-term observations may demonstrate that they have a common home range. Terms such as **dispersed social system** are sometimes applied in such cases.

Strepsirrhines constitute the majority of primate species that are characterized by pairs and pair-groups (see Chapter 3). The core of such a group is a male and a female in a long-term relationship minimally based on having roughly the same range and synchronizing their movements to some extent (Kappeler and van Schaik 2002). This leads us to a history of problems that began with reports of "monogamous" groups (Fuentes 1998). Setting aside anthropomorphisms, the main issue was the unsupported assumption that the social relationship between the male and female was also the sole sexual relationship. This misstep argues for the analytical separation of social and mating systems advocated by Kappeler and van Schaik.

Fuentes (2002) distinguished two kinds of long-term relationship: (1) the **sexual pair** (an exclusive copulatory relationship between two adults within a group that may or may not include other adults); and (2) the **social pair** (an association between two non-kin adults characterized by spatial proximity and affiliative behaviors and/or energetic investment in offspring or territorial defense). Fuentes used the word *bond* in both cases, but this is problematic because it suggests a proximate mechanism (an emotional connection) that has rarely been demonstrated for primate pairs.

A pair-group may contain adult offspring that have not emigrated and sometimes includes one or more unrelated adults. With regard to the central pair, the distinction between social pair and sexual pair (without the assumption of bonding) is useful. Thus, behavioral evidence for a central pair distinguishes a pair-group from a troop. In many species with this kind of organization, we see variation from pairs to groups that include two adult females, and sometimes two adult males.

The typical organization of some nocturnal species fits the definition of **noyau** (Rowe 1996): a society in which the home range of one male encompasses those of several females and the male defends his territory against other males. There are also some multi-male communities in which several males interact with each other and with several females. Mating in the multi-male community is promiscuous and competitive, while the noyau and the mated pair imply greater limitation on sexual behavior. All of these social units are dispersed in ordinary circumstances.

Changing concepts of strepsirrhine sociality represent increasingly sophisticated fieldwork under difficult conditions (Gursky 2005). Brief studies with limited equipment left impressions of solitary life. Then we learned that many strepsirrhines sleep in groups, where they display various affiliative behaviors that may include allogrooming and play. They were reconceptualized as "solitary-but-social" or solitary foragers, i.e. spending most or all of their waking time foraging alone, even if they are social sleepers. Telemetry had facilitated the research, but had its pitfalls. Fieldworkers tended to follow one focal animal at a time and could easily miss interactions with nearby individuals in the dark. Now there is increasing evidence for socializing during the active period, and the proportion of social interactions may fall well within the range of anthropoids (Nekaris and Bearder 2011).

Strepsirrhines use chemical signals for individual recognition, sexual recognition, dominance interaction, and mating coordination (delBarco-Trillo and Drea 2014). Most strepsirrhines, like many other mammals, engage in **urine marking** of the environment. Chemicals in urine can identify an individual who deposits it and can contribute to mutual sexual stimulation. The main function of the vomeronasal organs in the roof of the mouth seems to be to detect chemicals in female urine, which allows a male to ascertain the reproductive status of a female from the scent she has left behind. The most common form of communication with urine is *punctuated urine marking* (also called *rhythmic micturition*) in which the animal deposits small quantities of urine while it is moving. On the other hand, only nocturnal strepsirrhines perform **urine washing** in which the individual urinates on the hands and feet. Urine washing seems to be associated with aggressive or stressful situations, such as moving into a new territory or approaching an **estrus** female.

Strepsirrhines also use specialized cutaneous glands to scent mark environmental features and each other. These chemical signals are vigorously investigated and over-marked by conspecifics. They encode information about species and individual identity, genetic similarity, sex and social status, reproductive state, and health. A comparative study of 12 species suggests that a shift in emphasis from urine marking to glandular marking is associated with increased activity in daylight (delBarco-Trillo et al. 2011). The researchers found greater chemical complexity in the urine of markers versus non-markers. The chemical profiles of the markers were more discrete, suggesting a function as species signatures.

The researchers suggested four factors in the shifting emphasis from urine to glandular marking. First, the relative simplicity of urine's chemical composition suggests that it is a less costly mode of communication; thus, stronger arguments for the benefits of glandular marking are required. Second, cooler night temperatures may facilitate the continued integrity of the simpler urine chemicals, while glandular secretions resist degradation in daytime temperatures. Third, glandular marks often entail distinctive postures and visible marks so that this becomes a multi-modal form of communication. Fourth, urine marking seems designed for delayed detection while glandular marking is more conspicuous.

Facial communication is limited by the fact that the upper parts of the mouth are firmly tethered to the gum. However, a variety of other visual signals occur. At a high level of excitement the animal may display erection of the ears, widening of eyes and mouth, and spreading of arms and hands. The average vocal repertoire is smaller than those of most anthropoids in size and complexity. Some of these signals are widely shared among strepsirrhines (Petter and Charles-Dominique 1979). One example is contact rejection calls that function to prevent the approach of a conspecific, which include staccato spitting sounds and grunting divided into inhaled and exhaled phases. Distress calls that express fear and/or pain, comparable to those of other mammals, are also consistent throughout the strepsirrhines. Experiments show that these calls attract others. They arrive quickly, observe for a few seconds, and then leave, often uttering alarm calls. This reaction, when the threat is real, may facilitate predator recognition. Unlike the ubiquitous distress calls, alarm calls are typical of particular species.

Strepsirrhines sometimes groom each other, licking and using the toothcomb, but less often than they groom themselves (Goosen 1987). Allogrooming is most likely to be sexual or maternal, though it may be more common among strepsirrhine species than has been thought. In any case, self-grooming with the toothcomb seems to serve a social as well as a hygienic function. It stimulates glands in the skin and distributes the scents that they produce all over the body, which sends social signals to other individuals.

Sex is a small part of strepsirrhine life because it typically occurs during one brief period each year. During these mating periods a female undergoes one or more **ovarian cycles** in which fertile eggs become available. In strepsirrhines these are **estrus cycles** like those of other mammals, which means that sexual behavior is tightly regulated by hormonal changes (Dixson 2012). The **peri-ovulatory period** (the time around ovulation) is typically the main time for sexual activity.

Key components of sexual interaction in all mammals are **attractivity** and **receptivity**. Attractivity refers to signals from the female that stimulate sexual interest from the male. Receptivity denotes the female's readiness (physical and behavioral) to accept the male's sexual attempts. A female strepsirrhine's attractivity increases shortly before she becomes receptive, probably as a result of hormonal effects on vaginal, urinal, or cutaneous cues. The main function of the vomeronasal organ seems to be to detect chemicals in female urine. This allows a male to ascertain the reproductive status of a female from the scent she has left behind. Males approach females, try to sniff their genitals, and often follow when the females move away. Females usually react aggressively to these intrusions, and males usually back off (at least for a while), but the females eventually prove to be receptive during a period lasting one to three days (estrus). This is the only time that copulation takes place.

Copulation commonly takes the form of a **single prolonged intromission** (SPI) with a duration that ranges from four minutes to four hours (contrasting with under one minute for brief-intromission species). As to function, SPI could be the most direct form of **mate guarding**, preventing other males from copulating with the female. It could also provide the best opportunity for multiple ejaculations, enhancing the male's status in sperm competition. These hypotheses indicate mating competition despite the tendency toward prolonged male–female social relationships in pair-groups or noyaus. Neighboring or roaming males may try to steal copulations.

These hypotheses assume that sexual selection governs sexual behavior, but sexual behavior can be affected by natural selection (Dixson 2012). For example, predator pressure may affect copulation length. All things being equal, long copulations increase risk. On the other hand, there is less risk for small, cryptic, nocturnal species. However, low predator pressure can only explain why SPI is *possible* for some species. To understand how it might be *adaptive*, we have to return to sexual selection hypotheses.

The social arrangements of strepsirrhines provide little opportunity for sexual monopolization of females by males beyond an SPI. One reason is that females tend to be widely dispersed, as individuals or in small groups. Another is that highly seasonal breeding means a limited time during which fertile females are available. Physical features in many species indicate the importance of sperm competition, including large testes, complex penile structures, and complex copulatory patterns.

The relative lack of direct competition is probably a factor in the low levels of sexual dimorphism. Only about half the species in a survey displayed any significant sexual dimorphism in the dimensions of the canine teeth, and this included 6 of 13 species where such competition was to be expected. These results fail to support sexual selection theory (Kappeler 1996). Male **roaming** contributes to sexual competition. That is, some males leave their usual home ranges during the mating season and travel widely in search of estrus females.

Strepsirrhine Life Cycle

Multiple offspring (usually two or three) recur in a number of taxa. Some species can deliver twice in one year because of either a short lactation period or a resumption of estrus before lactation ends (Whitten and Brockman 2001). However, most females in most strepsirrhine species give birth to a single infant most of the time, and the overall average for litter size is barely more than one. Females of all strepsirrhine species have at least two pairs of nipples, but this seems to be a relict from ancestors in which multiple births were typical. Birth mass is smaller for offspring in litters, and weaning age is relatively young.

Neonates of most nocturnal strepsirrhines are kept in nests for several days following birth, and sometimes for a week or more (Ross 2001). The mother may carry them from one nest to another, which could be a defense against predation and/or male infanticide. However, it might also prevent the build-up of parasites in the nest, and perhaps excrement or other detritus as well. The young are commonly transported in the mother's mouth, as in many other mammals. Nesting is associated with the production of small, multiple offspring. A closer comparison across species shows that some non-nesters also have small offspring; therefore, multiple-birth seems to be the key factor.

Nesting mothers eventually begin **parking** their infants, while some other species do it virtually from birth: the mother leaves her infant clinging to a branch or twig while she forages. This probably protects the infant from parasites and perhaps other nest problems, while freeing the mother for more effective foraging. Parking might also compensate for a lack of paternal assistance. However, strepsirrhine neonates are typically less than 22 percent of the mother's body weight; comparison with monkeys shows that a primate mother can readily carry an infant of that relative size (Whitten and Brockman 2001).

The pace of life history in strepsirrhines has engendered controversy. Some primatologists held that the durations of gestation and lactation are similar between strepsirrhines and anthropoids of the same size (Whitten and Brockman 2001). Others maintained that particular strepsirrhines, at least, have faster life histories than anthropoids of the same size (Purvis et al. 2003). According to Kappeler et al. (2003), the fast–slow continuum that characterizes life history among mammalian orders is repeated within the primates: strepsirrhine life history is slow compared to that of most other mammals, but faster than that of anthropoids.

Strepsirrhine Cognition

Anatomy suggests that strepsirrhines may be little different from other mammals in cognitive functions. First, their brain/body ratios are about the same as those of most mammalian orders (Byrne 2000). Further, because most are small animals, their brains are small in absolute terms. Some experimental evidence (early and sparse) indicated limited mechanical intelligence

(Jolly 1985). Strepsirrhines solved standard learning problems more slowly than monkeys. In manipulation tests they approached unfamiliar objects in the same way that they approached their regular food. They oriented their manipulations to cues of general direction rather than the details of the objects. Adult strepsirrhines made learning mistakes similar to those of infant anthropoids. Finally, in social situations strepsirrhines did not display differentiated reactions to different individuals, nor do they use third parties to attain goals in social interactions. Further research has revealed recurrent strepsirrhine capabilities, including positive results from tests on learning sets, delayed response, hidden objects, and object manipulation (Fichtel and Kappeler 2010). The most recent work has placed lemuriforms far beyond lorisiforms in their cognitive abilities, especially with regard to social skills (see Chapter 6).

Van Woerden et al. (2010) tested two views of brain size with a comparative study of strepsirrhines. According to the Expensive Brain Hypothesis, difficult environments limit brain size due to the difficulty of obtaining sufficient energy to support a large brain. According to the Environmental Buffer Hypothesis, natural selection favors large brains in difficult environments because they provide needed flexibility in behavior. The comparative study provided clear support for the first hypothesis and only weak support for the second. However, the second result may be more significant than it seems. Positive evidence came mainly from the lemuroid clade, which seems to contain more taxa in which behavioral adaptations are achieved by intelligence.

Summary

Most strepsirrhines are nocturnal and depend heavily on their noses, ears, and vibrissae for information. Nevertheless, large eyes and the tapetum lucidum give them good night vision. Strepsirrhine hands are flexible enough for grasping. Small size and high metabolic rate are supported by insectivory in most species. Individual strepsirrhines are solitary foragers, but live in a variety of social systems. Chemical signals are important for social and sexual functions. Sexual behavior is controlled by an estrus cycle like that of non-primate mammals. Nests and parking are prominent in infant care. Strepsirrhines are more intelligent than previously thought, but there is a gap between lorisoids and lemuroids.

Sources

Basic Bibliography

Charles-Dominique, P. (1977). *Ecology and behaviour of nocturnal primates*. NY: Columbia University Press.

Fleagle, J.G. (2013). *Primate adaptation and evolution* (3rd ed.). NY: Academic Press.

Kappeler, P.M. (2012). The behavioral ecology of strepsirrhines and tarsiers. In J.C. Mitani, et al. (Eds.), *The evolution of primate societies* (pp. 17–42). Chicago: UCP.

Masters, J., et al. (Eds.) (2013). *Leaping ahead: advances in prosimian biology*. NY: Springer.

Mittermeier, R.A., et al. (Eds.) (2013). *Handbook of the mammals of the world, Vol. 3. Primates* (pp. 28–235). Barcelona: Lynx Edicions.

Additional References

Byrne, R.W. (2000). The evolution of primate cognition. *Cognitive Science, 24*, 543–570.

Carter, A.M., & Enders, A.C. (2013). The evolution of epitheliochorial placentation. *ARo Animal Bioscience, 1*, 443–467.

Chaplin, G., et al. (2014). The role of piloerection in primate thermoregulation. *FP*, *85*, 1–17.

Colquhoun, I.C. (2011). A review and interspecific comparison of nocturnal and cathemeral strepsirrhine primate olfactory behavioural ecology. *IJZ*, *vol. 2011*, 362976 (11pp.).

delBarco-Trillo, J., & Drea, C.M. (2014). Socioecological and phylogenetic patterns in the chemical signals of strepsirrhine primates. *AB*, *97*, 249–253.

delBarco-Trillo, J., et al. (2011). Night and day: the comparative study of strepsirrhine primates reveals socioecological and phylogenetic patterns in olfactory signals. *JEB*, *24*, 82–98.

Dixson, A.F. (2012). *Primate sexuality*. NY: OUP.

Eaglen, R.H. (1980). Toothcomb homology and toothcomb function in extant strepsirrhines. *IJP*, *1*, 275–286.

Fichtel, C., & Kappeler, P.M. (2010). Human universals and primate symplesiomorphies: establishing the lemur baseline. In P.M. Kappeler & J.B. Silk (Eds.), *Mind the gap: tracing the origins of human universals*. NY: Springer.

Finstermeier, K., et al. (2013). A mitogenomic phylogeny of living primates. *PLONE*, *8*, e69504.

Fuentes, A. (1998). Re-Evaluating primate monogamy. *AA*, *100*, 890–907.

Fuentes, A. (2002). Patterns and trends in primate pair bonds. *IJP*, *23*, 953–978.

Garrett, E.C., et al. (2013). The vomeronasal complex of nocturnal strepsirrhines. *AR*, *296*, 1881–1894.

Gaulin, S.J.C. (1979). A Jarman/Bell model of primate feeding niches. *Human Ecology*, *7*, 1–20.

Gebo, D.L. (2011). Vertical clinging and leaping revisited. *AJPA*, *146*, 323–335.

Goosen, C. (1987). Social grooming in primates. In G. Mitchell & M.G. Erwin (Eds.), *Comparative primate biology*, *vol. 2B: behavior, cognition and motivation* (pp. 107–131). New York: Liss.

Gursky, S. (2005). Associations between adult spectral tarsiers. *AJPA*, *128*, 74–83.

Joffe, B., et al. (2014). Diurnality and nocturnality in primates: an analysis from the rod photoreceptor nuclei perspective. *EB*, *41*, 1–11.

Jolly, A. (1985). *Evolution of primate behavior*. NY: Macmillan.

Kappeler, P.M. (1996) Causes and consequences of life-history variation among strepsirrhine primates. *AN*, *148*, 868–891.

Kappeler, P.M. (2012). The behavioral ecology of strepsirrhines and tarsiers. In J.C. Mitani, et al. (Eds.), *The evolution of primate societies* (pp. 17–42). Chicago: UCP.

Kappeler, P.M., et al. (2002). Evolution of primate social systems. *IJP*, *23*, 707–740.

Kappeler, P.M., et al. (2003). Primate life histories and socioecology. In P.M. Kappeler & M.E. Pereira (Eds.), *Primate life histories and socioecology*. Chicago: UCP.

Martin, R.D. (1990). *Primate origins and evolution*. Princeton: Princeton University Press.

Muchlinski, M.N., Durham, E.L., Smith, T.D., & Burrows, A.M. (2013). Comparative histomorphology of intrinsic vibrissa musculature among primates. *AJPA*, *150*, 301–312.

Nekaris, K.A.I., & Bearder, S.K. (2011). The lorisiform primates of Asia and mainland Africa. In C.A. Campbell, et al. (Eds.), *Primates in perspective*. NY: OUP.

Petter, J.-J., & Charles-Dominique, P. (1979). Vocal communication in prosimians. In G.A. Doyle & R.D. Martin (Eds.), *The study of prosimian behavior*. NY: Academic Press.

Prugh, L.R., & Golden, C.D. (2013). Does moonlight increase predation risk? Meta-analysis reveals divergent responses of nocturnal mammals to lunar cycles. *Journal of Animal Ecology*, *83*, 504–514.

Purvis, A., et al. (2003). Primate life histories and phylogeny. In P.M. Kappeler & M.E. Pereira (Eds.), *Primate life histories and socioecology* (pp. 25–40). Chicago: UCP.

Ross, C. (2001). Park or ride? Evolution of infant carrying in primates. *IJP*, *22*, 749–771.

Rowe, N. (1996). *The pictorial guide to the living primates*. Charlestown, RI: Pogonias Press.

Santini, L., et al. (2015). Evolving through day and night: origin and diversification of activity pattern in modern primates. *BE*, doi: 10.1093/beheco/arv012.

van Woerden, J.T., et al. (2010). Effects of seasonality on brain size evolution: evidence from strepsirrhine primates. *AN*, *176*, 758–767.

Whitten, P.L., & Brockman, D.K. (2001). Strepsirrhine reproductive ecology. In P. Ellison (Ed.), *Reproductive ecology and human evolution* (pp. 321–350). NY: A. de Gruyter.

5 The Lorisiform Infraorder

Strepsirrhines in the Dark

The flashlight beam illuminates a tiny animal clinging upright to a thin tree in the dark forest. Its eyes glow, its long tail twitches, and it pushes away from the vine with its long legs into a leap that ends on another tree five meters away.

The lorisiforms are one of the two major subdivisions of the strepsirrhines (Box 5.1). They are distributed across southern Asia and sub-Saharan Africa. The taxon encompasses vertical clingers and relatively slow-moving quadrupeds. Common features of the lorisiforms are reviewed here and variations are considered in the next chapter.

Box 5.1 Lorisiform Taxonomy

The infraorder Lorisiformes contains a single living family, the Lorisoidea. Since the two taxonomic groupings coincide, the latter term once sufficed. However, in accord with recent taxonomic thinking, the infraorder classification recognizes the time depth of the clade (Groves 2004).

The shared derived traits that define the clade are not obvious at a glance. Some are cranial details, including the location of the tympanic ring (fused to the lateral wall of the bulla rather than suspended within it). Also internal, the blood supply to the anterior brain is provided by a different artery than is the case in other primates. The combined study by Masters et al. (2013) resulted in one troublesome genus, the potto, clustering with the galagids rather than the lorisids. However, they concluded that the strength of the morphological similarities should prevail and the lorisid clade, including pottos, should be retained.

Charles-Dominique and Bearder (1979) considered 14 prominent behavioral characteristics and concluded that 13 are retentions from the group's ancestry: mammalian, primate, and strepsirrhine. The only shared derived features they identified were vocalizations. Nevertheless, the shared derived traits of the lorisiforms are sufficient to distinguish them, especially in light of the distinctive traits displayed by their closest living relatives.

Lorisiform Behavior

Lorisiform Ecology

Lorisiforms are typically small (see Table 5.1 online for physical traits of all lorisiforms), nocturnal, and almost entirely arboreal. Faunivory is more prominent among lorisiforms than in many other primate taxa (see Table 5.2 online for ecological patterns of all lorisiforms). Animal matter (mostly insects) constituted more than half of the diet in 8 of 15 species in one survey. Charles-Dominique (1977) argued that insectivory is basic to the lorisiform lineage in the functional, proximate, and phylogenetic senses. First, lorisiforms of different sizes and genera eat about the same quantity of insects in absolute weight, which suggests that other foods are added to the insect diet in accord with body size. Second, laboratory tests on five species, including fruit and gum eaters, showed that all fed almost exclusively on insects when both insects and fruit were provided in large quantities. Finally, one frugivorous species displayed high levels of a digestive enzyme that attacks the chitin in insect exoskeletons. This is best explained as a retention from an insectivorous past.

Faunivorous behavior includes modal action patterns that are probably innate. Reaching, for example, is stereotyped and distinctive (A. Bishop, cited by Martin 1990): an individual stretches out its arm with the hand closed and then, when nearing the object of interest, opens the hand with the fingers spread. The closed hand may give the prey less visual warning, while the suddenly spread digits are ready to encompass the target quickly. The primate immediately brings the prey to its mouth and bites.

The daily cycle is probably photosensitive. Two very different species of captive lorisiforms always began nightly activities when sunlight levels were within a certain range, and this response adjusted to seasonal changes in the time of sunset (Kavanau and Peters 1976). Timing for cessation of activity was less consistent, but was least variable in clear weather (that is, when the rising sun was most visible). Lorisiforms are capable of some daytime activity. They forage in daylight if food is scarce and escape into daylight if disturbed by a predator while in the sleeping site. They sleep in nests or clumps of vegetation. Heavy rain may drive them from these shelters, and they sometimes change locations for thermoregulatory reasons.

Lorisiform Sociality

Lorisiform societies are dispersed to varying degrees, so social organization is largely inferred from ranging and sleeping patterns (see Table 5.3 online for social patterns of all lorisforms). The basic social unit in most of the well-studied taxa is a sleeping group composed of a female with her offspring. In some species the sleeping group includes one or more additional adults of one or both sexes. In other species adults without young sleep alone. Some species engage in enough social interaction during nightly foraging that they have been called *semi-dispersed*. Individual home ranges overlap in varying ways from one species or population to another.

Whatever the details, lorisiforms conform to the Socioecological Model in that females distribute themselves in relation to food, and males distribute themselves in relation to the females. When 20 captured bushbabies and lorisids were released into unfamiliar areas, females quickly settled into new ranges while the males moved on if there were no females present. The model is also supported by the fact that variation in the size of female home ranges seems to be related to basic ecological needs (Bearder 1987).

Males typically have individual home ranges and there is usually little overlap. They avoid each other in the small common areas, but mark them with urine. Marking seems to notify others of a continuing presence. Any interaction is likely to be aggressive. Wounds and behavioral observations suggest physical fighting, presumably over territory and/or females (Wiens and Zitzmann 2003b). Two species provided evidence that the largest and heaviest males were the most successful.

Fights may not be common. Animals retrapped over four years in one study area showed few new wounds. Male success, as in most primates, probably depends more on dominance and threat than actual fighting. A lorisiform threatens with a stereotyped posture in which body and tail are extended and rejection is signaled by widely spread ears, open mouth, and a two-phased growl. At some point an opponent usually adopts a submissive posture or flees.

Taking excessive risk for immediate mating gain is probably not adaptive for subordinate males. It may be more productive for younger ones to wait for a vacancy, since the tenure of most dominant males is limited to a few years. They are more exposed to predators because of their conspicuous behavior and the amount of time they spend in travel to visit associated females.

Male–female relationships can extend beyond any estrus period or breeding season. Though a male usually sleeps alone, he regularly visits associated females. These visits involve affiliative behaviors such as allogrooming and mutual scent marking, as well as checking the female for signs of estrus. A male is likely to court any new female that he encounters, regardless of age or sexual condition, if she is unattached. The current reproductive state of a female is not necessarily important, because courtship can establish a social bond that may facilitate future mating. The female usually reacts to courtship with reluctance at first, but she generally responds positively to persistent pursuit. Acceptance of the male is followed by a period of intense mutual grooming that may last for hours or days. Thereafter the two become more independent of one another, but the male will visit his partner with some regularity.

The *baculum* (penis bone), which is elongated in the lorisiforms, may provide support for long copulations, resisting pressure on the urethra. In combination with the long baculum, prolonged intromission might increase the probability of fertilization by keeping the ejaculate in contact with the *os cervix*. This combination might also stimulate female responses that increase fertility (Dixson 2012).

Lorisiform Life Cycle

Twin births occur in more than half of the lorisiform taxa studied, and females may reproduce more than once a year. These facts might explain the lack of infanticide reports. High reproductive output provides many opportunities for male success, which diminishes the importance of infanticide as a sexual strategy.

The following description of infancy leans heavily on captive studies of a few but varied species. Lorisiform mothers typically begin parking their infants away from the sleeping shelter during the first week of life, and eventually leave them for extended periods. Most or all suckling takes place during the daytime sleeping period (Tilden and Oftedal 1997; Wiens 2002). Apparently as an adaptation to this pattern, lorisiforms produce the richest milk of any primate. It is dense (high in solids rather than dilute) and rich in fat and protein. When the mother comes to collect her infant at the end of the night, she emits a special vocalization. The infant responds with a high-pitched clicking sound that helps her to locate its hiding place. It may vocalize spontaneously if disturbed during the night or if the mother does not return before the dawn becomes too bright (see Table 5.4 for reproductive parameters of all lorisiforms).

By about three weeks of age the infant engages in some independent movement around the parking site. It also begins to take solid food and in some cases snatch bits of food from the mother's hand or mouth, which has been called **scrounging**. This behavior has also been called "dietary conditioning". The term assumes the function of the behavior, which is a questionable practice, but in this case it seems reasonable: infants probably do learn food choices from scrounging. When two young human-reared individuals captured noxious moths, one suffered severe swelling of the snout and the other died. Although mothers never give food to their offspring, they do not resist scrounging. Maternal aggression and infant tantrums are absent from weaning (see Table 5.5 online for all life cycle stages of lorisiforms).

Following the mother becomes routine when the offspring is ambulatory, and it can be independent by about six months. Shortly after that a young male leaves the area of his birth, and will probably lead a nomadic life until he can find a place to establish a home range of his own. A female is likely to settle in a range near or overlapping with that of her mother.

Lorisiform Cognition

A few experiments bear on lorisiform psychology. The species involved are few, but they represent both branches of the taxon, bushbabies and the lorisids. Jaenicke and Ehrlich (1972) based their experiment on the conditions of natural nocturnal life, postulating that their subjects would respond to movement, smell, and sound more than to vision. They presented the animals with a transparent box in three conditions: empty, containing a snake, and containing a domestic cat. The two species displayed similar levels of interest in each of the various stimuli: 12 percent of a test session for the empty box and 42 percent of a test session for the box when containing either live animal. This research contradicted an earlier view that lorisiforms (and other strepsirrhines) differed from anthropoids in lacking curiosity about novel objects. The negative results of the earlier work only showed that the subjects had no interest in inanimate objects, such as balls and dolls. This improvement on earlier work represents two important principles of behavioral experimentation: (1) do not overgeneralize from test conditions; and (2) relate the research to the subjects' natural behavior.

Lorisiform solutions to other experimental problems illustrate the concept of **cognitive style**, which holds that different species may approach problems in different ways according to their natural adaptations. Lorisiforms seem to apply natural behavior involved in prey capture, which is a vital part of their subsistence (Jolly 1985). They display long bouts of attention, and depend on hand and eye (rather than nose and mouth) to investigate and manipulate objects. They are particularly good at solving delayed-response problems because of their tendency to wait, watch, and then pounce. These characteristics are less pronounced in lorisiform species that depend less on insects for food.

Summary

Lorisiforms live in Africa and Asia. All are nocturnal, relatively small, and depend heavily on insects and gums for food. Female sleeping groups are the basic social units. The social systems are dispersed and many display elements of the noyau. A male courts a non-estrus female to establish a long-term association with her. Dominant males regularly visit female associates, which exposes them to danger and makes waiting a viable reproductive strategy for subordinates. Males are typically non-affiliative with each other. The high reproductive rate seems to obviate infanticide. Most mothers park infants while they forage, sometimes for long periods, and compensate by producing rich milk. Infants obtain their first solid food by scrounging. Little is known about lorisiform cognition. The cognitive style seems to be derived from prey capture.

Sources

Basic Bibliography

Brandon-Jones, D., et al. (2004). Asian primate classification. *IJP*, *25*, 97–164.
Burrows, A.M., & Nash, L.T. (Eds.) (2007). Evolution, morphology, and behavior of lorisiform primates. *AJP*, *69*, 1–125 (special issue).
Charles-Dominique, P. (1977). *Ecology and behaviour of nocturnal primates*. NY: Columbia University Press.
Masters, J., et al. (Eds.) (2013). *Leaping ahead: advances in prosimian biology*. NY: Springer. [Chapters 9, 10,11, 19, 30]
Mittermeier, R.A., et al. (Eds.) (2013). *Handbook of the mammals of the world, Vol. 3. Primates* (pp. 184–235). Barcelona: Lynx Edicions.
Nekaris, K.A.I., & Bearder, S.K. (2011). The lorisiform primates of Asia and mainland Africa. In C.A. Campbell et al. (Eds.), *Primates in perspective* (pp. 34–54). NY: OUP.

Additional References

Bearder, S.K. (1987). Lorises, bushbabies, and tarsiers: diverse societies in solitary foragers. In B.B. Smuts, et al. (Eds.), *Primate societies*. Chicago: UCP.
Bearder, S.K., et al. (2002). Dangers in the night: are some nocturnal primates afraid of the dark? In L. Miller (Ed.), *Eat or be eaten: predator sensitive foraging in primates* (pp. 21-40). NY: CUP.
Bearder, S., et al. (2003). Species-typical patterns of infant contact, sleeping site use and social cohesion among nocturnal primates in Africa. *FP*, *74*, 337–354.
Charles-Dominique, P. (1977). *Ecology and behaviour of nocturnal primates*. NY: Columbia University Press.
Charles-Dominique, P., & Bearder, S.K. (1979). Field studies of Lorisid behavior: methodological aspects. In G.A. Doyle, & R.D. Martin (Eds.), *The study of prosimian behavior*. NY: Academic Press.
Das, N., et al. (2014). Medicinal plant exudativory by the Bengal slow loris, *Nycticebus bengalensis*. *ESR*, *23*,149–157.
Dixson, A.H. (2012). *Primate sexuality: comparative studies of the prosimians, monkeys, apes, and human beings*. NY: OUP.
Ehrlich, A. (1974). Infant development in two prosimian species: greater galago and slow loris. *Developmental Psychobiology*, *7*, 439–454.
Ehrlich, A., & Macbride, L. (1989). Mother–infant interactions in captive slow lorises (*Nycticebus coucang*). *AJP*, *19*, 217–228.
Ehrlich, A., & Musicant, A. (1977). Social and individual behaviors in captive slow lorises. *Behaviour*, *60*, 195–220.
Fisher, H.S., et al. (2003). Countermarking by male pygmy lorises (*Nycticebus pygmaeus*): do females use odor cues to select mates with high competitive ability? *BE&S*, *53*, 123–130.
Fitch-Snyder, H., & Ehrlich, A. (2003). Mother–infant interactions in slow lorises (*Nycticebus bengalensis*) and pygmy lorises (*Nycticebus pygmaeus*). *FP*, *74*, 259–271.
Fitch-Snyder, H., & Jurke, M. (2003). Reproductive patterns in pygmy lorises (*Nycticebus pygmaeus*): behavioral and physiological correlates of gonadal activity. *Zoo Biology*, *22*, 15–32.
Groves, C. (2004). The what, why and how of primate taxonomy. *IJP*, *25*, 1105–1126.

Izard, M.K., & Rasmussen, D.T. (1985). Reproduction in the slender loris (*Loris tardigradus malabaricus*). *AJP*, *8*, 153–165.

Izard, M.K., & Weisenseel, K. (1989). Comparative reproduction of the lorisidae. *AJP*, *18*, 140.

Izard, M.K., et al. (1988). Reproduction in the slow loris (*Nycticebus coucang*). *AJP*, *16*, 331–339.

Jaenicke, C., & Ehrlich, A. (1972). Effects of animate vs. inanimate stimuli on curiosity behavior in greater galago and slow loris. *Primates*, *23*, 95–104.

Jolly, A. (1985). *The evolution of primate behavior* (2nd edn.). NY: Macmillan.

Kavanau, J.L., & Peters, C.R. (1976). Activity of nocturnal primates: influences of twilight zeitgebers and weather. *Science*, *191*, 83–86.

Martin, R.D. (1990). *Primate origins and evolution*. Princeton: Princeton University Press.

Munds, R.A., et al. (2013). Living together in the night: abundance and habitat use of sympatric and allopatric populations of slow lorises and tarsiers (*Nycticebus* and *Tarsius*). *ESR*, *22*, 269–277.

Munds, R.A., et al. (2013). Taxonomy of the Bornean slow loris, with new species *Nycticebus kayan* (Primates, Lorisidae). *AJP*, *75*, 46–56.

Nekaris, K.A.I. (2001). Activity budget and positional behaviour of the Mysore slender loris: implications for slow climbing locomotion. *FP*, *72*, 228–241.

Nekaris, K.A.I. (2003). Observations on mating, birthing and parental care in three taxa of slender loris in India and Sri Lanka (*Loris tardigradus* and *Loris lydekkerianus*). *FP* supp. (Eds. S. Gursky, & K.A.I. Nekaris), *74*, 312–336.

Nekaris, K.A.I. (2003). Spacing system of the Mysore slender loris (*Loris lydekkerianus lydekkerianus*). *AJPA*, *121*, 86–96.

Nekaris, K.A.I. (2005). Foraging behaviour of the slender loris: implications for theories of primate origins. *JHE*, *49*, 289–300.

Nekaris, K.A.I. (2006). The social lives of Mysore slender lorises. *AJP*, *68*, 1–12.

Nekaris, K.A.I., & Jayewardene, J. (2004). Distribution of slender lorises in four ecological zones in Sri Lanka. *Journal of Zoology*, 262, 1–12.

Nekaris, K.A.I., & Rasmussen, DT (2003). Diet of the slender loris. *IJP*, *24*, 33–46.

Nekaris, K.A.I., & Stephens, N.J. (2007). All lorises are not slow: rapid arboreal locomotion in the newly recognised red slender loris (*Loris tardigradus tardigradus*) of southwestern Sri Lanka. *AJP*, 69, 112–120.

Nekaris, K.A.I., et al. (2007). Anti-predator behaviour of lorises and pottos. In S.G. Gursky & K.A.I. Nekaris (Eds.), *Primate anti-predator strategies* (pp. 220–238). New York: Springer.

Nowack, J., et al. (2010). Torpor on demand: heterothermy in non-lemur primates *Galago moholi*. *PLONE*, 5, e10797. DOI: 10.1371/journal.pone.0010797.

Nowack, J., et al. (2013a). Nonshivering thermogenesis in the African lesser galagid *Galago moholi*. *JEB*, *216*, 3811–3817.

Nowack, J., et al. (2013b). Surviving the cold, dry period in Africa: behavioral adjustments as an alternative to heterothermy in the African lesser galagid (*Galago moholi*). *IJP*, *34*, 49–64.

Nowack, J., et al. (2013c). Torpor as an emergency solution in *Galago moholi*: heterothermy is triggered by different constraints. *Journal of Comparative Physiology B*, *183*, 547–556.

Pontzer, H., et al. (2014). Primate energy expenditure and life history. *PNAS*, *111*, 1433–1437.

Radhakrishna, S. (2004). Sociality in a solitary primate: how gregarious is the slender loris? *Resonance*, 9, 64–81.

Radhakrishna, S., & Kumara, H.N. (2010). Behavioural variation in the Mysore slender loris *Loris lydekkerianus lydekkerianus*. *Current Science*, 99, 1226–1232.

Radhakrishna, S., & Singh, M. (2002). Social behaviour of the slender loris (*Loris tardigradus lydekkerianus*). *FP*, *73*, 181–196.

Radhakrishna, S., & Singh, M. (2004). Infant development in the slender loris (*Loris lydekkerianus lydekkerianus*). *Current Science*, 86, 1121–1127.

Starr, C., & Nekaris, K.A.I. (2013). Obligate exudativory characterizes the diet of the pygmy slow loris *Nycticebus pygmaeus*. *AJP*, 75, 1054–1061.

Starr, C.R., et al. (2012). Hiding from the moonlight: luminosity and temperature affect activity of Asian nocturnal primates in a highly seasonal forest. *PLONE 7*, DOI: 10.1371/journal.pone.0036396.

Tan, C.L., & Drake, J.H. (2001). Evidence of tree gouging and exudate eating in pygmy slow lorises (*Nycticebus pygmaeus*). *FP*, *72*, 37–39.

Tilden, C.D., & Oftedal, O.T. (1997). Milk composition reflects pattern of maternal care in prosimian primates. *AJP*, *41*, 195–211.

Wiens, F. (2002). Behavior and ecology of wild slow lorises (*Nycticebus coucang*): social organization, infant care system, and diet. PhD dissertation, University of Bayreuth.

Wiens, F., & Zitzmann, A. (2003a). Social dependence of infant slow lorises to learn diet. *IJP*, *24*, 1007–1021.

Wiens, F., & Zitzmann, A. (2003b). Social structure of the solitary slow loris *Nycticebus coucang* (Lorisidae). *JZL*, *261*, 35–46.

Wiens, F., et al. (2006). Fast food for slow lorises: is low metabolism related to secondary compounds in high-energy plant diet? *Journal of Mammalogy*, *87*, 790–798.

Zimmermann, E. (1985). Vocalizations and associated behaviours in adult slow loris (*Nycticebus coucang*). *FP*, *44*, 52–64.

Zimmermann, E. (1989). Reproduction, physical growth and behavioral development in slow loris (*Nycticebus coucang*, Lorisidae). *Human Evolution*, *4*, 171–179.

6 Lorisiform Variation

Leapers and Creepers

Moonlight illuminates the moth fluttering on a twig. A round face with a tiny snout and large eyes slowly emerges from the darkness. The loris moves slowly, one foot at a time, along a small branch. Its legs are long and thin, but the hands and feet are powerful and hold the slender body without swaying or lurching. The loris lunges, grasps the insect, and kills it with several bites.

The lorisiform taxon can be divided at the family level into Galagidae and Lorisidae. The former are largely vertical clingers and leapers; the latter are mainly quadrupeds that move slowly most of the time. Lorisidae are rather sharply divided between Asian and African genera. This chapter describes the distinctive features of the families and some of the genera they contain.

Galagids

Galagids inhabit all of sub-Saharan Africa except for the southernmost tip and exist in a wide range of forest and scrub environments (Box 6.1). They typically have slender bodies, long legs, and long bushy tails. Many galagids, especially the smaller species, engage in vertical clinging and leaping. However, leaping is primarily used to span gaps between trees rather than as the preferred mode of locomotion.

Box 6.1 Galagid Taxonomy

Seven experts at an international workshop reached a consensus on classification within the galagids independent of whether they are a family (Galagidae) or a subfamily (Galaginae) (Grubb et al. 2003). They recognized five genera, distinguished by vocalizations as well as morphological features:

- *Galagoides* (small and dwarf galagids) are forest-living species that weigh less than 200g. They have relatively short limbs and light build.
- *Galago* (lesser galagids) are somewhat larger and stockier with longer limbs.
- *Euoticus* are the needle-clawed galagids. They specialize in gum consumption, using their claws to cling near their food sources and their distinctive craniodental features for gouging bark (Dumont 1997).
- *Sciurocheirus* (squirrel galagids) are larger than *Galago* at 250–450g.
- *Otolemur* (greater galagids) are also large, and rarely leap.

The experts recognized at least 24 species among these genera, based to a significant degree on bioacoustic studies interpreted according to the relatively new mate recognition concept of species. According to this view, distinct species should be identified by whether or not the opposite sexes recognize each other as potential mates. The resultant classification has been supported by more traditional morphological, genetic, and behavioral criteria.

Galagid Ecology

With ears that are large even by strepsirrhine standards, galagids hear the movements of active insects in the dark and rotate toward such sounds. A galagid can follow the movement of an insect on the other side of a plywood screen, as shown by the movement of the primate's head. A galagid also hunts by sight. It often catches an insect as it takes flight by lunging at it while anchored to a branch by its feet. It uses a power grip with one or both hands.

Quantitatively, consumption of insects is not quite what theory would predict for such small primates. In the smallest species insects comprise only 70 percent of the diet. Insect consumption may be somewhat limited by dependence on evasive prey that can escape by jumping, flying, or at least rapid scuttling. Almost 80 percent of the prey consists of grasshoppers, moths, and beetles.

Dependence on insects is affected by consumption of gums, which is ubiquitous. This may be a fundamental adaptation of the taxon. Galagids need their sense of smell to find these exudates because they are unable to penetrate trees and depend on openings caused by breakage or boring insects. The primates scrape up gums with the toothcomb. An elongated cecum contains microorganisms that facilitate digestion of the complex sugars in gums. Mid-sized species make the most use of exudates to augment their diets, while the largest depend more on fruit. Some species eat nectar and serve the ecosystem as pollinators.

Fruit may be more important to some smaller species than previously thought. Scheun et al. (2014) observed the lesser bushbaby (*Galago moholi*), thought to be an insect/gum specialist, eating fruit during winter. Survey of the habitat revealed that the prime gum-producing tree was largely absent and other species produced only small amounts. Analysis of the fruits eaten by the galago showed them to be rich in protein, fat, and energy content. Previously unknown dietary flexibility allowed this species to use an important fallback food.

Flexibility also emerged in an experimental study of galagos by McArthur et al. (2012). They used artificial food patches to investigate galago response to shifts in the balance between levels of plant toxins and predator danger. The researchers demonstrated that the animals quantified the relative costs of toxins versus potential predation and sought food patches with the lower net cost. This work may have broader implications for galago cognition and for primate foraging patterns in general.

The lesser bushbaby illustrates the strepsirrhine need for thermoregulation as a result of limited capability for piloerection (see Chapter 4). **Torpor** involves both metabolic and behavioral adaptation. The animal assumes a location and posture in which the metabolism can be temporarily slowed to save energy. However, the bushbabies rarely use this tactic. Purely behavioral thermoregulation is more common: less activity at night, more use of enclosed sleeping places, more huddling with others, and more consumption of gum that increases in energy content in winter (Nowack et al. 2013a). In this species torpor is an emergency response to extreme temperatures (Nowack et al. 2013b).

Galagid Sociality

There are few studies of *Galago senegalensis*, the most widely distributed nocturnal primate in Africa. Recent field research illustrates our growing knowledge of sociality in galagids (Svensson and Bearder 2013). Subjects were alone in 40 percent of the sightings, but in groups of three or more in 23 percent of the encounters. They slept alone in 30 percent of daytime observations, but in groups in 70 percent.

Galagid societies are typically dispersed multi-male systems. One of the most consistent features is the **matrilocal cluster**, an association of females marked by spatial proximity and friendly or tolerant behavior. Members of such an association sleep together in groups, usually four to six individuals, and/or share a single home range or live in overlapping ranges. Members of a female cluster tend to be aggressive toward members of other such associations.

Early studies inferred that associated females are all related, based on the apparent fact that females rarely dispersed from the natal area. However, few of these studies exceeded two years in length and they recorded only small samples of maturing males and females that were potential emigrants (Nash 2004). Many dyads are certainly mother and daughter, but other relationships remain unclear. Sharing of a sleeping place, a potentially limited resource, may be a kin-selected behavior.

Male home ranges are larger than those of females and overlap them. In some species a dominant male sleeps with females at least some of the time. A dominant male may tolerate smaller and/or younger males within his home range, but any other males are peripheral because galagid males are mutually aggressive. Although these systems are widely characterized as "multi-male", it is important to note that they differ significantly from the multi-male troops of anthropoids that we will encounter later on. In some species there are **floaters**, males who have no permanent home ranges and may travel widely. Some floaters may be young males who become solitary during dispersal from the natal area.

Aggressive competition for females might stem from the fact that galagids display the greatest degree of sexual dimorphism among the strepsirrhines. Typically males are larger than females in body size (up to 20 percent larger) and in all three dimensions of the canine teeth (width, height, and length). However, evidence from one species (*G. moholi*) suggests a more complicated situation (O'Mara et al. 2012). Males are larger and add even more body mass during the mating season. However, testicle size also increases, and sperm competition is indicated by the fact that two-thirds of females mate with more than one male. It may be that both contest and scramble competition are involved, with perhaps some degree of female choice.

Both male and female territories are signaled by vocalizations and by scent marks. Dominant males engage in urinary and glandular marking more often than other males, and they countermark the traces of subordinates. An emphasis on vocal communication may be related to auditory capabilities that originally evolved for prey location (Rasmussen and Nekaris 1998). Various species have vocal repertoires of 8 to 25 calls, some of which are discrete and others graded. They can mix these into rapidly changing sequences, some of which last for more than 30 minutes. Even when seemingly alone, a galagid can communicate effectively with others of its kind.

Every species has one particular loud call that is given by both sexes. Because it helps bring mates together, it is different in each species. The typical sexual advertisement of galagids is a series of rhythmically repeated, lower frequency barks. These calls travel further than higher pitched calls would, because they are less prone to degrade in the forest environment. There is some danger that they will attract predators as well as potential mates, but this is offset by galagid speed for a sudden escape (Zimmermann 1995).

Galagid Life Cycle

The wild infant spends its first week or two in a nest or tree hole, the same shelter used by adults for sleeping. Kappeler (1996) suggested a connection with the galagid tendency to bear twins. Most species produce one or two offspring once each year. Early developmental stages are similar across the studied species (Doyle 1979; Zimmermann 1989). Within half an hour the neonate can crawl awkwardly, grip with its hands, and pull itself into a suckling position (this precocial behavior pattern probably occurs in all lorisoids). The infant can walk within a week or so, but may be kept in the shelter for somewhat longer. During the initial stages of parking the galagid mother may stay close to her offspring. When foraging requires significant movement, the young may be parked at several different sites during the course of a night. This pattern has been labeled **cache and carry** (Gursky 2000).

Kessler and Nash (2010) presented evidence for complexity in the social life and life cycle of *Galago senegalensis*. They recorded allonursing in two captive **matrilines** (groups of individuals related through females). Grandmothers nursed three infants in about one-fourth of their feeding bouts. This was not explained by reciprocity, because the mothers rarely nursed younger siblings. The researchers suggested that this behavior might be the result of kin selection in which the older females, some of whom had lost neonates, increased their reproductive success.

Within a month the infant starts jumping, a vital behavior for most galagids (Doyle 1979). Play begins during the first few days, with nest material and with a sibling if available. By the end of the first month the infant plays with other infants, juveniles, adolescents, adults, and inanimate objects. Around this time it begins to interact more with adolescents than with the mother, and soon afterward she begins to ignore its approaches. Nursing begins to decline after two or three weeks and weaning is commonly completed by the end of the second month.

The young of most galagid species reach sexual maturity between eight and ten months of age. Females give birth after a gestation period that varies across species from 16 to 20 weeks. Smaller galagids can live for at least eight years in the wild while larger species may live twice as long (Nowak 1999). The pace of life stages in most galagids is typical for primates in their size range.

Big Galagids

The largest galagids are the needle-clawed genus (*Euoticus*) and the greater galagids (*Otolemur*). Weighing up to about two kilograms, these giants have limited jumping ability or none at all. In both genera a mother often carries her infant on her fur and rarely parks it. Single birth may be one factor, and having a large enough body to accommodate the infant may be another. These features resemble monkey traits (Figure 6.1).

Exploitation of gums by needle-clawed galagids illustrates concomitant specialization in morphology and behavior. Two anatomical features give them access to gum sources unavailable to sympatric primates. First, each nail is keeled and ends in a sharp point (hence the common name of the genus). This allows the animal to move like a squirrel, sometimes head downwards, on broad tree trunks and smooth branches that grasping primates cannot negotiate. Second, the tongue and toothcomb are relatively long and narrow, which probably

Figure 6.1 The monkey-like features of a greater galago are apparent on the ground.
© Y.A. de Jong and T.M. Butynski, www.wildsolutions.nl

allows the harvesting of gum from narrow crevices. Needle-clawed galagids spend entire nights in search of gums. Each individual follows the same itinerary from night to night, accumulating detailed knowledge of gum locations. An individual averages 100 collections in an hour, including in some cases a single drop. Allowing for resting time, this suggests that a needle-clawed galagid may visit 500 to 1000 sites in a single night.

Captive greater galagids provided evidence for innovation and social learning. When offered basins containing swimming fish, several individuals learned to prey on them (Welker 1976). Experiments were performed 18 months later on two groups with diverse membership, including some with fishing experience and some naive. Wild-born animals readily recognized fish as potential food and improved their capture ability from one trial to the next. Captive-born animals had to learn that fish are food by taking some from others. Predation on fish had to start with one animal, and participants had to learn the method of grasping from above.

Experiments by Watson et al. (1994) elaborated on the sources of learning. Subjects were faster to approach fishbowls after seeing either fishing or non-fishing models, as opposed to those who saw no model. The behavior of parents rather than the models had the greatest effect on learners. Those with fishing parents were quicker to make the initial fishing attempt and made the greatest number of attempts.

Lorisid Behavior

Members of the lorisid family (Box 6.2) inhabit widely scattered areas of Asia (slow lorises and slender lorises) and Africa (pottos and angwantibos). The Asian forms are the most distinctive and have received the most study. Highly specialized in a variety of ways, they have been characterized as the "most extreme primates" (Nekaris 2014). They are the basis for most of the following discussion.

Box 6.2 Lorisid Taxonomy

There is widespread recognition of a lorisid family that contains the Asian genera *Loris* (slender lorises) and *Nycticebus* (slow lorises) (Brandon-Jones et al. 2004). They are commonly joined in this assignment by the African genera *Arctocebus* (angwantibos) and *Perodicticus* (pottos) (Grubb et al. 2003). However, lorisids may be divided at the subfamily level as Perodicticinae and Lorisinae (or Lorinae). Fleagle (2013) made no distinction at the subfamily level. On the other hand, Nekaris and Bearder (2011) did away with the distinction at the family level and classified the Lorisiformes into three subfamilies: Galaginae, Lorisinae, and Perodicticinae. The combined study by Masters et al. (2013) resulted in one troublesome genus, *Potto*, clustering with the galagids rather than the lorisids. However, they concluded that the strength of the morphological similarities should prevail and the lorisid clade including pottos should be retained.

To summarize for the lorisids:

Family Lorisidae

[Subfamily Lorisinae? or Lorinae?]
- Genus *Loris* (slender lorises)
- Genus *Nycticebus* (slow lorises)

[Subfamily Perodicticinae?]
- Genus *Arctocebus* (angwantibos)
- Genus *Perodicticus* (pottos)

Lorisid Ecology

Despite their "extreme" specializations, lorisids display considerable ecological flexibility (Nekaris 2014). In addition to tropical rainforests, these primates live in montane forests, dry scrublands, and peat swamps. They survive at altitudes from sea level to 2000m and in monthly mean temperatures that vary from 7 to 50 degrees Centigrade. They can use all levels of the forest canopy, although they are most often found below 10m.

These primates are characterized by a preference for slow climbing in which three limbs are firmly fastened to the substrate while the fourth one moves (Nekaris 2001). Physical specializations for this form of quadrupedalism include unusually strong hands and feet with shortened second digits, highly mobile wrists and ankles, and blood vessels in the arms and legs that allow blood to flow freely even when there is no limb movement. Emphasis on the forelimb contrasts with the hindlimb dominance of other primates and results in more stress on the long arms than on the hindlimbs (Schmitt and Lemelin 2004). All studied species have the greatest elbow yield and the least knee yield of any primates (Larney and Larson 2004). Having no need for aerial balance, these animals have greatly reduced tails.

Lorisids are capable of rapid movement (Nekaris 2001). Fieldworkers suggest that some observers have been fooled by the animals' fluid locomotion. Lorisids often move easily from tree to tree because they can exert a secure grip on fine twigs and can extend their long bodies to bridge small gaps. Caution may also create an illusion of slowness. Lorisids often test the substrate ahead and then back up and find another route. This is because they never leap more than a few inches.

Lorisids have very low rates of basal metabolism even by strepsirrhine standards, less than half of what would be expected in animals their size. This is seen in other mammals that specialize in eating toxic insects—for example, anteaters and aardvarks. Slender lorises eat toxic insects while slow lorises consume large amounts of toxic plant foods (Nekaris 2014). It seems that low metabolism is related to physiological mechanisms for detoxification of the food (Rasmussen and Nekaris 1998). Greatly slowed reproductive rates and life history schedules also seem to be related to low BMR, since they cannot be accounted for by body size, sociality, or ecological stability. Lorisids are among the most K-selected primates for their size.

Lorisid Sociality

In contrast to galagids, female lorisids are strongly territorial towards one another and males are sometimes mutually tolerant. Lorisids seem to spend less time in social interaction than galagids, associating with conspecifics during less than 10 percent of each night (Wiens and Zitzmann 2003b). However, a body of evidence for lorisid social life is increasing (Nekaris 2014).

A suggestion that slow locomotion limits the time available for social contacts was countered by the finding that these primates are capable of rapid locomotion. Furthermore, slender lorises travel about 1km each night and "slow" lorises average more than 5km. The area of lorisid home ranges is about the same as those of galagids of similar body size. The latter fact seems to be the more telling one, since it indicates habitual or average speed rather than top speed under unusual conditions.

Some unusual mating behavior may have been shaped by predator pressure. Lorisid sexual advertisement calls are higher pitched than those of galagids, which makes them harder to locate because they degrade more quickly in a forest environment. Since this applies to both potential predators and potential mates, such calls are probably a compromise between reproductive function and danger. The danger may be of greater consequence to lorisids than to galagids because they are slower and more dependent on crypsis (Zimmermann 1995). Lorisids in general seem to be less vocal than galagids; however, this may be illusory since some lorisid calls are ultrasonic.

When mates find each other, they copulate in a position that is virtually unique among the primates. The male is in a fairly conventional dorsoventral position in relation to the female, but both are suspended underneath a branch. This may aid in concealment from predators (especially raptors, presumably), and, if danger is imminent, they can drop to the ground and flee. This is known to be an escape behavior for adults in at least one species in the group. The lorisid pattern illustrates Dixson's point that sexual behavior may be shaped by natural selection as well as sexual selection (see Chapter 2).

Lorisid Life Cycle

The fact that lorisids never use nests reinforces the view that nesting is mainly an adaptation to multiple births rather than to small neonates (Ross 2001). The baby clings to the mother's abdomen for a period of two to four months. Then the mother begins parking it while she forages. In contrast to the cache and carry policy of the galagids, a lorisid mother is likely to leave her infant in one place throughout the night. It calls for her as dawn approaches.

Lorisid life histories are among the longest of any primates of their body size. Gestation takes 4 to 6 months (compared with 3.5 to 4.5 in galagids). Weaning is typically completed between 4 and 6 months of age, compared with 1.4 to 4 months in galagids. At about 3 or 4 months, the young may ride on the mother's back or follow her (Bearder et al. 2003).

Lorisid Cognition

In the previous chapter we reviewed an experiment by Jaenicke and Ehrlich (1972) in which a lorisid and a galagid species displayed equal curiosity about novel objects (live animals in a box). This comparative study also revealed significant differences, making the point that the same mental state may be expressed differently in different species. Lorisid responses seemed to reflect vulnerability to predators in the wild. They moved slowly and silently, and made contact with the box less often. They took longer to approach within about a meter, and made shorter visits and spent less total time within that zone. The galagids were less cautious in their investigation, possibly because of their capability for rapid escape. Here again we encounter the importance of interpreting experimental results in relation to natural behavior.

Emotion (or temperament) may also be important here. According to Schulze and Meier (1995), both slow and slender lorises are affected by psychic stress during tests of learning and intelligence. This can mask the fact that they do learn. Sometimes a loris will suddenly display a new behavior that could only be due to a learning experience that took place months before. These range from tameness to performance on discrimination tests. In evaluating the intelligence of the slender loris, especially in tests, it is important to consider personality and temperament because this is an "excitable" species (Schulze and Meier 1995).

Slow Lorises

Slow lorises (genus *Nycticebus*) all live in southern and southeastern Asia (Figure 6.2). In contrast to slender lorises they subsist primarily on exudates and nectar (Nekaris 2014). They are "gouging machines", capable of producing a large hole in tree bark in seconds with a specialized stout toothcomb. A slow loris can cling to a tree trunk for 45 minutes with its strong hands and u-shaped hind legs, while camouflaged from predators by the stripe along its back. The fundamental nature of this adaptation is indicated by the beginning of gouging at the early age of three months.

While brutal to trees, slow lorises are delicate with the flowers from which they obtain nectar. The animal carefully pulls a flower toward itself, extracts nectar with its long tongue, and then gently replaces the flower without any damage. In the process its face becomes covered with pollen, which suggests that its progress from flower to flower results in pollination.

These feeding habits lead to encounters with high levels of toxic substances in some barks, exudates, and nectars. Slow lorises are resistant to substances that can blister or even kill humans. Many of the exudates that they prefer are used in the traditional medicines of local humans and may kill parasites (Das et al. 2014). In addition to providing a diet with little competition, toxic plant materials may contribute to the loris's generation of its own poison.

Slow lorises are unique among primates in producing a toxin that probably functions as a predator defense. The animal secretes potentially poisonous substances from brachial glands under the arms, which are activated by contact with saliva when the animal rubs its arms across its head and neck. Slow loris bites can cause anaphylactic shock in humans (Krane

Figure 6.2 A slow loris demonstrates its positional versatility and displays the lack of a tail that is characteristic of lorisids. © thousandlies via iStock

et al. 2003). The slow loris tends to hold its bite, which may allow time for more poison to be injected. Alterman (1995) suggested that grooves between the incisors of the toothcomb facilitate this process, and that this might have been the original adaptive function of the toothcomb for all lorisoids. Few other mammals produce toxins and none are known to use them routinely against other species.

In contrast to other nocturnal primates for which information is available, slow lorises become less active in moonlight (Rode-Margono and Nekaris 2014; Starr et al. 2012). At first predation risk was considered an important factor, but further research showed no connection with predator activity. Cold temperatures continued to be implicated, but the key factor for the lorises appears to be prey availability.

Slow lorises in one field study displayed a low level of social cohesion, devoting only 3 percent of nightly active time to social interactions (Wiens and Zitzmann 2003b). The most common "social" behavior (recorded only six times in 1000 hours) was simply following another group member. Group members slept together less than 40 percent of the time. The researchers wondered why this tenuous social system existed at all, because none of the standard primate explanations applied: no cooperative predator defense, no defense of resources, and no alloparenting.

Anecdotal evidence suggested an answer: allogrooming. The one loris in the study population without a social group had an unusually high ectoparasite load. Only one instance of allogrooming was seen in the field, but captive studies clarified the issue. Captive slow lorises engage in intensive allogrooming just before retiring and just after waking (Ehrlich and Musicant 1977). This corresponds to dusk and dawn in the field, times when observing conditions are very poor. Allogrooming may be important to wild lorises, even if not visible to a human observer.

Additional fieldwork provided more evidence for social life among slow lorises (Nekaris 2014). In one species up to 10 percent of the activity budget consisted of social behavior, including grooming, playing, and feeding together. In another species individuals spent up to 65 percent of their time (average 18 percent) in spatial proximity, including body contact.

Captive slow lorises provided more evidence for social cohesion. They spent the majority of waking time close to each other, and seemed to seek each other out for affiliative behavior (Ehrlich and Musicant 1977). Rare instances of aggressive behavior did not seem serious and wounds were unusual. Antagonists often engaged each other in grooming or play rather than separating after the quarrel ended. This appears to be **reconciliation**, a behavior that reflects the importance of particular social relationships and which is not often seen in strepsirrhines. Other captive studies also show low levels of aggression in slow loris groups and affiliative behaviors such as allogrooming and social play (Fitch-Snyder and Schulze 2001). Comparison with galagids suggests that these behaviors are characteristic of slow lorises rather than being an artifact of captivity. Greater galagids kept in a laboratory environment identical to that of the lorises were less affiliative and more aggressive toward one another.

Slender Lorises

The genus *Loris*, scattered across India and Sri Lanka, differs from the slow lorises and from primates in general in several important ways (Figure 6.3). Most of our current information comes from studies of the common grey loris. Slender lorises are highly insectivorous, with the proportion in the diet rising as high as 96 percent. They avoid competition with other small predators by concentrating on invertebrates that repel most predators with their noxious smell and/or taste, or some irritant quality such as spines. These prey are abundant and easily captured (Rasmussen and Nekaris 1998). Lorisids do not choose such prey in captive tests, indicating that they are tolerated rather than preferred. In natural circumstances lorisids also eat birds and other small vertebrates.

Figure 6.3 A slender loris displays its ability to grip thin supports. © Dirk Freder via iStock

Slender lorises are the most social lorisiforms (Nekaris 2001, 2002), spending 10 to 20 percent of their waking time in each other's company, as well as sleeping communally. The typical configuration of slender lorises during their daytime rest is a sleeping ball like that of greater galagids. Sleeping groups usually contain 3 to 6 individuals, but there may be as many as 14 (members sometimes sleep apart from the group).

Slender lorises are also highly sociable when awake at the sleeping site, and they often gather in pairs or larger associations during the nightly active period. Social behaviors include feeding together, allogrooming, and play wrestling. Though singles and pairs are more common, a traveling group might contain as many as nine individuals. Comparably sized groups can live peacefully in captive conditions (Fitch-Snyder and Schulze 2001).

Senior females play an unusual, if not unique, role in the social organization of slender lorises. Each sleeping group has a single adult female as its focus (Radhakrishna and Singh 2002). The group usually includes several of her offspring and in some cases other adult females. Unrelated adult males may also sleep with the group, most often during the mating season. The senior female dominates the group, gets first access to the largest insects, and decides whether or not to reciprocate grooming. These prerogatives are in addition to the typical lorisoid pattern of females firmly rejecting unwanted suitors. Senior females have home ranges that are largely separate from one another. Each adult male has a home range that overlaps those of several females, which leads to the debatable characterization of these societies as multi-male.

The mating system of slender lorises (Nekaris 2003a) is typically multi-male. Several males may follow and compete for an estrus female. Although the dominant female chooses her partner, males still fight among themselves and losers may harass the copulating couple. Males also pay a great deal of attention to infants, which may be another tactic to impress potential sex partners. Males regularly play with infants from sleeping groups other than their own. Females only spend time with their own infants. Copulation involves multiple intromissions, unusual among strepsirrhines, but consistent with the pattern of competition. Multiple intromissions can probably achieve most of the functions attributed to the prolonged intromissions that characterize other strepsirrhines.

Reproductive schedule differentiates the slender lorises. Slender loris gestation and lactation periods, at about five months each, are long in relation to the size of the animals (Izard and Rasmussen 1985). However, reproduction rate is not low because twinning is more common in slender lorises and because females often have fertile estrus cycles before lactation ends (Radhakrishna and Singh 2004). A female can have three or four offspring in one year. The high reproductive output provides males with many opportunities to sire offspring, which accounts for the fact that infanticide is rare or absent.

African Lorisids

The Perodicticinae are an African taxon composed of two genera (Box 6.2). Pottos (*Perodicticus*) are widely distributed across sub-Saharan Africa while angwantibos (*Arctocebus*) occur only in the rainforests of central Africa. The stocky pottos feed mainly on fruit and gums, while the smaller angwantibos are more faunivorous (about 85 percent of the diet). Several lines of evidence suggest that the potto has diverged from an insectivorous ancestry: (1) the potto has an enzyme for the digestion of insect integument; (2) captive pottos prefer to eat insects; (3) pottos and angwantibos use the same stereotyped behavior to capture larger prey (the victim is clamped down with one hand and bitten until dead); and (4) both genera continuously sniff around when active and find most prey by sense of smell.

The use of scent and the one-handed grab are common lorisid patterns. The potto's shift to a more omnivorous diet seems likely to be related to its increased size.

Both African genera can freeze for hours when frightened (a behavior that occurs in other lorisids but is less pronounced). Perhaps related, they seem to lack audible alarm signals. However, some lorisids have ultrasonic vocalizations so the African genera may be more vocal than we think. When confrontation is unavoidable, the potto lowers its head between the forelimbs and exposes its neck, which is protected by highly developed cervical spines. With the head down the potto is aware of the enemy's movements through long sensory hairs on the neck. This allows it to lunge forward at opportune moments and an enemy that lacks the potto's powerful grip may be knocked off its perch. The angwantibo posture is similar, but the mouth is open and directed beneath one of the armpits. When touched in the rear of the body the defender immediately bites in that direction. These variants of the same posture probably evolved as responses to the most frequent attacks encountered by each genus.

In some situations a potto may drop to the ground. Even an infant may drop from its mother, which suggests an innate response. Dropping was the predominant response to an experimentally presented large snake. The key may be that a large snake is more likely than a cat or mongoose to swallow prey whole and alive, which would render the other defenses useless.

Caterpillars are a large part of the angwantibo diet. The primate holds a caterpillar's head in its mouth and rubs its hands along the prey's body for up to 20 seconds, which removes most of the hairs from the body. The massage is more extensive for noxious caterpillars, and afterward the angwantibo wipes its hands and snout along branches. Young captive individuals massage proffered insects when they are two months old, before they can capture their own prey, which suggests that this is an innate modal action pattern.

Summary

Galagids tend toward vertical clinging and leaping, which facilitates escape from predators and capture of elusive prey. Related females form clusters. Neonates stay in shelters until ready for parking. Mothers carry infants in the mouth when moving.

Lorisids use slow, silent quadrupedalism to stalk insect prey and perhaps to avoid detection by predators. Neutralization of toxins from insect prey is probably a function of the low lorisid metabolic rate. The precocious neonate is carried on the mother's abdomen until old enough to park. Some lorisids seem to live in pair-groups; the slender loris system resembles interlocking noyaus. African lorisids are highly distinctive in their predator defense mechanisms.

Sources

Basic Bibliography

Charles-Dominique, P. (1977). *Ecology and behaviour of nocturnal primates*. NY: Columbia University Press.

Masters, J., et al. (Eds.) (2013). *Leaping ahead: advances in prosimian biology*. NY: Springer. [Chapters 9, 10, 11, 19, 30]

Mittermeier, R.A., et al. (Eds.) (2013). *Handbook of the mammals of the world, Vol. 3. Primates* (pp. 184–235). Barcelona: Lynx Edicions.

Nekaris, K.A.I. (2014). Extreme primates: ecology and evolution of Asian lorises. *EA*, *23*, 177–187.
Nekaris, K.A.I., & Bearder, S.K. (2011). The lorisiform primates of Asia and mainland Africa. In C.A. Campbell et al. (Eds.), *Primates in perspective* (pp. 34–54). NY: OUP.

Additional References

Alterman, L. (1995). Toxins and toothcombs: potential allospecific chemical defense in *Nycticebus* and *Perodicticus*. In L. Alterman, et al. (Eds.), *Creatures of the dark*. NY: Springer.
Bearder, S., et al. (2003). Species-typical patterns of infant contact, sleeping site use and social cohesion among nocturnal primates in Africa. *FP*, *74*, 337–354.
Brandon-Jones, D., et al. (2004). Asian primate classification. *IJP*, *25*, 95–164.
Das, N., et al. (2014). Medicinal plant exudativory by the Bengal slow loris, *Nycticebus bengalensis*. *ESR*, *23*, 149–157.
Doyle, G.A. (1979). Development of behavior in prosimians with special reference to the lesser bushbaby, *Galago senegalensis moholi*. In G.A. Doyle & R.D. Martin (Eds.), *The study of prosimian behavior*. NY: Academic Press.
Dumont, E.R. (1997). Cranial shape in fruit, nectar and exudates feeders: implications for interpreting the fossil record. *AJPA*, *102*, 187–202.
Ehrlich, A. (1974). Infant development in two prosimian species: greater galago and slow loris. *Developmental Psychobiology*, *7*, 439–454.
Ehrlich, A., & Macbride, L. (1989). Mother–infant interactions in captive slow lorises (*Nycticebus coucang*). *AJP*, *19*, 217–228.
Ehrlich, A., & Musicant, A. (1977). Social and individual behaviors in captive slow lorises. *Behaviour*, *60*, 195–220.
Fisher, H.S., et al. (2003). Countermarking by male pygmy lorises (*Nycticebus pygmaeus*): do females use odor cues to select mates with high competitive ability? *BE&S*, *53*, 123–130.
Fitch-Snyder, H., & Ehrlich, A. (2003). Mother–infant interactions in slow lorises (*Nycticebus bengalensis*) and pygmy lorises (*Nycticebus pygmaeus*). *FP*, *74*, 259–271.
Fitch-Snyder, H., & Jurke, M. (2003). Reproductive patterns in pygmy lorises (*Nycticebus pygmaeus*): behavioral and physiological correlates of gonadal activity. *Zoo Biology*, *22*, 15–32.
Fitch-Snyder, H., & Schulze, H. (Eds.), Larson, L. (compiler) (2001). *Husbandry manual for Asian lorisines* (*Nycticebus* & *Loris* SSP). Zoological Society of San Diego, San Diego, CA.
Fleagle, J.G. (2013). *Primate adaptation and evolution* (3rd edition). NY: Elsevier Science.
Grubb, P., et al. (2003). Assessment of the diversity of African primates. *IJP*, *24*, 1301–1357.
Gursky, S. (2000). Sociality in the spectral tarsier, *Tarsius spectrum*. *AJP*, *51*, 89–101.
Izard, M.K., & Rasmussen, D.T. (1985). Reproduction in the slender loris (*Loris tardigradus malabaricus*). *AJP*, *8*, 153–165.
Izard, M.K., & Weisenseel, K. (1989). Comparative reproduction of the lorisidae. *AJP*, *18*, 140.
Izard, M.K., et al. (1988). Reproduction in the slow loris (*Nycticebus coucang*). *AJP*, *16*, 331–339.
Jaenicke, C., & Ehrlich, A. (1972). Effects of animate vs. inanimate stimuli on curiosity behavior in greater galago and slow loris. *Primates*, *23*, 95–104.
Kappeler, P.M. (1996) Causes and consequences of life-history variation among strepsirrhine primates. *AN*, *148*, 868–891.
Kessler, S.E., & Nash, L.T. (2010). Grandmothering in *Galago senegalensis braccatus*. *African Primates*, *7*, 42–49.
Krane, S., et al. (2003). "Venom" of the slow loris. *Naturwissenschaften*, *90*, 60–62.
Larney, E., & Larson, S.G. (2004). Compliant walking in primates: elbow and knee yield in primates compared to other mammals. *AJPA*, *125*, 42–50.
McArthur, C., et al. (2012). The foraging tightrope between predation risk and plant toxins: a matter of concentration. *Functional Ecology*, *26*, 74–83.

Nash, L.T. (2004). Kinship and behavior among nongregarious nocturnal prosimians: what do we really know? In B. Chapais & C.M. Berman (Eds.), *Kinship and behavior in primates* (pp. 200–222). NY: OUP.

Nekaris, K.A.I. (2001). Activity budget and positional behaviour of the Mysore slender loris: implications for slow climbing locomotion. *FP*, *72*, 228–241.

Nekaris, K.A.I. (2002). Slender in the night. *Natural History*, *111(2)*, 54–59.

Nekaris, K.A.I. (2003a). Observations on mating, birthing and parental care in three taxa of slender loris in India and Sri Lanka (*Loris tardigradus* and *Loris lydekkerianus*). *FP* supp. (Eds. S. Gursky & K.A.I. Nekaris) *74*, 312–336.

Nekaris, K.A.I. (2003b). Spacing system of the Mysore slender loris (*Loris lydekkerianus lydekkerianus*). *AJPA*, *121*, 86–96.

Nekaris, K.A.I. (2014). Extreme primates: ecology and evolution of Asian lorises. *EA*, *23*, 177–187.

Nekaris, K.A.I., & Stephens, N.J. (2007). All lorises are not slow: rapid arboreal locomotion in the newly recognised red slender loris (*Loris tardigradus tardigradus*) of southwestern Sri Lanka. *AJP*, *69*, 112–120.

Nekaris, K.A.I., et al. (2007). Anti-predator behaviour of lorises and pottos. In S.G. Gursky & K.A.I. Nekaris (Eds.), *Primate anti-predator strategies* (pp. 220–238). New York: Springer.

Nekaris, K.A.I., et al. (2013). Mad, bad and dangerous to know: the biochemistry, ecology, and evolution of slow loris venom. *Journal of Venomous Animals and Toxins including Tropical Diseases*, *19*, DOI: 10.1186/1678-9199-19-21 (last accessed 4 August 2015).

Nowack, J., et al. (2013a). Surviving the cold, dry period in Africa: behavioral adjustments as an alternative to heterothermy in the African lesser galagid (*Galago moholi*). *IJP*, *34*, 49–64.

Nowack, J., et al. (2013b). Torpor as an emergency solution in *Galago moholi*: heterothermy is triggered by different constraints. *Journal of Comparative Physiology B*, *183*, 547–556.

Nowak, R.M. (1999). *Walker's primates of the world*. Baltimore: Johns Hopkins University Press.

O'Mara, M.T., et al. (2012). Growth and development of sexual size dimorphism in lorises and galagos. *AJPA*, *147*, 11–20.

Pontzer, H., et al. (2014). Primate energy expenditure and life history. *PNAS*, *111*, 1433–1437.

Radhakrishna, S. (2004). Sociality in a solitary primate: how gregarious is the slender loris? *Resonance*, *9*, 64–81.

Radhakrishna, S., & Kumara, H.N. (2010). Behavioural variation in the Mysore slender loris *Loris lydekkerianus lydekkerianus*. *Current Science*, *99*, 1226–1232.

Radhakrishna, S., & Singh, M. (2002). Social behavior of the slender loris (*Loris tardigradus lydekkerianus*). *FP*, *73*, 181–196.

Radhakrishna, S., & Singh, M. (2004). Infant development in the slender loris (*Loris lydekkerianus lydekkerianus*). *Current Science*, *86*, 1121–1127.

Rasmussen, D.T., & Nekaris, K.A.I. (1998). Evolutionary history of lorisiform primates. *FP*, *69*, 250–285.

Rode-Margono, E.J., & Nekaris, K.A.I. (2014). Impact of climate and moonlight on a venomous mammal, the Javan slow loris (*Nycticebus javanicus* Geoffroy, 1812). *Contributions to Zoology*, *83*, 217–225.

Ross, C. (2001). Park or ride? Evolution of infant carrying in primates. *IJP*, *22*, 749–771.

Scheun, J., et al. (2014). Spicing up the menu: evidence of fruit feeding in *Galago moholi*. *Primates*, *55*, 359–363.

Schmitt, D., & Lemelin, P. (2004). Locomotor mechanics of the slender loris (*Loris tardigradus*). *JHE*, *47*, 85–94.

Schulze, H., & Meier, B. (1995). Behavior of captive *Loris tardigradus nordicus*. In L. Alterman et al. (Eds.), *Creatures of the dark, the nocturnal prosimians* (pp. 221–249). NY: Plenum.

Starr, C., & Nekaris, K.A.I. (2013). Obligate exudativory characterizes the diet of the pygmy slow loris *Nycticebus pygmaeus*. *AJP*, *75*, 1054–1061.

Starr, C.R., et al. (2012). Hiding from the moonlight: luminosity and temperature affect activity of Asian nocturnal primates in a highly seasonal forest. *PLONE*, *7*, e36396.

Svensson, M.S., & Bearder, S.K. (2013). Sightings and habitat use of the northern lesser galago (*Galago senegalensis senegalensis*) in Niumi National Park, The Gambia. *African Primates*, *8*, 51–58.

Watson, S.L., et al. (1994). Effects of modeling and lineage on fishing behavior in the small-eared bushbaby (*Otolemur garnettii*). *IJP*, *15*, 507–519.

Welker, C. (1976). Fishing behaviour in *Galago crassicaudatus* E. Geoffroy, 1812 (Prosimiae; Lorisiformes; Galagidae). *FP*, *26*, 284–291.

Wiens, F., & Zitzmann, A. (2003a). Social dependence of infant slow lorises to learn diet. *IJP*, *24*, 1007–1021.

Wiens, F., & Zitzmann, A. (2003b). Social structure of the solitary slow loris *Nycticebus coucang* (Lorisidae). *JZL*, *261*, 35–46.

Wiens, F., et al. (2006). Fast food for slow lorises: is low metabolism related to secondary compounds in high-energy plant diet? *Journal of Mammalogy*, *87*, 790–798.

Zimmermann, E. (1985). Vocalizations and associated behaviours in adult slow loris (*Nycticebus coucang*). *FP*, *44*, 52–64.

Zimmermann, E. (1989). Reproduction, physical growth and behavioral development in slow loris (*Nycticebus coucang*, Lorisidae). *Human Evolution*, *4*, 171–179.

Zimmerman, E. (1995). Acoustic communication in nocturnal prosimians. In L. Alterman, et al. (Eds.), *Creatures of the dark*. NY: Springer.

7 The Lemuriform Infraorder

Island Refuge

It is a winter morning in Madagascar. A little grayish-brown ball of fur is curled up in a hole in a tree. The dwarf lemur will sleep for several months. On the ground below, a troop of ringtailed lemurs wave their bushy tails as they travel through the sparse forest, foraging and occasionally bickering.

Lemuriforms (Figure 7.1) are strepsirrhines that exist only on the island of Madagascar and some small adjacent islands. Madagascar is about 1000km long and about 400 kilometers distant from the east coast of Africa. The ancestors of living lemuriforms probably arrived there by rafting on vegetation from Africa when the currents were favorable (Ali and Huber 2010), a view that was supported by a broader analysis of vertebrate fauna on the island (Samonds et al. 2012). This places their arrival between 50 and 60 million years ago, which is consistent with molecular calculations for their origin in a common ancestor (McLain et al. 2012; Montagnon 2013). Lemuriforms never faced competition with anthropoids until

Figure 7.1 The nocturnal mouse lemur is one of the smallest primates. © dennisvdw via iStock

the arrival of humans a few thousand years ago. During their isolation they proliferated into numerous and enormously diverse species (Tattersall 2013).

Lemuriforms are often called Malagasy primates because they share their island with the political entity called the Malagasy Republic. More problematic is the use of the word *lemur* for all of them. It is a somewhat slippery term because it is also used in the names of some families, genera, and species within the larger taxon, but not for all of them (e.g. *mouse lemur* vs. *sifaka*). Furthermore, the aye-aye is in a sense not a lemur at all. However, *lemuriform* encompasses all the Malagasy primates (Box 7.1).

Box 7.1 Lemuriform Classification

More than 100 species have been recognized among the Malagasy primates (McLain et al. 2012; Yoder 2013), although a conservative evaluation would place the number closer to 50 (Tattersall 2013). In any case, there is enormous primate diversity on the island, represented by general agreement about classification into five distinct families.

Ongoing molecular studies with new technologies have confirmed most of the preceding 50 years of research on the phylogeny of Malagasy primates (Yoder 2013). This includes the conclusion, consistent with morphological analyses, that this primate group is monophyletic (McLain et al. 2012). The main changes have been in taxonomy at the highest and lowest levels. Many subspecies have been elevated to the level of full species, with some dissent (Tattersall 2013). Perhaps less controversially, the lemuroid superfamily has been placed in a higher taxon (infraorder or series) designated Lemuriformes (Fleagle 2013; Masters et al. 2013). In this scheme the Lemuroidea are the sole superfamily in Lemuriformes. This parallels the elevation of the lorisoids (see Box 5.1).

Lemuriform Behavior

Many lemuriforms are nocturnal animals, but others are active to varying degrees in daylight. Despite this profound difference, there are many similarities across the entire taxon. Two opposing theories account for this fact. The Energy Conservation Hypothesis portrays the Madagascar environment as unusually difficult for primates and explains many common features of the lemuriforms as unique responses to the problems of extracting resources and conserving energy in the demanding Madagascar environment. This hypothesis is an amalgam of several predecessors and can also be called the Energy Frugality Hypothesis (Wright 1999).

The main alternative theory is based on phylogenetic inertia, the idea that some established traits change more slowly than the environment. This Evolutionary Disequilibrium or Mismatch Hypothesis explains common lemuriform traits as nocturnal adaptations that did not change when some lemuriforms became active in daylight. The scenario begins with giant eagles that lived 2000 years ago. The extinction of these birds allowed nocturnal primates to expand their activities into daylight, increasing their foraging opportunities. Because this happened so recently, there is a *mismatch* between the new diurnal environment and the persisting nocturnal traits.

The issue of monomorphy illustrates the distinction. Typically there is little or no sex difference in the body mass of lemuriforms (see Table 7.1 online for physical traits of all lemuriforms). Where the teeth differ (about half of studied species), male canines are only slightly larger. According to the Energy Conservation Hypothesis, larger-bodied males would have been at a selective disadvantage during frequent and stringent food shortages on Madagascar. In addition, because large male infants are a greater cost to nursing mothers, females would have been selected to have monomorphic offspring (possibly by choosing smaller males as sexual partners).

According to the Mismatch Hypothesis, monomorphy is an adaptation to the pair-groups that are typical of nocturnal lemuriforms and other nocturnal primates (such as many lorisoids). Lemuriforms that are active in daylight remain monomorphic because they have not had time to evolve different adaptations (van Schaik and Kappeler 1996). Wright (1999) answered that several non-pair species of nocturnal lemuriforms are also monomorphic (which makes this a recurrent trait in our terms).

Lemuriform Ecology

Madagascar habitats range from dry forests in the west and south to rainforests in the east, but all of these environments are characterized by poor soils, low plant productivity, severe climate, and catastrophic weather (Dunham et al. 2010; Ganzhorn 1999). Cyclones, droughts, and frosts depress overall productivity and create periodic food shortages. Lemuriforms seem to face greater problems of food availability than do primates in most other tropical habitats. This is the basis for the Energy Conservation Hypothesis, which amalgamated several ideas related to resource acquisition and the use of energy. Low basal metabolic rates of lemuriforms seemed to be an energy-conserving adaptation to Madagascar, but it turned out that all strepsirrhines have low BMRs compared to other primates (Harcourt 2008). However, the hypothesis still applies well to other lemuriform traits.

Fruit is the leading component of lemuriform diets, eaten by 23 of the 25 species in a survey, and a large percentage of the diet for most of them. Some fruiting plants seem to depend on the primates for seed propagation. Dependence on fruit seems maladaptive in the Madagascar environment where fruit tree growth is slower and productivity is lower than in other primate habitats. Madagascar fruit cycles tend to be unusually prolonged, asynchronous, and irregular. Even if relatively regular, a particular fruit may only be available in alternate years. Even when fruit is available, the supply is limited because tree crowns are smaller than elsewhere (possibly an adaptation to prevent felling by cyclonic winds). One compensating benefit may be that there are few frugivorous birds on Madagascar to compete with the primates (Smith and Ganzhorn 1996). The high level of energy return from fruit could be another factor (see Table 7.2 online for ecological patterns of all lemuriforms).

On the other hand, although fruit provides a substantial portion of the diet for many Malagasy primates, there are no lemuriform species that are frugivorous to the same extent as many primates elsewhere. Ganzhorn et al. (2009) noticed an especially large gap between Madagascar and the Americas: there were many more frugivorous genera in New World primate communities than in Madagascar. They hypothesized that this was due to a significant difference in protein content and tested the idea with a literature survey and their own analyses. Nitrogen concentrations, representing protein content, supported the hypothesis.

Lemuriforms survive irregular availability and the low nutritional value of fruit by using various fallback foods. Leaves and associated structures (stems, shoots, and petioles) are almost as important as fruit to most species. Though still in the minority, primarily folivorous species are more numerous among the lemuriforms than in other primate ecological communities elsewhere in the world. Some of them show a high tolerance for secondary compounds.

Lemuriforms also eat flowers, buds, petals, and nectar. The use of seeds has probably been underestimated because seeds are often swallowed along with fruit and some studies give no indication as to whether those seeds are digested or not. Lemuriforms use a large number of plant taxa and a great variety of plant parts. This unusual plasticity can be viewed as an adaptation to the erratic environment.

The typical action patterns of reaching in lemuriforms contrast with those of lorisiforms in ways consistent with diets placing more emphasis on plant foods than on insects. A lemuriform extends its hand with digits two through five parallel and the fingertips leading (Martin 1990). This is functional when most food items do not try to fly or leap away.

Lemuriforms suffer the highest predation rates among the primates, perhaps due to the dearth of other mammalian prey on Madagascar. Raptors seem to play the leading role. In playback experiments lemuriforms responded to raptor calls with appropriate vigilance and escape tactics (Karpanty and Grella 2001). The Mismatch Hypothesis would explain this as a genetic relict of a past when a greater variety of avian predators existed in Madagascar. However, there are formidable raptors in the present and some seem to have a preference for lemuriform prey, at least during certain seasons. Two hawk species in a national park took up to 21 percent of the lemuriform populations per year (Karpanty 2006).

The fossa (*Cryptoprocta ferox*) is another formidable enemy. This species is a viverrid (related to civets and genets), but is the Madagascar equivalent of a cat (Figure 7.2). It hunts during day or night and combines agility in the trees with a size of about 7kg. A fossa can take down adults of the largest lemuriform species. In some locales lemuriforms provide

Figure 7.2 The fossa is a cat-like viverrid that preys on lemuroids night and day. © Rhett A. Butler / www.mongabay.com

30 to 50 percent of the fossa's prey, as measured by weight. Even where the per capita rate of predation is low, this can be deceptive for several reasons. First, impact is multiplied by the clustering of kills due to the fossa's habit of hunting out an area before moving to another. Second, the small size of lemuriform social units means that fossa predation can extinguish an entire group. Boa constrictors can also take adults of the largest lemuriform species.

Lemuriform Sociality

Pair-groups occur in more than 25 percent of the lemuriforms and perhaps over 50 percent, compared to 13 percent of all primates and 3 percent of mammals. An ecological function of such small groups might be effective use of resources that are scarce and uniformly distributed at any time during the daily cycle. The Energy Conservation theory would see this as an especially stringent factor in Madagascar. Other forms of social organization include troops and multi-male–multi-female networks that in some cases display aspects of the noyau.

Pairs can last for several years, but there is significant variation in the parameters of the relationship, even within populations. Sleeping arrangements vary from solo to pair or pair-group. Cohesiveness during the active period varies from solitary foraging to proximity and vocal contact to complete coordination of travel. Social interactions may involve frequent affiliative acts or frequent hostility. Male allocare might explain pairing in some lemuriforms but not in others. Social monogamy is predominant, but a male or female may be added to the group. Maturing offspring might account for some extra adults, but this is demonstrably untrue in certain cases (see Table 7.3 online for social patterns of all lemuriforms).

Kappeler (2014) advanced a typology for living species to suggest two paths for the evolution of pairs from solitary and social ancestors. In one of these "routes" or "conduits" pairs resulted from the union of solitary individuals in accord with the Socioecological Model: females established independent territories, males associated with them, and males began to display paternal care. In the other conduit, pairs differentiated from troops in response to female competition and interspecific competition. The details are best discussed in connection with particular taxa in the following chapters.

Female dominance is a recurrent, if not typical, feature of lemuriform societies. In a number of species, across various social systems, individual females win almost all agonistic encounters with individual males. Females can threaten or slap males without retaliation, often eliciting submissive displays. Males sometimes signal submission spontaneously. In some species the outcomes are more mixed, but female success is greater than in many other primates. This success is all the more striking in view of the similarity between the sexes in body and canine size. The majority of aggressive interactions involve food, and many primatologists consider female feeding priority to be a manifestation of dominance. A recurring pattern of female leadership can be linked to feeding priority: females that lead group movements arrive at food sources first.

Alternatively, the Male Deference Hypothesis states that males *allow* females better access to food. The Wimpy Male Hypothesis (Richard 1992) made it a matter of sexual selection: females only mate with males that are compliant (or at least not successfully assertive). An underlying assumption (reasonable, but undemonstrated) is that this aspect of male personality is heritable (the phylogenetic gambit). The Wimpy Male Hypothesis can be extended to sexual selection of smaller males, leading to monomorphy. A possible objection arises from the fact that food conflicts often result in females charging, slapping, and biting males, and the males often retreat with submissive vocalizations. This behavior raises a question as to why deferential/wimpy males would place themselves in such situations in the first place unless they are trying to compete. They may be wimpy in conflicts, but the fact that they provoke them suggests that they are not deferential.

Agonistic superiority and male deference are proximate explanations for female feeding priority. They leave us with the question of function or ultimate causation. Most proponents of the Energy Conservation Hypothesis hold that the sparse and seasonal supply of food in Madagascar puts so much stress on females that they could not reproduce effectively without this advantage. Males would have to accept domination (whatever the proximal mechanism) because they too would fail to reproduce if the females they fertilized did not bear enough viable offspring. Challengers have argued that detailed investigation of lemuriform reproduction reveals no special burden and no significant difference from other primates (Curtis 2004; Pereira et al. 1999).

The Cost Asymmetry Hypothesis (Dunham 2008) proposed a synthesis based on game theory, which predicts that a conflict between opponents of comparable fighting ability will be resolved by the value to each contestant of the resources at stake. Female lemuriforms are like other primates in their greater need for food due to the costs of reproduction. The difference is monomorphy. Being physically equal to males, females are likely to win at least 50 percent of conflicts. Given the differential value of the resources, the cost of fighting is greater for males and their best strategy is to defer to females and find other food sources. One problem with this hypothesis is that, as noted above, males often seem to lose to females rather than defer to them. Another is that it leaves open the cause of monomorphy.

Dunham and Rudolf (2009) proposed "passive mate guarding" as an explanation for monomorphy. This entails avoidance of direct confrontations with other males by devices such as **copulatory plugs**, i.e. congealed sperm left in the vaginal tract of a mate to impede fertile mating by other males. The researchers used data from 62 primate species to show that copulatory plugs are negatively associated with degree of sexual dimorphism and the length of female sexual receptivity. They concluded that lemuriform monomorphy could plausibly be explained by such post-copulatory competition, as opposed to selection for male body size and weaponry.

Nevertheless, aggressive behavior and sperm competition both occur in lemuriforms. Males in most species display progressive enlargement of the testes over two or three months leading up to the breeding season. This even occurs in supposedly monogamous species, presumably because paired individuals also breed with outsiders. When the season arrives, males augment their voluminous sperm production with fights that often cause long canine gashes and other serious injuries. Active mate guarding also occurs in a number of species.

Each species has its own strict mating season that lasts from one to two weeks and is triggered by daily sunlight totals that stimulate the pituitary gland. An individual female is in estrus for no more than about 30 hours, which might explain why lemuriform males practice an unusual diversity of competitive mechanisms. The time limit resulting from synchronization of estrus among females in a population makes it impossible for one or a few males to monopolize many of the females. This gives females more control over the mating process, which may contribute to year-round dominance and feeding priority. A shortcoming of the Mismatch Hypothesis is that it does not explain the uniquely strict seasonality of virtually all lemuriforms.

Sexually selected infanticide is unlikely in most lemuriforms because of the strict annual breeding seasons. Killing an infant would not ordinarily provide the male with a quick reproductive opportunity. However, Rasoloharijaona et al. (2000) made detailed observations on an infanticide that took place after the mother's partner had deserted her.

Lemuriform Life Cycle

Restriction of the mating season results in a narrow period for births (see Table 7.4 online for reproductive parameters of all lemuriforms). The Predator Satiation Hypothesis explains this phenomenon: every mother benefits statistically because the simultaneous presence of all infants decreases the odds that hers will be taken. Seasonal breeding can also be explained in terms of maximized food availability during either pregnancy or nursing. These interpretations are mutually contradictory for a given population but might apply across a species or higher taxon.

The greatest significance of lemuriform seasonality may be related to synchronized weaning (see Table 7.5 online for life cycle stages of all lemuriforms). Although different species give birth at different times, spread out over six months of each year, all wean their young during the same three-month period. This is the time when the food supply, with fruit augmented by leaves and insects, is most certain. The rapid growth of many species ensures that the infants will reach the weaning stage by the time this period begins.

The earliest stages of reproduction are equivocal with regard to the energy needs of female lemuriforms. They display higher rates of prenatal growth than lorisoids, which indicates a higher maternal investment per unit of time. However, lactation is the more costly stage of reproduction and the cost is no greater for lemuriforms than for lorisoids: the quantity of lemuriform milk is no greater and it is more dilute (van Schaik and Kappeler 1996).

The low quality of milk and the small size of neonates suggest that lemuriform females have been selected for low maternal investment. This can be explained by high infant mortality, ranging from 40 to 80 percent over periods of ten years. Bearing larger neonates and giving better milk would do little to counteract the causes: predation, natural disasters, and infanticide. Keeping resources in reserve for the next cycle probably maximizes lifetime reproductive success. Maternal nutrition seems be a key proximate factor in the wild. Captive females display earlier first birth, heavier neonates, and shorter inter-birth intervals.

Lemuriform maternal difficulties might suggest a role for allocare, which has been advanced as an important factor in primate life histories. However, Tecot et al. (2012) argued that allocare is not significant for lemuriforms. Rather, the early developmental variables that shape life history are nesting and parking. Comparative analysis showed that allomaternal care had no significant effect on reproductive output. Parking and nesting were positively associated with faster fetal and postnatal growth, and nesting was associated with shorter inter-birth intervals. The researchers noted that these factors applied particularly to species that produced litters. This may be consistent with the suggestion by Dewar and Richard (2007) that the extreme variability of the Madagascar environment has elicited different life history adaptations in different taxa.

Lemuriform Cognition

There is a growing body of evidence that lemuriform cognitive abilities overlap with those of anthropoids (Fichtel and Kappeler 2010). Lemuriforms are claimed to demonstrate the following capabilities in more than one of their zoological families: spatial memory; negotiation of detours and mazes; solution of simple box problems; reversal learning; reconciliation; dominance relationships; social learning; and referential alarm calls. These cognitive traits are all shared with anthropoids. In many cases where evidence is lacking for lemuriforms, relevant observations and experiments have not yet been made. Evidence for cognitive skills is especially prominent in species active in daylight, so this topic will be taken up again in Chapter 9.

Summary

Although lemuriforms are divided between nocturnal taxa and those that are active to varying degrees in daylight, they have many common features. These are accounted for by opposing hypotheses: Energy Conservation and Evolutionary Disequilibrium (Mismatch). The common traits include monomorphy, low basal metabolism, frequency of pair-groups, and recurrent female dominance. Despite the last, males compete intensely for mating opportunities during the brief mating seasons. Different species give birth at different times, but all wean their young during the same period of the year when the food supply is most certain. Low milk quality and small neonates suggest that low maternal investment is adaptive in the dangerously fluctuating environment. Recent years have seen a growing body of evidence for unsuspected cognitive abilities in some lemuriforms.

Sources

Basic Bibliography

Goodman, S.M., & Benstead, J.P. (Eds.) (2004). *The natural history of Madagascar*. Chicago: UCP.

Gould, L., & Sauther, M.L. (Eds.) (2006). *Lemurs, ecology and adaptation*. NY: Springer.

Kappeler, P.M., & Watts, D.P. (Eds.) (2012). *Long-term field studies of primates*. NY: Springer. [Part II]

Masters, J., et al. (Eds.) (2013). *Leaping ahead, advances in prosimian biology*. NY: Springer. [Chapters 1–5, 12, 34, 36]

Mittermeier, R.A., et al. (Eds.) (2010). *Lemurs of Madagascar*, 3rd ed. Conservation International, Washington, DC.

Mittermeier, R.A., et al. (Eds.) (2013). *Handbook of the mammals of the world, Vol. 3. Primates* (pp. 1–183). Barcelona: Lynx Edicions.

Rakotosamimanana, B., et al. (Eds.) (1999). *New directions in lemur studies*. NY: Plenum Press.

Wright, P.C. (1999). Lemur traits and Madagascar ecology: coping with an island environment. *Yearbook of Physical Anthropology 42*, 31–72.

Additional References

Ali, J.R., & Huber, M. (2010). Mammalian biodiversity on Madagascar controlled by ocean currents. *Nature, 463*, 653–656.

Curtis, D.J. (2004). Diet and nutrition in wild mongoose lemure (*Eulemur mongoz*) and their implications for the evolution of female dominance and small group size in lemurs. *AJPA, 124*, 234–247.

Dewar, R., & Richard, A. (2007). Evolution in the hypervariable environment of Madagascar. *PNAS, 104*, 13723–13727.

Dunham, A.E. (2008). Battle of the sexes: cost asymmetry explains female dominance in lemurs. *AB, 76*, 1435–1439.

Dunham, A.E., & Rudolf, V.H.W. (2009). Evolution of sexual size monomorphism: the influence of passive mate guarding. *JEB, 22*, 1376–1386.

Dunham, A.E., et al. (2010). Global climate cycles and cyclones: consequences for rainfall patterns and lemur reproduction in southeastern Madagascar. *Global Change Biology, 17*, 219–227.

Fichtel, C., & Kappeler, P.M. (2010). Human universals and primate symplesiomorphies: establishing the lemur baseline. In P.M. Kappeler & J.B. Silk (Eds.), *Mind the gap* (pp. 395–426). NY: Springer.

Fleagle, J.G. (2013). *Primate adaptation and evolution* (3rd edition). NY: Elsevier Science.
Ganzhorn, J.U. (1999). Body mass, competition and the structure of primate communities. In J.G. Fleagle et al. (Eds.), *Primate communities* (pp. 141–157). Cambridge: CUP.
Ganzhorn, J.U., et al. (2009). Possible fruit protein effects on primate communities in Madagascar and the Neotropics. *PLONE*, *4*, e8253.
Harcourt, A.H. (2008). Are lemurs' low basal metabolic rates an adaptation to Madagascar's unpredictable climate? *Primates*, *49*, 292–294.
Irwin, M.T., et al. (2009). Spatial and temporal variability in predation on rainforest primates. *Animal Conservation*, *12*, 220–230.
Kappeler, P.M. (2014). Lemur behavior informs the evolution of social monogamy. *TREE*, *29*, 591–593.
Karpanty, S.M. (2006). Direct and indirect impacts of raptor predation on lemurs in southeastern Madagascar. *IJP*, *27*, 239–261.
Karpanty, S.M., & Grella, R.A. (2001). Lemur responses to diurnal raptor calls in Ranomafana National Park, Madagascar. *Folia*, *72*, 100–103.
Martin, R.D. (1990). *Primate origins and evolution*. Princeton: Princeton University Press.
Masters, J., et al. (2013). What's in a name? Higher level taxonomy of the prosimian primates. In J. Masters et al. (Eds.), *Leaping ahead, advances in prosimian biology* (pp. 3–9). NY: Springer.
McLain, A.T., et al. (2012). An *Alu*-based phylogeny of lemurs (infraorder: Lemuriformes). *PLONE*, *7*, e44035.
Montagnon, D. (2013). Strepsirrhine divergence dates estimated from mitochondrial gene sequences and the status of *Daubentonia*. In J. Masters et al. (Eds.), *Leaping ahead, advances in prosimian biology* (pp. 21–32). NY: Springer.
Pereira, M.E., et al. (1999). Metabolic strategy and social behavior in Lemuridae. In B. Rakotosamimanana, et al. (Eds.), *New directions in lemur studies*. NY: Springer.
Rasoloharijaona, S., et al. (2000). Infanticide by a male Milne-Edwards' sportive lemur (*Lepilemur edwardsi*) in Ampijoroa, NW-Madagascar. *IJP*, *21*, 41–45.
Richard, A.F. (1992). Aggressive competition between males, female controlled polygyny and sexual monomorphism in a Malagasy primate *Propithecus verreauxi*. *JHE*, *22*, 395–406.
Samonds, K.E., et al. (2012). Spatial and temporal arrival patterns of Madagascar's vertebrate fauna explained by distance, ocean currents, and ancestor type. *PNAS*, *109*, 5352–5357.
Smith, A.P., & Ganzhorn, J.U. (1996). Convergence in community structure and dietary adaptation in Australian possums and gliders and Malagasy lemurs. *Austral Ecology*, *21*, 31–46.
Tattersall, I. (2013). Species-level diversity among Malagasy lemurs. In J. Masters et al. (Eds.), *Leaping ahead, advances in prosimian biology* (pp. 11–20). NY: Springer.
Tecot, S., et al. (2012). Infant parking and nesting, not allomaternal care, influence Malagasy primate life histories. *BE&S*, 66, 1375–1386.
van Schaik, C.P., & Kappeler, P.M. (1996). The social systems of gregarious lemurs: lack of convergence due to evolutionary disequilibrium? *Ethology*, *102*, 915–941.
Wright, P.C. (1999). Lemur traits and Madagascar ecology: coping with an island environment. *YPA*, *42*, 31–72.
Yoder, A.D. (2013). The lemur revolution starts now: the genomic coming of age for a non-model organism. *Molecular Phylogenetics and Evolution*, 66, 442–452.

8 Lemuriform Variation

The Night Life

A nightmare creeps out of the darkness, tapping the branch on which it moves with a long skeletal finger. It stops and gouges at the branch until it comes up with a meal of grubs. The aye-aye is not pretty to human eyes, but it is the survivor of an ancient lineage and the most widely distributed primate on the island of Madagascar.

There are three zoological families of nocturnal lemuriforms. This chapter surveys each of them, including some of the most interesting (or only) genera in each.

Cheirogaleidae

All the genera in this family are in about the same size range as mice. They are quadrupeds with varying degrees of leaping ability and inclination. Cheirogaleids display high reproductive rates with short gestation, litters of two to four, and short IBIs. This offsets a high mortality rate that is probably due largely to predation. Owls are a particular danger to these small nocturnal animals. Mortality per year was measured at 25 percent in one population.

> **Box 8.1 Cheirogaleid Taxonomy**
>
> Cheirogaleidae are very small quadrupeds. The mouse lemurs (*Microcebus*) and the dwarf lemurs (*Cheirogaleus*) are the most speciose in the family. They are the most closely related of the five genera and share the dormancy trait, though the dwarf lemurs take it further.

Dormancy: Torpor and Hibernation

Mouse lemurs (*Microcebus*) and dwarf lemurs (*Cheirogaleus*) are found in forests all across Madagascar. Both genera are capable of **heterothermy**, the ability to make substantial adjustments in body temperature (Dausmann 2014). They do this through forms of **dormancy**, a state in which body temperature is close to the ambient level and metabolic rate gets as low as 5 percent of normal. These are generally viewed as adaptations for energy conservation during times of food shortage, but they may also function to conserve water and/or to avoid predators during peaks of danger (Schmid and Kappeler 2005). Torpor is a form of dormancy that usually lasts less than 24 hours, though it may occur on a daily basis. **Hibernation** extends over a longer period, up to seven months in some cases.

Daussman (2014) argued that there is a spectrum of heterothermy (variable body temperature) that ranges in gradations from daily torpor through prolonged torpor to obligate hibernation. The last includes underground hibernation in one population (Blanco et al. 2013).

Mouse Lemurs

Mouse lemurs (Figure 7.1 in the previous chapter) are among the smallest primates, weighing as little as 30g. They eat insects and sugary insect secretions, but subsist largely on fruit. This seems to violate the metabolic rule linking small size and consumption of animal matter. Conservation of energy through dormancy may account for their dietary flexibility. Gray mouse lemurs (M. *murinus*) undergo daily torpor during the dry season and engage in sexually skewed hibernation. Females fatten their bodies and tails and then hibernate for up to five months. Males undergo no change in weight, rarely are in torpor for more than a few days at a time, and are active for weeks before the females emerge from hibernation. In brown mouse lemurs (M. *rufus*) a high proportion of both sexes may hibernate for at least a month (based on temporary absence from researchers' traps). Some brown mouse lemurs fluctuate in weight while others do not, which suggests that their hibernation pattern is influenced by variables such as age and social status (Schmid and Kappeler 2005). Dausmann (2014) considers it likely that all 20 or so mouse lemur species engage in some degree of heterothermy. In most cases this is inferred from gaps in sightings by reliable observers. However, in at least one species torpor is obviated by flexible reductions in energy expenditure in response to environmental situations and individual conditions (Kobbe et al. 2010; Kobbe et al. 2014).

Anti-predator behavior in nocturnal lemuriforms has received more attention with the realization that these primates are not as cryptic as was assumed (Rahlfs and Fichtel 2010). Experiments with mouse lemurs demonstrated appropriate predator-specific responses. Confronted with raptor models, they froze and otherwise behaved cryptically, which is appropriate to the suddenness of raptor attacks. The lemurs reacted to fossa and snake models by monitoring them, presumably to assess risk. However, the lack of response to playbacks of alarm calls indicated that they did not rely on conspecific warnings. In one case gray mouse lemurs mobbed a boa constrictor that had caught a conspecific and the snake released the lemur (Eberle and Kappeler 2008).

Mouse lemur social organization resembles that of galagos in that the basic units are female sleeping groups, some groups including one male. Golden mouse lemurs (M. *ravelobensis*) display more variation, including multiple males. Comparison suggests an ecological explanation. Gray mouse lemurs sleep in relatively well-insulated tree holes while golden mouse lemurs sleep in nests or on branches. The addition of extra individuals among the latter probably contributes to heat retention (Weidt et al. 2004).

Dammahn and Kappeler (2009) tested the relevance of the Socioecological Model to mouse lemurs by comparing two species. Dependence on small, dispersed food patches by M. *berthae* led to high levels of scramble competition and dispersion among females and low levels of association. M. *murinus* used large patches of high quality resources that caused contest competition, leading to clustering of females. This study supported basic predictions of the Socioecological Model and also showed that it can be applied to solitary foragers.

The mouse lemur mating system is multi-male/multi-female. Each individual has access to a number of potential partners during the breeding season and males display a variety of scramble and contest tactics. By the time mating season arrives the male mouse lemur is prepared for sperm competition by testes that have become larger than his brain. During the breeding season males roam in search of mates. This form of scramble competition favors familiarity with the area, including the location of female sleeping sites. Older males copulate with a female earlier in her estrus period and those larger than average will guard her against other males. A study that included genetic data found that most litters had mixed paternity, but older and heavier males had greater reproductive success. Female choice supports these patterns: most females prefer males with calling activity that reflects male–male dominance (Craul et al. 2004). They also tend to prefer older males for copulation, which might correlate with dominance or might represent fitness (longer survival). Estrus female gray mouse lemurs are less aggressive toward dominant males, and dominant males sire more offspring.

Mouse lemurs lack scent glands and achieve olfactory communication with urine, feces, and genital excretions. However, they are highly vocal and display a repertoire of seven or eight calls. One important function is probably the identification of mating partners, since different mouse lemur species are sympatric in many locales. There is evidence that subordinate immigrant males imitate the calls of resident males, perhaps to enhance social acceptance and/or reproductive success.

Vocalizations of gray mouse lemurs contain kinship signatures that result in functional responses related to breeding and agonism. Advertisement calls of gray mouse lemurs contain patrilineal signatures while alarm calls do not (Kessler et al. 2013). Females paid more attention to advertisement calls from unrelated males as compared to related males, but made no such distinction in response to the alarm calls. Female calls contain matrilineal signatures that may reduce costly aggressive incidents between relatives during solitary foraging (Kessler et al. 2014). Recognition might result from innate mechanisms or it may be learned in sleeping associations, which are composed of related females.

Dwarf Lemurs

Dwarf lemurs (*Cheirogaleus*) are among the largest species in their family, weighing up to 50g. They have dark eye rings and a pink rhinarium, and they often walk in a distinctive quadrupedal creep along stout branches. The social system has been described as families that defend their home ranges. High-pitched whistles from both sexes announce territory to other groups.

Dwarf lemurs are the only primates that "hibernate". This is an obligate pattern of seasonal dormancy necessary for survival in cool, dry winters (Dausmann 2014). The period varies across species from about three to seven months. It is not hibernation in the same sense as mammals such as bears, because lemur body temperature closely tracks ambient temperature rather than remaining at a consistently low level. At least one species is known to prepare by eating large quantities of sugary fruits. This results in weight gains of up to 50 percent, stored mainly in the tail (hence the appellation "fat-tailed lemur").

Nectar is an especially important restorative food at the end of hibernation. Returning strength prepares the dwarf lemurs for mating season. Both sexes emit squeaking copulation calls. Infants, after they pass the mouth-carrying stage, cling to the mother's back. Though dormancy is usually achieved in tree holes, some dwarf lemurs hibernate for as long as seven months in underground chambers that they construct for themselves (Blanco et al. 2013).

Lahann and Dausmann (2011) studied the fat-tailed dwarf lemur (*Cheirogaleus medius*) to investigate the relative effects of environmental variation and evolutionary history on life history and physiological traits. They compared populations in a littoral rainforest and a dry deciduous forest and found geographical differences in the length of hibernation and in life history traits. Animals in the rainforest can spend more time in the active state due to a shorter hibernation period, but tend to have a shorter lifespan due to higher mortality rates. They seem to have higher reproductive rates (larger litters, more litters). Home ranges and social organization did not vary between the populations, which suggested that some requirements of the species are independent of environmental conditions. An important implication is that some life history traits that were assumed to be genetically fixed are actually very flexible. Different populations of the same species can vary their behavior to maximize individual reproductive output.

Blanco and Godfrey (2013) considered the relationship between hibernation and life history from a broader comparative perspective. They noted that hibernating mammals differ from others in that their life history pattern is not related to size. Instead the stages revolved around hibernation: short gestation and accelerated early growth to prepare for hibernation; retardation of development in connection with hibernation; and delayed achievement of full size and first reproduction. The researchers studied dwarf lemurs to determine whether primates displayed the same variation as other mammals. They found that dwarf lemurs follow the hibernating pattern while non-hibernators of similar size (e.g. *Galago*) do not. In this area primates are constrained in the same way as other mammals.

A phylogenetic reconstruction by Masters et al. (2014) led to the conclusion that the current size range of dwarf lemurs is the result of evolutionary size reduction in several dwarf lineages from ancestors about the size of *Lepilemur*. Comparison with the closely related lepilemurids indicated that this is an adaptation to a high level of unpredictable variation in the environment. Among the *Lepilemur* species, body size is correlated with environmental predictability. Beyond providing a better understanding of dwarf lemurs, the research is significant in casting doubt on portrayals of dwarf lemurs as a model for an ancient ancestral primate.

Other Cheirogaleids

The drolly named "giant mouse lemurs", *Mirza*, are now in a separate genus from mouse lemurs (*Microcebus*) because they do not resemble them as much as was originally thought. Their elongated bodies weigh three times as much and dentition is markedly different. Vocalization is relatively infrequent and characterized by soft social calls (except for the high-pitched mating whistles of both sexes). Female genitals swell and redden. There is no real torpor in *Mirza*, although they are likely to return to their nests during the second half of a winter night. Presumably their greater size provides some insulation.

Rode et al. (2013) combined data from nest examination and relatedness to infer the social organization of *M. zaza*. The nests were well covered by canopy, even during the dry season. As many as four individuals shared nests exclusive to their group for as long as 50 days. Two groups included two and three males with fully developed testes. Relatedness data revealed that the adult males sharing nests were either unrelated or closely related. These data suggest that *M. zaza* is sleeping in social nest groups including multiple males, which is unusual among nocturnal strepsirrhines. Some of these cheirogaleids display daily torpor.

"Hairy-eared" dwarf lemurs (*Allocebus*) resemble mouse lemurs in size and color, but other morphological traits are highly distinctive. In addition to the tufts around their ears, they have stronger legs that facilitate frequent leaping. Enlarged upper incisors suggest bark scraping to obtain exudates and long tongues seem adapted to feeding on nectar. Pairs are the only adult social grouping that has been observed. Dausmann (2014) assumed torpor in this taxon because of its phylogenetic position between mouse lemurs and dwarf lemurs.

Fork marked lemurs (*Phaner*) are specialized for exudates, a major part of the diet throughout the year (Schülke 2001, 2003). Exudates (rather than insects) provide protein that allows these primates to remain active despite extremes of climate and food supply. They are the only cheirogaleids that do not display dormancy (Schülke and Ostner 2005, 2007). However, they suffer a 20 percent seasonal loss of body mass. During the dry season there is intense competition with mouse lemurs for gums. *Phaner* and *Mirza* displayed active predator response in one incident, mobbing a snake (Schülke 2001).

In one population paired *Phaner* seemed to be bonded: they were near each other during each night, maintained vocal contact, and allogroomed regularly. At another site avoidance was usual, and accidental meetings at food sources stimulated hostility. The latter pattern may result from intense competition for exudates. Males were responsible for proximity and affiliation, which suggests that protection for females does not explain pairing in this species.

Males often fought each other in territorial conflicts, and also fought with females of other pair-groups. The fact that the females always won seems contrary to the Male Deference Hypothesis, since a male gains no fitness advantage by deferring to a female in a rival group. Neighboring females never fought and often engaged in affiliative behavior, which suggests that they were related (the opposite of the situation in mouse lemurs).

Lepilemuridae

Lepilemur (Figure 8.1) is the only surviving genus in its family (Box 8.2). The nickname sportive lemur comes from a tendency to stand up when disturbed in what resembles a boxing stance. Small animals at less than 1kg, sportive lemurs display several features that are anomalous for their size. One is the ability to jump about 1.5m (Warren and Crompton 1997).

Figure 8.1 A sportive lemur clings to a tree with its pair partner below. They are larger than mouse lemurs and differ in positional behavior. © jalvarezg via iStock

Box 8.2 Lepilemurid Taxonomy

Lepilemur is the only genus in the family Lepilemuridae, but it contains almost 30 species. Also known as sportive lemurs for their distinctive defensive stance, they are medium-sized lemurs with short tails. A conical head and short snout are among the more distinctive physical features.

Sportive Lemur Ecology

Sportive lemurs are the smallest folivorous primates, straddling Kay's Threshold of 500g. Folivores are rare among nocturnal mammals in general, probably because leaves are low in sugars at night due to the cessation of photosynthesis. One permissive factor is that sportive lemurs use relatively little energy in traveling short distances to harvest their low-quality foods, including old leaves (Warren and Crompton 1997). Their metabolic rate is low, even for a strepsirrhine, and among the lowest of all mammals for this body size (Schmid and Ganzhorn 1995).

A species of sportive lemur displayed functional responses to playbacks of predator vocalizations (Fichtel 2007). Like mouse lemurs, they froze for raptor calls and also scanned the sky. Reactions to fossa calls included flight or climbing, scanning the ground, and increasing the scan rate. Another sportive lemur species showed similar responses to the same predators (Seiler et al. 2013a). Like mouse lemurs, they did not give alarm calls. However, the sportive lemurs reacted in a similar way to the alarm calls of other species (Seiler et al. 2013b).

Sportive Lemur Sociality

Pair-groups seem to be typical and relatively monogamous sexually. These dispersed units display a variety of proximate mechanisms and functions. A study of five competing hypotheses about the functions of *Lepilemur* loud calls found evidence to support mate attraction, mate defense, and resource defense functions (Rasoloharijaona et al. 2006). Duetting seems to coordinate the activities of *L. edwardsi* pairs and may also strengthen their relationship (Mendez-Cardenas and Zimmermann 2009). Mate guarding by males seems to be paramount in *L. ruficaudatus* (Hilgartner et al. 2012). Interactions between mates are often hostile and males are responsible for any proximity that occurs. The male stops sharing the sleeping hole after his mate gives birth and has nothing to do with the infant (Hilgartner et al. 2008). Latrines are important among *L. leucopus* (Dröscher and Kappeler 2014b). All members of a social unit used latrines communally, creating a source of olfactory communication that seems to achieve a degree of social bonding. Visits were more frequent when intruders were perceived, suggesting a mate-guarding function. The researchers considered the reason for dispersed social units in this species. Finding no evidence for feeding competition, even in the lean season, they suggested that spatial avoidance is a cryptic anti-predator strategy (Dröscher and Kappeler 2014a).

Sportive Lemur Life Cycle

A single infant is born after about five months of gestation. This is much longer than in other lemuroids of the same body mass, even in other lemuroid folivores. The delay may be due to a relatively high level of toxins in the diet.

Daubentoniids

The aye-aye, *Daubentonia* (Figure 8.2), is another case of one surviving genus in a family. Nevertheless, it maintains a striking presence in its habitat: it is the largest nocturnal lemuriform, weighing up to about 3kg, and the most widely distributed primate on Madagascar (Sterling 2003). The aye-aye is also strikingly unique in a number of ways. It is the only primate with abdominal mammary glands and claws on all its digits except the thumb. The digits themselves are highly specialized. Some aye-aye features are paralleled only among marsupials, some are like those of anthropoids, and some resemble those of lorisoids. In a sense the aye-aye may not be a lemuroid. Its lineage emerged before the other families differentiated, and the tendency among taxonomists has been to distinguish it at a high level (Box 8.3).

Figure 8.2 The lemur that isn't: the aye-aye is so different from other Malagasy primates that it has been placed in its own higher taxon. © javarman3 via iStock

Box 8.3 Daubentoniid Taxonomy

Many of the aye-aye's distinctive physical and behavioral features have been described in the text. These are such that extant aye-ayes constitute a single species, genus, family, and superfamily. With more information and analysis the gap from other primates has grown to the point that some would place the aye-ayes in their own infraorder, called the Chiromyiformes (Masters et al. 2013; see also Montagnon 2013).

Aye-aye Ecology

Despite its size, the aye-aye is usually active in the highest levels of the canopy, where it builds its leafy nests. It specializes in **structurally defended resources**; that is, food items with hard or dangerous casings. Other vertebrates on the island cannot access these resources. The aye-aye is among the few vertebrates (including bats and woodpeckers) that use self-generated acoustic cues in foraging. It locates wood-boring larvae by means of **percussive foraging**, moving along a wood surface with nose down and large ears pointed forward while it taps the wood with its long middle finger. The size and mobility of the ears are greater than in any other Malagasy primate.

When an aye-aye locates a subsurface channel, it pries open the tree or liana with chisel-like incisors that grow continuously (like those of rodents) and then uses the mobile finger to reach the exposed grubs. Rather than skewer the prey (an erroneous assumption made before close-up observations were available), the aye-aye uses its claw to hook each grub out of its refuge. This ensures that the whole insect is retrieved, which maximizes return for effort.

Lhota et al. (2008) observed a functional distinction between the third and fourth fingers. The third was used preferentially for tapping and probing (the latter for nectar and kernels as well as insect prey). It seems to be specialized for tasks that require mobility, sensitivity, and precision. The robust fourth finger appears to be specialized for strength and a scooping action, as in consumption of certain fruits. The use of curled thumbs to secure food leads to a very different conclusion. Since the digits are not physically specialized for the function, the researchers considered this a case of brain mechanisms overcoming morphological limitations (Pellis and Pellis 2012).

Aye-aye specializations expand the diet rather than limiting it. For example, an aye-aye may spend more than 30 percent of its feeding time on seeds. It penetrates the shell with its incisors and scrapes out the heart with the middle finger. It also uses its incisors to scrape fungus from stems and its middle finger to dip nectar.

A less obvious feature of the aye-ayes is the retention of a gene that allows them to see blue (Melin et al. 2012). The primatologists compared aye-ayes to other terrestrial vertebrates with similar optics with regard to intensity thresholds for color vision. With this perspective they considered the twilight activities of aye-ayes and found that twilight is enriched in short-wavelength (bluish) light. They also presented evidence that the intensity of twilight and full moonlight is sufficient to support cone-mediated color vision. They speculated that the gene in question is a crepuscular adaptation. Blue vision is significant for some potential visual targets of the aye-aye, such as scent marks and the arils of *Ravenala madagascariensis*.

Aye-aye Sociality

Female home ranges do not overlap, females never sleep near one another, and proximity usually results in fighting or chasing. Females and males associate, display more affiliative than agonistic behavior, and sometimes forage together temporarily. Males interact with each other more often than with females, including occasional joint foraging and nest sharing. Male home ranges are larger than those of females and overlap with others of both sexes.

Estrus apparently occurs only once each year for an individual, but the cycles of different females are not synchronized into a mating season. Estrus lasts for several days, rather than a few hours as in many lemuriforms. Behaving somewhat like a male loris, the estrus female sits for about an hour and emits loud calls that result in a convergence of males. They fight with and chase each other and attempt to mate with the female. She exerts choice by repelling some attempts. When she accepts a male they copulate for about an hour (single or multiple intromissions) while the other males continue to chase each other and try to dislodge the successful male. When the copulation ends, the female quickly moves across a distance of about 500 to 1000 meters and begins the mating sequence again. The main mode of male–male competition is long searches for estrus females (roaming). This form of scramble competition probably selects for attributes such as mobility and spatial memory.

Aye-aye Life Cycle

In captivity the aye-aye infant develops slowly compared to other strepsirrhines, staying in or near the sleeping structure for its first two months. Up to the age of one year a captive infant obtains much of its food by begging from the mother. The long dependence of the infant aye-aye can be explained in terms of the unique foraging pattern of the species. Food items such as nut kernels and embedded insect larvae would be difficult for a young individual to locate, extract, and process. Development of species-specific patterns seems to start as play. For instance, tapping goes through a clumsy stage before it becomes the efficient *tap-scanning* of an adult.

Aye-aye Cognition

Field reports suggested that aye-ayes use lianas as tools with a high level of comprehension; however, experiments indicated that such manipulations are based on trial and error (Sterling and Povinelli 1999). Kaufman et al. (2005) analyzed aye-aye brain structures by various techniques. They concluded that the volumes of sensory structures are consistent with nocturnal extractive foraging—in particular an enhanced auditory capability that is probably related to percussive foraging. The frontal cortex seems large in proportion to the whole brain, which supports the idea that the complex foraging behavior of this primate entails a relatively high level of sensorimotor intelligence.

Summary

Some species of cheirogaleids are characterized by torpor and "hibernation", forms of dormancy that are absent from all but one other primate. This is one of several strategies for heat retention in these small primates. Mouse lemurs completely lack scent glands and depend on excretions for marking behavior, but larger cheirogaleids have a wider array of chemical signals. Sportive lemurs (lepilemurids) are the smallest folivorous primates. This is apparently made possible by low metabolism and low energy expenditure in foraging. Aye-ayes seem to represent an ancient lineage that is completely separate from the other Malagasy primates. They are distinguished by numerous physical and behavioral adaptations.

Sources

Basic Bibliography

Goodman, S.M., & Benstead, J.P. (Eds.) (2004). *The natural history of Madagascar*. Chicago: UCP.

Masters, J., et al. (Eds.) (2013). *Leaping ahead, advances in prosimian biology*. NY: Springer. [Chapters 7, 8, 13, 14, 22–24, 33, 37]

Mittermeier, R.A., et al. (Eds.) (2013). *Handbook of the mammals of the world, Vol. 3. Primates* (pp. 1–88, 176–183). Barcelona: Lynx Edicions.

Schülke, O., & Ostner, J. (2005). Big times for dwarfs—social evolution, sexual selection and cooperation in Cheirogaleids. *EA*, *14*, 170–185.

Additional References

Blanco, M.B., & Godfrey, L.R. (2013). Does hibernation slow the "pace of life" in dwarf lemurs (*Cheirogaleus* spp.)? *IJP*, *34*, 130–147.

Blanco, M.B., et al. (2013). Underground hibernation in a primate. *SR*, *3*, 1768.

Colquhoun, I.C. (2011). A review and interspecific comparison of nocturnal and cathemeral strepsirrhine primate olfactory behavioural ecology. *IJZ*, v. 2011, ID 362976, 1–11.

Craul, M., et al. (2004). First experimental evidence for female mate choice in a nocturnal primate. *Primates*, *45*, 271–274.

Dammahn, M., & Kappeler, P.M. (2009). Females go where the food is: does the socio-ecological model explain variation in social organisation of solitary foragers? *BE&S*, *63*, 939–952.

Dausmann, K.H. (2010). Effective predation defence in *Cheirogaleus medius*. *Lemur News*, *15*, 18–20.

Dausmann, K.H. (2014). Flexible patterns in energy savings: heterothermy in primates. *JZL*, *292*, 101–111.

Dröscher, I., & Kappeler, P.M. (2014a). Competition for food in a solitarily foraging folivorous primate (*Lepilemur leucopus*)? *AJP*, *76*, 842–854.

Dröscher, I., & Kappeler, P.M. (2014b). Maintenance of familiarity and social bonding via communal latrine use in a solitary primate (*Lepilemur leucopus*). *BE&S*, *68*, 2043–2058.

Eberle, M., & Kappeler, P.M. (2008). Mutualism, reciprocity, or kin selection? Cooperative rescue of a conspecific from a boa in a nocturnal solitary forager the gray mouse lemur. *AJP*, *70*, 410–414.

Eichmueller, P., et al. (2013). The lack of female dominance in golden-brown mouse lemurs suggests alternative routes in lemur social evolution. *AJPA*, *150*, 158–164.

Farris, Z.J., et al. (2011). Comparing aye-aye (*Daubentonia madagascariensis*) presence and distribution between degraded and non-degraded forest within Ranomafana National Park, Madagascar. *FP*, *82*, 94–106.

Fichtel, C. (2007). Avoiding predators at night: antipredator strategies in red-tailed sportive lemurs (*Lepilemur ruficaudatus*). *AJP*, *69*, 611–624.

Ganzhorn, J.U., et al. (2004). Selection of food and ranging behaviour in a sexually monomorphic folivorous lemur: *Lepilemur ruficaudatus*. *JZL*, *263*, 393–399.

Hilgartner, R., et al. (2008). Life history traits and parental care in *Lepilemur ruficaudatus*. *AJP*, *70*, 2–11.

Hilgartner, R., et al. (2012). Determinants of pair-living in red-tailed sportive lemurs (*Lepilemur ruficaudatus*). *Ethology*, *118*, 466–479.

Kaufmann, J.A., et al. (2005). Anatomical analysis of an aye-aye brain (*Daubentonia madagascariensis*, primates: Prosimii) combining histology, structural magnetic resonance imaging, and diffusion-tensor imaging. *ARA*, *287*, 1026–1037.

Kessler, S.E., et al. (2013). Paternal kin recognition in the high frequency/ultrasonic range in a solitary foraging mammal. *BMCE*, *12*, doi: 10.1186/1472-6785-12-26.

Kessler, S.E., et al. (2014). Modeling the origins of mammalian sociality: moderate evidence for matrilineal signatures in mouse lemur vocalizations. *Frontiers in Zoology*, *11*, doi: 10.1186/1742-9994-11-14.

Kobbe, S., et al. (2011). Extreme individual flexibility of heterothermy in free-ranging Malagasy mouse lemurs (*Microcebus murinus*). *JCPhB*, *181*, 165–173.

Kobbe, S., et al. (2014). Torpor is not the only option: seasonal variations of the thermoneutral zone in a small primate. *JCPhB*, *184*, 789–797.

Lahann, P., & Dausmann, K.H. (2011). Live fast, die young: flexibility of life-history traits in the fat-tailed dwarf lemur (*Cheirogaleus medius*). *BE&S*, *65*, 381–390.

Lahann, P., et al. (2006). Geographic variation in life history traits of *Microcebus murinus* in Madagascar. *IJP*, *27*, 983–999.

Lhota, S., et al. (2008). Specialized use of fingers in free-ranging aye-ayes (*Daubentonia madagascariensis*). *AJP*, *70*, 786–795.

Masters, J., et al. (2013). What's in a name? Higher level taxonomy of the prosimian primates. In J. Masters et al. (Eds.), *Leaping ahead, advances in prosimian biology* (pp. 3–9). NY: Springer.

Masters, J.C., et al. (2014). The red island and the seven dwarfs: body size reduction in Cheirogaleidae. *Journal of Biogeography*, doi:10.1111/jbi.12327.

Melin, A.D., et al. (2012). Why aye-ayes see blue. *AJP*, *74*, 185–192.

Mendez-Cardenas, M.G., & Zimmermann, E. (2009). Duetting—a mechanism to strengthen pair bonds in a dispersed pair-living primate (*Lepilemur edwardsi*)? *AJPA*, *139*, 523–532.

Montagnon, D. (2013). Strepsirrhine divergence dates estimated from mitochondrial gene sequences and the status of *Daubentonia*. In J. Masters et al. (Eds.), *Leaping ahead, advances in prosimian biology* (pp. 21–32). NY: Springer.

Pellis, S.M., & Pellis, V.C. (2012). Anatomy is important, but need not be destiny: novel uses of the thumb in aye-ayes compared to other lemurs. *BBR*, *231*, 378–385.

Radespiel, U., et al. (2009). Sociogenetic structures, dispersal and the risk of inbreeding in a small nocturnal lemur, the golden-brown mouse lemur (*Microcebus ravelobensis*). *Behaviour*, *146*, 607–628.

Rahlfs, M., & Fichtel, C. (2010). Anti-predator behavior in a nocturnal primate, the grey mouse lemur (*Microcebus murinus*). *Ethology*, *116*, 429–439.

Rasoloharijaona, S. (2010). Does nonnutritive tree gouging in a rainforest-dwelling lemur convey resource ownership as does loud calling in a dry forest-dwelling lemur? *AJP*, *72*, 1062–1072.

Rasoloharijaona, S., et al. (2006). Loud calling, spacing, and cohesiveness in a nocturnal primate, the Milne Edwards' sportive lemur (*Lepilemur edwardsi*). *AJPA*, *129*, 591–600.

Rode, E.J., et al. (2013). Social organisation of the northern giant mouse lemur *Mirza zaza* in Sahamalaza, northwestern Madagascar, inferred from nest group composition and genetic relatedness. *Contributions to Zoology*, 82, 71–83.

Schmid, J., & Ganzhorn, J.U. (1995). Resting metabolic rates of *Lepilemur ruficaudatus*. *AJP*, *38*, 169–174.

Schmid, J., & Ganzhorn, J.U. (2009). Optional strategies for reduced metabolism in gray mouse lemurs. *Naturwissenschaften*, 96, 737–741.

Schmid, J., & Kappeler, P.M. (2005). Physiological adaptations to seasonality in nocturnal prosimians. In D.K. Brockman & C.P. van Schaik (Eds.), *Seasonality in primates* (pp.129–155). NY: CUP.

Schmid, J., & Speakman, J.R. (2009). Torpor and energetic consequences in free ranging gray mouse lemurs (*Microcebus murinus*): a rain–dry forest comparison. *Naturwissenschaften*, 96, 609–620.

Schülke, O. (2001). Social anti-predator behaviour in a nocturnal lemur. *FP*, *72*, 332–334.

Schülke, O. (2003). The natural history of *Phaner furcifer*. In S.M. Goodman, & J.P. Benstead (Eds.), *The natural history of Madagascar*. Chicago: UCP.

Schülke, O., & Ostner, J. (2005). Big times for dwarfs—social evolution, sexual selection and cooperation in Cheirogaleids. *EA*, *14*, 170–185.

Schülke, O., & Ostner, J. (2007). Physiological ecology in cheirogaleids: hibernation and torpor. *Acta Ethologica*, *10*, http://dx.doi.org/10.1007/s10211-006-0023-5.

Seiler, M., et al. (2013a). Anti-predator behaviour of Sahamalaza sportive lemurs, *Lepilemur sahamalazensis*, at diurnal sleeping sites. *Contributions to Zoology*, *82*, 131–143.

Seiler, M., et al. (2013b). Interspecific semantic alarm call recognition in the solitary Sahamalaza sportive lemur, *Lepilemur sahamalazensis*. *PLONE*, 8, e67397.

Sterling, E. (2003). *Daubentonia madagascariensis*, aye-aye. In S.M. Goodman, & J.P. Benstead (Eds.), *The natural history of Madagascar*. Chicago: UCP.

Sterling, E.J. & Povinelli, D.J. (1999). Tool use, aye-ayes, and sensorimotor intelligence. *FP*, *70*, 8–16.

Tattersall, I. (2013). Species-level diversity among Malagasy lemurs. In J. Masters et al. (Eds.), *Leaping ahead, advances in prosimian biology* (pp. 11–20). NY: Springer.

Thoren, S., et al. (2011). Different competitive potential in two coexisting mouse lemur species in northwestern Madagascar. *AJPA*, *145*, 156–162.

Warren, R., & Crompton, R. (1997). A comparative study of ranging behaviour, activity rhythms and sociality of *Lepilemur edwardsi* (Primates, Lepilemuridae) and *Avahi occidentalis* (Primates, Indriidae) at Ampijoroa, Madagascar. *JZL*, *243*, 397–415.

Weidt, A., et al. (2004). Social organization of the golden brown mouse lemur (*Microcebus ravelobensis*). *AJPA*, *123*, 40–51.

Zimmermann, E., & Radespiel, U. (2014). Species concepts, diversity, and evolution in primates: Lessons to be learned from mouse lemurs. *EA*, *23*, 11–14.

Zohdy, S., et al. (2011). Elevated testosterone levels in female brown mouse lemurs (*Microcebus rufus*). *AJPA*, *144*, 318.

9 Lemuriform Variation in the Light of Day

The sun is rising and a small troop of lemurs are foraging. A male finds an enticing cluster of fruit, but a smaller female swats him aside. He displays a submissive signal and departs.

This chapter explores two lemuriform families that have adapted to life in the sun, at least part of the time: lemurids (containing the well-known ringtailed lemurs) and indriids. Color vision is recurrent among these primates, found in at least two lemurid genera and one indriid genus. All of them use sight to identify their low quality foods, but their sensory patterns for discovery of preferred foods differ (Rushmore et al. 2012, and see below).

Lemurids

The study of this family introduced primatology to **cathemerality** (Tattersall 1987). Broadly defined, this means that animals can be active for significant periods at any time of the 24-hour cycle. The most common form continues throughout the year while others are seasonal. Diurnal activity occurs in every lemurid taxon and nocturnal activity has now been reported for almost all of them (LaFleur et al. 2014).

Several functions have been attributed to cathemerality. These hypotheses are not mutually exclusive, especially considering the diverse taxa and varied ecological conditions involved. They include: (1) thermoregulation; (2) mitigation of feeding competition; (3) predator avoidance; and (4) increased feeding time.

Thermoregulation is plausible in view of the extremes in Madagascar weather: nocturnal activity peaks during the colder months, so the hypothesis proposes that the primates seek to be warm during cold nights by being active. Mitigation of feeding competition is also plausible in light of the formidable lemuriforms seeking limited resources during the day: large-bodied indriids and relatively large groups of aggressive ringtailed lemurs (*Lemur catta*). Night provides refuge from most raptors (except owls), especially in dry western forests that are seasonally bare. **Crepuscular** activity (at dawn and dusk) reduces danger from the fossa, which hunts during full day or night. A study of fossa ecology revealed that the *crepuscular* species in the study area was underrepresented in the fossa diet while other lemuriforms, both nocturnal and diurnal, were overrepresented.

The cited hypotheses assume that cathemerality originated as a shift from daytime activity. However, according to the Mismatch Hypothesis, diurnality is a recent development and cathemerality is a prior intermediate stage (Kappeler and Erkert 2003). A phylogenetic reconstruction of the genus *Eulemur* tested this argument and concluded that the most recent common ancestor was probably cathemeral, indicating that this is a stable evolutionary strategy (Griffin et al. 2012). Indications that cathemerality occurs in all lemurid genera suggest that it may be ancestral for the entire lemurid family (Donati et al. 2013).

Compared to other primate taxa, female–female agonistic rates among daylight lemurs are relatively low (Erhart and Overdorff 2008). However, competition increases during reproductive periods in some taxa and occasionally results in **targeted aggression**: one or more females focus on another for severe unprovoked attacks that persist until the victim is driven out of the social unit. This behavior probably maintains small groups in which females are closely related.

"True" Lemurs

Three of the five lemurid genera (Box 9.1) are broadly similar, resulting in their casual lumping together as "true lemurs". These are *Eulemur*, *Lemur*, and *Varecia*. They are all quadrupeds with a size range similar to that of small domestic cats (~1.5–3kg). They typically live in troops and tend to be frugivorous or generalized in feeding habits. Bamboo lemurs (*Hapalemur* and *Prolemur*) differ in that they are smaller, display vertical clinging and leaping, and commonly eat bamboo or grass.

Box 9.1 Lemurid Classification

Though once subsumed by *Eulemur*, *Lemur catta* is now universally accorded a genus of its own. With the elevation of several subspecies, *Eulemur* still contains about 12 species. Molecular phylogenetic studies seem to have resolved issues of genus relationships within the family Lemuridae (Horvath 2008; Yoder 2003): (1) *Eulemur* is monophyletic; (2) *Lemur* and *Hapalemur* form a clade that is fairly closely related to *Eulemur*; and (3) *Varecia* is basal to the family, i.e. became a separate lineage before any of the others differentiated.

Hapalemur includes the smallest of the lemurids. Many taxonomists have elevated the largest species, triple the size of the smallest, to the genus level as *Prolemur*. This is one of the many cases where an informal name creates less confusion for the non-taxonomist. All of these species together, regardless of generic assignment, can still be called "bamboo lemurs" and contrasted with the "true lemurs".

Eulemur

Sexual dichromatism is ubiquitous in *Eulemur*. This means that there are consistent differences between the sexes with regard to color. In some species the entire pelage is differently colored. In others it is parts of the body, such as abdomens, heads or facial markings. The male red-bellied lemur, for example, is simply distinguished by white spots next to the eyes (Figure 9.1). In the crowned lemur (*E. coronatus*) the female is predominantly gray; the male is reddish brown and topped with a black and orange crown. An experiment with females of one species found that they preferred to view photographs of more colorful males, which suggests that some form of sexual selection is involved (Cooper and Hosey 2003).

Most of the species live in troops of 5 to 10 individuals. Larger troops tend to fission into temporary subgroups for foraging and to divide permanently when membership exceeds 20. Pair-groups, typically cohesive and territorial, occur in two species (*E. mongoz* and

Figure 9.1 A pair of red-bellied lemurs. The male has white eye-markings, illustrating sexual dichromatism in *Eulemur*. © Stacey Tecot

E. rubriventer). Males carry and care for infants in both species (Figure 9.2). Units with an additional adult male and/or female may represent adult pair-group dynamics or near-adult offspring on the verge of emigration, both known in other lemuriforms. Mongoose lemurs display high levels of aggression between adult females of different groups, often resulting in chases on the ground. Some researchers inferred that females rather than males seek to monopolize mates (Curtis and Zarmody 1999).

Colquhoun (2011) compared the roles of olfaction in cathemeral *Eulemur* and nocturnal strepsirrhines. Despite the different environments, olfaction was similarly multifunctional. The main distinction was the absence of urine washing in *Eulemur*. Also interesting was the fact that the pair-grouping species displayed behaviors not observed in troop-living *Eulemur*: (1) they marked in boundary zones and during territorial encounters; and (2) each sex marked the other with anogenital glands rather than just male marking female. DelBarco-Trillo et al. (2012) studied the chemical composition of olfactory signals in *Eulemur*. They found that the chemical complexity of secretions varied with social organization and suggested that olfactory complexity, like vocal complexity, is favored by social complexity.

The brown lemur (*E. fulvus*) is the largest species in the genus, weighing up to 4kg or more. It is the most widely distributed species of the genus, overlapping all the others. Brown lemurs live in a variety of forest environments while smaller *Eulemur* species live in drier forests where food is less abundant. Tolerance for tannins and alkaloids from unripe fruit and mature leaves may explain their extensive geographic range and their ability, unique in the genus, to thrive in both wet and dry forests. Brown lemurs in dry forests depart from the typical cathemeral pattern with a seasonal shift from diurnal to 24-hour activity, perhaps to expand their food supply. Use of smaller patches in vertical microhabitats may also extend their resource base.

Figure 9.2 A young red-bellied lemur carried by a male. This illustrates the common pattern of dorsal carriage and the recurring phenomenon of male allocare. © Stacey Tecot

Strong dyadic relationships exist *within* some troops. These most commonly involve a male and a female (although other age/sex combinations also occur). Such dyads are visible more often in times of food scarcity and in feeding contexts during the mating season. The pattern may be related to the absence or lack of clarity in dominance hierarchy, to the distribution and density of food patches, and/or to vulnerability to predators. The male–female dyads are unlike pair relationships in nocturnal lemuriforms in that the connection is not stable throughout the year (Overdorff 1998). Females often fed at higher rates when near their partners; however, since no mating preference was established, the benefit to the males was unclear.

Another study of the same (sub)species (Ostner and Kappeler 1999) found no evidence for dyads and females copulated with all resident males, though most often with a central dominant male. Paternity tests on yet another subspecies demonstrated a correlation between dominance and reproductive success, but showed that subordinate males also sired offspring (Gachot-Neveu et al. 1999). These variations suggest that the wide distribution of brown lemurs is due to social as well as ecological flexibility.

Ringtailed Lemurs

The genus *Lemur* contains a single living species, *L. catta*. Ringtails are mostly diurnal and the most terrestrial of the lemuriforms. When they walk, ringtails hold aloft the long tail for which they are named. The display may function as a signal for troop cohesion and a caution to other troops. Ringtail behavior and ecology converge with anthropoids in a number of ways.

About 70 percent of group travel is on the ground and each individual is on the ground for about a third of an average day (Figure 9.3). However, ringtails spend some time in all strata of the forest. They are **opportunistic omnivores**, which means that they eat most of the digestible foods that the forest offers. These include exudates, arthropods, spider webs, cocoons, birds, and reptiles. They also eat dirt from termite mounds, which might be geophagy or a way to feed on termites. Tamarind trees are the most important food source because some are producing fruit and leaves at any time of year. During certain periods they provide half of the ringtails' food. These primates can obtain water, a problem in the driest habitats, from succulent plants, dew, and puddles in tree holes and other crevices. In contrast to other lemurs they can use either sight or olfaction independently to find their preferred foods (Rushmore et al. 2012).

Ringtails typically live in troops of 10 to 15 individuals that are composed of two or more matrilines, subgroups of related females spanning more than one generation. Large troops usually fission, an event governed by the dominance system. Matrilines are arranged in a hierarchy and so are individuals. At the top of the troop's hierarchy is the senior female of the dominant matriline. She initiates troop movement and is otherwise the focal point for other members. Most female-female interactions are aggressive, but mother–daughter and sister–sister pairs are usually friendly, marked by grooming and proximity.

Males have a separate hierarchy, divided between central and peripheral individuals. Two or three central males, usually in their physical prime, stay close to the dominant females. They may benefit from better access to food and reproductive females, and perhaps by less vulnerability to predation. Peripheral males are young, old, or recently transferred from another troop. They stay on the edges of the troop and bring up the rear during troop movements. Immigrant males have to fight the resident males for dominance.

Unlike many other primates with extensively overlapping home ranges, ringtails frequently engage in aggressive confrontations. The females of each troop face each other and stare, which may lead to a fight with cuffing and biting. Escalation can result in

Figure 9.3 Part of a ringtailed lemur troop foraging on the ground. Small troops like this also occur in many species of *Eulemur*. © Matej Hudovernik via iStock

serious injuries and even death. Both males and females mark substrates in overlap areas with their anogenital glands, sometimes standing on their hands to mark a tree as high as possible.

Ringtails are among the most vocal nonhuman primates, presumably related to their complex social lives. They have a repertoire of 28 distinct types of call. Two aggressive displays are distinctive. In the jump fight one lemur stands on its hindlimbs with arms spread and jumps or hops around the opponent. This can lead to serious fighting. In the stink fight, practiced only by males because they have the glandular equipment, a male wipes his tail on the scent glands of his wrists and chest and then waves it over his head toward the opponent. The other either responds in kind or flees. A stink fight can last up to an hour.

During the six months before mating season, many males begin to roam and perform spur marking. A male drags his wrist gland and the associated nail across a sapling, leaving scratches and secretions. This informs other males of his presence and perhaps deters them from entering the marked area. Males continue to roam and mark during the mating season. The mating season lasts for more than a month in a given area, but the mating period for a particular troop is only about two weeks and each female is in estrus for four to six hours. Males can mate with many females by migrating during the season, but they face intense competition for each female.

Central males of a troop can stay close to estrus females and the top two are likely to mate first or most frequently. Nevertheless, there is reason for serious competition because the high-ranking resident males are not always the most successful. In addition a female is likely to mate with several males during her estrus, although she is most likely to conceive with the first. Males try to guard mates with jump fights. Females sometimes exercise choice by leaving the fighters. They also reject mating attempts by relatives and sometimes seek out males from other troops. An estrus female solicits a male by turning her hindquarters to him, lifting her tail, and looking at him over her shoulder.

According to one account, female ringtails prepare themselves for gestation by accumulating fat reserves. Pregnant during the cold and dry winter, they do little moving or feeding during the five months of gestation. However, several researchers interpret ringtails as **income breeders** rather than **capital breeders** (Gould et al. 2003). That is, the females use the maximum amount of resources during reproduction rather than relying on previously stored fat.

Ringtails are inclined toward daytime parturition, which seems dangerous. Although no attacks by predators or neighboring groups were observed, there were two instances in which mother and neonate were attacked by females of their own troop (Takahata et al. 2001). The researchers postulated that this is a non-adaptive pattern explained by the Mismatch Hypothesis.

Although captive ringtails often have twins, the great majority of wild births are singletons. The infant clings ventrally at first, but shifts to the jockey position in less than a month. The motivational/emotional state of ringtail mothers is suggested by their responses to dead infants (Nakamichi et al. 1996). They sniff, lick, and touch them. Six of seven mothers returned to their dead infants several times. This behavior suggests an attachment like that of many anthropoid mothers.

Females in a troop show great interest in a mother and live infant. They direct more affiliative acts toward the mother than at other times, and they lick the infant. Licking infants may contribute to stable social relationships among adults. All age/sex categories provide the infant with some kind of care. Occasional alloparental behavior by males may give infants a chance to enter the central part of the troop, where they may benefit from increased predator protection and eventual mating possibilities.

Some adult males continue to have affiliative relationships with older immatures during lactation and post-migration periods, a pattern that occurs in some monkey species. These juveniles and adolescents receive better protection and more practice in social skills. Despite the efforts of various caretakers, about half of all infants die by the age of one year during non-drought years (Gould et al. 2003). Development after infancy is somewhat slow by lemurid standards. Males and females reach adult size at about three years of age, but may not reproduce for another year. This seems to be a constraint of the environment, since captive females consistently give birth at the age of two years.

Maturing males leave the natal troop, and move again every three years or so (Parga and Lessnau 2008). Data from a Caribbean colony indicate that many adult males stay in their troops beyond the sexual maturation of potential daughters. This argues against the inbreeding avoidance explanation for transfer. Sexual competition is more likely, since males were significantly more likely to join groups with fewer adult males and a more favorable sex ratio.

Ruffed Lemurs

Ruffed lemurs (*Varecia*) are the largest extant lemurids, weighing up to 5kg. They are primarily diurnal and mainly occupy the higher strata in eastern rainforests. Ruffed lemurs expand their available feeding space with suspensory behavior (Figure 9.4). This point was originally associated with use of the arms or tails in certain anthropoids, but ruffed lemurs demonstrated that it can be done with quadrupedal or hindlimb suspension (Meldrum et al. 1997). These dedicated canopy dwellers come to the ground at certain times of the year to ingest soil, which underlines the importance of geophagy.

Ruffed lemurs are the most frugivorous of living lemurids, ordinarily spending 75 to 90 percent of their feeding time on fruit. They can identify their preferred foods by smell alone but not by sight alone (Rushmore et al. 2012). The frugivory of *Varecia* makes them important seed dispersers and experiments showed that defecated seeds of one ruffed species had a higher germination rate than controls (Razafindratsima and Martinez 2012). However, the ecological significance of seed dispersal by *Varecia* (as well as *Eulemur*) is complicated (Razafindratsima et al. 2014). Primates are especially important for seed dispersal in Madagascar because of the dearth of frugivorous birds and bats. The larger daylight lemurs are probably important for the dispersal of large seeds that are beyond the smaller nocturnal forms. On the other hand, the distribution range of the larger lemurs is small and this casts doubt on their long-term role in the preservation of Madagascar forests.

Ruffed lemur society is a kind of fission–fusion system in which females defend a common territory, but every individual belongs to a subgroup associated with its own area. Females limit their activities to these core areas on a seasonal basis. At other times they may move anywhere in the community territory. At any time of the year females associate in small parties that can change membership on a day-to-day basis. Scent marking is most frequent during territorial conflicts. It is also important in the very different context of greeting, when females engage in mutual marking as they leapfrog over each other.

Males have individual home ranges within a community territory and are generally antagonistic toward one another. They associate with females in small parties that often consist of just one female and one male with the female's offspring. These can be mistaken for pair-groups if the study does not last long enough to follow the more general and complex pattern of movements and affiliations throughout the community.

Figure 9.4 A ruffed lemur in suspensory position. © evenfh via iStock

Infants are kept in a nest, constructed by the mother a few weeks before parturition, for their first 10 to 15 days. These are the only primates that build birthing nests. Some females join with nesting pairs that allow each mother more feeding time, and apparently increase survival probability for offspring. Nearly all (13 of 14) infants with cooperating mothers survived, versus 2 of 5 others (Baden et al. 2013). The proximate basis for this behavior is a puzzle. Though some of the pairs were kin, others were not.

Beyond this, ruffed lemurs have unusual life history traits compared to other large diurnal lemurids: shorter gestation (about 100 days), larger litters (typically two or three and up to five), rapid development to an early age of first reproduction, and delayed age of last reproduction. Rapid infant development is supported by milk that is richer than that of any other primate, females have two sets of nipples rather than one set like other diurnal lemuriforms, and allonursing occurs. The lifespan of *Varecia* (15 to 20 years) seems short for its body size. This may be related to the energy expended in reproduction and compensated in evolutionary terms by the production of litters rather than single births. Baden et al. (2013) suggested that all this is the response of an obligate frugivore to the uncertainties of the Madagascar environment. In their own fieldwork they found that the natural IBI of the genus can be as long as six years.

The Female Need Hypothesis applies at least in part to this genus. Prenatal energy investment by the *Varecia* mother is greater than that of any other primate, and the same may be true of postnatal investment (Vasey 2005b). Ruffed lemur mothers need help in raising their young, and even though they usually get it, they seem to wear down at a relatively young age. Given their need for energy, why are these primates so committed to fruit, an irregular source of food? Vasey (2006) suggested that the unusual social organization of *Varecia* evolved to sustain a frugivorous way of life. The flexible grouping pattern allows them to respond effectively to the shifting patterns of fruit availability. In this respect they converged with spider monkeys (Chapter 13) and chimpanzees (Chapter 25).

Ruffed lemurs have an extensive vocal repertoire and their calls appear to have several ecological functions. Some loud calls seem to coordinate movements within a social group when the members are foraging individually or in subgroups. For example, such calls can mediate sequential use of the most important fruit trees. Vocalizations are also important in female greetings and in male mating displays. Vocalizations, including several loud calls, convey information about spacing between and within troops, dominance relations, travel/contact, and aerial and ground predators.

Bamboo Lemurs

The genus *Hapalemur* includes the smallest of the lemurids. The largest species has been elevated to the genus level as *Prolemur* (Box 9.1). This is a case where an informal name creates less confusion for the non-taxonomist because all of these species have been called "bamboo lemurs", in reference to the fact that all but one are heavy consumers of bamboo (Tan 2006). The other popular name for this group, "gentle lemurs", is a somewhat misleading label for these fiercely territorial primates.

Considering all species, head/body length ranges from 25cm to 45cm and the tail adds about the same amount. Weights vary from less than a kilogram to 2kg. All of these species are vertical clinger-leapers, the only ones in the lemurid family. They have the typical limb proportions and jumping style of clinger-leapers, but they are also capable of rapid quadrupedal locomotion on the ground. The larger forms may be awkward leapers. Fieldworkers saw one individual (of a medium-sized species) miss a jump and fall into a lake (Petter and Peyrieras 1975). Recounting this incident may seem trivial, but it reminds us of an important theoretical point. Natural selection does not design perfect machines; it is a blind process that works on variation.

For all but one of these species bamboo is a huge component of the diet, comprising as much as 95 percent (Tan 2006). This is unique among the primates and comparable to the pandas (Mutschler and Tan 2003; Tan 2006). Three of these lemur species can live sympatrically because they concentrate on different parts of the bamboo plants (Yamashita et al. 2009). However, all of them are able to ingest cyanide in the plants, which is a remarkable adaptation and unique among the primates (Ballhorn et al. 2009; Gould and Sauther 2007).

Taken as a whole, the activity patterns of the bamboo lemurs (Mutschler 2002; Tan 2006) are intermediate among the lemurids. They vary from diurnal to cathemeral to seasonally nocturnal. Social organization resembles that of many other lemuriforms. Dyadic pair-groups predominate but there is a significant percentage of triads in many if not all populations (Grassi 2006). Bamboo lemurs might be a little more flexible than their nocturnal relatives. Observers have reported what may be small troops with as many as 13 members. However, the significance of these units is debatable because the larger ones appear in more disturbed habitats (Grassi 2006). There are also some units with one female and more than one male. However, it seems that only one male is a breeder and the males may be father and son (Mutschler 2002).

The bamboo lemur social unit is cohesive, in contrast to similar nocturnal lemurs. Members of a group travel and forage together, moving from one clump of bamboo to another (Petter and Peyrieras 1975). Males and females do tend to separate during feeding, however. In one observed group the male fed at significantly higher levels than the female and two juveniles (Grassi 2006). The simplest explanation is that the male plays a primary role in predator detection (males give alarm calls in response to raptors). An alternative is that separation mitigates feeding competition between members of the same group. Both patterns occur in other primates.

Other hypotheses are more specifically lemuriform. The female requires the richer foods that are nearer to ground level in order to sustain her high reproductive costs, and the male cooperates because of female dominance and/or male deference to females. From a functional point of view these interpretations are not mutually exclusive. At this point we have no evidence as to what the ultimate cause of the pattern might be, or even if it has any innate basis.

Nievergelt et al. (2002) used DNA technology to clarify the social and mating systems in the Aloatran taxon (species or subspecies). They found that reproducing females were the cores of the social units. Each female stayed in the same area over a study period of three years while resident males shifted rather frequently. Desertion and disaster accounted for some male changes; expulsion probably caused others, though takeovers were not seen. In all units containing two breeding females, they were mother and daughter or perhaps sisters. This is in accord with the prediction from inclusive fitness theory that close relatives are most likely to share resources.

The mating system in this population (and perhaps in similar ones) coincides closely with the social system. Genetic data showed that current or previous resident males sired more than 90 percent of the offspring. Extra-group copulations were special cases. Three involved inbreeding avoidance, and in two other instances the resident pair had produced no offspring during the previous two years (Nievergelt et al. 2002). The Childhood Familiarity Hypothesis of inbreeding avoidance seems especially applicable in these small groups. A mechanism that stimulates extra-group copulation after reproductive failure may be a thornier ethological problem.

Two other important points emerge from these data. First, there is no evidence for infanticide, despite the frequent changes in resident males. This is to be expected in annual seasonal breeders where killing the infant does not hasten the mother's return to cycling. Second, the discreteness of these social units counters the prediction from the Disequilibrium (Mismatch) Hypothesis that daytime groups of cathemeral lemuroids are composed of pairs temporarily united into larger groups to counter predator pressure.

The mating and birth seasons of one highly localized population, the Alaotran population, are unusually long and flexible, extending over four to six months each. These bamboo lemurs are comparable to the nocturnal aye-ayes with regard to both extended mating season and use of continuously reliable resources. Taken together the special situations of the Alaotrans and the aye-ayes support the idea that other lemuriforms are constrained to strict mating and birth seasons by the seasonality and low productivity of their main food sources (Mutschler 2002). This is in accord with the Energy Conservation Hypothesis.

Bamboo lemur gestation is slightly shorter than that of other lemurids (4.5 to 5 months) and the early life of most bamboo lemurs seems to be different from other lemurids and more like that of nocturnal lemuriforms. Newborns are sheltered in dense vegetation, carried by mouth, and parked while the mother forages (Mutschler and Tan 2003). However, Petter and Peyrieras (1975) reported that the *H. griseus* infant rides on its mother's back from the beginning.

The largest species of bamboo lemurs differs from the others in several life history patterns. Gestation is longer (about 150 days) and weaning is later (eight months). The mother carries her infant rather than parking it. There is also evidence for male dominance over females.

Indriids

Indriids are basically folivorous, as shown by enlarged salivary glands, large stomach, and a long intestine. Leaves sustain them through harsh dry seasons. Indriids are vertical clinger-leapers with elongated hands and long legs. They can also move quadrupedally in the trees and hop or jump bipedally on the ground. Indriids live in small, cohesive social groups. Single birth is typical. Mothers carry their infants ventrally at first and then allow them to shift to the jockey position.

There are three genera in the family (Box 9.2). Diurnal indriids (*Propithecus* and *Indri*) are the largest strepsirrhines, weighing up to about 10kg. Some can jump as far as 10m. Woolly lemurs (*Avahi*) have evolved back to nocturnal life, live in pairs, and weigh less than 2kg. Sifakas are by far the most thoroughly studied.

Box 9.2 Indriid Classification

The taxonomic grouping that we are calling indriids has been rendered technically as a zoological family called either the Indridae or Indriidae (Gould and Sauther 2007). Its members are very uniform in cranial morphology and share a distinctive pattern of dental reduction marked by two premolars in each quadrant and four teeth in the toothcomb rather than six. They also share specializations for folivory (enlarged cecum and large intestine) and for vertical clinging-leaping (long hindlimbs and short, dorsally oriented ischium) (Fleagle 2013). There seems to be a consensus about the family level of this taxon and about the genera it contains: *Propithecus*, *Indri*, and *Avahi*. There also seems to be agreement that *Indri* contains only one species, *I. indri*.

Two species (or perhaps superspecies) of *Propithecus* (the sifakas) are well established: *P. verreauxi* (Verreaux's sifaka) and the significantly larger *P. diadema* (the diademed sifaka). These correspond to the east–west split that is common in Malagasy fauna and flora, with *P. verreauxi* living in drier western forests and *P. diadema* inhabiting eastern rainforests (Mayor et al. 2004). *Avahi* is commonly viewed as having three species (e.g. Gould and Sauther 2007), inhabiting three separate geographic areas.

Sifakas (*Propithecus*) are lanky indriids named for their typical warning call (Figure 9.5). They are distributed around the perimeter of Madagascar, one group of species in the dry western forests and another in eastern rainforests (Irwin 2006). Folivory is vital during the dry seasons, but in better times sifakas eat a greater variety of plant parts than other lemuriforms. Fruit, flowers, and seeds constitute more than half of the year-round diet for many populations, and fruit consumption may rise above 60 percent during the wet season, especially in the rainforests. Some sifakas eat plant parts that are underground and further obscured by leaf litter on the ground, apparently finding them by smell (Irwin et al. 2007). There is little evidence for such use of olfaction by other primates.

Sifaka behavior suggested a new explanation for geophagy. They ate soil from termite mounds that might contain bacteria that would facilitate digestion of cellulose. These bacteria could enter the soil as the termites exchange nutrients with each other or use saliva to cement the nest structure (Norscia et al. 2005). Some females in western populations seem to make medicinal use of tannin, which is usually avoided by primates. They increased

Figure 9.5 A sifaka in the vertical clinging posture. © Stacey Tecot

tannin levels during pregnancy by briefly feeding on a particular plant. Tannin consumption can increase body weight and stimulate milk secretion. It can also counter parasites, hemorrhage, and spontaneous abortion (Carrai et al. 2003).

Sifakas show that strepsirrhine allogrooming can be social as well as utilitarian (Lewis 2008, 2010). Half of all allogrooming was directed toward body parts that could easily be reached by a recipient, and the rate of the behavior did not increase during the rainy season when a higher ectoparasite load was to be expected. Rank and sex influenced allogrooming rates and the rates increased during the mating season. Females with infants received more grooming than females without them. Immigrants used grooming to facilitate entry into new groups. Males groom other individuals at higher rates than do females, and when females groom males they are likely to choose subordinate males. Sifaka grooming patterns are similar to those of anthropoids.

During mating season, aggression increases and male subordination to females may be abandoned. Males sometimes fight each other, and both sexes harass copulating couples. There is also more aggression between groups, perhaps reflecting the fact that the promiscuous mating system crosses social boundaries. Most females mate with several males, including visitors, but they are discriminating; they choose particular males according to complex criteria. Even so, the main function of multiple mating may be to confuse paternity and forestall infanticide.

Sifaka infanticide (Erhart and Overdorff 1998; Lewis et al. 2003) is frequent compared to other strepsirrhines, and both sexes are perpetrators when they join new groups. Infanticide by males conforms to the Sexual Selection Hypothesis. In contrast to other strepsirrhines, the sifaka inter-birth interval is often two years. Thus, infanticide more than doubles the probability that a female will give birth in the following year. The function of infanticide by female immigrants is different. A resident female (usually the only breeding female in a group) emigrates after the loss of her infant and the newcomer becomes the breeding female. Researchers think that food competition in the difficult Madagascar environment is the probable cause of female infanticide.

Neither sex emigrates more frequently or goes farther than the other. This raises the possibility that relatives will meet and breed, a problem that may be solved by using scent to distinguish relatives. Natal individuals sometimes become reproductive soon after an opposite-sex immigrant displaces the opposite-sex parent. Immigrants scent mark more often than natal group members, perhaps signaling their sexual eligibility. Distinguishing among potential sex partners is especially important because some individuals do not leave the natal group until they are seven or eight years old. Given the dangers of dispersal and the rarity of reproductive opportunities, these individuals may be waiting for a breeding vacancy in their own group or a neighboring one.

Sifaka social units are unusually variable in composition, containing every possible combination of adult male(s) and female(s). Pochron et al. (2004) explained variability in sifaka social organization in terms of life history. They saw the absence of female bonding as the key issue. Due to high mortality and low fertility, a sifaka female is likely to have just one surviving daughter by the time of her death. As a result of these variables, social composition and mating patterns vary from year to year within a single group and there is no chance for related females to establish social bonds. In these conditions it could be that the mated pair is the only viable bond. Other forms of organization are shaped by environmental compulsion or individual opportunism. The researchers pointed out that the life history variables involved are similar to those found in lemurid species that have been subjected to long-term studies. They emphasized sharp contrasts in female life history between strepsirrhines and anthropoids, which suggest that their conclusions might apply widely across the lemuriforms and even the strepsirrhines in general.

Cognition in Daylight Lemuriforms

Various combinations of daylight lemuriform taxa have been subjected to numerous and varied cognitive studies in recent years. This section reviews some of those studies. One of the broadest studies sampled *Propithecus* and the "true lemur" genera to explore the cognitive implications of group size, an issue of great interest with regard to all primates. MacLean et al. (2013) tested social cognition with an experiment in which the subjects chose to take food in front of a watching researcher and another who was not. Non-social cognition was represented by a task that required the inhibition of reaching for food. Success in the social task was correlated with group size, while success in the non-social task was not. This suggests some sort of domain separation, as opposed to general intelligence. Also significant, brain size was not correlated with success in either task.

Another study with a similar range of subjects compared three aspects of spatial memory: learning mechanisms supporting memory; recall after a long delay; and recall of multiple locations in a complex environment (Rosati et al. 2014). The results conformed to variation in subsistence pattern. Ruffed lemurs, the most frugivorous of the subjects, displayed more robust spatial memory across the tasks while the folivorous sifakas were the least adept.

Experiments comparing "true lemurs" yielded the predicted result that they preferred to minimize risk rather than try to obtain a larger quantity of food (MacLean et al. 2012). This is consistent with the unpredictable Madagascar environment, but it is a result that has been obtained with many other animals. The researchers also caution that the underlying mechanisms might be quite simple and specific.

Our last example of recent work sets the ringtailed lemurs apart from other "true lemurs" and suggests comparison with anthropoids (Sandel et al. 2011). One experiment was a competitive task that involved avoidance of food watched by a researcher; the other involved non-competitive gaze following. The ringtails performed above the chance level and better than the other lemurs. The experimenters suggested that ringtails have converged with monkeys as a result of living in comparably large and complex societies. They note that the convergence of behavior does not necessarily imply that the underlying cognitive mechanisms are identical.

Summary

Two lemuriform families display regular daylight activity. They are generally larger than the nocturnal species and typically manifest longer life stages. They have displayed a variety of mental skills that belie previous assumptions of cognitive simplicity.

Lemurids tend to be cathemeral, but display flexibility. "True lemur" genera in this family are similar in several common behavior patterns. Multi-male troops are typical. Recurrent female patterns are targeted aggression against other females and varying degrees of dominance over males. Brown lemurs (*Eulemur*) are the largest and most widely distributed lemurids. Sexual dichromatism is prominent. They are ecologically flexible and display some tendencies toward male–female pairing. Ringtailed lemurs (*Lemur*) are the most diurnal and terrestrial of the lemurids, subsisting as opportunistic omnivores. They display several traits that converge with anthropoids, including matriline social organization and a higher level of social cognition than other strepsirrhines. Ruffed lemurs (*Varecia*) live in fission–fusion troops. Some temporary subgroups include only one adult of each sex. Despite large size, ruffed lemurs display traits like those of nocturnal lemuriforms that are seemingly related to a high rate of reproduction. Bamboo lemurs are aptly named for their typical diet. They differ from most other lemurids in being vertical clinger-leapers that live in small groups that may be pair-based.

Two genera of the indriid family are the largest strepsirrhines and strongly diurnal. They are vertical clinger-leapers with physical adaptations for folivory. Sifakas (*Propithecus*), the most studied genus in the family, display unusually complex and dynamic social patterns. One of their most unusual features is infanticide by both sexes.

Sources

Basic Bibliography

Curtis, D.J., et al. (Eds.) (2006). Cathemerality. *FP*, *77*, 1–194.

Goodman, S.M., & Benstead, J.P. (Eds.) (2004). *The natural history of Madagascar*. Chicago: UCP.

Jolly, A., et al. (Eds.) (2006). *Ringtailed lemur biology*. NY: Springer.

Masters, J., et al. (Eds.) (2013). *Leaping ahead, advances in prosimian biology*. NY: Springer. [Chapters 15, 16, 18, 21, 35]

Mittermeier, R.A., et al. (Eds.) (2013). *Handbook of the mammals of the world, Vol. 3. Primates*. Barcelona: Lynx Edicions. [pp. 90–175]

Vasey, N. (Ed.) (2005a). Behavioral ecology and conservation of ruffed lemurs. *AJP*, 66, 1–94 (special issue).

Additional References

Baden, A.L., et al. (2013). Communal nesting, kinship, and maternal success in a social primate. *BE&S*, *67*, 1939–1950.

Ballhorn, D.J., et al. (2009). Quantitative variability of cyanogenesis in *Cathariostachys madagascariensis* – the main food plant of bamboo lemurs in southeastern Madagascar. *AJP*, *71*, 305–315.

Blair, M., et al. (2013). Ecological divergence and speciation between lemur *Eulemur* sister species. *JEB*, *26*, 1790–1801.

Bolt, L.M. (2013a). The function of howling in the ring-tailed lemur (*Lemur catta*). *IJP*, *34*, 157–169.

Bolt, L.M. (2013b). Squealing rate indicates dominance rank in the male ring-tailed lemur (*Lemur catta*). *AJP*, *75*, 1174–1184.

Brockman, D.K., et al. (2009). Birth season glucocorticoids are related to the presence of infants in sifaka (*Propithecus verreauxi*). *PRSB*, *267*, 1855–1863.

Carrai, V., et al. (2003). Increase in tannin consumption by sifaka (*Propithecus verreauxi verreauxi*) females during the birth season: a case for self-medication in prosimians? *Primates*, *44*, 61–66.

Colquhoun, I. (2011). A review and interspecific comparison of nocturnal and cathemeral strepsirrhine primate olfactory behavioural ecology. *IJZ*, *v.2011*,362976. DOI: 10.1155/2011/362976.

Cooper, V.J., & Hosey, G.R. (2003). Sexual dichromatism and female preference in *Eulemur fulvus* subspecies. *IJP*, *24*, 1177–1188.

Curtis, D.J. (2004). Diet and nutrition in wild mongoose lemurs (*Eulemur mongoz*) and their implications for the evolution of female dominance and small group size in lemurs. *AJPA*, *124*, 234–247.

Curtis, D.J., & Zaramody, A. (1999). Social structure and seasonal variation in the behaviour of *Eulemur mongoz*. *FP*, *70*, 79–96.

Curtis, D.J., et al. (2006). A re-evaluation of the role of vision in the activity and communication of nocturnal primates. *FP*, *77*, 50–71.

delBarco-Trillo, J., et al. (2012). *Eulemur*, me lemur: the evolution of scent-signal complexity in a primate clade. *PTRSB*, *367*, 1909–1922.

Donati, G., et al. (2009). Proximate cues and ultimate determinants of brown lemur cathemerality. *AB*, *77*, 317–325.

Donati, G., et al. (2013). (Un-)Expected nocturnal activity in "diurnal" *Lemur catta* supports cathemerality as one of the key adaptations of the lemurid radiation. *AJPA*, *150*, 99–106.

Engqvist, A., & Richard, A. (1991). Diet as a possible determinant of cathemeral activity patterns in primates. *FP*, *57*, 169–172.

Eppley, T.M., et al. (2011). Coping with low quality diets: a first account of the feeding ecology of the southern gentle lemur, *Hapalemur meridionalis*, in the Mandena littoral forest, southeast Madagascar. *Primates*, *52*, 7–13.

Erhart, E.M., & Overdorff, D.J. (1998). Infanticide in *Propithecus diadema edwardsi*: an evaluation of the sexual selection hypothesis. *IJP*, *19*, 73–81.

Erhart, E.M., & Overdorff, D.J. (2008a). Population demography and social structure changes in *Eulemur fulvus rufus* from 1988 to 2003. *AJPA*, *136*, 183–193.

Erhart, E.M., & Overdorff, D.J. (2008b). Rates of agonism by diurnal lemuriforms: implications for female social relationships. *IJP*, *29*, 1227–1247.

Fleagle, J.G. (2013). *Primate adaptation and evolution* (3rd edition). NY: Elsevier Science.

Gachot-Neveu, H., et al. (1999). Paternity determination in two groups of *Eulemur fulvus mayottensis*: Implications for understanding mating strategies. *IJP*, *20*, 107–119.

Gordon, A.D., et al. (2013). Females are the ecological sex: sex-specific body mass ecogeography in wild sifaka populations (*Propithecus* spp.). *AJPA*, *151*, 77–87.

Gould, L., & Sauther, M. (2007). Lemuriformes. In C.J. Campbell et al. (Eds.), *Primates in perspective* (2nd edition). NY: OUP.

Gould, L., et al. (2003). Demographic and life-history patterns in a population of ring-tailed lemurs (*Lemur catta*) at Beza Mahafaly Reserve, Madagascar: a 15-year perspective. *AJPA*, *120*, 182–194.

Grassi, C. (2006). Variability in habitat, diet, and social structure of *Hapalemur griseus* in Ranomafana National Park, Madagascar. *AJPA*, *131*, 50–63.

Griffin, R.H., et al. (2012). Evolutionary disequilibrium and activity period in primates. *AJPA*, *147*, 409–416.

Horvath, J.E., et al. (2008). Development and application of a phylogenetic toolkit: resolving the evolutionary history of Madagascar's lemurs. *Genome Research*, *18*, 489–499.

Irwin, M.T. (2006). Ecologically enigmatic lemurs: the sifakas of the eastern forests (*Propithecus candidus*, *P. diadema*, *P. edwardsi*, *P. perrieri*, *and P. tattersalli*). In L. Gould & M.L. Sauther (Eds.), *Lemurs, ecology and adaptation*. NY: Springer.

Irwin, M.T., et al. (2007). Diademed sifakas (*Propithecus diadema*) use olfaction to forage for the inflorescences of subterranean parasitic plants (Balanophoraceae: Langsdorffia sp., and Cytinaceae: Cytinus sp.). *AJP*, 69, 471–476.

Irwin. M.T., et al. (2014). Nutritional correlates of the "lean season": effects of seasonality and frugivory on the nutritional ecology of diademed sifakas. *AJPA*, *153*, 78–91.

Kappeler, P.M., & Erkert, H.G. (2003). On the move around the clock: correlates and determinants of cathemeral activity in wild redfronted lemurs (*Eulemur fulvus rufus*). *BE&S*, *54*, 359–369.

Kappeler, P.M., & Fichtel, C. (2012). A 15-year perspective on the social organization and life history of sifaka in Kirindy Forest (pp. 101–121). In P.M. Kappeler & D.P. Watts (Eds.), *Long-term field studies of primates*. NY: Springer.

Karpanty, S.M. (2006). Direct and indirect impacts of raptor predation on lemurs in southeastern Madagascar. *IJP*, *27*, 239–261.

Karpanty, S.M., & Grella, R.A. (2001). Lemur responses to diurnal raptor calls in Ranomafana National Park, Madagascar. *FP*, *72*, 100–103.

LaFleur, M., et al. (2014). Cathemerality in wild ring-tailed lemurs (*Lemur catta*) in the spiny forest of Tsimanampetsotsa National Park. *Primates*, *55*, 207ff.

Lewis, R.J. (2008). Social influences on group membership in *Propithecus verreauxi verreauxi*. *IJP*, *29*, 1249–1270.

Lewis, R.J. (2010). Grooming patterns in Verreaux's sifaka. *AJP*, *72*, 254–261.

Lewis, R.J., et al. (2003). Observed infanticide in a seasonal breeding prosimian (*Propithecus verreauxi verreauxi*) in Kirindy Forest, Madagascar. *FP*, *74*, 101–103.

MacLean, E.L., et al. (2012). Variance sensitive choice in lemurs. *AC*, *15*, 15–25.

MacLean, E.L., et al. (2013). Group size predicts social but not nonsocial cognition in lemurs. *PLONE*, *8*, e66359.

Mahajan, N., et al. (2009). Enumeration of objects and substances in non-human primates: experiments with brown lemurs (*Eulemur fulvus*). *Developmental Science*, *12*, 920–928.

Mayor, M., et al. (2004). Species status of *Propithecus* spp. *IJP*, *25*, 875–900.

Meldrum, D.J., et al. (1997). Hindlimb suspension and hind foot reversal in *Varecia variegata* and other arboreal mammals. *AJPA*, *103*, 85–102.

Mertl-Millhollen, A.S., et al. (2011). Tamarind tree seed dispersal by ring-tailed lemurs. *Primates* *52*, 391–396.

Morelli, T.L., et al. (2009). The rules of disengagement: takeovers, infanticide, and dispersal in a rainforest lemur, *Propithecus edwardsi*. *Behaviour*, *146*, 499–523.

Mutschler, T. (2002). Alaotran gentle lemur: some aspects of its behavioral ecology. *EA Supplement*, *1*, 101–104.

Mutschler, T., & Tan, C.L. (2003). *Hapalemur*, bamboo or gentle lemurs. In S.M. Goodman, & J.P. Benstead (Eds.), *The natural history of Madagascar*. Chicago: UCP.

Nakamichi, M., et al. (1996). Maternal responses to dead and dying infants in wild troops of ring-tailed lemurs at the Berenty Reserve, Madagascar. *IJP*, *17*, 505–523.

Nievergelt, C., et al. (2002). Social system of the Alaotran gentle lemur (*Hapalemur griseus alaotrensis*): genetic characterization of group composition and mating system. *AJP*, *57*, 157–175.

Norscia, I., & Palagi, E. (2011). Do wild brown lemurs reconcile? Not always. *Journal of Ethology*, *29*, 181–185.

Norscia, I., et al. (2005). Termite soil eating in Kirindy sifakas (Madagascar): proposing a new proximate factor. *FP*, *76*, 119–122.

Ostner, J. (2002). Social thermoregulation in redfronted lemurs (*Eulemur fulvus rufus*). *FP*, *73*, 175–180.

Ostner, J., & Kappeler, P.M. (1999). Central males instead of multiple pairs in redfronted lemurs, *Eulemur fulvus rufus* (Primates, Lemuridae)? *AB*, *58*, 1069–1078.

Ostner, J., & Kappeler, P.M. (2004). Male life history and unusual sex ratio of redfronted lemur, *Eulemur fulvus rufus*, groups. *AB*, *67*, 249-259.

Ostner, J., et al. (2006). What males know about female reproductive status. *AJP*, *68*, 701–712.

Ostner, J., et al. (2008). Androgen and glucocorticoid levels reflect seasonally occurring social challenges in male redfronted lemurs (*Eulemur fulvus rufus*). *BE&S*, *62*, 627–638.

Overdorff, D.J. (1998). Are *Eulemur* species pair-bonded? *AJPA*, *105*, 153–166.

Overdorff, D.J., & Tecot, S. (2006). Resource defense and pair-bonding in red-bellied lemurs (*Eulemur rubriventer*) in southeastern Madagascar. *AJPA*, *129*, 143.

Palagi, E., et al. (2012a). Scratching around mating: factors affecting anxiety in wild *Lemur catta*. *Primates*, *53*, 247–254.

Palagi, E., et al. (2012b). Sexual signalling in *Propithecus verreauxi*: Male chest badge and female mate choice. *PLONE*, *7*, e37332.

Parga, J.A. (2006). Male mate choice in *Lemur catta*. *IJP*, *27*, 107–131.

Parga, J.A. (2009). Dominance rank reversals and rank instability among male *Lemur catta*: the effects of female behavior and ejaculation. *AJPA*, *138*, 293–305.

Parga, J.A. (2010a). Evaluation of male inter-troop transfer as a mating strategy in ring-tailed lemurs on St. Catherine's Island, USA. *FP*, *81*, 146–162.

Parga, J.A. (2010b). Post-ejaculatory mounting in the ring-tailed lemur (*Lemur catta*): a behavior solicited by females? *Ethology*, *116*, 832–842.

Parga, J.A. (2011). Nocturnal ranging by a diurnal primate: are ring-tailed lemurs (*Lemur catta*) cathemeral? *Primates*, *52*, 201–205.
Parga, J.A., & Henry, A.R. (2008). Male aggression during mating: evidence for sexual coercion in a female dominant primate. *AJP*, *70*, 1187–1190.
Parga, J.A., & Lessnau, R.G. (2008). Dispersal among male ring-tailed lemurs (*Lemur catta*) on St. Catherine's Island. *AJP*, *70*, 650–660.
Pereira, M.E., & McGlynn, C.A. (1997). Special relationships instead of female dominance for redfronted lemurs, *Eulemur fulvus rufus*. *AJP*, *43*, 239–258.
Petter, J.J., & Peyrieras, A. (1975). Preliminary notes on the behavior and ecology of *Hapalemur griseus*. In I. Tattersall, & R.W. Sussman (Eds.), *Lemur biology*. NY: Plenum Press.
Pochron, S.T., et al. (2004). Demography, life history, and social structure in *Propithecus diadema edwardsi* from 1986–2000 in Ranomafana National Park, Madagascar. *AJPA*, *125*, 61–72.
Razafindratsima, O.H., & Martinez, B.T. (2012). Seed dispersal by red-ruffed lemurs. *Ecotropica*, *18*, 15–26.
Razafindratsima, O.H., et al. (2014). Patterns of movement and seed dispersal by three lemur species. *AJP*, *76*, 84–96.
Rea, M.S., et al. (2014). Daily activity and light exposure levels for five species of lemurs at the Duke Lemur Center. *AJPA*, *153*, 68–77.
Richard, A.F. (1992). Aggressive competition between males, female controlled polygyny and sexual monomorphism in a Malagasy primate *Propithecus verreauxi*. *JHE*, *22*, 395–406.
Rosati, A.G., et al. (2014). The ecology of spatial memory in four lemur species. *AC*, *17*, 947–961.
Rushmore, J., et al. (2012). Sight or scent: lemur sensory reliance in detecting food quality varies with feeding ecology. *PLONE*, *7*, e41558.
Sandel, A.A., et al. (2011). Evidence from four lemur species that ringtailed lemur cognition converges with that of haplorrhine primates. *AB*, *81*, 925–931.
Takahata, Y., et al. (2001). Daytime deliveries observed for the ring-tailed lemurs of Berenty Reserve, Madagascar. *Primates*, *42*, 267–271.
Takahata, Y., et al. (2008). The relationship between female rank and reproductive parameters of the ringtailed lemur: a preliminary analysis. *Primates*, *49*, 135–138.
Tan, C.L. (2006). Behavior and ecology of gentle lemurs (genus *Hapalemur*). In L. Gould & M.L. Sauther (Eds.), *Lemurs: ecology and adaptation* (pp. 369–381). NY: Springer.
Tattersall, I. (1987). Cathemeral activity in primates: a definition. *FP*, *49*, 200–202.
Tecot, S. (2006). Activity budget and microhabitat in *Eulemur rubriventer*. *AJP*, *68*, 111–112.
Tecot, S., & Romine, N. (2012). Female leadership of group movements in a pair-living, co-dominant, monomorphic primate species. *AJP*, *74*, 591–601.
Tecot, S., et al. (2011). Patterns of mortality and group transfer explain differences in male and female longevity in *Propithecus edwardsi*, Milne-Edwards' sifaka. *AJPA*, *144*, 291–292.
Tecot, S., et al. (2013). Risky business: sex ratio, mortality, and group transfer in *Propithecus edwardsi*. *BE*, *24*, 987–996.
Thalmann, U. (2001). Food resource characteristics in two nocturnal lemurs with different social behavior: *Avahi occidentalis* and *Lepilemur edwardsi*. *IJP*, *22*, 287–324.
Vasey, N. (2005b). Activity budgets and activity rhythms in red ruffed lemurs (*Varecia rubra*) on the Masoala Peninsula, Madagascar: seasonality and reproductive energetics. *AJP*, *66*, 23–44.
Vasey, N. (2006). Impact of seasonality and reproduction on social structure, ranging patterns, and fission–fusion social organization in red ruffed lemurs. In L. Gould, & M.L. Sauther (Eds.), *Lemurs: ecology and adaptation*. NY: Springer.
Vasey, N. (2007). The breeding system of wild red ruffed lemurs (*Varecia rubra*): a preliminary report. *Primates*, *48*, 41–54.
Yamashita, N., Vinyard, C.J., & Tan, C.L. (2009). Food mechanical properties in three sympatric species of *Hapalemur* in Ranomafana National Park, Madagascar. *AJPA*, *139*, 368–381.
Yoder, A.D. (2003). Phylogeny of the lemurs. In S.M. Goodman, & J.P. Benstead (Eds.), *The natural history of Madagascar*. Chicago: UCP.

10 The Tarsioid Suborder

Common Features and Variation

In a dark tangle of vegetation near the floor of the nighttime rainforest, a small figure clings with long fingers and toes to a thin vertical stalk of bamboo. The tarsier scans its surroundings with large ears and eyes. Suddenly the round head swivels, and the tarsier focuses on its insect prey and pushes off with its long legs into a powerful leap.

The tarsioid lineage, another of the three ancient primate clades, was much more widespread in the past. Extant tarsiers occur only in insular Southeast Asia. All living tarsiers are small, weighing about 100–150 grams. Head/body length is about 10cm to 15cm and the tail 20cm to 25cm (see Table 10.1 online for physical dimensions of all tarsioids). Absence of the tapetum lucidum and a reduced sense of smell have led several authorities to believe that the nocturnal adaptation of the tarsiers is a secondary one, superimposed on a diurnal ancestry. The size of tarsier eyes is a derived trait.

Most tarsiers belong to one of three geographic populations that we will call the Sulawesi, Borneo, and Philippine tarsiers (Box 10.1). Several important traits seem to be clinally distributed across the major geographic populations. These include positional behavior, variety of prey, and social cohesion. Sulawesi tarsiers, perhaps the basal population, represent one end of each cline. Borneo tarsiers are at the other extreme and Philippine tarsiers are intermediate. The eastern group can be loosely called spectral tarsiers, but they probably include several distinct species.

Box 10.1 Tarsier Taxonomy

Tarsiiformes can be placed at the level of infraorder, suborder, or series. In any case, the name represents the same ancient lineage that originated perhaps 80 million years ago. The living representatives can all be placed in the single superfamily Tarsioidea, containing the lone family Tarsiidae. For approximately the last 50 years just one genus was recognized: *Tarsius*. Groves and Shekelle (2010) presented evidence for recognition of three genera, including morphological details of the living forms, molecular evidence for the age of the clades, and biogeographic separation.

Cephalopachus, informally known as the western tarsiers, occupies the island of Borneo and the adjacent area of Sumatra. *Tarsius*, called eastern tarsiers, inhabits the island of Sulawesi. *Carlito* occurs in the southern part of the Philippines.

Cephalopachus have the largest eyes and dentition and the longest hind limbs and hands. *Tarsius* have the smallest eyes and dentition and the shortest hind limbs and hands. *Cephalopachus* are the least social and least vocal. *Tarsius* are the most social and vocal. Philippine *Carlito* tarsiers are intermediate in all of these traits.

Groves and Shekelle refer to the eastern tarsiers as a taxonomically "cryptic" group that contains at least 8 different species and possibly as many as 16. This complex has been known as the single species *Tarsius spectrum*, the spectral tarsier, but *Tarsius tarsier* may be a more correct nomenclature.

Tarsier Behavior

Tarsier Ecology

Tarsiers typically inhabit lowland rainforest, more commonly in secondary than in primary growth. The tarsier's eyes glow in the dark, but with a dull orange color because of the fovea, which provides acute low-light vision. The animal scans the dark environment with huge eyes and large membranous ears. It can rotate its head almost 180 degrees to either side, which is necessary because the tiny primate cannot move its enormous eyes at all.

Tarsiers are active in the understory where there is an abundance of thin vertical supports, such as saplings, generally within 3m of the ground (Fogden 1974). Those seen at greater heights may have climbed higher when feeling endangered or otherwise insecure (Niemitz 1979). Tarsiers use vertical clinging and leaping in pursuit of prey (see Table 10.2 online for ecological characteristics of all tarsioids). A motionless tarsier typically clings to a vertical support in an upright position. The long fingers and toes end in wide pads for firm grasping, especially on smooth surfaces such as bamboo. To free its hands a tarsier can release its grip on the stalk and use only its feet and long tail for stabilization.

The tarsier uses acute night vision to search for prey, mostly on the ground in the form of insects. It also uses its hearing to locate prey under the leaf cover. The tarsier leaps down to attack, biting with its eyes closed (Niemitz 1979), perhaps because the bulging eyes are so vulnerable. It grabs the prey with its hands and kills it with several bites. Alternatively it may put a whole insect into its mouth, made possible by an unusually wide gape. Then the tarsier returns to a perch to feed. Its molars can crush an exoskeleton, releasing nutrients quickly and facilitating digestion by the simple gut. The distinctive upper teeth may also aid the process: the first six are all like premolars, tending to be pointed and somewhat conical.

Tarsiers are the only primates that are completely faunivorous. Feeding on vegetation has never been reported. Researchers have seen tarsiers biting on fungi and leaves, but without chewing or swallowing. Given comparable behavior in other primates, such actions are probably exploratory and/or shaped by medicinal value in some plants.

These tiny animals can easily leap three meters or more, though most of their jumps are shorter. In the air a tarsier uses its tail, which is nearly twice the length of its body, to control direction. Despite a greater maximum, jumps average 1.3m and constitute about 60 percent of the tarsier's locomotion. Climbing is the next most important mode (20 to 30 percent of locomotor bouts), followed by brief episodes of walking or hopping on the ground (Dagosto et al. 2001).

Tarsier locomotion and predation are closely interrelated. The primate must move quickly to find and seize sufficient prey, thus using a great deal of energy. The expenditure is made possible by the fact that the insects are high-energy food. This way of life is linked to the understory of the forest because the high moisture and low light produce leaves that are large, soft-textured, and low in toxicity. Such vegetation supports a greater proportion of larger-bodied arthropods than exist in other forest environments. Not only are these prey high in energy, they are also less subject to seasonal variation than plant resources in the environment.

Rosenberger (2010) argued that standard phrases like *visual acuity* are not sufficient to explain the unique eyes of tarsiers, which are far larger than those of nocturnal insect-hunting strepsirrhines. More than half of the globe of the tarsier eyeball projects beyond the rim of the orbit (Schultz 1940, cited by Rosenberger 2010). Rosenberger postulated that the adaptive result of eye enlargement was to increase surface area in the peripheries of the retinas, where the rods are located. These cells function in the recognition of spatial relations and movement and contribute to a high rate of success in prey capture, which was measured at 88 percent in one captive group (Roberts and Kohn 1993).

Other features may also facilitate detection of spatial relations and movement (Rosenberger 2010). Wide spacing of the ears on the round head increases ability to locate the source of a sound. Coordination of the senses is automatic because the ears and the immoveable eyes must both focus on the same target as the head swivels. The upshot is that the faunivorous tarsiers are highly adapted to prey capture by senses that enhance detection, spatial mapping, and calculation of interception.

Rosenberger (2010) linked these features to the subsistence strategy that he called **nonpursuit hunting**. Rather than foraging, the tarsier waits for the food to come to it. This approach requires a maximized level of success when opportunities arise, so tarsiers are highly specialized for detection and interception. Tarsier physical features sustain faunivory and faunivory sustains them. High quality food provides the energy needed for numerous leaping captures and for metabolic maintenance of the eyes (retinal tissue consumes oxygen more rapidly than brain tissue). Nonpursuit hunting and small home ranges minimize energy expenditure.

Niemitz (1979, 2010) suggested that the tarsier adaptation is ecologically analogous to that of owls. These primates and birds feed on similar prey and obtain them in similar ways. They are completely silent in locomotion, locate their prey by sound, and pounce from a short distance above ground level. Owls and tarsiers also display physical similarities that are arguably related to their feeding patterns. Most strikingly, both have very large eyes and can rotate their heads almost 180 degrees. Tarsiers have the advantage of operating in vegetation that is too dense for flying. Perhaps they can be understood as the owls of the undergrowth.

The daytime sleeping nests of tarsiers are mostly in tree hollows or in dense vegetation such as vine tangles and grass platforms. They are generally well above the ground, ranging from about 5m to the canopy. The animals emerge from their sleeping places around sunset and take up the vertical posts from which they will launch themselves in search of food. Activity occurs regularly throughout the night, with bouts of leaping at about one-hour intervals and rests in between (Niemitz 1979).

The known activity budgets of the various tarsiers are similar to one another. Foraging occupies 55 to 65 percent of their active time and traveling 25 to 30 percent. The remaining 10 to 15 percent is devoted to resting and to whatever social behavior they engage in. The relatively stereotyped nature of tarsier foraging (Rosenberger 2010) might explain this consistency across taxa and environments.

Some observers have suggested that tarsiers lack caution with regard to predators and proposed that they are protected by repellent odor and/or taste. It was also suggested that the capability for sudden long leaps is a more important defense than advance caution. However, accumulating evidence suggests that tarsiers are under significant predator pressure from the same enemies as other primates and react accordingly.

In experiments spectral tarsiers responded to models and replayed vocalizations of various raptors; to live feral and domestic cats; and to at least one species of civet, a catlike animal. They were more likely to freeze in response to a falcon model than to a civet or monitor lizard, and they maintained the freeze as long as the falcon was in sight. They were also more likely to mob the falcon model. The tarsiers regularly alarm-called in response to civets and sometimes mobbed them, but occasionally ignored them. Although local people report that monitor lizards eat spectral tarsiers, the reptiles only elicit occasional alarm calls and never freezing or mobbing. These varied responses, which differentiate among enemies, may have inspired the idea that tarsiers are indifferent to all predators. This is a case where speculation led to more effective research and experimentation.

Spectral tarsiers react strongly to boa constrictors in natural and experimental situations. They often mob snakes (40 of 90 experimental presentations in the field). Mobs for live snakes numbered from three to ten tarsiers, the larger numbers resulting from males of neighboring groups sometimes joining in. Resident females often attended, but males initiated mobbing, behaved more intensely, and persisted for longer. The snake retreated in 15 of 19 natural (non-experimental) events. This was more likely when five or more tarsiers participated.

Gursky (2005, 2006) used these snake responses to test competing hypotheses for mobbing, gathering data from 11 natural events and 31 experiments. The duration of mobbing was strongly correlated with the number of participants. Gursky rejected several explanations for the behavior: (1) site avoidance, because mothers continued to park their infants in the vicinity after an event; (2) protecting offspring, because mothers consistently brought infants to the events; (3) cultural transmission, because very young infants (even some less than a week old and with no mobbing experience) responded to snakes with alarm calls; (4) attracting stronger help, because outsider males who joined mobbing were no larger than resident males; (5) move-on, because the number of participants had no effect on whether snake or tarsiers retreated first. The surviving explanation was the Perception Advertisement Hypothesis. The data fulfilled its predictions: live snakes were rarely seen in the area following mobbing calls, while snakes continued to be in evidence after the primates ignored the snake or only gave alarm calls. This process is a good example of hypothesis testing and consideration of competing hypotheses.

Like nocturnal strepsirrhines, spectral tarsiers tend to become more active during full moons (called **lunar philia** by Gursky 2003). During full moons the tarsiers increased travel by all measures. They also increased foraging, which explains this behavior. Tarsiers captured almost three times as many insects on full moon nights as they did on new moon nights, either because the insects were more active or because the primates could see them better, or both. Increased activity might mean greater exposure to predators. According to one hypothesis, the risk is worthwhile because of the foraging benefit. According to another, there is no greater risk because the primates can see the predators as well as the predators can see them (Gursky 2003; Nash 1986). The potential importance of vision in these matters is underscored by the theory that tarsiers are secondarily nocturnal and may benefit more than other nocturnal animals from more light.

Tarsiers sometimes groom themselves with their teeth and specialized claws in the manner of strepsirrhines. They differ in using their lips more effectively, because the lips are more mobile (not being bifurcated and attached to the gum). Tarsiers use their fingers for grooming more than strepsirrhines do (Goosen 1987).

Tarsier Sociality

Pair bonding in the strict sense is characteristic of spectral tarsiers (MacKinnon and MacKinnon 1980). That is, the partners are strongly oriented toward one another and interact with affiliative behaviors (see Table 10.3 online for social features of all tarsioids). While ranging at night they tend to stay near each other, maintaining auditory and even visual contact. Sometimes they come together and sit in close contact. At these times one may hold the other's tail or the two may intertwine their tails. One partner may jump back and forth over the other, or scent mark the other. Either tarsier may rub its face in the partner's hair, presumably leaving scent from the **circum-oral gland**, and the male can mark the female with a gland on his abdomen. The mates seem to enhance their close contact communication with subtle facial expressions and body postures.

Despite the strong bond, the proximity of paired adults varies with the food supply. Measuring day-to-day availability of insects, Gursky (2002) found that about 0.35g per hour was pivotal. Below that threshold the average distance between adult members of the same family almost tripled, from about 20 to 55 meters. Seasonal change in food abundance produced the same pattern with a similar range of distance. This raises the question of why the tarsiers ever come close to one another. Theory predicts that the benefit(s) outweigh the cost. Sulawesi tarsiers that are near each other get less food, but they appear to gain in protection from predators. The subjects of an experiment maintained significantly shorter inter-individual distances after the presentation of snake models. Similar observations have been made on another Sulawesi species.

Pair-groups characterize all of the studied Sulawesi populations (MacKinnon and MacKinnon 1980; Merker 2010). A second adult female is present in some groups and occasionally a third. These socially polygynous populations constitute about 10 percent of a population (Driller et al. 2009; Gursky 2000). In at least one case an extra female is known to be a matured offspring from the previous year and it is possible that females tend to disperse later than males. However, in other cases the extra females may be additional unrelated partners of the one adult male. Gursky (1995) interpreted this as "facultative polygyny".

Common ranging patterns or sharing of a nest served to identify groups of up to nine individuals among the Sulawesi tarsiers. Group members relocated each other when the night's foraging ended and slept in the same tree. The maximum number probably included subadults, which can be characterized as tolerated or marginal because their social interactions with the adults are minimal. Gursky (2000) calculated that interaction between adults and subadults amounted to less than eight percent of total interaction time (compared to 23 percent for the adult pair and 43 percent for mother and offspring). On the other hand, subadults of both sexes provided some allocare and male subadults often participated in territorial conflicts and behaved with intensity in mobbing predators.

The solidarity in the Sulawesi family or group is represented in the treatment of infants that have been parked by their mothers: other group members of both sexes frequently visit and play with the infants. In at least one population of the species, adult males and subadult females play with babies. The young females also retrieve them when they fall and "babysit" while the mother is gone.

Several Sulawesi species perform duets, comparable to those of some strepsirrhines (Merker 2010). The duets are sung at dawn, before retiring for sleep, and sometimes at dusk. Similar duets are associated with hostile encounters between neighboring pairs. Subadults often join in and make the performance a chorus. The acoustic parameters of duets can be used to distinguish tarsier species from one another (Nietsch 1999). In at least one species the dawn duet takes place in a tree other than the sleeping tree, which obviates the possibility that a predator will be attracted to the hiding place.

During the course of her long field study, Gursky (2007) recorded three extra-pair matings by a female spectral tarsier, one with a stranger and two with males from neighboring groups. Now we have genetic evidence from the Lariang population on Sulawesi (Driller et al. 2009). The researchers captured and sampled 26 animals. They found evidence for one extra-pair offspring, while in all other cases where assignment was possible the observed social parents were probably the genetic parents. In these two species of Sulawesi tarsiers, at least, the mating system coincides closely with the social system and both are essentially monogamous.

Spectral tarsiers are territorial and males are hostile to intruding males. Gursky (2007) evaluated the alternative explanations of mate defense and resource defense. She found more support for the former hypothesis. Apparently, the low rate of extra-pair copulation is at least partly due to mate guarding. This is consistent with significant sexual size dimorphism in spectral tarsiers. Average weights are roughly 125g for males and 105g for females. This contrasts with relative monomorphism in Philippine and Borneo tarsiers (Gursky 2007).

Pygmy tarsiers live in the highland moss forests of Sulawesi at much higher altitudes than other tarsiers. They are about half the weight of other tarsiers and sexual dimorphism is less pronounced. They are the most divergent of the Sulawesi tarsiers in environment and behavior. What little we know about the social life of these primates comes mainly from a single study (Grow and Gursky-Doyen 2010). The one observed group contained two adult males and one adult female, which contrasts sharply with the social monogamy and facultative polygyny of other Sulawesi tarsiers. We cannot tell at this point whether this is a polyandrous pattern or simply one manifestation of openness to additional group members. In the colder environment of the highlands, the huddling of a larger group at the sleeping site could be important to thermoregulation. The problem is presumably exacerbated by their small size.

Tarsiers make somewhat more use of facial communication than do strepsirrhines because the external olfactory apparatus does not tether their facial features. However, as in strepsirrhines, deposition of urine and other forms of scent marking are important. Tarsiers have glands for olfactory communication in their lips, abdomens, and anogenital regions, which they rub on physical substrates and on other tarsiers. There is variation by sex in some species with regard to the possession and use of these glands. Tarsiers of various ages and sexes also mark with urine. Even humans, with their limited olfactory ability, can detect the presence of tarsiers by smell alone.

There are a number of types of vocalizations produced by tarsiers that are not part of duets (Niemitz 1984; Nietsch 2003). For example, spectral tarsiers have 15 different types of call, including distress calls, alarm whistles, alarm calls, mid-intensity alarm calls, female screams, contact trills, contact whistles, food calls, play whistles, infant squeaks and other types of whistles. Borneo tarsiers have at least four general types of vocalization and Philippine tarsiers display at least three main calls: (1) a loud piercing single note used as an alarm, during fights, and emitted by deserted babies; (2) locust-like chirping when several tarsiers are within calling distance of one another; and (3) a soft, sweet birdlike trill uttered when the animals were otherwise quiet and apparently contented (Wharton 1950, cited by Niemitz 1984).

Vocalizations and scent marks are rare among pygmy tarsiers (Shekelle 2008). The absence of duets may mean that these tarsiers are less territorial than the lowland species. Alternatively, they may use a cryptic system such as ultrasonic vocalizations. The significance of sound for the pygmy tarsiers is indicated by an enlarged auditory bulla, possibly an adaptation to sound attenuation in the moss forest. The environment may also account for the paucity of scent marks, which may be rendered less effective by the rain and the ubiquity of moss.

Tarsier Life Cycle

Gestation sharply distinguishes tarsiers from strepsirrhines. It lasts for about six months (175 to 195 days), which is very long for such tiny animals (see Table 10.4 online for the reproductive parameters of all tarsioids). A comparative study showed that tarsier gestation is longer than that of any of the 25 strepsirrhine species that were sampled (Roberts 1994). It is also longer than that of some anthropoids that are many times larger than tarsiers (Wright et al. 2003). Correspondingly, the fetal growth rate of tarsiers is the slowest known among mammals in general. The result is the birth of a single highly developed neonate. The single large-brained tarsier offspring seems to be the product of strong selection forces. Although the female has two or three pairs of nipples, they are apparently a vestigial trait since her reproductive output is strictly limited to one offspring per year.

Tarsier neonates are very heavy in proportion to their mothers, up to 30 percent of maternal weight. This is larger than in any of the 25 strepsirrhine species in the comparative survey mentioned above (Roberts 1994). Much of the neonate's unusual weight is due to its brain, which is 60 to 70 percent of adult size. This infant is prepared for rapid acquisition of the skills that are necessary for the tarsier's uniquely faunivorous way of life, especially in locomotion and foraging (Roberts 1994). The newborn is also precocious in other ways. It is fully furred, its eyes are open, some teeth have erupted, it can hold tightly, and it is capable of some independent locomotion (Niemitz 1979).

A striking feature of tarsier life history is the similarity between wild spectral tarsiers and captive Borneo tarsiers with regard to infant mortality. Infant mortality is concentrated in the first three weeks of life and is restricted to the first three months. Gursky cited the discussion of this point by Wright (1995), which alludes to maternal investment and to offspring deaths due to predation and infanticide. Gursky (2007) suggested that the birth of large infants is an evolutionary strategy for minimizing infant mortality by getting the offspring to the juvenile stage as quickly as possible. This is unusual among primates, which are generally characterized by long periods of infancy in relation to body size.

The ubiquity of complex pair-groups in the Sulawesi tarsiers creates distinctive conditions for development of the young and this affects maternal behavior to an undetermined degree (see Table 10.5 online for life cycle stages of all tarsioids). This begins with the mother's desire to protect the newborn from conspecifics. Though some simply keep their distance by frequent movement (MacKinnon and MacKinnon 1980), Gursky (2000) found that four of six pregnant females moved from the family sleeping site to a separate one shortly before the birth and maintained this pattern for the infant's first three weeks. These mothers chased away other family members who tried to join them.

After reunion with the family the situation is reversed. Other group members of both sexes visit and play with infants that have been parked by their mothers. Subadult females are particularly prominent in allocare (Gursky 2000). In addition to grooming and play, they retrieve fallen infants from the ground. They carry a smaller one by mouth, and with movements and vocalizations they urge a larger one to follow them to a tree. In at least one population of the species, adult males play with babies. All family members show increased vigilance when an infant is in the group, measured by the number of alarm calls.

Infant protection may be an important function of spectral tarsier gregariousness (Gursky 2007). A female stays closer to her male partner when she is lactating than at other times.

We may reasonably speculate that this accounts for the fact that lactating females are the most likely to participate in predator mobbing. It might seem paradoxical, but on the periphery of a mobbing event she knows where the predator is and is close to protection.

A cached infant responds to signs of a predator by changing its position, responding most strongly to model snakes (Gursky 2007). This can be distinguished from exploration by whether or not the infant returns to its original position. Vocalizations demonstrate that infants less than a week old recognize danger from predators.

Participation of other family members in allocare may explain the rapid decline in the mother's attentions over the first ten weeks of her offspring's life. Physical contact drops from about 60 percent of observations to as low as 5 percent. Being the nearest neighbor of the infant goes from 100 percent to less than 60 percent, and grooming drops from 15 percent to as little as 4 percent. Even more striking, mothers virtually never play with their infants at any time.

Despite the mother's unusual burden, tarsier fathers provide little care for their offspring compared to other primate species in which the infants are unusually large (Wright et al. 2003). It may be that greater benefits accrue to males from other activities, such as seeking other matings, scanning for predators, and patrolling territorial boundaries (Gursky 2000). Mothers and offspring may share in the benefits from the latter two activities.

Most of our knowledge about the interaction of infant and mother comes from wild Sulawesi tarsiers. The mother grooms her infant with her incisor teeth in a comb-like fashion, much as strepsirrhines do (Gursky 2007). This might be part of the unique mix of traits in the tarsiers. Alternatively, it might distinguish Sulawesi tarsiers from the others and represent their basal phylogenetic position. The mother carries her baby in her mouth or parks it during the first three weeks of its life. However, unlike most strepsirrhines, the mother stays close to her parked infant, often foraging in the same tree and keeping an average distance of about 4m. The average duration of a parking bout is only about half an hour. Rather than use the term parking, which is usually applied to similar behavior in strepsirrhines, Gursky (2002) prefers to call the tarsier system cache and carry (which can also be applied to some strepsirrhines).

Gursky addressed the issue of why tarsiers park their infants at all. One possibility is phylogenetic inertia. The extra nipples suggest that tarsiers retain remnants of an adaptation to multiple offspring. However, this is not testable. A testable hypothesis is that the infants are too big to carry continuously. Studies of energy acquisition and use show that continuous carrying by the tarsier mother would make it extremely difficult for her to obtain enough food to sustain herself and the nursing baby. Lactating mothers occasionally share solid food with the infant after the second week, but the precocious young play an active role in this haphazard occurrence. Every case involved the capture of an unusually large insect, and the mother simply allowed her infant to feed itself from the prey. This is a form of sharing called tolerated theft.

For development of the infant in the first month, much of our information comes from a few captive Borneo tarsiers (Niemitz 1979). During the first day, it is almost invisible because the mother engages in little activity other than holding it close and keeping it away from any other tarsiers with whom she is associated. On the second day it begins to creep and climb. Quadrupedal locomotion precedes the characteristic leaping of the taxon. By the end of two weeks the baby has begun to groom itself in several ways, first by licking, then with the front teeth, and finally with its grooming claws. It also displays protective behavior in the form of threats and biting. Jumping is elaborated in the third and fourth weeks, including the first leap from one vertical surface to another, and the first prey are caught.

Maternal carrying ends by the age of one month among spectral tarsiers, and play appears a week or so later (Niemitz 1979). The mother weans her infant before the age of three months, a very short nursing period. By this time the young is half the size of an adult. Having been born large and grown rapidly, a nursing infant would presumably place an intolerable physical and energy burden on the mother after the age of three months. However, juvenile growth slows and the individual is still smaller than an adult at the age of one year. Tarsier postnatal growth to maturity is slower than in 25 other strepsirrhine species (Roberts 1994).

Observations on apparent weaning conflict in spectral tarsiers are probably relevant to all tarsiers because of shared physical traits. A mother's disputes with her developing infant have more to do with her refusal to carry it than with nursing rejection. Carrying is the greater burden due to the weight of tarsier infants, which are born large and then undergo rapid postnatal growth. The infant registers its displeasure by refusing to follow, calling to the mother, and emitting alarm calls.

Though subadults can be considered marginal in relation to the interaction of the adult pair, they play a significant role in the lives of infants. Interaction between subadults and infants accounts for 20 percent of a group's total social behavior (Gursky 2000). In addition to allocare, subadult males contribute to infant welfare by participating in mobbing and territorial defense. On the other hand, subadult females do not participate in mobbing at all. Considering the aggressive behavior of lactating females, the avoidance of mobbing by subadult females may be related to the fact that they are pre-parous.

Fifteen years is a reasonable estimate for the natural lifespan of a tarsier, based on data provided by Shekelle and Nietsch (2008). Captive tarsiers behave like old animals at about that age (although they do not show clinical signs of aging) and the record in captivity is 17 years 7 months. The record for a wild tarsier is an estimated 10 years 9 months for a retrapped female that showed no signs of advanced age.

Tarsier Cognition

The anatomy of the brain seems unpromising as far as primate intelligence is concerned. It is small and smooth, like those of carnivores (Wright et al. 2003) and insectivores (Schwartz 2003). The relationship between brain and body size differs little from that of strepsirrhines (Fish and Lockwood 2003). This is something of a puzzle in that the high-energy faunivorous diet of tarsiers could support a larger brain. One explanation is that this is the primitive condition for primates and there has been no selection for a larger brain in tarsiers. An alternative is that the huge eyes and orbits constrain brain volume (Fish and Lockwood 2003).

Tarsiers seem to display language-like communication in that they distinguish different kinds of predators with different calls (Gursky 2003). They twitter for raptors and give a harsh loud triple call for snakes. Infants produce both calls when less than a week old, which suggests to Gursky that the calls are not really specific to particular predators. Following Hauser (1993), she interpreted the different vocalizations as representing the infant's assessment of different risk levels. A similar pattern has been reported in ground squirrels.

Infants sometimes gave "inappropriate" alarm calls in response to new things in the environment rather than to predators. This can be viewed as mistaken classification, but Hauser would interpret it as the infant's way of asking questions about the environment. To Gursky the latter hypothesis is supported by tarsier mothers' responses to their infants. They often follow a legitimate call by repeating the same call, but seldom do so when the call is inappropriate. Rather than being a language-like phenomenon, this could be an example of teaching. That would be noteworthy in itself, since teaching is rare or absent in nonhuman primates.

Summary

Tarsiers are nocturnal, but this may be a secondary adaptation. They are highly specialized for a faunivorous way of life based on vertical clinging and leaping. Though once thought to be immune to predation, evidence now indicates that they are much like other primates in this regard. Eastern tarsiers live in pair-groups, while the others are largely or entirely solitary. Though tarsiers lack the olfactory equipment of strepsirrhines, scent marking plays a large role in their social life. However, facial communication plays a somewhat greater role. Gestation is long and the neonate is heavy. The precocious infant develops rapidly. Mothers follow a pattern of "cache and carry", probably because infants are too big to be carried continuously. Tarsiers seem to display some language-like vocal communication.

Sources

Basic Bibliography

Gursky, S.L. (2007). *The spectral tarsier*. Upper Saddle River, NJ: Pearson Prentice Hall.

Gursky S.L. (2011). Tarsiiformes. In C.A. Campbell et al. (Eds.), *Primates in perspective* (pp. 79–90). NY: OUP.

Kappeler, P.M. (2012). The behavioral ecology of strepsirrhines and tarsiers. In J.C. Mitani et al. (Eds.), *The evolution of primate societies* (pp. 17–42). Chicago: UCP.

MacKinnon, J., & MacKinnon, K. (1980). The behavior of wild spectral tarsiers. *IJP*, *1*, 361–379.

Mittermeier, R.A., et al. (Eds.) (2013). *Handbook of the mammals of the world, Vol. 3. Primates*. Barcelona: Lynx Edicions. [pp.236–261]

Wright, P.C., et al. (Eds.) (2003). *Tarsiers: past, present, and future*. New Brunswick, NJ: Rutgers University Press.

Additional References

Crompton, R.H., & Andau, P.M. (1986). Locomotion and habitat utilization in free-ranging *Tarsius bancanus*: a preliminary report. *Primates*, *27*, 337–355.

Crompton, R.H., & Andau, P.M. (1987). Ranging, activity rhythms, and sociality in free-ranging *Tarsius bancanus*: a preliminary report. *IJP*, 8, 43–71.

Dagosto, M., et al. (2001). Positional behavior and social organization of the Philippine tarsier (*Tarsius syrichta*). *Primates*, *42*, 233–243.

Driller, C., et al. (2009). The social system of Lariang tarsiers (*Tarsius lariang*) as revealed by genetic analyses. *IJP*, 30, 267–281.

Fish, J.L., & Lockwood, C.A. (2003). Dietary constraints on encephalization in primates. *AJPA*, *120*, 171–181.

Fogden, M.P.L. (1974). A preliminary field study of the western tarsier, *Tarsius bancanus* Horsefield. In: R.D. Martin & 2 co-editors, *Prosimian biology* (pp. 151–165). Pittsburgh: University of Pittsburgh Press.

Goosen, C. (1987). Social grooming in primates. In G. Mitchell & M.G. Erwin (Eds.), *Comparative primate biology, vol. 2B: behavior, cognition and motivation* (pp. 107–131). New York: Alan R. Liss.

Groves, C., & Shekelle, M. (2010). The genera and species of Tarsiidae. *IJP*, *31*, 1071–1082.

Grow, N.B., & Gursky-Doyen, S.L. (2010). Preliminary data on the ecology, behavior, and morphology of *Tarsius pumilus*. *IJP*, *31*, 1174–1191.

Gursky, S.L. (1995). Group size and composition in the spectral tarsier. *Tropical Biodiversity*, *3*, 57–62.

Gursky, S.L. (2000). Allo-care in a nocturnal primate: data on the spectral tarsier, *Tarsius spectrum*. *FP*, *71*, 39–54.

Gursky, S. (2002). Determinants of gregariousness in the spectral tarsier. *JZL*, *256*, 1–10.

Gursky, S. (2003). Lunar philia in a nocturnal primate. *IJP*, *24*, 351–367.

Gursky, S. (2005). Predator mobbing in *Tarsius spectrum*. *IJP*, *26*, 207–221.

Gursky, S. (2006). Function of snake mobbing in spectral tarsiers. *AJPA*, *129*, 601–608.

Gursky, S.L. (2007). The spectral tarsier. Upper Saddle River, NJ: Pearson Prentice Hall.

Gursky-Doyen, S. (2010a). Dispersal patterns in *Tarsius spectrum*. *IJP*, *31*, 117–131.

Gursky-Doyen, S. (2010b). Intraspecific variation in the mating system of spectral tarsiers. *IJP*, *31*, 1161–1173.

Gursky-Doyen, S. (2011). Infanticide by a male spectral tarsier (*Tarsius spectrum*). *Primates*, *52*, 385–389.

Haring, D.M., & Wright, P.C. (1989). Hand-raising a Philippine tarsier, *Tarsius syrichta*. *Zoo Biology*, *8*, 265–274.

Hauser, M.D. (1993). Do vervet monkey infants cry wolf? *AB*, *45*, 1242–1244.

Izard, M.K., et al. (1985). Gestation length in *Tarsius bancanus*. *AJP*, *9*, 327–331.

Merker, S. (2006). Habitat-specific ranging patterns of Dian's tarsiers (*Tarsius dianae*) as revealed by radiotracking. *AJP*, *68*, 111–125.

Merker, S. (2010). The population ecology of Dian's tarsier. In S. Gursky, & J. Supriatna (Eds), *Indonesian primates*. NY: Springer.

Merker, S., & Yustian, I. (2008). Habitat use analysis of Dian's tarsier (*Tarsius dianae*) in a mixed-species plantation in Sulawesi, Indonesia. *Primates*, *49*, 161–164.

Nash, L.T. (1986). Influence of moonlight levels on traveling and calling patterns in two sympatric species of Galago in Kenya. In D.M. Taub, & F.A. King (Eds.), *Current perspectives in primate social dynamics*. NY: Van Nostrand Reinhhold.

Neri-Arboleda, I., et al. (2002). Home ranges, spatial movements and habitat associations of the Philippine tarsier (*Tarsius syrichta*) in Corella, Bohol. *JZL*, *257*, 387–402.

Niemitz, C. (1979). Results of a field study on the western tarsier (*Tarsius bancanus borneanus* Horsfield, 1821) in Sarawak. *Sarawak Museum Journal*, *27*, 171–228.

Niemitz, C. (1984). Vocal communication of two tarsier species (*Tarsius bancanus* and *Tarsius spectrum*). In C. Niemitz (Ed.), *Biology of tarsiers*. Stuttgart: Gustav Fischer Verlag.

Niemitz, C.T. (2010). Progreditur ordinara saltando et retrorsum ... Normally proceeds in a leaping fashion, and backwards... *IJP*, *31*, 941–957.

Nietsch, A. (1999). Duet vocalizations among different populations of Sulawesi tarsiers. *IJP*, *20*, 567–583.

Nietsch, A. (2003). Outline of the vocal behavior of *Tarsius spectrum*. In P.C. Wright et al. (Eds.), *Tarsiers: past, present, and possible future*. New Brunswick (NJ): Rutgers University Press.

Roberts, M. (1994). Growth, development, and parental care in the western tarsier (*Tarsius bancanus*) in captivity: evidence for a "slow" life-history and nonmonogamous mating system. *IJP*, *15*, 1–28.

Roberts, M., & Kohn, F. (1993). Habitat use, foraging behavior, and activity patterns in reproducing western tarsier, *Tarsius bancanus*, in captivity. *Zoo Biology*, *12*, 217–232.

Rosenberger, A.L. (2010). The skull of *Tarsius*: functional morphology, eyeballs, and the nonpursuit predatory lifestyle. *IJP*, *31*, 1032–1054.

Schwartz, J.H. (2003). How close are the similarities between *Tarsius* and other primates? In P.C. Wright et al. (Eds.), *Tarsiers: past, present, and possible future*. New Brunswick (NJ): Rutgers University Press.

Shekelle, M. (2008). Distribution and biogeography of tarsiers. In M. Shekelle et al. (Eds.), *Primates of the Oriental night*. Jakarta: LIPI Press.

Shekelle, M., & Nietsch, A. (2008). Tarsier longevity: data from a recapture in the wild and from captive animals. In M. Shekelle et al. (Eds.), *Primates of the Oriental night*. Jakarta: LIPI Press.

Tremble, M., et al. (1993). Field observations of *Tarsius dianae* at Lore Lindu National Park, central Sulawesi, Indonesia. *Tropical Biodiversity*, *1*, 67–76.

Wright, P.C., et al. (1986a). Courtship and copulation in *Tarsius bancanus*. *FP*, *46*, 142–148.

Wright, P.C., et al. (1986b). Reproductive cycles in *Tarsius bancanus*. *AJP*, *11*, 207–215.

11 The Anthropoid Suborder

Monkeys and Apes

As the day ends, two kinds of primates prepare for the period of sleep. An ape grasps branches and twigs with hands and feet as it works its way toward a favorable crook in a tree. Once there, it deftly builds a nest for itself and settles into it for the night. A troop of monkeys enters an adjacent tree, some climbing and some walking on top of thick branches. They spread out into clusters and lone individuals and sit back on their padded bottoms in preparation for sleep.

Box 11.1 Anthropoid Taxonomy

The suborder Anthropoidea can be characterized by a suite of traits that are diagnostic of its status as a clade; some may be found in other mammals, including tarsiers, but the combination is unique (Fleagle 2013). All anthropoids have two incisors in each quadrant and they are typically vertical. The lower incisors are procumbent in only a few species, and never to the same degree as in strepsirrhines. The two halves of the lower jaw (the mandible) are fused together, which seems to be associated with the vertical position of the incisors, whether genetically or just functionally. The canine teeth are never procumbent, nor are they ever drastically reduced or absent. The canines are always taller and wider than the incisors and retain large roots even if the crowns are reduced. Detailed morphology of the teeth also distinguishes the anthropoids.

The importance of vision is reflected in the joining of three bones to form a complete socket for each eye. The snout is typically small. Smaller details of the skull are diagnostic. One example is the location of the lacrimal bone within the orbit rather than on the snout. Another is the fusion of tympanic ring to the lateral wall of the auditory bulla in the ear. Anthropoids have relatively short trunks and the forelimbs are about the same length as the legs or are longer.

Anthropoids constitute the third of the three major primate lineages described in Chapter 1 (Box 11.1). Humans, apes, and monkeys are the living forms. Loosely speaking, monkeys are quadrupedal anthropoids with tails while apes are long-armed primates that lack tails (Box 11.2).

Anthropoids occur in both the Americas and in the Old World. Despite geographic separation, morphological differences, and deep phylogenetic divisions, anthropoids display some profound physical and behavioral similarities that distinguish them from strepsirrhines and tarsiiforms. This chapter explores the general features of the anthropoids (Box 11.3). Those that follow will consider the many variations.

Box 11.2 Misplaying the Name Game

Despite profound taxonomic and functional similarities, monkeys and apes are very distinct subgroups of the Anthropoidea. It is especially important to emphasize this because of popular confusion. The monkeys that live on the Rock of Gibraltar, for example, are often called "Barbary apes". Many large monkeys are mislabeled in the same way, even howlers in the New World. At the same time, chimpanzees and other apes are often called monkeys, even in news sources.

Anthropoids have often been called "higher primates". This seems to imply that they are more advanced forms of life than strepsirrhines and tarsiers. However, each of those lineages has a history of evolutionary change that is at least as long as that of the anthropoids. Each has its own specialized adaptations, in accord with modern theory about natural selection.

Box 11.3 Distinctive Features of Anthropoid Behavior

These traits are common among anthropoids and distinguish them from other primates, either quantitatively or qualitatively, as described in the text.

- diurnality
- visual acuity
- trichromacy
- digital control
- sexual dimorphism
- male predator defense
- social reconciliation
- flexible sexual behavior
- isosexual behavior
- slow life history
- cognitive traits (controversial)
- culture (controversial).

Anthropoid Behavior

Anthropoid Ecology

Extant anthropoids are typically diurnal, although one genus is primarily nocturnal and others may be more flexible than previously thought (Tan et al. 2013). Diurnal activity favored several visual enhancements. One is visual acuity exceeding that of all other mammals. The eye shape of mammals in general, regardless of daily cycle, is similar to that of *nocturnal* birds and lizards. The anthropoid clade is the only exception, which indicates that their eyes have evolved for diurnal life (Hall et al. 2012).

Another feature of anthropoid vision is **trichromacy**, the ability to see color in the red–green range as well as the blue–yellow range used by most strepsirrhines. Anthropoids and a few other primates are the only mammals with this capability. The functional significance of color vision may be similar for most or all anthropoids, but the genetic and phenotypic foundations vary.

The leading functional explanation for color vision is the Fruit Foraging Hypothesis: trichromacy facilitates discovery and evaluation of fruit in daylight. Color is related to ripeness, size, hardness, type of flesh, amount of edible tissue, and the number of seeds. It can also be used to distinguish between species of differing nutritive value. However, not all fruits eaten by primates are conspicuously colored, and in some cases color is not correlated with nutritive value (Dominy 2004; Valenta and Melin 2012). Thus, additional or alternative factors may be important.

The Predator Detection Hypothesis stems from the observation that cats that hunt primates are generally yellowish in color. It proposes that these predators are more conspicuous to trichromats against a background of green foliage (Caine 2002; Pessoa et al. 2014). This hypothesis is difficult to test because attacks on primates are rarely observed.

Anthropoids are most commonly arboreal quadrupeds (Beard 2013), but they display wide variation on that theme. Some include suspensory locomotion in their repertoires, holding branches above the head during movement. Feet may also be used for suspension. Anthropoids generally have more precise digital control than other primates and groom mainly with their hands (Goosen 1987; Figure 11.1). The groomer parts the subject's hair with the fingers, makes a close visual inspection of the skin, and removes small objects with the fingers. These include parasites, intrusive material, and body products such as dead skin. The anthropoid approach is facilitated by greater manual dexterity than that of strepsirrhines, but it is probably also related to the fact that anthropoids have coarser hair.

Figure 11.1 As these monkeys demonstrate, anthropoids groom mostly with their hands instead of orally like strepsirrhines. © Rhett A. Butler / www.mongabay.com

A higher degree of sexual dimorphism in body mass and canine size typically distinguishes anthropoids from other primates (Plavcan 2001). These differences are generally attributed to sexual selection. More detailed analysis indicated that selection is stronger for canine size than body size, which suggests that canines are the more important factor in male–male competition (Thoren et al. 2006). Male teeth are up to four times taller than those of females (Plavcan and Ruff 2008). Relative to skull length and body mass, the canines of male anthropoids equal or exceed those of similarly sized carnivores. The difference is usually greater in the canines of the upper jaw, which are used for fighting. The nearest premolar in the mandible is elongated and hones the rear edge of the upper canine. A study of canine enamel formation supports the view that anthropoid dimorphism results from prolonged male canine growth (Guatelli-Steinberg et al. 2009).

Key and Ross (1999) used the wide disparity in body size to test a common assumption that female mammals have greater energetic costs than males due to gestation and lactation. They postulated that intense inter-male competition might incur comparable energetic costs. Their survey disclosed that male energetic costs are greater than those of females when males exceed female body size by 60 percent or more.

Kamilar and Pokempner (2008) investigated the relationship between sexual dimorphism and food with a survey of anthropoids (36 of 38 species). They found no significant relationship between body mass sexual dimorphism and male–female differences in the percentage of fruit or leaves in the diet. Differences in fauna consumption correlated negatively with body mass dimorphism. Dimorphism correlated positively with male–female differences in the percentage of time spent feeding.

A functional hypothesis relates the canine pattern to predator defense by males, which is a recurrent trait among anthropoids. This behavior should be distinguished from mobbing, which may involve females and even young. In male predator defense one or more adult males separate themselves from the group and aggressively confront a predator (usually a mammal but occasionally a raptor). Adolescent and even juvenile males may participate. The primates can drive away or even kill predators, but they may also be injured or killed themselves. In terms of ultimate causation the benefit is (known in some cases and plausible in others) that the males are defending their own or related offspring. Thus male predator defense can be explained by a combination of direct genetic benefit and kin selection. The cost is obviously low enough, especially when males join forces.

In contrast to other primates, anthropoids are capable of piloerection (Chaplin et al. 2014). Full-body MAP was re-established in their early evolution. One function, and perhaps the original adaptive reason, has been to maintain thermoregulation as the anthropoids expanded into cold temperate zones that included snowy winters in some cases. However, piloerection has become ritualized in the ethological sense. Occurring at times of high arousal, it signals the emotional state of an individual to others. Most importantly, it is part of the aggressive displays of many anthropoid species.

Anthropoid Sociality

The complexity of social grooming is related to the fact that anthropoids typically live in relatively large troops (Kappeler 2012), most of them cohesive, which entails diverse and dynamic relationships. Though partly hygienic, allogrooming is more of a social activity in anthropoids than in other primates (Goosen 1987). Social grooming is mostly between adults, is prevalent during rest periods, and constitutes a major part of social interaction in many species. It is often invited, the soliciting individual assuming a posture as if it were already being groomed. The inviting animal may, for example, sprawl in front of the other or sit with its head bowed.

Grooming is often reciprocal, with the two participants changing roles at some point in the bout. The duration of a bout depends partly on the relationship between the participants. Social grooming is most likely to take place between individuals of similar rank. A subordinate individual is likely to perform more grooming than a dominant one. Females typically do more grooming than males and in many species adult males receive more grooming than any other age/sex class, with the possible exception of infants. Grooming frequency increases during mating seasons and males groom females more frequently during the ovulation period. Grooming has played a major role in the development of biological market theorizing about primate behavior, but many primatologists emphasize psychological interpretations (e.g. stress reduction).

Conflict is prominent in many anthropoids and it has costs apart from the risk of injury during a fight (Arnold et al. 2011). The loser may be subject to further attacks by the winner and by others in the group. He or she may leave the group and the benefits that it provides. A damaged relationship may impair future cooperation that would be advantageous to both parties involved. These considerations probably explain the widespread phenomenon of reconciliation in which opponents engage in affiliative interaction with each other shortly after their conflict. Some apparent reconciliation has been observed in lemuriforms (Palagi and Norscia 2011; Palagi et al. 2008), but it is primarily an anthropoid pattern.

The Valuable Relationships Hypothesis holds that reconciliation is most likely to occur between individuals whose relationship benefits the evolutionary fitness of both. This is an adaptive explanation that implies ultimate causation. The Uncertainty-Reduction Hypothesis addresses proximate cause. It postulates that both parties to a conflict feel uncertainty about their relationship and that reconciliation relieves this mental state. Anxiety is expressed in **self-directed behavior** such as scratching.

The form of reconciliatory behavior varies. **Implicit reconciliation** consists of affiliative behaviors that regularly occur in other contexts. **Explicit reconciliation** uses conspicuous behaviors that rarely occur in other contexts. The latter seem to be ritualized in the ethological sense and may have evolved for this particular function. Considering all manifestations, reconciliation is now known in about 60 anthropoid species (Arnold et al. 2011).

Facial expression is an important aspect of anthropoid communication (Burrows 2008; Dobson 2009). Greater facial mobility than in strepsirrhines is a result of changes in the external olfactory system (their faces are "untethered"). Anthropoids also have more complex musculature in the area of the nose and upper lip. Body size is a significant factor in the degree of facial mobility; smaller species are subject to allometric constraint. When body size is controlled, group size emerges as a good predictor of facial mobility, which indicates significance for more extensive and complex communication.

The Social Signal Hypothesis is an alternative to environmental explanations for trichromacy. It is based on the fact that primate skin on the face and rump undergoes color changes that reflect emotions, threats, health, and sexual status. Changizi et al. (2006) elaborated on this idea with evidence that anthropoid trichromacy is nearly optimal for recognizing modulations in skin color, which led them to conclude that selection for the trait came from sociosexual factors. They asserted that monochromats and dichromats have hairy faces, while those of trichromats are bare.

Fernandez and Morris (2007) countered the argument with a phylogenetic reconstruction based on comparison of more than 200 anthropoid species. They concluded that trichromacy originated before red hair and skin and thus could not be selected by those physical traits.

However, they acknowledged the functional connection and proposed that trichromacy was a "preexisting bias" that favored the evolution of red sociosexual signals.

Sexual behavior in anthropoids is more flexible than that of strepsirrhines. Rather than being limited to a typically mammalian estrus cycle, anthropoids can and do copulate when ovulation is not taking place. Dixson (2012) argued that the term *estrus* should not be used in relation to anthropoids because it implies rigidity. He preferred the term *menstrual cycle*. An apparent behavioral effect is a greater frequency of **proceptivity**, female behavior that invites sexual approaches from males. Eye contact is a recurring feature of proceptivity and it may also be an aspect of copulatory interaction. A common female trait is sexual skin that may change color or swell in connection with the sexual cycle.

In a review of nonhuman primates Vasey (1995) defined "homosexual behavior" as genital contact and/or manipulation between individuals of the same sex. He distinguished between non-sexual and "partially sexual" interactions, defining the latter as non-aggressive series mounting, consorting, and/or directed genital stimulation. The most common form is mounting with pelvic thrusting that is often indistinguishable from heterosexual behavior. Genital fondling, oral-genital contact, and anogenital intromission also occur in some cases. Because "homosexual" has anthropomorphic connotations, many primatologists now replace it with **isosexual**.

Isosexual behavior occurs in more than 30 anthropoid species (Vasey 1995). At least 20 species display such behavior in the wild, so it cannot be considered an artifact of captivity. Frequencies are generally low, but these are not isolated or idiosyncratic acts. Isosexual behavior is frequent or occasional (as opposed to rare) in more than half of the species in the sample. In several cases the frequency of isosexual interactions approaches or surpasses that of heterosexual interactions. In most species that display isosexual behavior, it occurs in both males and females. It happens in all major types of primate social systems, though it seems to be more common in multi-male/multi-female troops. Comparable behavior occurs in a variety of non-primate mammalian species, but it has never been reported for strepsirrhines.

Anthropoid Life Cycle

The life history of anthropoids is (to an even greater extent than other primates) characterized by single births, long inter-birth intervals, and slow maturation. To maintain a stable population (and successful reproductive competition) parents must produce a sufficient number of offspring one at a time to replace themselves. The long period of immaturity requires parental care. This led to the hypothesis that in any species with these characteristics the parent with the greater burden of tending the young will live longer. The hypothesis was sustained by a survey of diverse anthropoid species (Allman et al. 1998).

Anthropoid Cognition

Cunningham and Janson (2007) argued that the cognitive requirements of primate foraging have been underestimated, and criticized over-emphasis on social explanations for anthropoid mental abilities. A variety of anthropoids can integrate multiple attributes of a feeding site. They can find their way to a site according to their knowledge of "what–where–how much". Monkeys, for instance, will direct their travel toward the location of more productive trees and may skip less productive ones along the way. This ability to integrate multiple attributes is associated with episodic memory or a mental process much like it.

Episodic memory "binds together where, what, and when information in a recollection of a personal event" (Cunningham and Janson 2007).

There is an important emotional component to this kind of behavior, namely the ability to delay gratification. This is especially apparent in cases where primates skip a fruit tree on their way to a more productive one. Experiments with chimpanzees and monkeys verify that diverse anthropoids are capable of patience. We have noted reverse contingency experiments (Chapter 2) in which primates of all kinds are able to inhibit a first response in order to get a better result. These choices, like those made in nature, may be retained in memory for months and up to a year. The capacity for inhibition implies **mental time travel**, the ability to anticipate the future (Humle and Fragaszy 2011).

Another way that anthropoids interact with the environment is the use of tools. Although chimpanzees are famous for extensive tool use, such behavior also occurs to varying degrees in monkey species. It has been observed in natural as well as experimental situations. Tool-using performances by wild primates are often singular occurrences. For example, one female monkey used a leaf to wipe a sticky substance from her fingers (Worch 2001). Such anecdotes suggest that the mental capacities of many species are greater than they reveal in everyday life. Further weight is added by observations of similar performances in a number of species. Furthermore, some monkeys use tools regularly and selectively for purposes such as cracking the shells of marine animals (e.g. Gumert et al. 2009).

Anthropoids' greater ability to manipulate the physical environment is partly due to the structure of the hand in many species. However, it is also a result of differences in the brain. From a comparative study Kaas (2004) concluded that anthropoids are distinguished from strepsirrhines by more **serial processing**. Anatomically, he found four distinct fields in the anterior parietal cortex, and more areas in the posterior **parietal cortex**.

According to the Social Brain Hypothesis, large brains have evolved in response to social and ecological conflicts that arise from group living. Anthropoids differ from other vertebrates in that group size strongly affects brain size. An extension of the hypothesis holds that the anthropoids are a taxon in which the complexity of negotiated relationship has been generalized from pair bonding to all social partners. This accounts for larger brains and greater social intelligence in anthropoids (Schultz and Dunbar 2007).

Primates, like many other social mammals, have learned behavioral traditions. There is little evidence for traditions among strepsirrhines, while increasingly complex traditions have been discovered in some anthropoids. A growing number of primatologists have taken to calling these primate traditions "culture" and have established the subfield of **cultural primatology** (McGrew 2004). Application of the term "culture" to nonhuman primates is simply a matter of definition. Cultural primatologists use the word to denote relatively simple material and social behaviors, e.g. tool use and grooming postures. To anthropologists, "culture" is likely to denote a complex phenomenon based on language, including, for example, kinship structures and religious beliefs. No matter what terminology we use, these differences will remain. Extension of "culture" to include primates means we have to replace current usage with new words to make the distinction (e.g. linguistic culture vs. nonlinguistic culture). Then humans and other primates will still be profoundly different. Nonhuman primates cannot make a stone axe into a symbol of male dominance, forbid marriage between parallel cousins, or cast a spell to hold evil spirits at bay. We will use the word culture in its broad primatological sense because this usage is so popular in the discipline; however, you should keep the above reservations in mind.

Summary

Anthropoids, the third major primate lineage, contrast with strepsirrhines in being diurnal and sexually dimorphic. Most live in troops, where social grooming is typically important and reconciliation after conflicts is a recurrent pattern. Facial communication is more elaborate and important than in strepsirrhines. Heterosexual activities are not limited to an estrus period and isosexual behavior occurs. Life history revolves around single births and very slow maturation. Anthropoids exceed strepsirrhines in measures of cognitive ability, including tool use.

Sources

Basic Bibliography

Beard, K.C. (2013). Anthropoid origins. In D. Begun (Ed.), *A Companion to paleoanthropology* (pp. 358–375). Malden, MA: Wiley-Blackwell.
Campbell, C.J., et al. (Eds.) (2011). *Primates in perspective* (2nd ed.). NY: OUP. [Chapters 7–20]
Mitani, J., et al. (Eds.) (2012). *The evolution of primate societies*. Chicago: UCP. [Chapters 3–6]
Ross, C.F., & Kay, R.F. (Eds.) (2004). *Anthropoid origins* (2nd ed.). NY: Springer.

Additional References

Allman, J. et al. (1998). Parenting and survival in anthropoid primates: caretakers live longer. *PNAS*, *95*, 6866–6869.
Arnold, K., et al. (2011). Postconflict reconciliation. In C.J. Campbell et al. (Eds.), *Primates in perspective*, 2nd ed. (pp. 608–625). NY: OUP.
Beard, K.C. (2013). Anthropoid origins. In D. Begun (Ed.), *A Companion to paleoanthropology* (pp. 358–375). Malden, MA: Wiley-Blackwell.
Burrows, A.M. (2008). The facial expression musculature in primates and its evolutionary significance. *BioEssays*, *30*, 212–225.
Caine, N.G. (2002). Seeing red: consequences of individual differences in color vision in callitrichid primates. In L.E. Miller (Ed.). *Eat or be eaten: predator sensitive foraging among primates*. NY: CUP.
Changizi, M.A., et al. (2006). Bare skin, blood and the evolution of primate colour vision. *BL*, *2*, 217–221.
Chaplin, G., et al. (2014). The role of piloerection in primate thermoregulation. *FP*, *85*, 1–17.
Cunningham, E., & Janson, C. (2007). A socioecological perspective on primate cognition, past and present. *AC*, *10*, 273–281.
Dixson, A.F. (2012). *Primate sexuality*. NY: OUP.
Dobson, S. (2009). Allometry of facial mobility in anthropoid primates: implications for the evolution of facial expression. *AJPA*, *138*, 70–81.
Dominy, N.J. (2004). Fruits, fingers, and fermentation: the sensory cues available to foraging primates. *ICB*, *44*, 295–303.
Fernandez, A.A., & Morris, M.R. (2007). Sexual selection and trichromatic color vision in primates: statistical support for the pre-existing bias hypothesis. *AN*, *170*, 10–20.
Fleagle, J.G. (2013). *Primate adaptation and evolution* (3d ed.). NY: Elsevier Science.
Fleagle, J.G., & Reed, K.E. (1996). Comparing primate communities: a multivariate approach. *JHE*, *30*, 489–510.
Goosen, C. (1987). Social grooming in primates. In G. Mitchell & M.G. Erwin (Eds.), *Comparative primate biology, vol. 2B* (pp. 107–131). NY: Liss.
Guatelli-Steinberg, D., et al. (2009). Sex differences in anthropoid mandibular canine lateral enamel formation. *AJPA*, *140*, 216–233.

Gumert, M.D., et al. (2009). The physical characteristics and usage patterns of stone axe and pounding hammers used by long-tailed macaques in the Andaman Sea region of Thailand. *AJP*, *71*, 594–608.

Hall, M.I., et al. (2012). Eye shape and the nocturnal bottleneck of mammals. *PRSB*, *279*, 4962–4968.

Hiramatsu, C., et al. (2008). Importance of achromatic contrast in short-range fruit foraging of primates. *PLONE*, *3*, e3356.

Humle, T., & Fragaszy, D.M. (2011). Tool use and cognition in primates. In C.J. Campbell et al. (Eds.). *Primates in perspective* (2nd ed.) (pp. 637–651). NY: OUP.

Kaas, J.H. (2004). The evolution of the large, complex sensorimotor systems of anthropoid primates. *IJCP*, *17*, 34–52.

Kamilar, J.M., & Pokempner, A.A. (2008). Does body mass dimorphism increase male–female dietary niche separation? A comparative study of primates. *Behaviour*, *145*, 1211–1234.

Kappeler, P.M. (2012). The behavioral ecology of strepsirrhines and tarsiers. In J.C. Mitani et al. (Eds.), *The evolution of primate societies* (pp. 17–42). Chicago: UCP.

Key, C., & Ross, C. (1999). Sex differences in energy expenditure in non-human primates. *PRSB*, *266*, 2479–2485.

McGrew, W.C. (2004). *The cultured chimpanzee*. NY: CUP.

Melin, A.D., et al. (2013). Food search through the eyes of a monkey: a functional substitution approach for assessing the ecology of primate color vision. *Vision Research*, *86*, 87–96.

Melin, A.D., et al. (2014). The behavioural ecology of colour vision: considering fruit conspicuity, detection distance and dietary importance. *IJP*, *35*, 258–287.

Palagi, E., & Norscia, I. (2011). Scratching around stress: hierarchy and reconciliation make the difference in wild brown lemurs (*Eulemur fulvus*). *Stress*, *14*, 93–97.

Palagi, E., et al. (2008). Peacemaking on treetops: first evidence of reconciliation from a wild prosimian (*Propithecus verreauxi*). *AB*, *76*, 737–747.

Pessoa, D.M.A., et al. (2014). The adaptive value of primate color vision for predator detection. *AJP*, *76*, 721–729.

Plavcan, (2001). Sexual dimorphism in primate evolution. *AJPA*, *Suppl*. 33, 25–53.

Plavcan, J.M., & Ruff, C.B. (2008). Canine size, shape, and bending strength in primates and carnivores. *AJPA*, *136*, 65–84.

Shultz, S., & Dunbar, R.I.M. (2007). The evolution of the social brain: anthropoid primates contrast with other vertebrates. *PRSB*, *274*, 2429–2436.

Tan, C.L., et al. (2013). Into the night: camera traps reveal nocturnal activity in a presumptive diurnal primate, *Rhinopithecus brelichi*. *Primates*, *54*, 1–6.

Thoren, S., et al. (2006). Phylogenetic analyses of dimorphism in primates: evidence for stronger selection on canine size than on body size. *AJPA*, *130*, 50–59.

Valenta, K. & Melin, A.M. (2012). Protein limitation explains variation in primate color vision phenotypes. In M.D. Garcia (Ed.), *Zoology*. InTech, ISBN: 978-953-51-0360-8.

Vasey, P.L. (1995). Homosexual behavior in primates: a review of evidence and theory. *IJP*, *16*, 173–204.

Worch, E.A. (2001). Simple tool use by a red-tailed monkey (*Cercopithecus ascanius*) in Kibale Forest, Uganda. *FP*, 72, 304–306.

12 The Platyrrhine Infraorder

New World Monkeys

Lanky spider monkeys with sleek black coats are feeding in a fruit-laden tree. They use their long slender limbs to maintain a variety of positions as they enjoy the succulent flesh of the ripe fruits. Below them sits a bushy-furred uakari with a bald red head. Its main interest in the fruits is crushing and swallowing the seeds.

Platyrrhines (Figure 12.1) are the anthropoids of the Americas, usually called New World monkeys (Box 12.1). They are small to medium on the primate size scale, weighing between 100g and 10kg (see Table 12.1 online for physical dimensions of all platyrrhines). Platyrrhines typically have slender palms and long fingers that enhance arboreal locomotion. They are distributed through almost all of South America, and some extend through Central America and into part of Mexico.

Figure 12.1 Squirrel monkeys constitute one of the most widespread platyrrhine genera.
© temis via iStock

Box 12.1 Platyrrhine Taxonomy

The highest taxon that distinguishes New World monkeys is the Platyrrhini.

Diagnostic features include the broad flat shape of the external nose, relatively late fusion of cranial sutures, separation of the frontal and sphenoid bones by the joining of the parietal and zygomatic bones, and the loss of hypoconulids on the first two molar teeth in each quadrant. The platyrrhines are also distinguished by retention of primitive features, including three premolars in each quadrant and a tympanic ring that does not extend into a bony tube.

The platyrrhine taxon ranks high by virtue of the number of included taxa and the age of the clade. Hodgson et al. (2009), based on mtDNA, placed the last common ancestor at about 20 million years ago. Perez et al. (2013) used two molecular approaches to arrive at figures between about 20 and 30 million years ago. Finstermeier et al. (2013) used mitochondrial genomes to distinguish a separation of platyrrhines and catarrhines at about 45 million years ago (which they considered to be in line with other studies).

Platyrrhine Behavior

Platyrrhine Ecology

Platyrrhines display their greatest numbers and diversity in the rainforests of **Amazonia**, the basins of the Amazon and Orinoco rivers. Some genera exist in semi-deciduous forests and dry woodlands east and south of Amazonia. The altitude range runs from sea level to almost 3000m (see Table 12.2 online for ecological characteristics of all platyrrhines).

Arboreality is ubiquitous among platyrrhines. Some species come to the ground occasionally, but few spend much time there. Common reasons for these brief visits are access to water or mineral licks, foraging for insects, crossing gaps in the forest, escape from predators, and play. An additional reason is geophagy. The fact that geophagy brings these highly arboreal primates to the ground underlines its importance to primate health (Ferrari 2009; Ferrari et al. 2008).

Fruit and insects are the main foods. The seven primate genera in an ecological community all ate fruit, despite the other differences among them (Ramdarshan et al. 2011). Though leaves are abundant in the arboreal setting of platyrrhines, folivory is rare compared to tropical primates elsewhere in the world. Ganzhorn et al. (2009) provided evidence that New World fruits contain enough protein to satisfy primate needs. They inferred that selection pressure for specialization in harder-to-digest proteinaceous foods such as leaves would have been lower than elsewhere. According to an alternative hypothesis, fruiting and leafing are in phase with one another so that fruit (the preferred food) and leaves are always equally available. However, fieldwork by Heymann (2001) showed that the postulated synchronization did not occur. His speculation about different explanations, such as plant chemistry, was borne out by the more recent research.

Valenta and Melin (2012) proposed that figs are a preferred and limiting resource for platyrrhines, rather than the fallback food that many have assumed. Figs in the Americas are more diverse and abundant than those in the Old World. They contain moderate to high levels of protein compared to other fruits (about 4 to 6 percent by dry weight), supplemented by insects and larvae that they contain. The overall nutritional value of the figs seems to sustain some frugivores without other foods (Wendeln et al. 2000). Some monkeys curtailed foraging for insects, their main source of protein, on days when they consumed figs (Parr et al. 2011).

With regard to color vision platyrrhines display both dichromatic and trichromatic phenotypes, differentiated by sex. Functional research indicates that both can be advantageous in various circumstances (Jacobs 2009). A molecular study demonstrated that the two forms of color vision in platyrrhines are maintained by balancing selection (Hiwatashi et al. 2010). Several hypotheses have been advanced to explain balancing selection on platyrrhine color vision (Melin et al. 2013), but only two have substantial support. One is the Niche Divergence Hypothesis, which holds that different phenotypes have different advantages. The other is Heterozygote Advantage, which postulates that trichromatic individuals have a fitness advantage over other phenotypes because red–green chromatic discrimination is more useful than either of the homozygous phenotypes. This requires further explanation, which the Fruit Foraging Hypothesis provides.

The Fruit Foraging Hypothesis has been questioned on the grounds that many fruits are equally available to trichromats and dichromats because they can be distinguished in the blue–yellow range or by variation in luminance (Dominy et al. 2001; Hiramatsu et al. 2008). Furthermore, less conspicuous fruits are usually more odiferous and most platyrrhines retain a strong olfactory sense. Field studies of two species found that they rely more on their sense of smell when feeding on greenish fruits (Hiramatsu et al. 2009).

Melin et al. (2014) approached the problem by using wild capuchin monkeys (*Cebus capucinus*) to model the ability of dichromatic and trichromatic phenotypes to discriminate fruits from leaves. This task is part of the long-distance search for food patches in the tropical forest. The trichromatic models could correctly distinguish about three-fourths of the fruits in the monkey diet, while the dichromatic phenotypes could discriminate less than one-third of the fruits. The pattern held for separate analysis of the most heavily consumed diet items, preferred foods, and seasonally critical species. These results support the hypothesis that long-distance detection of fruit patches selects for trichromacy in platyrrhines.

The researchers made two further points. First, detection of fruit species with small patch sizes may provide an important advantage to trichromats because these resources can provide a high reward. Second, the fact that trichromats readily detected greenish-brown fruits suggests that these resources may have been more important to the evolution of primate color vision than is generally realized.

Valenta and Melin (2012) hypothesized that figs are responsible for the evolution of trichromatic color vision. According to their Protein Limitation Hypothesis, trichromacy is a response to the need for protein and the association of protein with the color red, which the platyrrhines see in reddish figs found in the New World.

The bottom line in evolutionary theory is differential reproduction, so Fedigan et al. (2014) tested the evolutionary significance of color vision with a study of female fitness in wild capuchins. Using long-term data (collected over 26 years), they found that trichromats had no advantage in three fitness measures: fertility rates, offspring survival, and maternal survival. This finding suggested that a selective mechanism other than heterozygote advantage is operating to maintain color vision polymorphism. The researchers proposed that field tests focus on alternative mechanisms of balancing selection that might explain opsin polymorphism: niche-divergence, frequency-dependence, and mutual benefit of association.

Research on the advantages of dichromatic vision seems to support the Niche Divergence Hypothesis, which holds that the different phenotypes are advantageous in different situations or suited to distinct tasks (Fedigan et al. 2014). Dichromats are superior in distinguishing camouflaged objects from background, probably because color does not distract them from perception of variation in texture or luminance. Dichromats may have an advantage when foraging in shade or low light (Caine et al. 2010; Perini et al. 2009) or for invertebrates that are camouflaged (Melin et al. 2007).

In the evolution of their sensory system the platyrrhines retained the vomeronasal organ that plays such an important role in the strepsirrhine olfactory sense. It is ubiquitous among living species. However, the organ varies in structure and it is reduced or even vestigial in some species. The sense of smell has declined somewhat in platyrrhines, but continues to be important. This may have more to do with social than ecological functions.

New World monkeys respond to raptors with simple, stereotyped calls—short, sharp whistles that seem to be innate (Ferrari 2009). Functionally, these seem to be true alarm calls that warn conspecifics of danger. Hearing the call, an individual either freezes or immediately takes cover. Rapid stereotyped responses are effective because raptor attacks always take the same form, a sudden swoop, and the only effective defenses are vigilance and avoidance. The reaction occurs in infants as soon as they can move independently. This suggests an innate basis for the response, though it may be reinforced by adult behavior.

Platyrrhine alarm calls have been less well studied than those of Old World monkeys, but the evidence indicates that they have similar characteristics (Cäsar and Zuberbühler 2012). In a number of cases they are functionally referential in that they designate particular environmental phenomena, especially raptors. They may also contain information about location. These messages are conveyed by different devices in different species, e.g. acoustically distinct calls or the assembly of calls into different sequences.

Platyrrhine arboreality somewhat reduces danger from terrestrial carnivores (Ferrari 2009). Nevertheless, they are hunted by at least seven different species of New World cats (Figure 12.2). The largest, pumas and jaguars, are formidable but smaller cats may be a greater danger. Ocelots are good climbers, and an analysis of ocelot scats indicated that primates were their third most favored prey (Bianchi and Mendes 2007). Another small carnivore that kills primates is the tayra (Bezerra et al. 2009). These weasel-like animals, typically weighing about 4.5kg, are agile in the trees as well as on the ground and can be active during day or night.

Among all primates, the ecological communities of the New World display the least diversity (Fleagle 2013; Fleagle and Reed 1996). Most species are diurnal, medium-sized, frugivorous quadrupeds. Only one genus is nocturnal and one tends toward folivory. One displays a significant amount of suspensory behavior, but is still largely quadrupedal. We have noted explanations for dietary limitation in terms of the characteristics of fruits and leaves. Another hypothesis suggested that competition with sloths limited the evolution of platyrrhine leaf consumption. The abundance of opossums may have restricted insectivory and also nocturnality. Competition among the primates themselves may have limited diversity, since competitive exclusion seems to be strongest among frugivores (Kamilar and Ledogar 2011). As to small size, the fossil record indicates that this is a recent development but does not indicate whether this is due to natural selection or some sort of historical cause.

Figure 12.2 The jaguar and various smaller cats prey on ceboids. © Rhett A. Butler/ www.mongabay.com

Platyrrhine Sociality

Most smaller-bodied platyrrhine species live in pair-groups while the larger ones live in troops (see Table 12.3 online for social features of all platyrrhines). However, greater flexibility in social grouping patterns and more diversity of dispersal patterns seem to distinguish platyrrhines from Old World primates. Female emigration takes place in most species, but the extent of male emigration varies.

Intergroup conflicts are recurrent and perhaps typical. Some seem to be territorial in that they occur at well-defined boundaries. Others may be for the defense of specific resources, including mates. There is a recurring pattern of invasion by outsiders with the intent of evicting same-sex residents and taking over their mates. Some of these takeovers lead to infanticides.

Platyrrhines are more like lorisids than they are like Old World anthropoids in the extent to which they make use of olfactory communication (Heymann 2006). The mammalian vomeronasal organ has been retained, though its presence and function vary (Smith et al. 2011), and most platyrrhines have sternal and anogenital scent glands. They also use urine (most commonly), saliva, and even nasal mucous to leave olfactory information on substrates or on the bodies of conspecifics. The importance of such information is indicated by the frequency and intensity with which the primates investigate these marks. The significance of olfactory communication throughout the platyrrhines can be attributed to the fact that they are all arboreal and most have historically lived in relatively dense rainforests (although many of their habitats are now impoverished).

Numerous functions have been attributed to platyrrhine scent marking, deriving from hypotheses about other mammals. These include the expression of territoriality, dominance, and sexuality. Overall there seems to be a trend toward interpretation of scent marking in terms of mate attraction and mating competition (Heymann 2006). However, some criticize that view and present new evidence to sustain older hypotheses, such as territoriality (Roberts 2012). Two perspectives seem important here. One is that many of the hypotheses are not mutually exclusive. The same scent mark by a male may convey sexual attraction to a female and a warning to other males. The other perspective is that functions may vary from one species to another.

Urine washing is a special kind of scent marking in which the monkey who is secreting the odiferous substance places it on itself. Urine washing can be defined as the deliberate deposition of urine on an individual's own body, extremities, or tail. Functional interpretations, especially sexual and territorial, overlap with those for scent marking. However, distinctive hypotheses include thermoregulation, laying down trails for spatial orientation, and improvement of grip.

The advance of technology in behavioral research is illustrated by an experiment in which fMRI was used to monitor the brain processes of female monkeys that were presented with samples of male urine (Phillips et al. 2011). Urine from adult males elicited responses that were absent when the urine came from juvenile males. The researchers view this as support for the Sexual Signaling Hypothesis.

Anointing, a recurrent behavior, consists of rubbing an odiferous foreign substance on the body, and it can be interpreted as a form of social communication (Lynch Alfaro et al. 2012; Laska et al. 2007). It occurs in one or two species of lemurs, but is more common in platyrrhines. Anointing substances come from crushed leaves or arthropods (ants and millipedes) and are often mixed with saliva. Some captive platyrrhines anoint themselves with substances such as garlic and onion. Weldon et al. (2003) elicited anointing from captive animals by presenting them with chemicals like those extracted by rubbing with millipedes in the wild. Control substances did not have the same effect, indicating the importance of the olfactory stimulus. However, effective substances vary from one species to another, even in the same genus, which suggests a cultural factor.

The behavior may have several different functions that vary from one species to another. As with urine washing, the application of fluid to the body suggests the possibility of thermoregulation. Some anointing compounds possess medicinal properties or can repel insects. However, interpretations such as thermoregulation, insect repellent, and medicinal value are rendered doubtful by: (1) application to only certain parts of the body; and (2) the fact that relatively few individuals engage in the behavior. The social status of these individuals suggests that the behavior functions as social communication. In a troop of spider monkeys, for example, only two adult males anointed themselves and one was the group alpha and father of the only infant (Laska et al. 2007).

Platyrrhine Life Cycle

Perinatal life history traits in platyrrhines are expressed as a function of maternal body size, as in other mammals. The typical gestation length of about five months correlates roughly with the relatively small maternal body masses of most New World species (see Table 12.4 online for reproductive parameters of all platyrrhines). The neonate weighs about a tenth as much as the mother, which is substantial. Mothers tend to neglect altricial young, so relatively large neonatal size is important to survival. Most species concentrate their births before the peak of food availability in order to support peak lactation or to wean before the lean season begins (Di Biteti and Janson 2000).

Dorsal carriage of infants is ubiquitous among platyrrhines, but the pattern is modified by relative birth weight (Nakamichi and Yamada 2009). In smaller species birth weight is a relatively high percentage of the mother's weight and infants are typically carried dorsally, even on the first day after birth. In larger bodied platyrrhines, with relatively low birth weight, infants are carried ventrally at first and then on the back (see Table 12.5 online for life cycle stages of all platyrrhines). This supported the researchers' hypothesis that dorsal carriage is better for managing relatively heavy infants.

Platyrrhine Cognition

An important feature of primate evolution is expansion of the brain. The Expensive Tissue Hypothesis (ETH) proposed that evolution of an energetically expensive large brain must be compensated by reduction of other expensive tissue such as the digestive tract. This reduction requires the adoption of a high-quality diet because food intake is reduced. Thus diet quality constrains brain growth.

According to Allen and Kay (2012), platyrrhines provide a good test for the hypothesis because they are so variably encephalized (more than any other primates except for the hominoids). They used diet quality as a proxy for gut size due to limited data for the latter. Across the platyrrhines, they found no correlation between diet quality and encephalization. Furthermore, they performed a phylogenetic reconstruction that indicated the existence of high-quality diets in ancestral platyrrhines before brain expansion began. They concluded that the ETH does not apply to platyrrhine brain expansion because the process was not constrained by diet. It must have begun when some positive selective factor or factors entered the situation.

Van Woerden et al. (2014) performed a comparative study in which they tested the Expensive Tissue Hypothesis and the Cognitive Buffering Hypothesis. Introducing a time dimension, they reviewed 21 platyrrhine species for diet composition and seasonality. Monthly variation in diet composition (representing energy intake) was negatively correlated with brain size, indicating support for the Expensive Tissue Hypothesis. The researchers also found support for cognitive buffering when they considered the difference between habitat seasonality and the experienced variation in net energy intake. These results were consistent with their work on catarrhines and lemuriforms.

Multiple explanations for encephalization are likely because it took place in different lineages. This is apparent from the fact that three of four major clades encompassed by a comparative study displayed a wide range of relative brain sizes (Hartwig et al. 2011). Thus brain size increased independently within each. The researchers noted that these episodes of encephalization were associated with varied ecological and social problems. Smaller brains were associated with social monogamy or other small-group adaptations within each of the clades.

Summary

Platyrrhines occur from Mexico through South America. They are highly arboreal and the prominence of loud calls and olfactory communication are probably related to dense forest habitats. Grouping and dispersal are typically flexible. Relatively large neonates affect reproductive patterns. Encephalization is associated with varied social and ecological challenges.

Sources

Basic Bibliography

Campbell, C.J., et al. (Eds.) (2011). *Primates in perspective* (2nd ed.). NY: OUP. [Chapters 7 through 11]

Defler, T.R. (2004). *Primates of Colombia.* Conservation International.

Fernandez-Duque, E., et al. (2012). The behavior, ecology, and social evolution of New World monkeys. In J.C. Mitani et al. (Eds.), *The evolution of primate societies* (pp. 43–64). Chicago: UCP.

Garber, P.A., et al. (Eds.) (2009). *South American primates.* NY: Springer.

Additional References

Allen, K.L., & Kay, R.F. (2012). Dietary quality and encephalization in platyrrhine primates. *PRSB, 279,* 715–721.

Aristide, L., et al. (2015). Modeling lineage and phenotypic diversification in the New World monkey (Platyrrhini, Primates) radiation. *MP&E, 82B,* 375–385.

Bezerra, B.M., et al. (2009). Predation by the tayra on the common marmoset and the pale-throated three-toed sloth. *Journal of Ethology, 27,* 91–96.

Bianchi, R.C., & Mendes, S.L. (2007). Ocelot (*Leopardus pardalis*) predation on primates in Caratinga Biological Station, Southeast Brazil. *AJP,* 69, 1173–1178.

Buchanan-Smith, H.M. (2005). Recent advances in color vision research. *AJP, 67,* 393–398.

Caine, N.G., et al. (2010). A foraging advantage for dichromatic marmosets (*Callithrix geoffroyi*) at low light intensity. *BL,* DOI: 10.1098/rsbl.2009.0591.

Cäsar, C., & Zuberbühler, K. (2012). Referential alarm calling behaviour in New World primates. *Current Zoology, 58,* 680–697.

Di Biteti, M.S., & Janson, C.H. (2000). When will the stork arrive? Patterns of birth seasonality in Neotropical primates. *AJP, 50,* 109–130.

Dominy, N.J., et al. (2001). The sensory ecology of primate food perception. *EA, 10,* 171–186.

Fedigan, L.M., et al. (2014). The heterozygote superiority hypothesis for polymorphic color vision is not supported by long-term fitness data from wild neotropical monkeys. *PLONE,* 9, e84872.

Ferrari, S.F. (2009). Predation risk and antipredator strategies. In P.A. Garber et al. (Eds.), *South American primates* (pp. 251–277). NY: Springer.

Ferrari, S.F. et al. (2008). Geophagy in New World monkeys (Platyrrhini). *FP, 79,* 402–415.

Finstermeier, K., et al. (2013). A mitogenomic phylogeny of living primates. *PLONE,* 8, e69504.

Fleagle, J.G. (2013). *Primate adaptation and evolution,* 3rd ed. Amsterdam: Academic Press.

Fleagle, J.G., & Reed, K.E. (1996). Comparing primate communities: a multivariate approach. *JHE, 30,* 489–510.

Ganzhorn J.U., et al. (2009). Possible fruit protein effects on primate communities in Madagascar and the Neotropics. *PLONE, 4,* e8253.

Hartwig, W.G., et al. (2011). Relative brain size, gut size, and evolution in New World monkeys. *AR, 294,* 2207–2221.

Hawes, J.E., & Peres, C.A. (2013). Ecological correlates of trophic status and frugivory in neotropical primates. *Oikos, 123,* 365–377.

Heymann, E.W. (2001). Can phenology explain the scarcity of folivory in New World primates? *AJP, 55,* 171–175.

Heymann, E.W. (2006). Scent marking strategies of New World primates. *AJP,* 68, 650–661.

Hiramatsu, C., et al. (2008). Importance of achromatic contrast in short-range fruit foraging of primates. *PLONE, 3,* e3356.

Hiramatsu, C., et al. (2009). Interplay of olfaction and vision in fruit foraging of spider monkeys. *AB, 77,* 1421–1426.

Hiwatashi, T., et al. (2010). An explicit signature of balancing selection for color-vision variation in New World monkeys. *Molecular Biology and Evolution*, *27*, 453–464.

Hodgson, J.A., et al. (2009). Successive radiations, not stasis, in the South American primate fauna. *PNAS*, *106*, 5534–5539.

Jacobs, G.H. (2009). Evolution of colour vision in mammals. *PTRSB*, *364*, 2957–2967.

Kamilar, J.M., & Ledogar, J.A. 2011. Species co-occurrence patterns and dietary resource competition in primates. *AJPA*, *144*, 131–139.

Kay, R.F. (2015). Biogeography in deep time – What do phylogenetics, geology, and paleoclimate tell us about early platyrrhine evolution? *MP&E*, *82B*, 358–374.

Laska, M., et al. (2007). Self-anointing behavior in free-ranging spider monkeys (*Ateles geoffroyi*) in Mexico. *Primates*, *48*, 160–163.

Lynch Alfaro, J.W., et al. (2012). Anointing variation across wild capuchin populations. *AJP*, *73*, 1–16.

Maiolino, S., et al. (2011). Morphological correlates of the grooming claw in distal phalanges of platyrrhines and other primates. *AR*, *29*, 1975–1990.

Melin, A.D., et al. (2007). Effects of colour vision phenotype on insect capture by a free-ranging population of white-faced capuchins, *Cebus capucinus*. *AB*, *73*, 205–214.

Melin, A.D., et al. (2010). Can color vision variation explain sex differences in invertebrate foraging by capuchin monkeys? *Current Zoology*, 56, 300–312.

Melin, A.D., et al. (2013). Food search through the eyes of a monkey: a functional substitution approach for assessing the ecology of primate color vision. *Vision Research*, 86, 87–96.

Melin, A.D., et al. (2014). The behavioural ecology of colour vision: considering fruit conspicuity, detection distance and dietary importance. *IJP*, *35*, 258–287.

Nakamichi, M., & Yamada, K. (2009). Distribution of dorsal carriage among simians. *Primates*, *50*, 153–168.

Palminteri, S., et al. (2011). Regional-scale heterogeneity in primate community structure at multiple undisturbed forest sites across south-eastern Peru. *JTE*, *27*, 181–194.

Parr, N.A, et al. (2011). Figs are more than fallback foods: *IJZ*, *vol. 2011*, Article ID 967274, 10 pages, 2011. doi:10.1155/2011/967274.

Perez, S.I., et al. (2012). Species tree estimation for a deep phylogenetic divergence in the New World monkeys (Primates: Platyrrhini). *MP&E*, 65, 621–630.

Perez, S.I., et al. (2013). Divergence times and the evolutionary radiation of New World monkeys (Platyrrhini, Primates). *PLONE*, 8, e68029.

Perini, E.S., et al. (2009). Detection of fruit by the *Cerrado*'s marmoset (*Callithrix penicillata*): modeling color signals for different background scenarios and ambient light intensities. *Journal of Experimental Zoology Part A: Ecological Genetics and Physiology*, *311*, 289–302.

Phillips, K.A., et al. (2011). Why do capuchin monkeys urine wash? An experimental test of the sexual communication hypothesis using fMRI. *AJP*, *73*, 578–584.

Ramdarshan, A., et al. (2011). The primate community of Cachoeira (Brazilian Amazonia): a model to decipher ecological partitioning among extinct species. *PLONE*, 6, e27392.

Roberts, S.C. (2012). On the relationship between scent-marking and territoriality in callitrichid primates. *IJP*, *33*, 749–761.

Rosenberger, A.L. (2010). Platyrrhines, PAUP, parallelism, and the long lineage hypothesis: a reply to Kay et al. (2008). *JHE*, 59, 214–217.

Rosenberger, A.L. (2011). Evolutionary morphology, platyrrhine evolution, and systematics. *AR*, *294*, 1955–1974.

Rosenberger, A.L. (2012). New World monkey nightmares. *AJP*, *74*, 692–695.

Rosenberger, A.L., et al. (2011). The making of platyrrhine semi-folivores: models for the evolution of folivory in primates. *AR*, *294*, 2112–2130.

Santana, S.E., et al. (2012). Adaptive evolution of facial colour patterns in Neotropical primates. *PRSB*, *279*, 2204–2211.

Schneider, H., & Sampaio, I. (2013). The systematics and evolution of New World primates – a review. *MP&E, 82B*, 348–357.

Schrago, C.G., et al. (2013). Combining fossil and molecular data to date the diversification of New World primates. *JEB*, doi:10.1111/jeb.12237.

Smith, T.D., et al. (2011). The vomeronasal organ of New World monkeys (Platyrrhini). *AR, 294*, 2158–2178.

Valenta, K., & Melin, A.D. (2012). Protein limitation explains variation in primate color vision phenotypes. In M.D. Garcia (Ed.), *Zoology*. InTech, ISBN: 978-953-51-0360-8.

Van Woerden, J.T., et al. (2014). Seasonality of diet composition is related to brain size in New World monkeys. *AJPA, 154*, 628–632.

Villanea, F.A., et al. (2012). ASPM and the evolution of cerebral cortical size in a community of New World monkeys. *PLONE 7*, e44928.

Weldon, P.J., et al. (2003). Benzoquinones from millipedes deter mosquitoes and elicit self-anointing in capuchin monkeys [*Cebus* spp.]. *Naturwissenschaften*, 90, 301–304.

Wendeln, M.C., et al. (2000). Nutritional values of 14 fig species and bat feeding preferences in Panama. *Biotropica, 32*, 489–501.

Wildman, D., et al. (2009). A fully resolved genus level phylogeny of neotropical primates (Platyrrhini). *MP&E, 53*, 694–702.

Youlatos, D., & Meldrum, J. (2011). Locomotor diversification in New World monkeys. *AR, 294*, 1991–2012.

13 Platyrrhine Variation

Atelids and Pitheciids

The tourists are distracted from their pina coladas by a loud, pulsing, grinding sound that seems to last forever. They look up into the trees and see a large brown monkey with his mouth rounded and his throat pouch inflated.

This chapter describes one platyrrhine family that has been well known for a long time (atelids) and one that is only now becoming well known (the pitheciids). Variation within the families is such that generalizations are unusually difficult. After brief introductions to the families, we will consider them mainly in terms of their constituent genera.

Atelid Behavior

Atelids (including howler and spider monkeys) have muscular **prehensile tails** that can support the full weight of the hanging body while the monkey feeds. The tail is elongated and has a hairless pad on the ventral tip with specialized tactile receptors like those in primate hands. The tip can grasp somewhat like a hand, so the tail sometimes provides additional support during arboreal locomotion. Howler monkeys differ substantially from the spider monkey group of genera (Box 13.1).

Box 13.1 Atelid Taxonomy

Atelidae are large (5–12kg) frugivore–folivores with long, strong prehensile tails. The lumbar region is shortened, which probably facilitates suspensory behavior.

In addition to *Alouatta* (howlers) and *Ateles* (spider monkeys), the Atelidae include *Brachyteles* (muriquis) and two genera of woolly monkeys, *Lagothrix* and *Oreonax*. Howler dentition is shaped by the degree of folivory in the genus, and the large hyoid bone supports the vocal apparatus. Together these features give howlers a large facial skeleton compared to other New World monkeys. The frugivorous spider monkeys have small and rounded heads with a flat face. Their long arms allow them to brachiate. Woolly monkeys are believed to retain more ancestral traits than the other atelids, including thumbs. Muriquis are called woolly spider monkeys because they have the robust build and thick hair of the woolly monkeys along with the long limbs, absence of thumb, and suspensory behavior that characterize spider monkeys.

Howler Monkeys

Howler monkeys, genus *Alouatta* of the atelids, are the most widespread primate genus in the Americas. Their distribution runs from eastern Mexico to northern Argentina.

Howler Ecology

Howler monkeys are among the most folivorous platyrrhines. More than half of their diet consists of leaves. Molars, salivary glands, and gut bacteria are adapted to this diet. Minimization of other energy use facilitates leaf digestion, so howlers spend 65 to 80 percent of every day resting. They typically move in a slow quadrupedal walk along the tops of branches. Average day range is about 525m and the longest less than 700m. Howlers are selective eaters, preferring young leaves to mature ones, and they eat a considerable amount of fruit. One species exemplifies phylogenetic constraint in this regard. These howlers eat more fruit than others and increase frugivory when possible, yet their activity levels remain close to the genus norm (Pavelka and Knopf 2004). These levels seem to be limited by phylogenetic heritage, a reminder that ecological explanations do not account for all primate behavior.

Even though folivory is unusual for platyrrhines, it should not leave the impression that howlers are highly specialized. Rosenberger et al. (2011) pointed out that the colobines, a taxon of Old World monkeys, constitute the model for primate folivores and howlers do not match their pattern. Rosenberger and colleagues proposed the term "semifolivores" to designate the howler pattern. The genus actually displays a great deal of variation and flexibility. They have survived in almost every kind of forest, from undisturbed rainforest to small fragments reduced by human activities. Even in forest fragments these monkeys exploit a variety of foods (Chaves and Bicca-Marques 2013), indicating that diversity is an important part of their foraging strategy (Dias et al. 2014).

Despite their relatively large size, howlers are subject to predation. Attacks by raptors and tayras have been observed and scat analysis confirmed felid predation. One observed kill by a boa constrictor at 7.5 meters above the ground demonstrated vulnerability to large snakes even in the trees (Quintino and Bicca-Marques 2013). An adult female ran at the snake and hit it with her hands, apparently trying to rescue the adult female victim.

Howler Sociality

The howler troop is cohesive and typically contains five to ten individuals. Troop composition is always in flux, largely because of substantial emigration rates for both sexes. Either males or females may stay if mating opportunities arise. Solitary individuals and small unisex bands live among the troops. These are unsettled emigrants: some form new troops, while others try to enter established units.

A troop may contain three males, but usually only one copulates with fertile females. Extra-troop mating seems rare, so male replacement attempts are common. With the reproductive need for risk-taking maximized, males fight viciously and sometimes lethally. Many wounds and scars are probably the result of takeover attempts (Cristobal-Azkarate et al. 2004).

A one-male troop becomes multi-male if: (1) two or three invaders take over; (2) an invader deposes the resident male but allows him to stay; or (3) one or more sons of a dominant male mature and stay in the troop (perhaps because the father tolerates mating by his sons). Any multi-male troop can quickly return to the one-male form as a result of emigration or deaths.

The howling vocalization sounds to some people like the roar of a lion. It can be heard for about 0.3km through jungle vegetation and twice as far across open water. Extra volume comes from inflation of the throat sac, which is larger in males (Figure 13.1). It exemplifies co-evolution of a specialized behavior and an associated morphological feature. Howling occurs in the morning on a regular basis and may convey location in order to advertise occupation and/or avoid confrontation with other troops. However, this interpretation raises a problem. Locational information might tell other troops that other parts of the range are unoccupied and subject to exploitation. Howling is also used in aggressive displays when troops meet, although some contacts are peaceful.

Continuing attempts to understand howling (or roaring or loud calls, as these vocalizations have been named) have had somewhat mixed results. Holzmann et al. (2012) tested three hypotheses of defense: space/food, mates, and infants. Their evidence supported mate defense and argued against the other explanations. Males participated in all howling bouts and females in less than a third. These bouts were only directed against conspecifics, even though the study group also shared food resources with another howler species. Roaring frequency did not depend on the presence of unweaned infants or spatial association with food or boundaries.

In contrast, Hopkins (2013) found that howlers were more likely to approach roaring neighbors when fruit and flower availability was low. A seeming peculiarity was that subjects were more likely to approach a higher-ranking group than a lower-ranking one. Hopkins explained this in terms of the probability that a more dominant group would be associated

Figure 13.1 Male and female howler monkeys display sexual dimorphism in color that occurs in some species of *Alouatta*. The male is howling (roaring). © roejoe via iStock

with higher quality resources. She also noted the implication that the vocalizations contained information about identity. Hopkins tested other variables and found no indications that howling was associated with mates or infants.

Van Belle et al. (2014) measured several dimensions of howling and also found links to food. Howling occurred more frequently when the monkeys were feeding, especially on fruit, and was concentrated in the vicinity of major feeding sites. The duration of spontaneous calls was correlated with feeding time at those localities. Howling lasted longer when neighboring groups or lone males were visible, which suggests that the vocalizations might also function in mate defense and/or infant defense. These explanations are supported by the fact that male participation was greater when howling occurred during an intergroup encounter, as opposed to spontaneous calling (Van Belle et al. 2014; see also Fernandez at al. 2013). Adult males initiated and participated in these bouts more than subadult males, especially if they had sired offspring in the group. Howling is apparently multi-functional.

Studies of leadership are producing similarly mixed results. Fernandez at al. (2013) reported that age was the only factor. Adults influenced group coordination in about 95 percent of the cases. Observation of a different species discerned a very different pattern (Van Belle 2013a). A single female was the habitual leader of each group. She was the most widely connected adult in terms of social relationships

Early reports indicated little or no fighting within howler troops, but Glander (1992) saw frequent aggression and suggested that fights are rarely seen because they are usually quick and silent. Subsequently, evidence accumulated for various forms of violence in a number of howler species (Cristobal-Azkarate et al. 2004). Howler fighting may stem from dominance competition in both sexes. Dominant females are aggressive toward subordinates and their offspring, an example of female reproductive competition. One complete fight between two males was observed, which ended with one of them fleeing from the group (Meyer and Young 2011).

Dominant males in one study group had the greatest mating success and, as demonstrated by genotyping, fathered all the offspring (Oklander et al. 2014). Similar skew was found in two neighboring troops. Research on another species produced less clear-cut results in that dominant males did not have the highest copulation rates (Corewyn 2014). It seemed that the central males were exercising tolerance.

An important feature of the mating system in all species is the active role of the females (Van Belle et al. 2015). Females solicit dominant males during the peri-ovulatory period and enter consortships with them. At other times they copulate with other males in the group, with males in neighboring groups, and with solitary males.

An alpha male, whether in a one-male or multi-male troop, leads responses to other troops and invading males. He seems to avoid outsiders when his group is outnumbered; otherwise a confrontation is likely. The dominant individual recruits other males as allies in these conflicts. Some evidence suggests that howling by the outsiders provides relevant information, including the number of males and perhaps their age.

Adult females do most of the grooming in the studied species. It is rare and brief, possibly because few troop members are kin due to emigration by both sexes. However, kin selection may play a role in male replacement. One study showed that takeover coalitions containing related males lasted more than three times longer than those composed of non-relatives.

Howler Life Cycle

According to available evidence, howler mothers are accommodating and tolerant but show few positive responses to their infants. Even grooming is rare. Perhaps these limits are connected with the brevity of the relationship. Offspring leave the troop as juveniles, possibly expelled by adults other than the mother. The mother reacts protectively to the interest of other females in her infant, avoiding interaction when possible. Dominant females are aggressive toward subordinates and their infants. In at least one species they pull at an infant and sometimes take it from the mother.

Allocare by juvenile sisters has a somewhat unusual effect on mothers. Rather than forage unencumbered, a mother usually rests or sleeps. This behavior is counter to the Foraging Efficiency Hypothesis, but seems in accord with the howler need for energy conservation. The Maternal Learning Hypothesis for allocare was questioned in regard to howlers because most first-born infants died in the first year. However, closer examination of the data revealed that troop size was a factor. Almost all of the infants in smaller troops survived, while 13 of 16 infants in larger troops died (Clarke et al. 1998). Successful care by young females may be confounded by the actions of numerous hostile adult females in a large troop.

Weaning begins around 2 months, but may not be completed until 15 months. Both sexes emigrate before first reproduction (between one and two years of age) and may go through a period of solitary life or residence in a unisex band. Female emigrants seem to be more numerous and travel farther. Male groups or single males may join with dispersing females to form a new troop. Most females have their first birth by about age five. Males become fertile at about the same time as females, but reproduction may be delayed another two years until they reach full adult weight.

Male behavior toward infants goes beyond tolerance in at least one species, possibly based on perceived paternity. Zandona (2011) observed an adoption and saw behavior like that previously reported for probable fathers: carrying, cuddling, watching, and sitting in proximity. The other adult male in Zandona's group never displayed such behaviors.

The "moo" vocalization used in at least one species undergoes an interesting transition during ontogeny. In young individuals it is a distress call, while the adult function is group cohesion (da Cunha and Byrne 2013). The researchers suggested that the different manifestations are linked by an underlying motivation: a feeling of insecurity.

In at least one species some young females are driven from the natal troop by the aggression of older females. Females in a new troop try to recruit their own daughters and force the offspring of others to leave, so female relatedness in a troop tends to increase over time. Reproductive output increases along with relatedness, which suggests a fitness benefit. The benefit might just be reduction in the competition that suppresses reproduction.

Howler Cognition

The play of juvenile mantled howlers varies from one site to another (Baldwin and Baldwin 1978) in ways that suggest primate culture. In one location young howlers play a kind of follow-the-leader game not seen elsewhere. In another example, players repeatedly run a tree from top to bottom and back.

Spider Monkeys

Spider monkeys (*Ateles*) have the largest numbers of species and individuals among the atelines (which includes three other genera). They are also the most widespread genus, ranging from southern Mexico to northern Bolivia. Males and females are virtually identical in size, color, and slender physique.

Spider Monkey Ecology

Spider monkey anatomy facilitates diverse locomotor patterns, including arboreal quadrupedalism and bipedalism. These primates are in suspensory mode during most of their feeding time (Figure 13.2), which facilitates gathering the mostly ripe fruits that constitute more than 70 percent of their food. Spider monkeys spend more than half of each day resting, a behavior typical of folivores. This may conserve energy needed to search for scattered fruit resources across an average day range of about 2100m. The strategy is feasible because ripe fruit, when discovered, offers a high return in energy. Alternatively, the resting pattern might represent phylogenetic inertia, as in one howler species noted above.

The fact that the highly arboreal spider monkeys mobbed large cats by shaking and breaking branches was a puzzle until further research revealed that they are sometimes killed on the ground (Matsuda and Izawa 2008). The pattern varies regionally (Campbell et al. 2005). Forest discontinuities at northern sites require terrestrial movement between fragments, and dry seasons cause drinking from streams. Southerly spider monkeys may come to the ground less often, likely due to more intact predator faunas, and only for essential nutrients. Anecdotally, these spider monkeys display lengthy hesitation before descending and appear fearful as they continually scan the environment.

Spider Monkey Sociality

Spider monkeys live in extreme fission–fusion societies. Subgroups, typically containing two or three individuals, can coalesce into larger groups (up to about 20 individuals) that approximate the entire society. In this dispersed troop (or community), each female has a separate home range and adult females travel independently of males. Female ranges overlap and females are not anchored to their individual home ranges. They may join with wider-ranging parties and they may also visit neighboring troops for several hours, a whole day, or overnight. Since females emigrate, temporary transfers may be visits to relatives.

Party size is correlated with the availability of high-quality food, especially ripe fruit. This suggests that fission is adaptive in limiting travel costs when high-quality foods occur only in small patches. Alternatively, under the same conditions, fission might function to limit food competition. Favoring the latter hypothesis, a lone community on an island maintained cohesion while exploiting an unusually large home range.

As in other fission–fusion societies, there is a question about the nature of community cohesion. Two studies found two different answers. In one of them the distances between parties were close to those predicted by a null hypothesis based on independent travel (Ramos-Fernandez et al. 2011). The other study inferred coordination based on loud calls (Spehar and Di Fiore 2013): over 80 percent of the calls were made within 30 minutes of a call from another location; parties that called were likely to increase in size within an hour; and monkeys approached loud calls more often than they ignored or avoided them.

Figure 13.2 Suspensory behavior in a spider monkey, aided by its prehensile tail. © rusm via iStock

The difference between the studies may be due to the measures chosen or the fact that the subjects were of different species.

Males, apparently the philopatric sex, display strong affiliations with each other. A male most often associates with other males and a male pair constitutes the most common adult grooming dyad. A distinctive expression of male affiliation is the embrace. Males jointly patrol the edges of community ranges that overlap to a relatively small degree. When a male sees outsiders, he roars. Nearby females join in and other parties follow the sound. The other troop responds in kind. Males continue to roar and shake branches while others reassure each other with touches and embraces. Of course, there are few absolutes in primate life. Recent fieldwork recorded male immigration and disclosed that males in the study community were no more closely related to each other than were the females (Aureli et al. 2013).

A typical male spends more time moving, moves farther each day, and covers a larger area than a female. Territorial patrol is one possible cause. Another might be a feeding difference reported for at least one population: females prefer large patches while males seek more scattered foods. The two explanations are not mutually exclusive, although the two goals might conflict for a particular male on a particular day. Male affiliation, territoriality, and patrols are striking parallels with chimpanzee behavior (see Chapter 25).

Van Roosmalen and Klein (1988) described certain females in one population leading parties in their own core areas, apparently applying superior knowledge of local food sources. Followers might have had increased survival chances in the highly seasonal environment of that population. Males showed some feeding aggression toward females and juveniles, but not toward leaders. The leader rarely aggressed against other females, but those few episodes were severe. Occasionally a subordinate female initiated an embrace. Sometimes the leader took her followers to meet other parties, which seemed to be social occasions.

Despite monomorphy, at least some spider monkey troops display an unbalanced sex ratio. This has been attributed to the absence of males that are in all-male groups or are solitary. These do exist, but all stay within the perimeter of the community so they are likely to be counted. The ratio may represent a real imbalance resulting from male mortality during food shortages, due to inferior knowledge of resources.

Female mate choice is indicated by the fact that receptive females sometimes bypass nearby males for more distant ones. There are reports of females mating with several males, sometimes on the same day. In at least one species females often mate outside the peri-ovulatory period. Males cooperate to dominate individual females and subject them to sexual investigation. This may be a tactic to counter female choice.

A female initiates a sexual encounter by rapidly approaching the male, presenting to him, and sitting in his lap. They remain seated for copulation in a unique posture with the male's legs over the female's thighs. The length of intromission varies from about 10 to 25 minutes. Spider monkeys try to copulate in seclusion, apparently to avoid harassment from other females and from offspring.

Spider Monkey Life Cycle

Spider monkey reproductive rates are slow. Gestation is about six months and the IBI is three years. An infant rides on its mother's back regularly until four to six months old, and up to the age of two years when tired or in need of help. Weaning is not finished until age three. Juveniles play with peers or older individuals, including adults. They show no preference for either sex in their play, but they do prefer particular individuals. Play bouts last as long as 35 minutes and a play period can last for 1.5 hours. Head shaking and panting are signals of playful intent (typical of chimpanzees). Subadults, beginning at about five years old, play with juveniles but also participate in adult loud calling. An individual develops his own call by trial and error. During this stage of life, a female stays close to her mother.

Spider Monkey Cognition

The behavior of a leader female suggests memory of local resources and a mental map of her home range or core area. Advance planning is implied by the fact that travel routes vary greatly from day to day, but always seem to make the shortest possible connections between food sources. With or without a leader, spider monkeys seem to use mental maps. *Ateles geoffroyi* in Mexico follow highly linear travel segments, typically ending at resources that were not in sight at the outset. They usually travel in a different direction after exploiting the resource. The pattern suggests that they use spatial memory to guide their travel and may plan in advance for arrival at more than one resource site (Valero and Byrne 2007).

Lindshield and Rodrigues (2009) reported sporadic tool use—three events in which females of *A. geoffroyi* scratched themselves with sticks. Two of them chewed the tip of the stick before application. The researchers speculated that an unusually wet environment facilitated a higher frequency of ectoparasites causing skin irritation. They also suggested that chewing the sticks might release a medicinal chemical. These anecdotes are significant because they question some proposed conditions for the evolution of tool use: opportunity for improved food exploitation and effective manual dexterity (Van Schaik et al. 1999). Spider monkey hands (suited to brachiation) lack dexterity for tools, and the only known tool use is not food related. Spider monkey brain morphology resembles that of other primate tool users, which suggests that cognitive skills may outweigh other factors.

Scheel and Edwards (2012) described an example of what they considered intentional communication in captive spider monkeys. Six of eight individuals in small groups raised their arms, a gesture not previously reported, to solicit grooming from others. Responses entailed increased rates and duration of grooming. This kind of social innovation raises the possibility of culture.

Wild spider monkeys display possible cultural traditions. Using methods similar to those employed with great apes, Santorelli et al. (2011) developed evidence for 22 behavioral traditions that varied from absent to "customary" across five long-term field sites. The monkeys differed from apes in that about half of these patterns pertained to social behavior while only 19 percent were food related.

Pitheciids

This family consists of four genera (Box 13.2), three of which are **seed predators**, which means they consume seeds for nutrition rather than passing them through the digestive tract as a byproduct of frugivory. The seeds are thus destroyed rather than going back into the environment to sprout new plants. Titis (genus *Callicebus*) are the exception in the family, emphasizing fruit and leaves in their diet and diverging in many physical traits from the seed predators.

Box 13.2 Pitheciid Taxonomy

The seed-eating genera vary in size from about 2kg to 4.5kg. They have prognathic snouts with a pronounced overbite. Their incisors are broad, tall, and procumbent. The canines are large and laterally splayed. The cheek teeth are relatively small. External features are enormously variable. The top of the head, for example, varies from bald to extremely bushy.

In striking contrast, the titis seem to have converged with the callitrichids (see next chapter) in many ways. They have short faces, long skulls, and large ears that are usually hidden by hair. Canine teeth are short while the incisors are tall and narrow. Enlarged hyoid bones are the basis for their loud territorial calls.

The Seed Predators

The pitheciid seed predators are uakaris, bearded sakis, and common sakis (Figure 13.3). The three genera are marked as seed predators by their dental specializations. These include procumbent lower incisors, thick and flaring canines, and square cheek teeth with cusps so low that the teeth are virtually flat. The molars resist cracking because the enamel is infused with distinctive bands. These are all adaptations for opening fruits that are mechanically protected. That is, the fruits are brittle and/or thick-husked. Often the fruits chosen by pitheciins are unripe, which is shown by their green or brown color. The combination of protective covering and lack of ripeness must make these fruits unappealing (or impossible) for other primates, lowering the level of competition.

After opening fruits, pitheciins use their procumbent incisors to scrape mesocarp from the inside of the husk (cf. strepsirrhines). The most important items for them are young seeds. Mature seeds and fruit pulp are a smaller part of the diet. Seeds often amount to a third of the diet, and sometimes more than two-thirds. The main advantage of heavy seed exploitation may be relief from the pressure of seasonal fruit shortages (Palminteri et al. 2012). Despite their physical specializations, sakis and uakaris are perfectly capable of eating a wide variety of other foods, which may be seasonally important. These range from leaves to nectar to insects. In many cases such fallback foods may comprise no more than 10 percent of the annual diet, but one study determined that insects alone comprised 20 percent.

Reporting on uakaris (*Cacajao*), Bowler and Bodmer (2011) emphasized that the specialization of these seed predators has gone beyond simply avoiding competition and has placed them in a divergent niche. The larger implication is that their behavior patterns are likely to be distinct from those of other primates of similar size. This is a very important point because of the relationship of body size to diverse aspects of primate life from feeding ecology to life history.

Thompson (2015) used sakis as a case study to examine the dynamics of pair-groups, a topic of importance for strepsirrhines as well as platyrrhines (Chapter 4). She found that same-sex aggression is a key component in the variable acceptance of "extra" males or females. Variation results from alternative strategies for balancing reproductive access with group defense. In addition, individual decisions might prevent others from maximizing fitness.

Figure 13.3 A white-faced saki monkey, one of the seed-eating taxa of the pitheciids.
© michaklootwijk via iStock

Titis

Titis are the best-studied and most widely distributed pitheciids, found mainly in the Amazon and Orinoco basins. The genus contains more species than any other in the New World except for *Saguinus*. Titis have little in common with other pitheciids apart from the physical details that define their phylogenetic relationship. The diet typically emphasizes fleshy fruit, supplemented by leaves. Titis occasionally join with tamarins but avoid larger primates that can chase them from food sources.

The most distinctive feature of titis is the pair bond (Figure 13.4). They seem to be strictly monogamous, at least socially, and a variety of social and physiological mechanisms promote attachment between male and female. Grooming occurs throughout the day and paired titis rest and sleep huddled together with tails intertwined. They also lip-smack, nuzzle, and gently grasp each other. Paired individuals closely coordinate activities and feed

Figure 13.4 A titi pair-group. The social organization of these pitheciids diverged from their closest relatives. The titi group is more cohesive and the central pair more strongly bonded than in most and perhaps all of the paired primates. © Sjo via iStock

on the same food items at the same time without conflict. Either partner can lead group movements. Duets are long sequences of various loud calls that can be produced by both sexes. Titis display distress and agitation when separated. However, despite the apparently emotional attachment, extra-group copulations take place.

The lactating mother leads her group into fruit trees and eats fruit and insects faster than others. Newborns cling ventrally to a carrier's neck and after a few weeks move to a dorsal position, usually across the shoulders. The adult male carries the infant almost exclusively in the first two months. More experienced mothers spend less time with their infants. While the father often shares food with the infant, the mother will not. Experiments with one species indicate that infants are more attached to the father than the mother. An infant separated from the father shows an increase in cortisol, an indicator of stress, while separation from the mother does not cause the same result. Lengthy retention of subadults as late as age four years may be a functional response to a high rate of juvenile mortality and can explain the larger groups that have been observed. Sons are more reluctant to leave than daughters, and they leave later.

Summary

Two platyrrhine families and six genera were covered in this chapter: the howlers and spider monkeys of the atelids, and the seed predators and titis of the pitheciids. Howlers take their name from their unusual distance call. They are distinctively folivorous among the platyrrhines and their behavior is strongly affected by energy conservation. Spider monkeys are strongly frugivorous and live in male-philopatric fission–fusion communities that parallel chimpanzees in several ways. Females display effective spatial memory and sporadic tool use.

Pitheciid seed predators have distinctive diets based on their ability to crack open hard-shelled seeds and fruits with their procumbent incisors. Their social systems vary across the taxa from pair-groups to large troops. Titis are small frugivores characterized by tightly knit pair-groups in which paired adults display strongly affiliative behaviors.

Sources

Basic Bibliography

Campbell, C.A. (Ed.) (2008). *Spider monkeys*. NY: CUP.

Campbell, C.A., et al. (Eds.) (2011). *Primates in perspective*. NY: OUP. [Chapters 8, 11]

Estrada, A., et al. (Eds.) (2005). *New perspectives in the study of Mesoamerican primates*. NY: Springer.

Kappeler, P.M., & Watts, D.P. (Eds.) (2012). *Long-term field studies of primates*. NY: Springer. [Part III]

Kowalewsi, M.M., et al. (Eds.) (2015). *Howler monkeys*. NY: Springer.

Veiga, L.M., et al. (Eds.) (2013). *Evolutionary biology and conservation of titis, sakis, and uakaris*. NY: CUP.

Additional References

Aureli, F., et al. (2013). Male philopatry in spider monkeys revisited. *AJPA*, *152*, 88–95.

Baldwin, J.D., & Baldwin, J.I. (1978). Exploration and play in howler monkeys. *Primates*, *19*, 411–422.

Bezerra, B.M., et al. (2011). Brevity is not always a virtue in primate communication. *BL*, *7*, 23–25.

Bowler, M., & Bodmer, R.E. (2011). Diet and food choice in Peruvian red uakaris (*Cacajao calvus ucayalii*). *IJP*, *32*, 1109–1122.

Briseño-Jaramillo, M., et al. (2015). Individual voice recognition and an auditory map of neighbours in free-ranging black howler monkeys (*Alouatta pigra*). *BE&S*, *69*, 13–25.

Campbell, C.J., et al. (2005). Terrestrial behavior of *Ateles* spp. *IJP*, *26*, 1039–1051.

Cäsar, C., et al. (2013). Titi monkey call sequences vary with predator location and type. *BL*, *9*, DOI: 10.1098/rsbl.2013.0535.

Chaves, O.M., & Bicca-Marques, J.C. (2013). Dietary flexibility of the brown howler monkey throughout its geographic distribution. *AJP*, *75*, 126–129.

Clarke, M.R., et al. (1998). Infant–nonmother interactions of free-ranging mantled howlers (*Alouatta palliata*) in Costa Rica. *IJP*, *19*, 451–472.

Corewyn, L.C., & Kelaita, M.A. (2014). Patterns of male–male association in mantled howlers (*Alouatta palliata*) at La Pacifica, Costa Rica. *Behaviour*, *151*, 993–1020.

Cristobal-Azkarate, J., et al. (2004). Causes of intraspecific aggression in *Alouatta palliata Mexicana*. *IJP*, *25*, 939–953.

Cunningham, E.P., & Janson, C.H. (2013). Effect of fruit scarcity on use of spatial memory in a seed predator, white-faced saki (*Pithecia pithecia*). *IJP*, *34*, 808–822.

da Cunha, R.G.T., & Byrne, R.W. (2013). Age-related differences in the use of the "moo" call in black howlers (*Alouatta caraya*). *IJP*, *34*, 1105–1121.

Defler, T.R. (2004). *Primates of Colombia*. Conservation International.

Di Fiore, A., & Suarez, S.A. (2007). Route-based travel and shared routes in sympatric spider and woolly monkeys: cognitive and evolutionary implications. *AC*, *10*, 317–329.

Dias, P.A.D., et al. (2014). Variation in dietary breadth among groups of black howler monkeys is not associated with the vegetation attributes of forest fragments. *AJP*, *76*, 1151–1162.

Fernandez, V.A., et al. (2013). Who is coordinating collective movements in black and gold howler monkeys? *Primates*, *54*, 191–199.

Glander, K.E. (1992). Dispersal patterns in Costa Rican mantled howling monkeys. *IJP*, *13*, 415–436.

Holzmann, I., et al. (2012). Roaring behavior of two syntopic howler species (*Alouatta caraya* and *A. guariba clamitans*): evidence supports the mate defense hypothesis. *IJP*, *33*, 338–355.

Hopkins, M.E. (2013). Relative dominance and resource availability mediate mantled howler (*Alouatta palliata*) spatial responses to neighbors' loud calls. *IJP*, *34*, 1032–1054.

Laska, M., et al. (2007). Self-anointing behavior in free-ranging spider monkeys (*Ateles geoffroyi*) in Mexico. *Primates*, *48*, 160–163.

Lindshield, S.M., & Rodrigues, M.A. (2009). Tool use in wild spider monkeys (*Ateles geoffroyi*). *Primates*, *50*, 269–272.

de Luna, A.G., et al. (2010). Predation and predation attempts on red titi monkeys (*Callicebus discolor*) and equatorial sakis (*Pithecia aequatorialis*) in Amazonian Ecuador. *FP*, *81*, 86–95.

Matsuda, I., & Izawa, K. (2008). Predation of wild spider monkeys at La Macarena, Colombia. *Primates*, *49*, 65–68.

Meyer, C., & Young, O.P. (2011). Observations of a fight between two adult male mantled howler monkeys. *NP*, *18*, 31–33.

Mourthé, I. (2011). Reactions of white-bellied spider monkeys to a predation attempt by a cougar. *NP*, *18*, 28–29.

Oklander, L.I., et al. (2014). Male reproductive strategies in black and gold howler monkeys (*Alouatta caraya*). *AJP*, *76*, 43–55.

Palminteri, S., et al. (2012). Advantages of granivory in seasonal environments. *Oikos*, *121*, 1896–1904.

Pave, R., et al. (2010). Adoption of an orphan infant in wild black and gold howler monkeys (*Alouatta caraya*). *Mastozoología Neotropical*, *17*, 171–174.

Pavelka, M.S., & Knopf, K.H. (2004). Diet and activity in howler monkeys (*Alouatta pigra*) in southern Belize: does degree of frugivory influence activity level? *Primates*, *45*, 105–111.

Quintino, E.P., & Bicca-Marques, J.C. (2013). Predation of *Alouatta puruensis* by *Boa constrictor*. *Primates*, *54*, 325–330.

Ramos-Fernández, G., et al. (2011). No evidence of coordination between different subgroups in the fission–fusion society of spider monkeys (*Ateles geoffroyi*). *IJP*, *32*, 1367–1382.

Rosenberger, A.L., et al. (2011). The making of platyrrhine semifolivores: models for the evolution of folivory in primates. *AR*, *294*, 2112–2130.

Santorelli, C.J., et al. (2013). Individual variation of whinnies reflects differences in membership between spider monkey (*Ateles geoffroyi*) communities. *IJP*, *34*, 1172–1189.

Scheel, M.H., & Edwards, D. (2012). Captive spider monkeys (*Ateles geoffroyi*) arm-raise to solicit allo-grooming. *BP*, *89*, 311–313.

Shaffer, C.A. (2013). Activity patterns, intergroup encounters, and male affiliation in free-ranging bearded sakis (*Chiropotes sagulatus*). *IJP*, *34*, 1190–1208.

Spehar, S., & Di Fiore, A. (2013). Loud calls as a mechanism of social coordination in a fission–fusion taxon, the white-bellied spider monkey (*Ateles belzebuth*). *BE&S*, *67*, 947–961.

Strier, K.B., et al. (2011). Low paternity skew and the influence of maternal kin in an egalitarian, patrilocal primate. *PNAS*, *108*, 18915–18919.

Thompson, C.L. (2015). To pair or not to pair: sources of social variability with white-faced saki monkeys (*Pithecia pithecia*) as a case study. *AJP*, doi: 10.1002/ajp.22360.

Valero, A., & Byrne, R.W. (2007). Spider monkey ranging patterns in Mexican subtropical forest: do travel routes reflect planning? *AC*, *10*, 305–315.

Van Belle, S., et al. (2013a). Collective group movement and leadership in wild black howler monkeys (*Alouatta pigra*). *BE&S*, *67*, 31–41.

Van Belle, S., et al. (2013b). Spatial and diurnal distribution of loud calling in black howlers (*Alouatta pigra*). *IJP*, *34*, 1209–1224.

Van Belle, S., et al. (2014). The function of loud calls in black howler monkeys (*Alouatta pigra*). *AJP*, *76*, 1196–1206.

Van Belle, S., et al. (2015). Insights into reproductive strategies and sexual selection in howler monkeys. In M.M. Kowalewski et al. (Eds.), *Howler monkeys* (pp. 57–83). NY: Springer.

van Roosmalen, M.G.M., & Klein, L.L. (1988). The spider monkeys, genus *Ateles*. In R.A. Mittermeier et al. (Eds.), *Ecology and behavior of neotropical primates*, Volume 2 (pp. 455–539). Washington DC: World Wildlife Fund.
Van Schaik, C.P., et al. (1999). The conditions for tool use in primates: implications for the evolution of material culture. *JHE*, *36*, 719–741.
Zandona, E. (2011). Male care in mantled howler monkeys (*Alouatta palliata palliata*). *NP*, *18*, 22–25.

14 Platyrrhine Variation
Cebids

Two small primates forage near one another. One clings to a tree trunk and licks exudate from the hole she has gouged. On the limb above the male pounces on a large insect. An infant rides on his back.

The cebid family as defined by Fleagle (2013) and others encompasses three diverse subfamilies. Two of them are considered below. The third subfamily, containing squirrel monkeys and capuchins, has a chapter to itself for two reasons. One is the special relevance of the two genera to humans. The other is the enormous amount of information about them.

Marmosets and Tamarins

Despite the fact that marmosets and tamarins are among the best-known platyrrhines, there is no popular name to encompass their taxonomic group (Box 14.1). They have been classified as Callitrichidae (family) and Callitrichinae (subfamily), so we will use the root **callitrich** as an informal and taxonomically noncommittal term for these monkeys. These

Box 14.1 Callitrich Taxonomy

Some taxonomists would demote the callitrichs below the family level and sink them into another family. Others, however, retain the family-level classification. Distinguishing features of the callitrichs are their size (100–700g, smaller than any other platyrrhines), "claws" on all digits except the big toes, two molar teeth in each quadrant instead of the three found in other ceboids, and a strong tendency to produce twins.

The typical callitrichs in these terms are the tamarins (*Saguinus*), lion tamarins (*Leontopithecus*), and marmosets (*Callithrix*, *Mico*, and *Cebuella*). Goeldi's monkey (*Callimico*) has been troublesome because it is callitrich size and has "claws" but resembles other platyrrhines with regard to the molar array and single birth. Recent work suggests that it is more closely related to marmosets than the tamarins are, which puts it firmly into the callitrich taxon.

Callitrichs have digestive specializations for consuming gum. Marmosets are distinguished by adaptations for obtaining gum, especially the gouge formed by their lower teeth. It is somewhat analogous to the strepsirrhine toothcomb.

small, monomorphic primates have relatively long braincases and short snouts. Their trunks, legs, and tails are all relatively long. Both sexes are brightly colored in many species.

Callitrich Ecology

Most callitrichs prefer dry secondary or disturbed forests, or edge habitats. They are unique among primates in having claw-like nails on most digits, which they use to cling to tree trunks. The hallux has a flat nail like those of other primates, probably because it is the most important digit for grasping. A gradient from claws to more nail-like structures is associated with greater frequency of small-branch foraging. Clinging to trees facilitates collection of exudates, a vital food for callitrichs. Availability throughout the year is important because callitrichs, being small and having a high metabolic rate, experience greater seasonal food pressures than most other primates. Exudate feeding is also attractive because competition is low.

Size and metabolism dictate dependence on another high-energy food source, animal matter (30 to 80 percent of feeding and foraging time). Mobile arthropods such as grasshoppers are typical prey. Callitrichs attack bigger insects than those taken by larger-bodied monkeys, perhaps because of greater need, and they also kill small vertebrates (mainly frogs and lizards). The typical predatory pattern is *scan and pounce.* The slowly moving primate scans among leaves, in holes, and on the surfaces of branches and trunks. Finding prey, it stalks and pounces.

Callitrichs put in short days compared to other primates and diurnal mammals in general. They leave their sleeping places about 45 minutes after full light and return an hour or two before sunset. There are at least four possible explanations, none mutually exclusive: (1) correlation with daily availability of large insect prey; (2) avoidance of competition with crepuscular mammals; (3) minimization of energy expenditure; (4) minimization of exposure to predators, especially raptors. The monkeys move to their sleeping trees quickly and quietly and huddle together, often in a dense tangle of vegetation (Dacier et al. 2006). During the day they are normally active in densely foliated areas and are cryptic when resting. If exposed, speed and agility provide a chance for escape.

Callitrichs seem to react to nonvisual stimuli, such as the fecal scents of predators (Buchanan-Smith et al. 1993). The responsiveness of inexperienced animals suggests innate recognition mechanisms. Marmosets can distinguish raptor calls from other sounds (Searcy and Caine 2003). Callitrichs generalize from their experiences. Predator-related stimuli, including sudden movements or loud noises, may result in behavioral alterations (such as unusual caution or avoidances) that can last for weeks (Dacier et al. 2006). Though small, callitrichs sometimes mob cats and snakes. In one incident two adult tamarins rescued a young member of the group from a boa constrictor by biting the snake (Tello et al. 2002). One rescuer was the victim's father and the other was a female relative who had nursed him after her own infant died (Huck et al. 2005).

Field observations on two species indicated that "alarm" calls did not serve as an immediate warning because monkeys that first perceived the threats moved away before giving the calls. For the same reason, the vocalizations could not function to confuse the predator in its initial approach. Instead the calls might alert distant group members, locate group members, localize and define the threat, or alert the predator that it has been detected.

Callitrich Sociality

Callitrich social units are highly variable in composition across taxa and time. The typical group of five to ten individuals is cohesive and affiliative. Members often groom each other during rest periods or just sit together in physical contact. Allogrooming is asymmetrical in some wild groups, with females receiving more than they give (Lazaro-Perea et al. 2004). Agonistic behavior is rare and usually happens during feeding. It is mild, consisting mainly of threats, chases, submissive signals, and the occasional slap. Male–female relations are either entirely egalitarian, or males can displace females at feeding sites.

Experiments implicate the hormone oxytocin in the proximal basis for callitrich pair bonding (Smith et al. 2010). Subjects treated with the hormone initiated huddling with partners more often than did controls. An oxytocin antagonist reduced huddling, proximity, and food sharing. In a tamarin species changes in affiliative behaviors (contact, grooming, and sex) were correlated with fluctuations in urinary oxytocin (Snowdon et al. 2010). Mutual contact and grooming explained most of the variance in female oxytocin levels while sexual behavior explained most of the male variance. Increased levels of the hormone affected both sexes, but differently: males were more likely to initiate physical contact and females were more likely to solicit sex.

Callitrich social organization is based on cooperative breeding and conforms to the Socioecological Model in that it is based on female distribution. The heart of the typical social unit is a single breeding female. She accepts long-term association with a male because (ultimate causation) she needs help in caring for her offspring (Figure 14.1). However, the pair needs further help due to the unusual burdens of callitrich reproduction: twins are the norm, infants are large in proportion to adults, and there is frequently an overlap between lactation and the next pregnancy due to postpartum ovulation. Therefore, one or more additional males may be accepted. Some may be adult offspring, but fieldworkers have determined that some are unrelated. An unrelated secondary male may copulate with the paired female (sexual polyandry), but conception rarely results. There is little overt competition among males in a polyandrous group, but mate guarding of ovulating females is recurrent if not common (Huck et al. 2004b; Porter and Garber 2004; Schaffner and French 2004).

Adult female offspring are also retained in the group as helpers, and the majority of social units contain more than one adult of each sex. However, breeding individuals of both sexes are the favored grooming partners of all members. In some cases dominant females seem to groom subordinates as an inducement to stay in the group (Lazaro-Perea et al. 2004). A single dominant female monopolizes breeding in most social groups, typically by suppressing reproduction in subordinate females. The proximal mechanisms vary across taxa and include suppression of ovulation and of sexual behavior. There has been debate about whether suppressed ovulation is the direct result of harassment by the dominant female, or is a spontaneous response to avoid serious aggression and preserve reproductive resources for the future.

In some groups a second female begins to breed and occasionally more than that (Hilário and Ferrari 2010; Smith et al. 2001). However, this rarely occurs unless an unrelated male joins the group, and the offspring of a secondary female rarely survive for more than a few months. They receive little or no care from the males and the dominant female may kill them—a recurring pattern with an uneven distribution across taxa. Infanticide is most probable when two females give birth within a limited time frame, which indicates that it is a form of competition for helpers. A female who loses her infants in this way often leaves the group within a few months.

Figure 14.1 A male silvery marmoset carries twin offspring, illustrating the birth pattern and allocare that are ubiquitous among callitrichids. © PaulMaguire via iStock

Genetic analysis has provided insight into male fitness in socially polyandrous groups. In a study of tamarins Diaz-Munoz (2011) determined that levels of relatedness among the males were consistent with brother–brother or father–son connections, so kin selection can provide part of the explanation for the cooperative system. In addition some "extra" males had breeding opportunities. In two of five study groups more than one male sired offspring.

Olfactory signals are essential to group cohesion. Though multifunctional, the callitrich olfactory system is especially effective for communication (Ziegler 2013). Both sexes have specialized scent glands in the anogenital region and females have large circumgenital scent glands. Glandular secretions mix with urine and possibly vaginal discharge to signal species and individual identity, sex, and dominance status. Olfactory communication ensures mating when fertility is greatest, which is particularly important in the context of the high rate of reproduction in these primates. A mother's scent may suppress ovulation in a maturing daughter, paving the way for her to stay in the group and assist with the care of her siblings.

Most intergroup encounters in the wild are mildly aggressive, characterized by vocalizations and chases, and physical aggression occurs only occasionally. Because this pattern is consistent across species, even where home ranges overlap extensively, it seems not to be defense of boundaries. It may be devoted to particular resources, such as exudate trees, or it may have more to do with mate defense than food resources. Extra-group copulations occur in some species and encounters between groups may provide an opportunity to assess potential partners.

Callitrich Life Cycle

Callitrich gestation time (typically under five months) is at the lower end of the platyrrhine range of variation. However, based on other placental mammals of similar size, callitrichs should have much shorter gestation periods. They may be constrained by a minimum threshold that affects anthropoids in general, but a simpler explanation is that additional developmental time is necessary for the twin births that constitute about 80 percent of the litters in the taxon (Goldizen 1987). Females have just one pair of nipples. In other anthropoids a single offspring alternates between the two, while strepsirrhines with litters have two or more pairs of nipples. This suggests that callitrich twinning is a derived trait.

Captive data show that all callitrichs can produce offspring twice each year. Not all species do so in the wild, which reminds us that environment has restraining effects. The bimodal pattern is possible because conception is not inhibited by lactation, though it may be delayed somewhat. Callitrichs can conceive again two to four weeks after giving birth (**postpartum estrus**). The energy derived from exudates probably makes twinning and semi-annual birth possible (Ah-King and Tullberg 2000).

High rates of postnatal growth add to the callitrich reproductive burden (Porter and Garber 2004). Mothers depend on help from mates and other group members, especially carrying and food sharing. Helpers may start carrying the infant on its first day. In captivity newborns may be scooped up even before they are cleaned (Welker and Schafer-Witt 1987). This enthusiasm from non-parents requires explanation because carrying is a costly behavior. Carriers lose up to 10 percent of their body weight over about three months of intensive care, and they experience deficits in locomotion and feeding (Snowdon and Cronin 2007). Infant stimuli elicit strong motivation in experienced male marmosets, and males who respond to infant vocalizations are the males likely to do the most carrying in a family (Zahed et al. 2008).

Allocare can be explained by two general hypotheses and two that are more specific to callitrichs: (1) young helpers gain experience; (2) related helpers increase inclusive fitness; (3) maturing helpers are allowed to stay in the natal unit until they can join another group; and (4) unrelated males are following a courtship strategy by helping the mother (Huck et al. 2004a; Snowdon and Cronin 2007). Higher infant mortality rates for inexperienced parents support the first hypothesis. High within-group relatedness in at least some species makes the second hypothesis plausible. Field observations that disclose few breeding vacancies support the third hypothesis. On the other hand, the courtship hypothesis generated three predictions that failed (Tardif and Bales 1997).

Another form of allocare is food sharing, beginning at about five weeks, which contrasts with the rarity of food transfer among other primates (Joyce and Snowdon 2007). The Nutritional Hypothesis holds that food sharing supports rapid maturation and survival at weaning (Joyce and Snowdon 2007). The Informational Hypothesis explains the behavior as providing knowledge about adult foods (Rapaport 2006). Obviously, these are not mutually exclusive. Allonursing occurs occasionally (Smith et al. 2001) and may be comparable to the more routine food transfers. Food sharing also occurs between adults but it may be either affiliative or coerced (Kasper et al. 2008; Rapaport 2001).

A widely cited study reported that callitrich fathers gained weight during their mates' pregnancies, presumably to prepare for the energetic cost of supportive behaviors (Ziegler et al. 2006). A follow-up study compared experienced and inexperienced fathers, with the expectation that experienced ones would gain more weight because of multiple hormonal changes (Sanchez Rodriguez et al. 2008). Instead, the inexperienced males gained while the experienced ones did not. The researchers concluded that other factors must be involved, such as feeding and resting by the males and communication between mates.

Research on the hormone prolactin started a similar roller coaster (Almond et al. 2008; Ziegler et al. 2006). A widely cited study implicated the hormone in callitrich paternal behavior, but a later study of tamarins failed to produce the effect in experienced fathers. Lowering prolactin in marmosets reduced the responsiveness of experienced males to infant stimuli, but raising prolactin did the same thing. This partially contradicted a previous study of marmosets, which found that lowering prolactin did not reduce infant care by experienced fathers. Research on marmosets by another group led them to conclude that increased prolactin results from physical effort associated with fatherhood, rather than being a hormonal trigger (Mota et al. 2006). Another series of studies suggested that paternal care is facilitated by a reduction in testosterone levels (Nunes et al. 2001; Prudom et al. 2008).

Callitrichs develop quickly and rapidly transition to adulthood. They reach sexual maturity between 12 and 18 months (although a male may not sire an infant until he is 25 months old). This rush to reproduction may be necessary because lifetime potential is limited by short reproductive tenure in a group, high infant mortality, and suppression of reproduction by dominant breeders.

Callitrich Cognition

Burkart and van Schaik (2010) suggested that callitrichs provide the strongest evidence for teaching among primates. The begging success of infant tamarins depends on whether or not a particular vocalization is uttered by the adult, which informs the infant about the food and the vocalization itself. These interactions are related to social learning on the part of the infant. Two aspects of this process are also represented in common marmosets and golden lion tamarins: adult consumption facilitates the acceptance of novel foods by infants, and infants obtain information as well as food by begging from adults. More suggestive of deliberate instruction, golden lion tamarin adults apparently guided their young to the locations of cryptic prey (Rapaport and Ruiz-Miranda 2002). A captive marmoset mother also seemed to actively promote learning in her offspring with regard to prey acquisition (Dell'mour et al. 2009). While solving experimental tasks for infants aged 11 to 15 weeks, she consumed less food and consumed it later than when foraging with older offspring or alone.

Plasticity in vocal signals, the ability to alter vocal structure after maturity, is documented in at least four callitrich species (Burkart and van Schaik 2010). For example, when new individuals were introduced to captive marmoset groups, call structure changed in all age/sex classes. Variation between populations also demonstrates plasticity, perhaps to a degree that is rare among primates. Contextual learning is the acquisition of knowledge as to how and when to use vocalizations after the vocalizations have developed. This applies to the chirps of cotton-top tamarins, for instance.

Common marmosets and cotton-top tamarins demonstrate some understanding of gaze by using the gaze of others to make object choices. Finally, these two species are both able to coordinate individual behaviors in space and time to solve mutual problems—for example, pulling two well-separated handles simultaneously in order to obtain food. Contrasting behavior in the presence and absence of a partner shows that callitrichs understand the role of the partner (Burkart and van Schaik 2010).

Kawai et al. (2014) experimented with the ability of marmosets to evaluate the behavior of others in social exchanges. The monkeys were able to distinguish between humans who reciprocated in social exchanges and those who did not. Such responses are often interpreted as indicating relatively sophisticated social cognition. However, the researchers in this case maintained that the response of the monkeys emerged from their natural tendency toward cooperation.

In an unusual sequence of experiments Gunhold et al. (2014a, b) taught wild marmosets two ways to open a box and then tested them with the task two years later. The initial learning experience was provided by videos of unfamiliar conspecifics performing the task. Despite the unusual presentation, the monkeys learned to perform the task. Tested again two years later, still in the wild, the original subjects showed that they remembered and still preferred the solution they had learned. The experiments demonstrate the power of social learning, even from a stranger in a video, and the potential for social traditions in these New World monkeys.

Aotids

Aotus (Figure 14.2) is the only living genus recognized in this family, but it has been divided into numerous species (Box 14.2). Common names for the genus include douroucoulis, night monkeys, and owl monkeys. The last probably alludes to the large eyes and hooting call as well as its nocturnal habits. Distribution of the genus ranges from Panama to Argentina and from the inland foothills of the Andes to eastern Brazil.

Figure 14.2 An owl monkey, representing the only anthropoid genus with entirely nocturnal species. © nicolasdecorte via iStock

Box 14.2 Aotid Taxonomy

Owl monkeys are small (between 0.5kg and 1.5kg) and sexual dimorphism is minimal. Their eyes display several morphological adaptations to nocturnal life in addition to sheer size. About 11 species have been recognized through the combination of morphological, karyotypic, and molecular data. Variations in body size, canine dimorphism, and activity patterns may be significant for species classification. The species are generally thought to fall into two "groups" based on karyotypes, susceptibility to malaria, and neck color. Gray-necked owl monkeys live north of the Amazon River and red-necked owl monkeys are south of the River.

Owl Monkey Ecology

Unique among anthropoids, owl monkeys are typically nocturnal. There are several explanations, none mutually exclusive. First, nocturnal insects up to about 8cm long provide a rich source of protein. Some make themselves conspicuous with loud calls. Second, potential competitors for food at night are small-bodied and/or live in small groups. Owl monkeys feed peacefully alongside kinkajous and opossums, while titis (diurnal monkeys of about the same size) are chased away from food sources by larger diurnal primates. Finally, there are fewer active predators at night. Especially important is the absence of raptors that are the main monkey killers in South American forests. Demography suggests the benefits of the nocturnal niche. All owl monkey groups in a survey contained offspring at every stage of life while half of the titi groups lacked an infant or a juvenile, probably due to predation (Wright 1994).

Nevertheless, several aspects of owl monkey behavior indicate a limited adaptation to nocturnal life. First, they tend to be more active at dusk and dawn than during the night. Second, they repeatedly follow the same travel routes, which suggests that the routes have been memorized in order to facilitate movement in relative darkness. Third, owl monkeys are more active and travel farther when the moon is bright. They are similar to tarsiers in this regard and nocturnality is probably a secondary adaptation in both.

Owl monkeys use all strata of their forest habitats except the ground, but prefer the higher parts of the canopy. They tend to occupy forests with dense canopy and high plant diversity, but other factors may be involved. The monkeys were absent from such an appropriate forest in one habitat, probably because the hardwood trees did not provide them with sleeping holes (Svensson et al. 2010). The proportion of leaves in the diet varies widely, rising to 46 percent of feeding time in at least one population during periods of fruit scarcity.

Owl Monkey Sociality

Owl monkeys are socially monogamous and probably reproductively monogamous. All field reports describe social units containing only two adults. There is rarely more than one female in a group, never more than one breeding female, and never more than one adult male engaged in infant care (Svensson et al. 2010). A pair may be accompanied by offspring of different ages, including adult-sized individuals that have delayed emigration. The maximum group of six is smaller than typical units among the other pair-grouping species of the New World. There are significant numbers of solitary individuals, apparently older monkeys that have been expelled from pair-groups or young emigrants that have yet to join a new group. The small size of groups may be related to the distribution of owl monkey food into small, predictable, uniform patches. These primates prefer to feed in relatively small-canopied trees.

Social interactions seem rare, but this might be due to variation in activity levels in relation to moonlight. Just as they travel more during the full moon, owl monkeys also engage in more social activity. These are generally affiliative behaviors; the level of aggression within the group is low. However, even when they are socially active, owl monkeys rarely engage in allogrooming (Wolovich et al. 2006). Owl monkey mates occasionally sit with their tails overlapping, deliberately holding them together.

Food sharing may be the most important affiliative behavior in owl monkeys. The proportion of food interactions between adult males and females that result in transfer of food is much higher than in other species in which such behavior has been studied. Furthermore, food transfer between pair members occurs in normal conditions with little or no begging. In these captive monkeys food transfers took place even though any individual could easily acquire food (Wolovich et al. 2008).

During sexual encounters the male may groom the female, but this does not occur in all cases. The only copulation described was a single brief intromission with no preliminaries. Afterward, the pair rested together for five minutes and then returned to foraging (Wright 1994). In one long-term study observers saw just eight instances of mating behavior over the course of 2000 hours.

As in many strepsirrhines, the nocturnal life of the owl monkeys is associated with extensive use of olfactory communication (MacDonald et al. 2008; Wolovich et al. 2010). Social interactions between captive mates include anogenital sniffing, nose sniffing, urine drinking, and marking the partner with the subcaudal gland (located below the tail). Owl monkeys also engage in scent marking with their muzzles, sternal glands, and urination. They have the largest olfactory bulbs among the platyrrhines, with apocrine glands on the lips, nose, and brow, as well as the perianal area. Chemical analysis indicates that perianal secretions could be used to distinguish sex, age, and family membership.

Sex identification is probably important because same-sex owl monkeys fight more intensely than those of opposite sex. Aggression in captivity is preceded by olfactory inspection. Making this distinction is probably important during the intergroup encounters of wild owl monkeys. In addition to the handicap of darkness, sex identification is made difficult by monomorphy, including the lack of any conspicuous difference in the appearance of external genitalia. Differentiation of older and younger individuals by scent might preclude aggression against sexually mature but non-reproducing members of a pair-group. Distinction of pair-groups by scent might facilitate the choice of a new pair-group by a dispersing individual. Differentiation of sexes and age groups by scent could be wholly or partly innate, but distinction of pair-groups formed from unrelated individuals must have an environmental origin.

Home range overlaps are slight and neighboring groups regularly meet at borders, where they usually engage in agonistic behavior that involves both sexes equally. Behaviors include whooping vocalizations produced by inflated throat pouches, piloerection, swaying and arching, stiff-legged jumping, chasing, and wrestling. These episodes can last anywhere from a few minutes to almost half an hour. Like other social interactions, these conflicts are more common when the moon is full or directly overhead.

Conflicts may be related to resources, as indicated by one population in which conflicts took place at fruiting trees on the borders (Wright 1994). However, reproductive competition may be involved. In captivity males and females only fight individuals of their own sex. This is not reported for the wild, but perhaps this is because the sexes are so difficult to distinguish in these monomorphic species. Aggression can be intense. Fights have been observed in several species and most of the adults in one population had broken or missing canine teeth. Captive experiments sometimes result in fighting so severe that the animals must be separated.

Owl Monkey Life Cycle

The gestation period is about the same as that of callitrichs, but relatively shorter in relation to the owl monkey's body size. At about 120 days this would be the shortest of any ceboid and suggests that neonates are extremely altricial. This could explain the fact that the mother carries her newborn for the first week or two in a distinctive position between abdomen and thigh, a warm spot that would compensate for undeveloped thermoregulation (Wolovich et al. 2008).

After the first week or so the male takes over infant care with greater intensity than among callitrichs. The father carries the infant on his back for more than 80 percent of waking time over the course of several months. He also plays with it, grooms it, protects it, and shares food with it. The mother rejects the infant after each nursing session, sometimes nipping it gently. The resulting protest brings the father for the transfer. In contrast to callitrichs, siblings do little to help with infants.

Carrying the infant can be a considerable burden for the father: he sometimes reaches food sources 10 to 15 minutes later than the others (Wright 1994). This suggests that the function of male carrying is to lessen the mother's energy expenditure. Protection from infanticide seems an unlikely alternative, since such attacks have never been seen in this genus. Eventually father and infant come into conflict. When the infant reaches four months of age the father tries to dislodge it by pushing and biting while the infant squeals and holds tight. Despite this troublesome transition, juveniles display greater attachment to the father than the mother (indicated by proximity).

In a captive group family members differed in sharing food with an infant aged between three and five months (Wright 1994). Out of 40 instances, the father accounted for 68 percent, a juvenile sibling 23 percent, and the mother 10 percent. When the infant approached the mother as she fed, she usually looked away or turned her back. This suggests that maternal responses are fixed in at least one species of the genus.

Food sharing between adults may play a role in successful reproduction (Wolovich et al. 2008). Females increase their rate of begging for food when they are lactating and males respond. Higher rates of such transfer are correlated with shorter inter-birth intervals. Insisting on an economic model, the researchers concluded that "males do not pay for sex but do buy the milk". On the other hand, females transfer food to males and do so without regard to their own reproductive condition. This suggests that social bonding is also a function of food sharing. The researchers postulated that females exchange food for future paternal care. However, in the absence of evidence that males gain material benefit from such exchanges, an ethological model of ritualized communication seems more appropriate.

The juvenile is weaned at about seven or eight months, but stays with its parents long after that. Most individuals emigrate around the time of sexual maturation at about three years of age. The maturing individual seems to be drawn away by his or her own motivation, mainly or entirely sexual. There is no aggression against young adults in captive groups and no evidence for it in the wild. Young adults emit hooting calls (much like owl vocalizations) that are not taken up by any others in the pair-group, often as they approach territorial boundaries. Individuals in other groups respond and may also approach the border. The young adult lags behind his or her group for a month or so, sometimes sleeping in a separate tree, and finally leaves.

The timing of owl monkey dispersal connects it with the birth season (a second or third sibling may be the proximate trigger) and with a season of food abundance (which may make it easier to be alone for a while). Dispersal may also be connected with the upcoming mating season. Some individuals stay in their natal groups until they are four or five years old, and parents tolerate them as long as they are non-reproductive. The reason for delayed emigration may be saturation of the habitat.

Summary

The two zoological families in this chapter are relatively small-bodied and tend toward pair-groups. Many callitrich groups include non-breeding adults. The putative father and others carry the infants, usually twins, and share food with them. Owl monkeys, the only nocturnal anthropoids, are socially monogamous but display little affiliative behavior. Male care for young is unusually intense.

Sources

Basic Bibliography

Campbell, C.J., et al. (Eds.) (2011). *Primates in perspective* (2nd ed.). NY: OUP. [Chapters 7, 9, and 11]

Fernandez-Duque, E. (2012). Owl monkeys *Aotus* spp in the wild and in captivity. *IZY, 46*, 80–94.

Fernandez-Duque, E., et al. (2012). The behavior, ecology, and social evolution of New World monkeys. In J.C. Mitani et al. (Eds.), *The evolution of primate societies* (pp. 43–64). Chicago: UCP.

Ford, S.M., et al. (Eds.) (2009). *The smallest anthropoids, the marmoset/callimico radiation*. NY: Springer.

Garber, P.A., et al. (Eds.) (2009). *South American primates*. NY: Springer.

Additional References

Ah-King, M., & Tullberg, B. (2000). Phylogenetic analysis of twinning in Callitrichinae. *AJP, 51*, 135–146.

Alba-Mejia, L., et al. (2013). Spatiotemporal interactions among three neighboring groups of free-ranging white-footed tamarins (*Saguinus leucopus*) in Colombia. *IJP, 34*, 1281–1297.

Almond, R.E.A., et al. (2008). Changes in prolactin and glucocorticoid levels in cotton-top tamarin fathers during their mate's pregnancy: the effect of infants and paternal experience. *AJP, 70*, 560–565.

Amora, T.D., et al. (2013). Use of alternative plant resources by common marmosets (*Callithrix jacchus*) in the semi-arid caatinga scrub forests of northeastern Brazil. *AJP, 75*, 333–341.

Babb, P.L., et al. (2011). mtDNA diversity in Azara's owl monkeys (*Aotus azarai azarai*) of the Argentinean Chaco. *AJPA, 146*, 209–224.

Barbosa, M.N., & Mota, M.T.S. (2013). Alloparental responsiveness to newborns by nonreproductive, adult male, common marmosets (*Callithrix jacchus*). *AJP, 75*, 145–152.

Buchanan-Smith, H.M., et al. (1993). Responses of cotton-top tamarins (*Saguinus oedipus*) to faecal scents of predators and non-predators. *Animal Welfare*, *2*, 17–32.

Burkart, J.M., & van Schaik, C.P. (2010). Cognitive consequences of cooperative breeding in primates? *Animal Cognition*, *13*, 1–19.

Dacier, A., et al. (2006). Rapid habituation of scan behavior in captive marmosets following brief predator encounters. *BP*, *71*, 66–69.

Dell'mour, V., et al. (2009) Social learning and mother's behavior in manipulative tasks in infant marmosets. *AJP*, *71*, 503–509.

Diaz-Munoz, S.L. (2011). Paternity and relatedness in a polyandrous nonhuman primate: testing adaptive hypotheses of male reproductive cooperation. *AB*, *82*, 563–571.

Fleagle, J.G. (2013). *Primate adaptation and evolution* (3rd ed.). NY: Academic Press.

French, J.A. (2013). The role of androgenic steroids in shaping social phenotypes across the lifespan in male marmosets (*Callithrix* spp.). *AJP*, *75*, 212–221.

Goldizen, A.W. (1987). Facultative polyandry and the role of infant-carrying in wild saddle-back tamarins (*Saguinus fuscicollis*). *BE&S*, *20*, 99–109.

Gosselin-Ildari, A.D., & Koenig, A. (2012). The effects of group size and reproductive status on vigilance in captive *Callithrix jacchus*. *AJP*, *74*, 613–621.

Gunhold, T., et al. (2014a). Memory, transmission, and persistence of alternative foraging techniques in wild common marmosets. *AB*, *91*, 79–91.

Gunhold, T., et al. (2014b). Video demonstrations seed alternative problem-solving techniques in wild common marmosets. *BL*, *10*, DOI: 10.1098/rsbl.2014.0439.

Hilário, R.R., & Ferrari, S.F. (2010). Four breeding females in a free-ranging group of buffy-headed marmosets (*Callithrix flaviceps*). *FP*, *81*, 31–40.

Huck, M., et al. (2004a). The many faces of helping: possible costs and benefits of infant carrying and food transfer in moustached tamarins (*Saguinus mystax*). *Behaviour*, *141*, 915–934.

Huck, M., et al. (2004b). Proximate mechanisms of reproductive monopolization in male moustached tamarins (*Saguinus mystax*). *AJP*, *64*, 39–56.

Huck, M., et al. (2005). Paternity and kinship patterns in polyandrous moustached tamarins (*Saguinus mystax*). *AJPA*, *127*, 449–464.

Jefferson, J.P., et al. (2014). Owl monkeys (*Aotus* spp.) perform self- and social anointing in captivity. *FP*, *85*, 119–134.

Joyce, S.M., & Snowdon, C.T. (2007). Developmental changes in food transfers in cotton-top tamarins (Saguinus oedipus). *AJP*, *69*, 955–965.

Kasper, C., et al. (2008). Tolerated mouth-to-mouth food transfers in common marmosets. *Primates*, *49*, 153–156.

Kawai, N., et al. (2014). Marmoset monkeys evaluate third-party reciprocity. *BL*, *10*, DOI: 0.1098/rsbl.2014.0058.

Kemp, C., & Kaplan, G. (2011). Individual modulation of anti-predator responses in common marmosets. *IJCP*, *24*, 112–136.

Lazaro-Perea, C., et al. (2004). Grooming as a reward? Social function of grooming between females in cooperatively breeding marmosets. *AB*, *67*, 627–636.

Macdonald, E.A., et al. (2008). Sex, age, and family differences in the chemical composition of owl monkey (*Aotus nancymaae*) subcaudal scent secretions. *AJP*, *70*, 12–18.

Mota, M.T.daS., et al. (2006). Hormonal changes related to paternal and alloparental care in common marmosets (*Callithrix jacchus*). *H&B*, *49*, 293–302.

Nunes, S. et al. (2001). Interactions among paternal behavior, steroid hormones, and parental experience in male marmosets (*Callithrix kuhlii*). *H&B*, *39*, 70–82.

Penate, L., et al. (2009). Reconciliation in captive cotton-top tamarins (*Saguinus oedipus*), a cooperative breeding primate. *AJP*, *71*, 895–900.

Petracca, M.M., & Caine, N.G. (2013). Alarm calls of marmosets (*Callithrix geoffroyi*) to snakes and perched raptors. *IJP*, *34*, 337–348.

Porter, L.M., & Garber, P.A. (2004). The Goeldi's monkey: a primate paradox? *EA*, *13*, 104–115.

Porter, L.M., & Garber, P.A. (2013). Foraging and spatial memory in wild Weddell's saddleback tamarins (*Saguinus fuscicollis weddelli*) when moving between distant and out-of-sight goals. *IJP*, *34*, 30–48.

Prudom, S.L., et al. (2008). Exposure to infant scent lowers serum testosterone in father common marmosets (*Callithrix jacchus*). *BL*, *4*, 603–605.

Rapaport, L.G. (2001). Food transfer among adult lion tamarins: mutualism, reciprocity or one-sided relationships? *IJP*, *22*, 611–629.

Rapaport, L.G. (2006). Provisioning in wild golden lion tamarins: benefits to omnivorous young. *BE*, *17*, 212–221.

Rapaport, L.G., & Ruiz-Miranda, C.R. (2002). Tutoring in wild golden lion tamarins. *IJP*, *23*, 1063–1070.

Rapaport, L.G. & Ruiz-Miranda, C.R. (2006). Ontogeny of provisioning in two populations of wild golden lion tamarins, *Leontopithecus rosalia*. *BE&S*, *60*, 724–735.

Sanchez Rodriguez, S.M., et al. (2008). Body weight increase in expectant males and helpers of cotton-top tamarin (*Saguinus oedipus*). *Psicothema*, *20*, 825–829.

Schaffner, C.M., & French, J.A. (2004). Behavioral and endocrine responses in male marmosets to the establishment of multimale breeding groups. *IJP*, *25*, 709–732.

Searcy, Y.M., & Caine, N.G. (2003). Hawk calls elicit alarm and defensive reactions in captive Geoffroy's marmosets (*Callithrix geoffroyi*). *FP*, *74*, 115–125.

Smith, A.C., et al. (2001). Multiple breeding females and allo-nursing in a wild group of moustached tamarins (*Saguinus mystax*). *Neotropical Primates*, *9*, 67–68.

Smith, A.S., et al. (2010). Manipulation of the oxytocin system alters social behavior and attraction in pair-bonding primates, *Callithrix penicillata*. *H&B*, *57*, 255–262.

Snowdon, C.T., & Cronin, K.A. (2007). Cooperative breeders do cooperate. *BP*, *76*, 138–141.

Snowdon, C.T., et al. (2010). Variation in oxytocin is related to variation in affiliative behavior in monogamous, pairbonded tamarins. *H&B*, *58*, 614–618.

Svensson, M.S., et al. (2010). Density estimates of Panamanian owl monkeys *Aotus zonalis* in three habitat types. *AJP*, *72*, 187–192.

Tardif, S.D., & Bales, K. (1997). Is infant-carrying a courtship strategy in callitrich primates? *AB*, *53*, 1001–1007.

Tello, N.S., et al. (2002). Boa constrictor attack and successful group defence in moustached tamarins, *Saguinus mystax*. *FP*, *73*, 146–148.

Thompson, C.L., et al. (2013). Spatial distribution and exploitation of trees gouged by common marmosets (*Callithrix jacchus*). *IJP*, *34*, 65–85.

Welker, C., & Schafer-Witt, C. (1987). On the carrying behaviour of basic South American primates. *Human Evolution*, *2*, 459–473.

Wolovich, C.K., et al. (2006). Social patterns of food sharing in monogamous owl monkeys. *AJP*, *68*, 663–674.

Wolovich, C.K., et al. (2008). Food transfers to young and mates in wild owl monkeys (*Aotus azarai*). *AJP*, *70*, 211–221.

Wolovich, C.K., et al. (2010). Insect-foraging in captive owl monkeys (*Aotus nancymaae*). *FP*, *81*, 63–72.

Wright, P.C. (1994). The behavior and ecology of the owl monkey. In J.F. Baer, & 2 co-editors, *Aotus, the owl monkey* (pp. 97–112). Elsevier.

Zahed, S.R., et al. (2008). Male parenting and response to infant stimuli in the common marmoset (*Callithrix jacchus*). *AJP*, *70*, 84–92.

Ziegler, T.E. (2013). Social effects via olfactory sensory stimuli on reproductive function and dysfunction in cooperative breeding marmosets and tamarins. *AJP*, *75*, 202–211.

Ziegler, T.E., et al. (2006). Pregnancy weight gain: marmoset and tamarin dads show it too. *BL*, *2*, 181–183.

15 Cebines

Squirrel Monkeys and Capuchins

The brown monkey with a black cap carefully positions a nut in a depression in the large stone that will function as an anvil. The monkey picks up a somewhat smaller rock in both hands, raises it to head level, and smashes it down on the nut. A meal begins.

This chapter completes our survey of the platyrrhines and our examination of the cebid family. It covers three of the best-known primate genera: the squirrel monkeys (*Saimiri*) and the capuchins, recently divided into the genera *Cebus* and *Sapajus* (Box 15.1). The former are called untufted because they lack the head ornaments of the other genus. They are also called the gracile capuchins because they are more lightly built. We divide the discussion of the increasingly complex subfamily Cebinae as follows:

1 a brief survey of the common features of the cebine subfamily
2 common features and major variants in the squirrel monkeys (*Saimiri*)
3 common features of the capuchins (formerly a single genus)
4 behavior of the gracile capuchins (*Cebus*)
5 behavior of the robust capuchins (*Sapajus*).

Box 15.1 Cebine Taxonomy

The Cebinae contains the squirrel monkeys and the capuchin monkeys. All have short faces, broad premolars, dimorphic canine teeth, tiny third molars, and relatively large brains. With short and unspecialized digestive tracts, they make insects a large part of their diet. Squirrel monkeys (*Saimiri*) are the smallest non-callitrich New World primates. They weigh between 600g and 1200g. Their rounded brain cases contain large brains with unusual structural features.

Molecular studies suggest that the capuchins separated from the squirrel monkeys about 14 or 15 million years ago. Extant capuchins are about five times larger than the squirrel monkeys. Their brains are even larger than those of squirrel monkeys relative to body mass. Semi-prehensile tails and more flexible fingers also distinguish them from their closest living relatives.

Capuchins can be divided into the genera *Sapajus* (tufted) and *Cebus* (untufted), based on morphology and supported by biogeographic analysis and behavioral observations. *Sapajus* is more compact and robust in the skull, jaws, limbs, and tail. *Cebus* is comparatively slender and longer-limbed.

The Cebine Subfamily

All three genera in this taxon are widely distributed in South and Central America and survive in a wide range of forest habitats. Their diets emphasize insects and fruit, and features of the skull, face, and dentition have been described as "prey based". The monkeys live in multi-male troops that maintain cohesion with constant calls when they travel. These calls convey information about the callers' emotional state and their immediate environment. Alarm calls are functionally referential, including a distinction between snakes that are on the ground and in the trees. Across the cebine species the encephalization quotient varies from 2.5 to 3.5 (chimpanzees range from 2 to 2.5).

Squirrel Monkeys

Squirrel monkeys (Figure 14.1) are widely distributed in the Amazon and Orinoco basins, and a separate population lives in Central America. Squirrel monkeys are small, within the weight range of nocturnal lemuroids, and sexual dimorphism varies from about 15 to 30 percent.

Squirrel Monkey Ecology

All squirrel monkey species prefer the lower and middle canopies in lowland rainforests. They are significantly dimorphic, but both sexes are small. Consistent with their size, squirrel monkeys depend on animal matter for a substantial portion of their diet. They spend up to 80 percent of some ordinary days foraging for arthropods, usually seeking visible prey but also unrolling leaves. Despite their agility, these monkeys prefer slowly moving caterpillars to other insects. They also favor small soft fruits, such as berries, and the main variable in fruit consumption is availability.

Anti-predator features pervade squirrel monkey morphology and behavior (Boinski 1999). Branch movements resulting from squirrel monkey activities alert raptors, perhaps because there are so many individuals in a troop. Alarm calls distinguish between medium and large raptors. Squirrel monkeys occasionally mob or attack raptors and various mammalian carnivores (Boinski et al. 2005).

Association with capuchins is a recurring pattern that provides defensive and subsistence benefits to squirrel monkeys. The larger and more vigilant capuchins produce elaborate alarm calls that squirrel monkeys respond to more readily than to their own. Capuchins know the location of fruiting trees when food is scarce and they leave behind bits of fruit from hard-shelled species that squirrel monkeys cannot open. Knowledge of fruiting trees suggests that capuchins might have better memories. Though the degree of encephalization is about the same, capuchin brains are larger than those of squirrel monkeys in absolute terms.

Squirrel monkeys may actively look for capuchins and maintain contact for days at a time. However, proximity to capuchins was at the chance level in a population where it would have imposed foraging costs on the squirrel monkeys (see Boinski 1999). This indicates that the squirrel monkeys assessed cost and benefit in particular situations. Capuchins apparently get no benefit from this association. They do not wait for lagging squirrel monkeys, and though generally tolerant, they sometimes chase them away.

Squirrel Monkey Sociality

Squirrel monkeys live in cohesive troops ranging in size from about 15 to 75, varying in mean size from one species to another. A recurring pattern is that residents avoid going closer than about 10 meters from the troop perimeter, apparently due to predation risk. An exception occurs where an unusually dense understory hides the monkeys.

Among recurring social patterns (Boinski 1999; Boinski et al. 2005), affiliative interactions between kin are frequent: proximity, feeding tolerance, and exchange of vocalizations. However, grooming among adults is almost nonexistent. Social aggression is rare in females, but common in males. Males threaten with a distinctive erect penis display. The unique pile-up display, in which males climb onto each others' backs, is associated with penile displays and fights and thus seems to be connected with dominance (Defler 2004). During anxious or exciting situations, squirrel monkeys engage in urine washing. The individual secretes a few drops of urine into the palm of the hand and then rubs the sole of the foot.

Montagu et al. (2014) combined behavioral observation with genetic analysis to illuminate some aspects of squirrel monkey troops. Females preferred to be within 5 meters of close female relatives and this may account for one aspect of subsistence: females captured insect prey at higher rates than males, possibly because the insects were flushed into view by nearby companions. Due to philopatry so an individual is likely to have at least a few relatives to fulfill this role.

Mating takes place during a two-month period each year and each female is available for about two days. The fatted male is a recurring pattern. Before and during the mating season, weight increases by as much as 20 percent from the accumulation of water between the skin and muscles, especially along the shoulders, arms, and back. The mechanism and degree of change are unique among primates. It seems to be a result of changing hormone levels, and the largest males are those with the highest levels of testosterone. This seems to be an honest signal, since females prefer the largest male. There is no courtship and the most common copulation pattern is a single intromission of less than 30 seconds.

Another recurring pattern is the affiliation of males into same-age coalitions for invasion of another troop. Resident males, numerous in many cases, attack intruders and bloody fights may result. Even if an invasion is repelled, one or more intruders may achieve a brief copulation with a resident female during the confusion.

Arching over these day-to-day aspects of everyday life, there are three drastically different social systems associated with three different species of squirrel monkey. These represent the greatest geographic variation to be found among closely related taxa. These social systems are compared in Box 15.2.

Squirrel Monkey Life Cycle

While the birth season for a population may last for two months, births within a particular troop often occur within a period of two weeks. It may be that the gestation period is adjusted to achieve birth synchrony. Females are subjected to intense genital inspection shortly before parturition. This could be a functionless response to pre-parturition odors, or the behavior might actually trigger the birth.

Evidence from one population suggests that the function of birth synchrony is related to predator defense (Boinski 1987). The birth period coincided with a sharp seasonal increase in raptors, and during this time females with neonates clustered together and displayed heightened vigilance. These associations loosened when infants became capable of

Box 15.2 Squirrel Monkey Social Systems

Central American	*Guianan*	*Black Capped*
egalitarian	linear dominance	females dominant
males integrated	males integrated	males peripheral
aggression rare	aggression level high	females aggressive
male–male affiliation	male–male affiliation	male–male affiliation
no female affiliation	no female affiliation	female kin relations
males philopatric	both sexes disperse	females philopatric*
intertroop neutral	intertroop hostile	sometimes friendly

*Dispersing males form the only known all-male groups among platyrrhines.

independent escape at about three months. At that point the cost of food competition among mothers probably outweighed the benefit of protecting offspring.

The neonate is precocious. It clings immediately after birth and may move to the mother's back at an early age. Up to five weeks, other females are interested in the infant and carry it. After that, the infant begins to leave its mother for exploration. The only significant allogrooming in squirrel monkeys takes place between mother and infant during the third and fourth months. At this time the baby begins to engage in peer interactions, especially play. Soon after that, adult males begin to interact more with the mother in anticipation of the coming mating season. The infant directs penile displays toward the males and is ignored.

Stone (2007a) tested juvenile perception of predation risk by setting up feeding platforms in different locations with different surroundings. In the rainy season, when natural food was relatively abundant, juveniles visited high-reward platforms only if they also offered a high level of concealment from predators. In the dry season they chose platforms based only on food quantity. Adults chose on the basis of food quantity throughout the year.

Comparison of natural behavior by Stone (2007b) did not support the hypothesis that primates need a prolonged juvenile phase to acquire complex foraging behaviors. Even young juveniles (8 to 12 months) were as efficient as adults in the capture and processing of prey, and the only limit on plant foraging was physical inability to open husked fruits. The data also provided a test of the Ecological Risk Aversion Hypothesis of Janson and van Schaik (slow growth of juvenile primates is a response to the risks of starvation and predation; juveniles avoid predation by proximity to adults and this is allowed because slow growth mitigates feeding competition). Adult squirrel monkeys did not have preferential access to fruit patches, and competition for prey was negligible (Stone 2007b).

Most females give birth by the age of three years, while males only reach reproductive maturity at five or six. Dispersal varies from species to species as to timing and the predominant sex. Males emigrate at or before sexual maturity. Females emigrate under varying conditions and encounter no resistance from residents of the troops that they enter. Typically young males try to join non-natal troops for reproductive opportunities, which is critical because females rarely engage in extra-group copulations.

Squirrel Monkey Cognition

Squirrel monkey mental processes are of special interest because the genus has a high ratio of brain to body weight. Various adaptive explanations include relatively large home ranges, variable diets, and/or complex foraging patterns. Two squirrel monkeys were given choices between arrays of food that varied in quantity (McKenzie et al. 2004). Both subjects showed a significant preference for larger quantities, such as 20 versus 10 peanuts. In the second part of the experiment the offered food was either replenished or reduced 15 minutes after the initial choice, such that the initial choice of a smaller quantity led to more food eventually. This caused a significant decline in the choice of the larger quantities, which suggests that the squirrel monkeys anticipated future consequences.

Smith et al. (2003) trained a squirrel monkey and an Old World monkey to represent the numerosities 1 to 4 in ascending order. When tested with novel stimuli of the same numerosities, both primates performed in a way that seemed to be based on numerical attributes. They both successfully ordered pairs of numerosities. Measurement of accuracy and latency indicated that both primates were responding to numerosity. The researchers inferred that both primates perceived ordinal relations, contrary to the assumption of greater intelligence in Old World monkeys.

Capuchin Monkeys

The geographic distribution of capuchins (Figure 15.1) rivals that of howler and spider monkeys. Capuchins are medium-sized by platyrrhine standards. The average male weighs 20 to 25 percent more than the average female, and his canine teeth are 15 to 20 percent

Figure 15.1 The white-faced capuchin monkey is an untufted species.
© Rhett A. Butler/www.mongabay.com

larger. However, there is considerable overlap between the sexes. Capuchins are divided into two genera (Box 15.1). *Sapajus* are called tufted because there is a tuft of hair on each side of the head, which *Cebus* lack. The latter are called gracile capuchins because they are more lightly built. Significant behavioral differences between the taxa will be considered after a general introduction.

Capuchin Ecology

The short tail is called "semi-prehensile" because it lacks the flexibility of atelid tails, but is used in much the same way in trees—as an anchor during foraging and for stability during quadrupedal locomotion. Adults rarely suspend themselves entirely by the tail (though the young often do), but the use of tail and feet together gives the hands complete freedom. The tail is unusual, but capuchins have a handgrip that is unique among New World monkeys. They can pick up small objects because the fingers and thumb can be moved independently and the thumb is pseudo-opposable.

Capuchins do well in varied environments from primary forests to swamps. They also survive heat stress and water scarcity in hot, dry habitats (Campos and Fedigan 2009). Though ordinarily frugivorous, they can subsist almost entirely on other plant products, including seeds, during times of stress. Capuchins spend more time on the ground than any other platyrrhine, largely in search of insect prey under leaf litter. They obtain most of their animal protein from insects that are cryptic or actively defensive. Along with insect bites and stings, the monkeys endure the bites of mammalian prey animals (e.g. squirrels) that may be a third of their weight. Some capuchins exploit shellfish and crabs (Port-Carvalho et al. 2004).

Capuchins are too large to obtain prey by stealth, so they emphasize the search for hidden animals in holes, under bark, and in rotten branches. To get access they bite, rip, twist, pull, and smash. This kind of foraging takes more time and energy, so capuchins can't use insects for fallback to the same extent as squirrel monkeys when fruit is scarce.

Diverse foraging strategies allow capuchins to minimize competition within the troop, between troops, and with other species. Dietary flexibility gives them an advantage over sympatric primates with regard to alternate food sources during times of scarcity. This does not mean that the diet is low quality. On the contrary, the small cecum shows that they require easily digestible foods. They can obtain such food under a variety of circumstances because of morphological and/or behavioral features: powerful jaws in some species and tool using in others (Wright et al. 2009).

All troop members are active in predator detection and alarm response, but adult males are the most prominent in both. There are probably several reasons (which apply in varying degrees to other primates). First, males are more often on the ground or the troop periphery. Second, males look outward to detect other troops or extra-troop males. Third, females have less time for vigilance because they eat more than males to support pregnancy and lactation.

Self-care includes anointing, in which the animal rubs its fur with pungent and sometimes topically irritating plant or animal materials. Many of these substances are known to have antibacterial or insect-repelling qualities, and anointing takes place during the rainy season when mosquitoes are most likely to be a problem. Some other primates engage in anointing, but capuchins do it with more intensity and perhaps greater frequency. Capuchins also engage in urine washing, which some strepsirrhines use in olfactory communication. However, at least some capuchin urine washing may function in thermoregulation: females do it during the part of their ovarian cycle when body temperature is elevated.

Capuchin Sociality

Capuchins are female-philopatric and affiliation among females is stronger than relationships with males except for attraction to the alpha male. Affiliation is represented by grooming and by proximity in traveling, resting, and sleeping. Grooming rates usually accord with patterns of dominance, coalition, aggression, kinship, sexuality, and the presence and age of infants. Females do most of the grooming and are most likely to groom other females, especially those with infants. Males groom less than females and rarely groom each other, although dyads may be marked by proximity.

A typical troop contains an obvious alpha male. In some troops an alpha female ranks just below him and can dominate subordinate males as well as females. Otherwise, hierarchies are vague and major rank changes are usually signaled by shifts in female proximity patterns. Alpha males vary across groups and taxa with regard to rigidity and tolerance. One factor is the presence or absence of desirable food in defensible patches.

Males generally dominate females, but females sometimes form coalitions to drive males from food trees, even the alpha male, overcoming sexual dimorphism with cooperation. Most males are tolerant of females feeding nearby (or perhaps wary of coalitions). Despite some conflicts, male–female relationships are usually affiliative and there is little physical aggression between the sexes. Particular individuals frequently groom each other and maintain proximity.

Male–male relationships are the least affiliative among adults, and aggression between males is more severe than between females. Details vary. In one species an alpha male keeps other males spatially and socially peripheral. He receives almost all mating invitations from females, and is believed to be the only male who produces offspring. Despite being peripheral, subordinate males cooperate in troop defense. They are younger and smaller than the alpha, which suggests that these are age-graded troops in which the alpha tolerates the continued presence of his own maturing offspring.

Another species is more variable. Although it is likely that an alpha male fathers more infants than other resident males, the latter get some mating opportunities. This probably accounts for the fact that they reliably cooperate in troop defense. Finally, there is a species in which males form a cohesive aggregation in the middle of the troop and take turns feeding. There is little agonistic behavior, mating is equitable, and the males are highly cooperative in defending the troop.

Defense against outsider males, which may be fatal (Gros-Louis et al. 2003), means more than preventing the occasional stolen copulation. Coalitions of intruders invade troops and attack the resident males (Fedigan and Jack 2004; Ramirez-Llorens et al. 2008). If they are successful, infanticide often follows. An alpha male may tolerate central location, feeding, and mating by subordinate males because he needs them to defend against these invasions. Data from one site indicate that such intrusions occur every three or four years.

Sexual behavior distinguishes capuchins from all other New World primates. Ovulation is signaled overtly by means of prolonged and insistent proceptive behavior. Females use facial expressions, vocalizations, gestures, and body postures to communicate their interest. A female typically courts one male during most of her estrus period. Often, but not always, this is the alpha male. The male does not respond to her initial advances, perhaps conserving sperm for a time closer to ovulation. It seems that capuchin males may be limited to one ejaculation per day.

When the male does respond, the interaction seems to reflect the amity of the sexes in everyday social life. Male and female perform an elaborate courtship ritual together. This often includes a "dance" in which the couple performs a series of coordinated movements. Copulation consists of multiple mounts, which means a low ejaculation rate. The male continues courting the female after ejaculation, which may be a form of mate guarding. Sperm competition also seems to be a factor, judging from the fact that males have moderately large testicles for their body size. A female may remove solidified ejaculate from the vagina after copulation, which suggests female choice.

Capuchin Life Cycle

The capuchin's life history schedule is slow for its body size. It is similar to that of spider monkeys, which are four times larger. Gestation lasts more than five months and the inter-birth interval ranges from 1.5 to 2 years. Births occur throughout the year, but there are peaks that correspond to food abundance. The capuchin neonate has only about 9 percent of its mother's body mass, but it grows rapidly at first. Much of the early weight gain comes from the brain, which quickly reaches 35 to 50 percent of adult size. Survivorship of infants is most significantly affected by whether or not there is a turnover in the troop's male cohort (Fedigan et al. 2008) because infanticide is likely. About 20 percent of infants die in the first six months, although there may be a great deal of variation in that figure.

A new mother is likely to allow only relatives to approach, but all troop members will probably carry the infant at one time or another. Some other mothers will probably nurse the baby. This unusual degree of alloparenting might be functionally significant because the troop is often spread out over a wide area while foraging, and the increasingly ambulatory infant may become separated from its mother. The pattern might be explained by kin selection.

Reciprocity is an alternative. A small amount of milk for the baby of another female could be a small investment to make in order to get the same in return. Reciprocity is plausible because cheaters are likely to be exposed: infants often throw tantrums when denied suckling, whether by their own mothers or other females. Offspring of dominant females may receive more allonursing than those of subordinates, suggesting that ingratiation is a proximate motivation (Baldovino and Di Bitetti 2008).

Weaning begins and is completed during the second year, with adults rejecting clinging as well as nursing. The sparse literature on juveniles suggests that they tend to stay close together in age-class clusters. However, they may try to develop relationships with adults by allogrooming, allomothering, and providing coalitionary support. In at least one species juveniles seem to be highly attracted to adult males and watch them intently.

A female begins to display sexual interest in males by the age of four years, but the first birth is likely to take place at age seven or eight. Male development is even slower. Sexual maturity comes after seven years, but the male does not attain full body size until after ten years. A male emigrates around the time of sexual maturity and joins another troop, usually non-aggressively. He will probably transfer several more times during his life, aggressively when he is in his prime.

Although they are the philopatric sex, females sometimes emigrate (Jack and Fedigan 2009). This is probably not due to incest avoidance because the majority of dispersers had already given birth and father–daughter mating was not possible because male tenure did not exceed female age at first birth. Feeding competition may be a factor since female emigrants tended to leave during the dry season and they had fewer relatives in the troop than those who stayed. The data are also consistent with avoidance of infanticide because females are more likely to emigrate during years of male takeovers.

Capuchins lead longer lives than other primates of their size. They can live for more than 50 years in captivity, and the average lifespan for wild individuals has been estimated at 30 years. Older capuchins (16 years and up) remain physically and socially active. One possible factor in delayed maturation is the unusually slow postnatal growth of the capuchin brain. A greater proportion of brain development takes place postnatally in this genus than in others.

Capuchin Cognition

Organization of some capuchin brain structures is unique among primates, except for the closely related squirrel monkeys. These structures seem to be related to sensory functions and their connection with other brain functions. Capuchins also have a large ratio of brain to body weight. The same explanations have been applied to both genera: relatively large home ranges, variable diets, and complex foraging patterns.

From the time that an infant can first reach out and touch a surface or grasp an object, it explores its environment with hands and teeth with a persistence that suggests innate curiosity. From one to three years old the infant frequently manipulates non-edible objects. The ontogeny of manipulation seems to start with innate components in perception and motivation.

Critical motor skills that distinguish capuchins from other ceboids are related to producing changes in surfaces or objects, or in the relationships between objects. They use their bodies in innovative ways; for example, in an experiment, surrounding multiple cups with tail and feet while working on them. These skills form the basis for tool use, which we find mainly in tufted capuchins. Beyond simply using any manipulable object at hand, capuchins choose objects for tools that are suitable for particular tasks. They learn to avoid irregularities in a tool surface that make it difficult to succeed. They can bring relevant objects together, such as carrying nuts to a rock pile. They can modify an object to accomplish a task. They can often generalize the use of an object from one context to another. They can solve the same task using different objects. Capuchins learn to perceive spatial relations among objects and to achieve specific outcomes through action with objects. This kind of knowledge can lead to an abstract appreciation of causal relations.

Capuchins are good candidates for social learning because they are relatively tolerant of each other, interested in each other, and interested in food and objects held by another capuchin. While eating, capuchins permit companions to remain close and collect dropped food or even take food directly. Infants are particularly tolerated in this regard.

Capuchin social learning is based on observation, but not imitation. Watching another monkey solve a problem does not result in learning a sequence of actions; rather, it promotes the observer's interest in the object and places that are associated with the work. Capuchins learn *with* another monkey rather than *from* another. A possible cultural dimension is represented by the fact that conspecifics at different sites may use different substances even though the same ones are available at both locales.

Gracile Capuchins

Untufted capuchins are like most other monkeys in many ways. They range widely to obtain high-quality foods, especially fruit, they live in large and cohesive troops, and female sexual behavior is consistent with paternity confusion (Lynch Alfaro et al. 2014). An intriguing observation on foraging behavior is that sex differences may be due to the typical platyrrhine pattern of sex differences in color vision. Gracile capuchins differ from the robust in that they are often seen on the ground. Perhaps their longer limbs facilitate terrestrial locomotion.

The larger troops of gracile capuchins contain more males and occupy larger home ranges, which means that an alpha male cannot monopolize food sources or females. He benefits more from the presence of other adult males who assist in agonistic encounters with neighboring troops and potential usurpers. The same benefits accrue to the philopatric females. Matthews (2009) studied one troop to test the Socioecological Model. He confirmed the model as to ranging and feeding, but found that direct ecological and demographic effects, as well as local conditions, influenced group size.

Cebus display strong bonds between males, which are manifested in close cooperation during hunts (Leca et al. 2007). The alpha male tolerates sexual behavior by other males, and troop members in general are tolerant of the close proximity of others during foraging. Aggression is typically bidirectional and followed by intense reconciliation. Leadership is distributed; any individual may initiate troop movement. Immigrant males form alliances with resident males for intertroop encounters and have affiliative and tolerant relationships within the troop (Schoof and Jack 2014). Dyads composed of subordinate males have the strongest and most equitable bonds. At the same time, much of the vigilance displayed by the alpha male is directed toward other males as probable sexual competitors.

Capuchin mammal hunts differ from their other faunivorous behavior. Adult males often travel away from the troop to places where the density of squirrels or coatis is high. Some hunts seem cooperative. For example, one monkey slaps at a female squirrel or coati while another approaches from behind and grabs the infants in the nest beneath her. Occasionally meat is shared by tolerated theft. There is little harassment of an individual who is eating meat, regardless of the possessor's age/sex class. Capuchins routinely act synchronously in the wild and their hunting behavior can be taken as a natural form of cooperation. In experiments they can learn to work together to achieve a shared goal, although there seem to be stringent limits. In pulling tasks, for instance, each individual has to see and feel the partner acting for cooperation to take place.

Studies of gracile capuchins indicated that females were responsible for the direction of troop movement. Petit et al. (2009) studied troop movements in an outdoor enclosure. They found that any individual could be the initiator and that the response was a collective decision. If three or more individuals followed the initiator, the group moved. Otherwise, the initiator gave up. According to the researchers this is a complex decision-making feedback system, such as had previously been studied only in insects. They interpreted the proximate basis for following as a mimetic process. The initiator's decision to stop arises from either a motivational state or a response threshold.

Cebus anointing fits with the sociability of the genus. They tend to do it in fluid subgroups of two to nine participants. In almost a third of the cases the initiator tries to recruit other participants and may stop quickly when left alone. Participants join in without regard for dominance, kinship, or other affiliation. This form of fur rubbing may create and reinforce social bonds. It may also be a test of those bonds. *Cebus* use a wide variety of leaves and flowers in anointing. In captivity they prefer onions and citrus fruits but readily use other substances ranging from hot peppers to wet-wipes. Onions have antifungal properties and some of the other materials might also have therapeutic effects.

Cebus occasionally use tools, as reported in 15 percent of all wild and captive studies (Lynch Alfaro et al. 2014). The tools used in wild populations are branches for defense and leaves to absorb liquid. An anecdote describes a male attacking a snake with a club after it was trapped under a heavy branch dislodged during mobbing. Some individuals used an orange peel and a leaf to obtain water and a branch as a prop to lean on while standing upright to watch birds (Urbani 1999). The only known systematic use of a "tool" (some would dispute this designation) involves a stationary and unmodified substrate. The monkeys open hard fruits and seedpods by rubbing, rolling, or pounding them against hard substrates such as trees (O'Malley and Fedigan 2005).

Robust Capuchins

Tufted capuchins (*Sapajus*) use robust jaws to eat hard nuts from small understory trees. These trees are densely and uniformly distributed, which allows the monkeys to live in small home ranges (Matthews 2009). In many of the concomitantly small troops, the alpha male is the only adult male. He monopolizes rich food sources but tolerates juveniles, who are likely to be his sons, in an age-graded troop.

The alpha male has the central role in the troop. He is closely monitored by others and receives the most grooming of all the males. However, the small contingent of adult males increases susceptibility to takeovers and the infanticide that follows. According to Verderane et al. (2013), tufted capuchins are consistent with the Socioecological Model in that females are philopatric, and they display linear dominance hierarchies and coalitions.

In tufted capuchins subordinate females exchange grooming for tolerance during feeding. This finding is important in several ways (Tiddi et al. 2011, 2012). It documents reciprocal exchange in which partner choice is based on benefits received, and it documents a platyrrhine case that conforms to a model developed for Old World monkeys.

Mixed results from research on grooming and female rank led Schino et al. (2009) to address the topic with a larger database than in previous studies and the genealogy of their captive group. Their subjects did not exchange grooming for rank-related benefits such as agonistic support or reduced aggression. Such exchanges might have been unprofitable because of the group's small size and strong kin bias.

Reunions are a dramatic aspect of tufted capuchin life. Male reunions are marked by a special display, the **scream-embrace** (Lynch Alfaro 2008; Phillips and Shauver Goodchild 2005). One explanation holds that the scream-embrace reduces tension between individuals with conflicting interests. Another interprets the behavior as a reaffirmation of affiliation. Field observations indicated that the two components of the display had different functions. The embrace risks attack by the other participant and thus strongly expresses affiliation. The scream (the loudest vocalization in the repertoire) calls attention to the act of embracing and thus advertises the alliance to other members of the troop.

Sapajus anointing fits a pattern of limited social bonding. *Sapajus* engage in fur rubbing mainly while alone and give no sign that they want to be joined. Dominance and kinship significantly affect any socializing that occurs in connection with fur rubbing (Leca et al. 2007). Initiation of fur rubbing by one individual results in greater proximity of others, but aggression then escalates and affiliative behavior decreases (Paulkner and Suomi 2008). The cause of these events is not clear, but they provide a significant contrast with the behavior of untufted capuchins.

Expressive behavior begins to appear in the infant at about one month of age, beginning with lip-smacking (an affiliative signal) and concluding with the open-mouth threat face. These facial displays are often exchanged with peers, less often with adults, and rarely with the mother (De Marco and Visalberghi 2007). Early in its first year the infant finds itself the subject of lip-smacking and carrying attempts by male and female adults and juveniles, peaking early and declining to low levels by the end of the year. Facial inspection of others by the infant begins when it is three to four months old and increases over time (Byrne and Suomi 2009).

Development of tool use begins when infants beat stones and other objects against substrates, and one-year-olds try to crack nuts. Learning, practice, and maturation are clearly implicated since the young watch experienced individuals and are unable to achieve the necessary coordination to crack nuts until their third year (Ottoni and Izar 2008). Scrounging may be the motivation that obliquely motivates the development of this potentially adaptive behavior: scrounging young watch nut crackers for the opportunity to gather tidbits, and they preferentially watch the most skillful nut crackers because they generate the most tidbits in the shortest amount of time.

Tiddi et al. (2010) investigated the applicability of Biomarket Theory to infant handling opportunities. Specifically, they studied the relationship between the relative availability of infants and the amount of grooming that mothers required before allowing access to their infants. Contrary to supply-and-demand, the number of infants in the group at a given time did not affect the amount of grooming necessary to gain access. The researchers suggested two hypotheses that are not mutually exclusive. Grooming may be a signal of "benign intent" that is not subject to market forces; or dominance relations among would-be handlers might take precedence over exchange competition.

Tool using appears in 75 percent of all *Sapajus* studies (Lynch Alfaro et al. 2014) and at least two species use tools regularly in the wild (Canale et al. 2009; Jalles-Filho and Teixera de Cunha 2008). The most widespread form is breaking hard shells (nuts or fruits) to obtain encapsulated food (Figure 15.2). *Sapajus* use stones to dig tubers out of soil and to cut open wood to obtain insects and larvae, and they bang stones in defense. They use sticks to probe for food and water in cracks, holes, and insect nests. In one population the monkeys use two different techniques to fish for termites: nest tapping and stick rotation (Souto et al. 2011).

Figure 15.2 A tufted capuchin monkey lifts a pounding stone, one of several tool-using behaviors that occur regularly in these New World primates. © MikeLane45 via iStock

Males use tools more frequently than females, especially in probing for prey with sticks. Moura and Lee (2010) suggested that this represents males signaling their investment quality as a mating attraction. Falotico and Ottoni (2014) connected probing for prey with the fact that males hunt more often and suggested that females might have fewer opportunities for social learning about probing tools.

Noting that all reports of natural capuchin tool use came from dry habitats, Ottoni and Izar (2008) proposed an ecological explanation. Their hypothesis is consistent with research on high-resistance foods in dry habitats: shell thickness is positively correlated with the volume of the contents, and the tools are essential to penetrate the shells (Visalberghi et al. 2008). Ottoni and Izar also noted that the most intensive tool use occurs in semi-free groups that are provisioned to some extent, experience low predation risk, and spend a great deal of time on the ground. Thus they favored terrestriality over food as the strongest predictor of tool use.

Further discussion concerned the Opportunity Hypothesis versus the Necessity Hypothesis (or the similar Energy Bottleneck Model). The former explains tool use in terms of the availability of desirable foods that require such processing; the latter sees tool use as a response to food scarcity. In the equivocal results of one study, tool use and consumption of fruit increased in the dry season, but the energetic return was low (Emidio et al. 2012). Another study clearly supported the Opportunity Hypothesis (Spagnoletti et al. 2012): the rate of tool use was not diminished by provisioning or natural availability of food, and the monkeys cracked the most common nuts when they were especially abundant.

Tool use in these capuchins has cognitive implications. Field experiments showed that wild *Sapajus* choose among potential stone tools and carry more effective ones to the work site (Visalberghi et al. 2009a, b). They bring heavier and harder ones to crack palm nuts and lighter stones to open less resistant items. When the effective tool is visually obvious, a monkey touches it first. Otherwise, it tests weights by moving, lifting, and/or tapping the stones. The monkeys always use their first choice and never discard it after the first strike. They draw on experience to recognize the affordances of objects and plan their use ahead of time. Remains of spontaneous nut cracking demonstrated that these experimental findings reflect natural behavior (Ferreira et al. 2010). Stone pounders found on anvils ranged in size from 0.2kg to more than 3kg, and weight was correlated with the associated remains of different kinds of nuts.

Selectivity was confirmed in the first study of another population (De Moraes et al. 2014). These monkeys used tools on five different food items, and the weight of their hammers and the characteristics of their anvils varied accordingly. In a field experiment by Massaro et al. (2012) monkeys made individual choices in which they balanced the weight of stones and the transport distance.

Another aspect of selectivity emerged from an experiment. Researchers observed that the monkeys consistently placed nuts into pits in the anvils (Fragaszy et al. 2013). Their consistency provided no control case for comparison, so the scientists enlisted a human to pound nuts in the pits and on the flat surface of the anvil. The human displaced the nuts in pits on proportionally fewer strikes, demonstrating that the monkeys use the more effective location.

Sapajus that use stick tools sometimes alter them to be more effective. They remove them from trees, resize them when necessary, and sometimes trim leaves and side branches. They also use the sticks in sequence with stones to open dead trunks and then probe for insects (Ottoni and Izar 2008). Another kind of sequence involves the use of a secondary tool to obtain a primary one. On three occasions monkeys used smaller stones to dislodge bigger ones from conglomerate rock and then used the latter as tools (Mannu and Ottoni 2009). Monkeys in an experiment used different tools sequentially to obtain a reward (Sabbatini et al. 2013). Some techniques that are present in one population and absent in others indicate social transmission (Sirianni and Visalberghi 2013). One long-term study provided evidence for a tool-using cultural tradition (Coelho et al. 2015). In the early observations of nut cracking juveniles were the most frequent performers, but eventually the behavior became commonly used by adults. Now the young observe older and more proficient individuals, reinforced by the opportunity to scrounge.

Summary

The morphology and behavior of the small-bodied squirrel monkeys is heavily affected by predation. They live in large troops with social organization that varies sharply across species. The fatted male is an unusual mating pattern. Capuchin monkeys adjust to a wide variety of environmental conditions, partly with the help of regular tool use. Grooming is important in their social behavior.

Capuchins engage in a dramatic reunion display comparable to that of the spider monkeys. Also like spider monkeys, they have a very slow life history for their body size. The complexities of capuchin tool use indicate cognitive skills that are unusual among primates. Tufted and untufted capuchins differ in tool use and other important aspects of behavior. Tufted capuchins especially engage in a variety of tool-using performance, some of which have profound implications for cognitive skills. These may include the ability to develop long-lasting behavioral traditions.

Sources

Basic Bibliography

Campbell, C.A., et al. (Eds.) (2011). *Primates in perspective*. NY: OUP. [Chapters 8, 11]
Estrada, A., et al. (Eds.) (2005). *New perspectives in the study of Mesoamerican primates*. NY: Springer.
Fragaszy, D.M., & 2 co-editors (2004). *The complete capuchin*. NY: CUP.
Kappeler, P.M., & Watts, D.P. (Eds.) (2012). *Long-term field studies of primates*. NY: Springer. [Part III]
Mittermeier, R.A., et al. (Eds.) (2013). *Handbook of the mammals of the world, Vol. 3. Primates*. Barcelona: Lynx Edicions. [pp. 348–413]
Perry, S. (2008). *Manipulative monkeys*. Cambridge, MA: HUP.

Additional References

Baldovino, M.C., & Di Bitetti, M.S. (2008). Allonursing in tufted capuchin monkeys (*Cebus nigritus*): milk or pacifier? *FP, 79*, 79–92.
Boinski, S. (1987). Birth synchrony in squirrel monkeys (*Saimiri oerstedi*): a strategy to reduce neonatal predation. *BE&S, 21*, 393–400.
Boinski, S. (1999). The social organization of squirrel monkeys: implications for ecological models of social evolution. *EA, 8*, 101–112.
Boinski, S., et al. (2002). An expanded test of the ecological model of primate social evolution: competitive regimes and female bonding in three species of squirrel monkeys (*Saimiri oerstedii, S. boliviensis*, and *S. sciureus*). *Behaviour, 142*, 525–632.
Boinski, S., et al. (2005). Dispersal patterns among three species of squirrel monkeys (*Saimiri oerstedii, S. boliviensis*, and *S. sciureus*): I. Divergent costs and benefits. *Behaviour, 142*, 525–632.

Byrne, G., & Suomi, S.J. (2009). Intimate social behavior in infant interactions in *Cebus apella*. *AJP*, *71*, 77–85.
Campos, F.A., & Fedigan, L.M. (2009). Behavioral adaptations to heat stress and water scarcity in white-faced capuchins (*Cebus capucinus*) in Santa Rosa National Park, Costa Rica. *AJPA*, *138*, 101–111.
Canale, G.R., et al. (2009). First record of tool use by wild populations of the yellow-breasted capuchin monkey (*Cebus xanthosternos*) and new records for the bearded capuchin (*Cebus libidinosus*). *AJP*, *71*, 366–372.
Coelho, C.G., et al. (2015). Social learning strategies for nut-cracking by tufted capuchin monkeys (*Sapajus* spp.). *AC*, DOI: 10.1007/s10071-015-0861-5.
De Marco, A., & Visalberghi, E. (2007). Facial displays in young tufted Capuchin monkeys (*Cebus apella*). *FP*, *78*, 118–137.
De Moraes, B.L.C., et al. (2014). Adaptability in stone tool use by wild capuchin monkeys (*Sapajus libidinosus*). *AJP*, *76*, 967–977.
Defler, T.R. (2004). *Primates of Colombia*. Bogota: Conservation International.
Emidio, R.A., et al. (2012). Energetic payoff of tool use for capuchin monkeys in the caatinga. *AJP*, *74*, 332–343.
Falotico, T., & Ottoni, E.B. (2014). Sexual bias in probe tool manufacture and use by wild bearded capuchin monkeys. *BP*, *108*, 117–122.
Fedigan, L.M., & Jack, K.M. (2004). The demographic and reproductive context of male replacements in *Cebus capucinus*. *Behaviour*, *141*, 755–775.
Fedigan, L.M., et al. (2008). Predictors of reproductive success in female white-faced capuchins (*Cebus capucinus*). *AJPA*, *137*, 82–90.
Ferreira, R.G., et al. (2010). Three stones for three seeds: natural occurrence of selective tool use by capuchins (*Cebus libidinosus*) based on an analysis of the weight of stones found at nutting sites. *AJP*, *72*, 270–275.
Fragaszy, D., et al. (2010a). Bearded capuchin monkeys' and a human's efficiency at cracking palm nuts with stone tools: experiments. *AB*, *79*, 321–332.
Fragaszy, D., et al. (2010b). How wild bearded capuchin monkeys select stones and nuts to minimize the number of strikes per nut cracked. *AB*, *80*, 205–214.
Fragaszy, D.M., et al. (2013). Wild bearded capuchin monkeys (*Sapajus libidinosus*) strategically place nuts in a stable position during nut-cracking. *PLONE*, *8*, e56182, DOI: 10.1371/journal.pone.0056182.
Gros-Louis, J., et al. (2003). Violent coalitionary attacks and intraspecific killing in wild white-faced capuchin monkeys (*Cebus capucinus*). *Primates*, *44*, 341–346.
Gunst, N., et al. (2010). The ontogeny of handling hard-to-process food in wild brown capuchins. *AJP*, *72*, 960–973.
Jack, K.M., & Fedigan, L.M. (2009). Female dispersal in a female-philopatric species, *Cebus capucinus*. *Behaviour*, *146*, 471–497.
Jalles-Filho, E., & Teixera de Cunha, R.G. (2008). Manipulation and tool use in captive yellow-breasted capuchin monkeys (*Cebus xanthosternos*). *IJCP*, *21*, 12–18.
Leca, J.B., et al. (2007). Social aspects of fur rubbing in *Cebus capucinus* and *C. apella*. *IJP*, *28*, 801–817.
Lynch Alfaro, J. (2008). Scream-embrace displays in wild black-horned capuchin monkeys. *AJP*, *70*, 551–559.
Lynch Alfaro, J., et al. (2014). Activity budget, diet, and habitat use in the critically endangered Ka'apor capuchin monkey (*Cebus Kaapori*) in Pará State, Brazil. *AJP*, *76*, 919–931.
Mannu, M., & Ottoni, E.B. (2009). The enhanced tool-kit of two groups of wild bearded capuchin monkeys in the Caatinga. *AJP*, *71*, 242–251.
Massaro, L., et al. (2012). Wild bearded capuchin (*Sapajus libidinosus*) select hammer tools on the basis of both stone mass and distance from the anvil. *AC*, *15*, 1065–1074.

Matthews, L.J. (2009). Intragroup behavioral variation in white-fronted capuchin monkeys (*Cebus albifrons*): mixed evidence for social learning from new and established analytical methods. *Behaviour*, *146*, 295–324.

McKenzie, T., et al. (2004). Can squirrel monkeys (*Saimiri sciureus*) plan for the future? Studies of temporal myopia in food choice. *Learning and Behavior*, *32*, 377–390.

Montague, M.J., et al. (2014). Population genetics, dispersal, and kinship among wild squirrel monkeys (*Saimiri sciureus macrodon*). *IJP*, *35*, DOI: 10.1007/s10764-013-9723-7.

Moura, A.C.deA., & Lee, P.C. (2010). Wild capuchins show male-biased feeding tool use. *IJP*, *31*, 457–470.

O'Malley, R.C., & Fedigan, L.M. (2005). Evaluating social influences on food-processing behavior in white-faced capuchins (*Cebus capucinus*). *AJPA*, *127*, 481–491.

Ottoni, E.B., & Izar, P. (2008). Capuchin monkey tool use: overview and implications. *EA*, *17*, 171–178.

Paulkner, A., & Suomi, S.J. (2008). The effects of fur rubbing on the social behavior of tufted capuchin monkeys. *AJP*, *70*, 1007–1012.

Petit, O., et al. (2009). Collective decision-making in white-faced capuchin monkeys. *PRSB*, *276*, 3495–3503.

Phillips, K.A., & Shauver Goodchild, L.M. (2005). Reunion displays in captive male brown capuchins (*Cebus apella*). *Primates*, *46*, 121–125.

Port-Carvalho, M., et al. (2004). Predation of crabs by tufted capuchins (*Cebus apella*) in Eastern Amazonia. *FP*, *75*, 154–156.

Ramirez-Llorens, P., et al. (2008). Infanticide in black capuchin monkeys (*Cebus apella nigritus*) in Iguazú National Park, Argentina. *AJP*, *70*, 473–484.

Sabbatini, G., et al. (2013). Sequential use of rigid and pliable tools in tufted capuchin monkeys (*Sapajus* spp.). *AB*, *87*, 213–220.

Schino, G., et al. (2009). Grooming, rank, and agonistic support in tufted capuchin monkeys. *AJP*, *71*, 101–105.

Schoof, V.A.M., & Jack, K.M. (2014). Male social bonds: strength and quality among coresident white-faced capuchin monkeys (*Cebus capucinus*). *Behaviour*, *151*, 963–992.

Sirianni, G., & Visalberghi, E. (2013). Wild bearded capuchins process cashew nuts without contacting caustic compounds. *AJP*, *75*, 387–393.

Smith, B.R., et al. (2003). Numerity of a socially housed hamadryas baboon (*Papio hamadryas*) and a socially housed squirrel monkey (*Saimiri sciureus*). *JCP*, *117*, 217–225.

Souto, A., et al. (2011). Critically endangered blonde capuchins fish for termites and use new techniques to accomplish the task. *BL*, *7*, 532–535.

Spagnoletti, N., et al. (2012). Stone tool use in wild bearded capuchin monkeys, *Cebus libidinosus*. Is it a strategy to overcome food scarcity? *AB*, *83*, 1285–1294.

Stone, A.I. (2007a). Age and seasonal effects on predator-sensitive foraging in squirrel monkeys: a field experiment. *AJP*, *69*, 127–141.

Stone, A.I. (2007b). Ecological risk aversion and foraging behaviors of juvenile squirrel monkeys. *Ethology*, *113*, 782–792.

Stone, A.I. (2008). Seasonal effects on play behavior of immature squirrel monkeys. *IJP*, *29*, 195–205.

Tiddi, B. (2011). Grooming for tolerance? Two mechanisms of exchange in wild tufted capuchin monkeys. *BE*, *22*, 663–669.

Tiddi, B., et al. (2010). Grooming for infant handling in tufted capuchin monkeys: a reappraisal of the primate infant market. *AB*, *79*, 1115–1123.

Tiddi, B., et al. (2012). Grooming up the hierarchy: the exchange of grooming and rank-related benefits in a New World primate. *PLOne*, *7*, e36641.

Urbani, B. (1999). Spontaneous use of tools by wedge-capped capuchin monkeys (*Cebus olivaceus*). *FP*, *70*, 172–174.

Verderane, M.P., et al. (2013). Socioecology of wild bearded capuchin monkeys (*Sapajus libidinosus*): an analysis of social relationships among female primates that use tools in feeding. *Behaviour*, *150*, 659–689.

Visalberghi, E., et al. (2008). Physical properties of palm fruits processed with tools by wild bearded capuchins (*Cebus libidinosus*). *AJP*, *70*, 884–891.

Visalberghi, E., et al. (2009a). Selection of effective stone tools by wild bearded capuchin monkeys. *CB*, *19*, 213–217.

Visalberghi, E., et al. (2009b). Distribution of potential suitable hammers and transport of hammer tools and nuts by wild capuchin monkeys. *Primates*, *50*, 95–104.

Wright, B., et al. (2009). Fallback foraging as a way of life: using dietary toughness to compare the fallback signal among capuchins and implications for interpreting morphological variation. *AJPA*, *140*, 686–699.

16 The Catarrhine Infraorder

Old World Monkeys and Apes

It is the dry season in a dry forest. In the trees monkeys squabble over the few fruits that remain or fill themselves with leaves. Apes pass by on the ground, moving slowly and plucking herbs that provide a nutritious fallback food.

Old World monkeys and apes are more closely related to each other than either is to New World primates. Together they constitute the catarrhines (Box 16.1). Some of the most important features of the catarrhines are related to behavior (Box 16.2). Sexual dimorphism is more common than in platyrrhines and goes to greater extremes. Males are in some cases twice the size of females and all have a **CP_3 complex** in which long upper canines are honed against the lower premolars behind them. A lone large cusp on the premolar performs that function (a parallel occurs in some New World monkeys, but it does not characterize the entire infraorder).

Box 16.1 Catarrhine Taxonomy

A number of anatomical specializations distinguish all catarrhines, Old World monkeys and apes, from the platyrrhines (Fleagle 2013; Harrison 2013). The most obvious is the trait that gives them their name, narrow nostrils that face downward. They have two molars in each quadrant of the mouth instead of the three found in platyrrhines. The upper canine teeth are kept sharp by honing from the lower premolar teeth behind them. The honing premolar is sectorial, i.e. it has a single cusp with a long and narrow crown. This functional relationship between the canine and adjacent premolar is called the CP_3 complex (P_3 designates the first premolar because the original first two have been lost in the course of primate evolution). More technical aspects of the skull are also diagnostic. For example, extensions of the frontal bone separate the zygomatic (cheek bone) from other parts of the skull, and the tympanic bone of the ear forms a tube. Molecular evidence suggests that the catarrhine clade may have originated before 40 million years ago, a time depth that indicates a high level of classification such as in an infraorder.

Box 16.2 Catarrhine Features Related to Behavior

- generally larger than platyrrhines
- greater sexual dimorphism
- more species folivorous
- terrestrial tendencies
- olfactory sense less important
- honed canine teeth
- facial coloration
- seasonal breeding.

Catarrhine Ecology

Catarrhines are more terrestrial than platyrrhines in several ways. More species spend significant amounts of time on the ground; some species spend more time on the ground than any platyrrhines; and a few species spend most of their active time on the ground. Terrestrial habits widen the array of available resources and in at least some cases can facilitate travel.

The realm of terrestriality may contain the explanation for a serendipitous discovery from research on artificial sweeteners. Catarrhines perceive aspartame as sweet while other primates do not. In this respect catarrhines differ from a variety of mammals that seem to experience sugar as sweet (Li et al. 2011). Noting that a genetic variant underlies the trait, the researchers hypothesize that it evolved in response to some natural equivalent of aspartame in a specialized feeding niche. That distinctive niche might well be in the terrestrial milieu that distinguished catarrhines from other primates.

Catarrhines engage their habitats with other senses that differ significantly from other primates. Most importantly, they are **routinely trichromatic**. This means that full color vision occurs equally in both sexes and is ubiquitous across the taxon in that form. The consistency between the sexes is the most important difference from the platyrrhines. The underlying physiology of catarrhine trichromacy is particularly consistent with the Fruit Foraging Hypothesis (see Chapter 11): spectral separation of the L and M cones is nearly optimal for distinguishing fruits from mature leaves (Melin et al. 2012). However, young leaves may also be significant because they differ from mature leaves in being reddish in color. They are a critical fallback food because they are relatively tender and are rich in protein and amino acids (Lucas et al. 2003). The "flush" of the leaves is seasonal and tends to coincide with low fruit availability (Dominy and Lucas 2001).

This is a case where two hypotheses may be complementary rather than competing. Taken together with the reddish figs of the New World (see Chapter 12), these young leaves formed the basis for the Protein Limitation Hypothesis of primate diet and vision (Valenta and Melin 2012). Platyrrhines and catarrhines seem to have undergone parallel evolution in order to use different resources to solve similar problems.

A further reduction in the sense of smell, beyond that which characterizes the anthropoids generally, goes with the consistent presence of trichromatic vision. The vomeronasal organ is non-functional or absent in all catarrhines examined for the trait (Smith et al. 2014). Its continuing but vestigial occurrence in some catarrhines can be explained by its role in embryonic development as a conduit to the brain for neurons. Another factor in the reduction of the VNO is the minimal role of pheromones in the complex societies of catarrhines (Swaney and Keverne 2009).

Opposable thumbs facilitate manual manipulation by catarrhines (Napier/Tuttle 1993). In this context opposability means the capacity to touch the pulp of the thumb to the pulp of another finger. Opposability is made possible by the thumb rotation that distinguishes the catarrhines. The limited contact between thumb and other digits that occurs in other primates is called pseudo-opposability. An opposability index applies only to catarrhines: length of thumb × 100/length of forefinger. This measurement differs significantly among catarrhine taxa. Opposability presumably functions in relation to the environmental diversity that catarrhines have confronted in expanding across two continents and spending more time on the ground.

The terrestrial tendencies of many catarrhines place them in greater danger from a wider array of predators than other primates, including lions, hyenas, crocodiles, and monitor lizards. However, the leopard seems to stand out among terrestrial predators. It is a large and versatile cat that can kill primates of any size throughout Africa and Asia, including gorillas. Leopards usually hunt monkeys on the ground during the day (Zuberbühler and Jenny 2002), but their climbing ability allows these cats to take primates from sleeping places in trees or other elevated locations. The degree of predation is positively related to the population density of the primates in question.

An extensive survey of leopard habits (Hayward et al. 2006) demonstrated a prey range of 10 to 40 kilograms with a strong tendency toward 25kg. Numerous primate species fall into this range. Within this range the cats usually take prey in accord with abundance in a habitat. Ungulates are always the bulk of the diet, but primates are significant in many locales. Langur monkeys in Asia were among the proportionally most common prey, at the same level as the small antelopes called duikers.

Jenny and Zuberbühler (2005) pointed out that most leopard studies had been done in savanna habitats. Their research indicated that forest leopards may differ in being diurnal/crepuscular, seasonal in their activities, and individualized in their prey preferences. While these variables probably affect the particulars of predator–prey interaction, the abundance of primates in forests and their availability (diurnal and often terrestrial) suggest that they will be no less important to leopards in the forest than in the savanna. Henschel et al. (2011) studied four rainforest sites and found that leopards preyed on four genera of monkeys, as well as chimpanzees and gorillas.

Leopards are more inclined to hunt primates that are larger in body size, live in larger groups, and have more males per group. Zuberbühler and Jenny (2002) took this to contradict the common view that those characteristics function as defenses against predators. They concluded that cognitive evolution has been the main primate response to leopard predation in some catarrhines.

Schel and Zuberbühler (2009) followed up with an experiment that compared experienced monkeys with those in an area where leopards had been extinct for generations. The responses to test stimuli by the two populations were "indistinguishable", leading to the conclusion that evolved cognitive mechanisms underlie recognition of leopard presence as dangerous. The implications may extend to other feline predators and other primates. The facial configuration of the Felidae is obviously very consistent across the many species in the taxon, and highly distinctive compared to other mammals (King, n.d.).

Catarrhine Sociality

Pronounced sexual dimorphism allows males to dominate females on an individual basis, although females form effective coalitions against males in some species. However, the prevailing explanation for sexual dimorphism is mating competition among males. Sexual dimorphism also has implications for ecological issues, such as energy requirements and feeding patterns. In general, dominance is a more prominent feature of social life among catarrhines than it is among platyrrhines.

Catarrhine faces exhibit every possible hue in the spectrum of mammal colors. Communication, physiology, and ecology select for animal coloration, but little was known about their proportional effects on catarrhine faces. Comparative analyses of catarrhine facial diversity (Santana et al. 2013) revealed two major axes, sociality and ecology, and their underlying correlates. Facial color patterns are linked to social factors, such that gregarious and highly sympatric species have evolved more colors in their faces. Species living in tropical, densely forested and humid habitats in Africa have evolved darker faces. Thus, both sociality and ecology have played a role in producing the highest diversity of faces within the mammals.

The catarrhine pattern is the opposite of the one in platyrrhines in that the more gregarious species have the more complex facial patterns. This may be related to greater dependence on facial expressions and displays for social communication (Santana et al. 2013). Dobson (2009) used a facial action coding system designed for humans to study video recordings of facial expressions in 12 anthropoid species housed in zoos. Informed by earlier work, he controlled for body size in his analysis. The results showed a positive correlation with group size in catarrhines.

Breeding seasonality is commonly more pronounced in catarrhines than in platyrrhines, because catarrhines live in more habitats with substantial seasonal fluctuation. However, seasonality is never as rigid as it is in most strepsirrhines. Furthermore, female sexual flexibility is greater in the catarrhines because they have a potential for extended estrus (displayed by some in the wild and more in captivity). Significant menstruation is more common in catarrhines and there is a sexual skin in the adult female genital region in many species. This skin may change color or even swell in association with sexual receptivity.

Female copulatory vocalizations are a recurring feature of catarrhine sexuality (Dixson 2012; Maestripieri and Roney 2005). They occur during and after copulation. Various explanations exist and functions may vary from one taxon to another. Two possibilities seem to stand out. One is that the vocalizations are a vehicle of female choice. As honest signals of fertility or ovulation, they induce a preferred male to linger and engage in mate guarding that will reduce the likelihood of sperm competition from another male. The other general possibility is social facilitation: the vocalizations stimulate copulation in others. Insofar as this means that other males are attracted to the vocalizing female for later copulation, its function may be to encourage sperm competition. The idea that stimulation of other pairs might be the function seems to suggest group selection, which is not a well accepted theory. After reviewing evidence and comparing hypotheses, Maestripieri and Roney (2005) suggested that the evolution of female copulation calls in primates is linked to the evolution of other female mating signals such as exaggerated sexual swellings; the potential for sperm competition; and the opportunity for precopulatory female mate choice.

Isosexual behavior is more common in catarrhines than in New World primates, probably due to: (1) the elaboration of sociosexual behaviors for communication in more complex societies; and (2) the discovery by primates of the potential for pleasure (Dixson 2012). New

World primates display such behavior only in the most basic forms and only in play and dominance interactions. Catarrhine interactions are more frequent and complex and more obviously sexual in nature.

Taking all the species together, catarrhine isosexual behavior includes consortship bonding, reconciliation, tension regulation, and alliance formation (Vasey 1995). Catarrhines choose their isosexual partners and engage in competition for them with members of their own sex and the opposite sex. These elaborations may have something to do with needs posed by higher levels of aggression in catarrhines. However, the evidence for sexual satisfaction in isosexual behavior and the parallels with heterosexual behavior suggest that a higher level of sexual motivation may also be a factor.

Catarrhine Cognition

The phylogenetic separation from the platyrrhines correlates with expansion of catarrhine brain size (Vallender 2012). Using human brain volume as a standard, it is about eight times larger than the platyrrhine average and six times larger than the catarrhine average. This follows a general primate pattern that connects encephalization with major divergence events. Particular causes may be ecological or social or both.

Large brains buffer energetic effects of seasonal habitats in catarrhine primates. Van Woerden et al. (2012) examined two different ways in which seasonality may be related to brain size. First, seasonality may impose energetic constraints on the brain because it forces animals to deal with periods of food scarcity (the Expensive Brain Hypothesis). Second, seasonality may act as a selective pressure to increase brain size, as behavioral flexibility helps to overcome periods of food scarcity (the Cognitive Buffer Hypothesis). Controlling for phylogeny, they found a strong negative relationship between brain size (relative to body mass) and the degree of experienced seasonality, as estimated by the variation in net energy intake. However, they also found a significant positive relationship between relative brain size and the effect of so-called cognitive buffering, proxied by the difference between environmental seasonality and the seasonality in net energy intake actually experienced by the animals. These results show that both energetic constraints of seasonal habitats as well as cognitive buffering affect brain size evolution, leaving environmental seasonality uncorrelated to brain size. The researchers emphasized the importance of simultaneously considering both costs and benefits in models of brain size evolution.

Some recent experiments have narrowed a presumed cognitive gap between Old World monkeys and apes, thereby suggesting that there are important features of cognition that are characteristic of the catarrhines. For example, Schmitt et al. (2012) tested 18 Old World monkeys (long-tailed macaques and olive baboons) in the so-called Primate Cognition Test Battery (PCTB) (Herrmann et al. 2007). The tests revealed largely comparable results between Old World monkeys and great apes. Single comparisons showed that chimpanzees performed better than macaques in experiments on spatial understanding and tool use, but in none of the socio-cognitive tasks. These results question the clear-cut relationship between cognitive performance and brain size and support the view of an accelerated evolution of social intelligence in humans. One limitation, however, is that the initial experiments were devised to tap into human-specific skills in the first place, thus potentially underestimating both true nonhuman primate competencies as well as species differences. Another is that the monkeys in the experiment belong to closely related species. Similarities to apes might be the result of limited convergence rather than representing widely distributed catarrhine traits.

The same caveat applies to another experiment involving the same species of monkeys. Researchers (Schmitt et al. 2013) compared six primate species, including humans, chimpanzees, bonobos, gorillas, olive baboons, and long-tailed macaques, in a task on fine-grained size discrimination. Except for gorillas, subjects of all taxa (i.e. humans, apes, and monkeys) were able to discriminate three-dimensional cubes with a volume difference of only 10 percent (i.e. cubes of 50 and 48 mm side length) and performed only slightly worse when the cubes were presented successively. The minimal size discriminated declined further with increasing time delay between presentations of the cubes, highlighting the difficulty of memorizing exact size differences. The results suggest that differences in brain size, as a proxy for general cognitive abilities, did not account for variation in performance, but that differential socioecological pressures may better explain species differences. The study highlights the fact that differences in cognitive abilities do not always map neatly onto phylogenetic relationships and that in a number of cognitive experiments monkeys do not fare significantly worse than apes, casting doubt on the assumption that larger brains confer an advantage in such tests.

An investigation of anthropoid brains demonstrated distinctive features of the catarrhines. Anthropoid primates differ from other mammals by having relatively large primary visual cortices (V1) and complex facial expressions. Dobson and Sherwood (2011a) tested the hypothesis that facial expression processing coevolved with the expansion of V1 by analyzing previously published data with phylogenetic comparative methods. Results suggested a pattern of correlated evolution linking social group size, facial motor control, and cortical visual processing. However, this held only for catarrhines and not for platyrrhines. Catarrhines that live in relatively large social groups tended to have relatively large facial motor nuclei and relatively large primary visual cortices. The researchers concluded that catarrhine brains are adapted for producing and processing complex facial displays.

Facial motor nucleus volume coevolves with both social group size and primary visual cortex volume in catarrhine primates as part of a specialized neuroethological system for communication using facial expressions. Dobson and Sherwood (2011b) investigated whether facial nucleus volume also coevolves with functionally unrelated brainstem motor nuclei due to developmental constraints. Using phylogenetically informed multiple regression analyses of previously published brain component data, they demonstrated that facial nucleus volume is not correlated with the volume of other motor nuclei after controlling for medulla volume. They concluded that brainstem motor nuclei can evolve independently of other developmentally linked structures in association with specific behavioral ecological conditions. This finding provides additional support for the mosaic view of brain evolution.

Recent comparative evidence suggests that anthropoid primates are the only vertebrates to exhibit a quantitative relationship between relative brain size and social group size. Dobson (2012) tried to explain this pattern with regard to facial expressivity and social bonding. He hypothesized that facial motor control increases as a secondary consequence of neocortical expansion owing to cortical innervation of the facial motor nucleus. This is supported by new analyses demonstrating correlated evolution between relative neocortex size and relative facial nucleus size. He also hypothesized that increased facial motor control correlates with enhanced emotional expressivity, which provides the opportunity for individuals to better gauge the trustworthiness of group members. This is supported by previous evidence from human psychology, as well as new analyses demonstrating a positive relationship between allogrooming and facial nucleus volume.

Gaze following is the act of following another individual's direction of attention. This occurs in a number of catarrhine species under natural conditions (Humle and Fragaszy

2011), as well as in captive ringtailed lemurs (Shepherd and Platt 2008). Experiments indicate that captive catarrhines can follow the gaze of humans as well as conspecifics. This is a significant degree of identification with another individual and some scientists think that it represents **perspective taking**, i.e. having the same mental processes as the other. Others think that gaze following does not necessarily imply an understanding of what the other viewer sees, knows, or intends to do. However, recent studies of rhesus monkeys and apes support the idea that catarrhines at least understand "something of the visual and auditory perceptions of others" (Humle and Fragaszy 2011).

Research with rhesus monkeys and apes also indicates that catarrhines are capable of **abstract conceptual representation**, i.e. the use of general criteria to categorize objects (Humle and Fragaszy 2011). For instance, chimpanzees can conduct a search for a real object based on criteria of form and composition derived from a two-dimensional image. This is contrasted with the simpler process of **stimulus generalization**, the association of objects based on perceptual features or cues. Ape and rhesus monkey experiments also indicate a recurring catarrhine system for adding and subtracting quantities (Humle and Fragaszy 2011).

Summary

The catarrhine infraorder consists of Old World monkeys and apes, found in a wide variety of species across Africa and Asia. Distinctive features include greater size, greater sexual dimorphism, opposable thumbs, and the CP_3 complex in the dentition. Catarrhines are routinely trichromatic and vision is favored even more than in other primates. Stronger terrestrial tendencies have probably caused increased predator pressure on many species.

Sexual dimorphism is probably an adaptation to competition among males, but it also functions to enhance male dominance over females. This can be partially offset in some species by female coalitions. Facial coloration is prominent and highly diverse in catarrhines, patterned by social and ecological factors. Breeding seasonality is more pronounced than in platyrrhines due to greater seasonal fluctuations, but never as rigid as in strepsirrhines. Individual females display a capacity for extended estrus. Two recurrent features of female sexuality are post-copulatory vocalizations and pronounced sexual swellings. Catarrhine isosexual patterns are more common and elaborated than in platyrrhines. There is evidence for greater cognitive skills in catarrhines than in other primates. Old World monkeys and apes may be more similar in these skills than previously thought.

Sources

Basic Bibliography

Harrison, T. (2013). Catarrhine origins. In D.R. Begun (Ed.), *A companion to paleoanthropology* (pp. 376–396). Wiley-Blackwell.

Additional References

Breuer, T. (2005). Diet choice of large carnivores in northern Cameroon. *AJE*, *43*, 97–106.

Dixson, A.F. (2012). *Primate sexuality: comparative studies of the prosimians, monkeys, apes, and human beings*. NY: OUP.

Dobson, S. (2009). Socioecological correlates of facial mobility in nonhuman anthropoids. *AJPA*, *139*, 413–420.

Dobson, S. (2012). Face to face with the social brain. *PTRSB*, *367*, 1901–1908.

Dobson, S., & Sherwood, C.C. (2011a). Correlated evolution of brain regions involved in producing and processing facial expressions in anthropoid primates. *BL*, *7*, 86–88.

Dobson, S.D., & Sherwood, C.C. (2011b). Mosaic evolution of brainstem motor nuclei in catarrhine primates. *Anatomy Research International*, *Volume 2011*, Article ID 236894, 5 pages.

Dominy, N.J., & Lucas, P.W. (2001). Ecological importance of trichromatic vision to primates. *Nature*, *10*, 363–366.

Fleagle, J.G. (2013). *Primate adaptation and evolution* (3rd ed.). New York: Elsevier Science.

Harrison, T. (2013). Catarrhine origins. In D.R. Begun (Ed.), *A companion to paleoanthropology* (pp. 376–396). Wiley-Blackwell.

Hayward, M.W., et al. (2006). Prey preferences of the leopard (*Panthera pardus*). *JZL*, *270*, 298–313.

Henschel, P., et al. (2011). Leopard prey choice in the Congo Basin rainforest suggests exploitative competition with human bushmeat hunters. *JZL*, *285*, 11–20.

Hermann, E., et al. (2007). Humans have evolved specialized skills of social cognition: the cultural intelligence hypothesis. *Science*, *317*, 1360–1366.

Humle, T., & Fragaszy, D.M. (2010). Tool use and cognition in primates. In C.J. Campbell, et al. (Eds.), *Primates in perspective*, 2nd ed. NY: OUP.

Jenny, D., & Zuberbühler, K. (2005). Hunting behaviour in West African forest leopards. *AJE*, *43*, 197–200.

King, G.E. (n.d.). An attentional hypothesis of chimpanzee responses to felids. https://www.researchgate.net/profile/Glenn_King/contributions.

Li, X., et al. (2011). Sweet taste receptor gene variation and aspartame taste in primates. *Chemical Senses*, *36*, 453–475.

Lucas, P.W., et al. (2003). Evolution and function of routine trichromatic vision in primates. *Evolution*, *57*, 2636–2643.

Mastripieri, D., & Roney, J.R. (2005). Primate copulation calls and post-copulatory female choice. *BE*, *16*, 106–113.

Melin, A.D., et al. (2012). Why aye-ayes see blue. *AJP*, *74*, 185–192.

Napier, J., revised by Tuttle, R.H. (1993). *Hands*. Princeton: Princeton University Press.

Pradhan, G.R., et al. (2006). The evolution of female copulation calls in primates: a review and a new model. *BE&S*, *59*, 333–343.

Santana, S.E., et al. (2013). Adaptive response to sociality and ecology drives the diversification of facial colour patterns in catarrhines. *Nature Communications*, *4*, 2765. DOI: 10.1038/ncomms3765.

Schel, A.M., & Zuberbühler, K. (2009). Responses to leopards are independent of experience in Guereza colobus monkeys. *Behaviour*, *146*, 1709–1737.

Schmitt, V. et al. (2012). Old World monkeys compared to apes in the primate cognition test battery. *PLONE*, *7*, e32024.

Schmitt, V. et al. (2013). Monkeys perform as well as apes and humans in a size discrimination task. *AC*, *16*, 829–838.

Shepherd, S.V., & Platt, M.L. (2008). Spontaneous social orienting and gaze following in ringtailed lemurs (*Lemur catta*). *AC*, *11*, 13–20.

Smith, T.D., et al. (2014). The shrinking anthropoid nose, the human vomeronasal organ, and the language of anatomical reduction. *AR*, *297*, 2196–2204.

Stanyon, R., et al. (2012). Evolutionary molecular cytogenetics of catarrhine primates: past, present and future. *Cytogenetic Genome Research*, *137*, 273–284.

Swaney, W.T., & Keverne, E.B. (2009). The evolution of pheromonal communication. *Behav. Brain Res.* *200*, 239–247.

Valenta, K. & Melin, A.M. (2012). Protein limitation explains variation in primate color vision phenotypes. In *Zoology*. M.D. Garcia (Ed.), InTech, ISBN: 978-953-51-0360-8.

Vallender, E.J. (2012). Genetic correlates of the evolving primate brain. *Progress in Brain Research*, *195*, 27–44. doi:10.1016/B978-0-444-53860-4.00002-7.

Van Woerden, J.T., et al. (2012). Large brains buffer energetic effects of seasonal habitats in catarrhine primates. *Evolution*, 66, 191–199.

Vasey, P.L. (1995). Homosexual behaviour in primates: a review of evidence and theory. *IJP*, *16*, 173–203.

Vasey, P.L., & Sommer, V. (2006). Homosexual behaviour in animals: topics, hypotheses and research trajectories. In V. Sommer & P.L. Vasey (Eds.), *Homosexual behaviour in animals: an evolutionary Perspective*, (pp. 3–44). New York: Cambridge University Press.

Zuberbühler, K. (2007) Predation and primate cognitive evolution. In S.L. Gursky & K.A.I. Nekaris (Eds.), *Primate anti-predator strategies* (pp. 3–26). NY: Springer.

Zuberbühler, K. & Jenny, D. (2002). Leopard predation and primate evolution. *JHE*, *43*, 873–886.

17 The Cercopithecoid Superfamily
Old World Monkeys

A large male monkey sits in the shade of a tree. Nearby are three females, one of them with an infant clinging to her abdomen. The male has just defeated the former leader of this one-male troop and put him to flight. Two females forage with a few darting glances at the newcomer, but the mother nervously edges away. The midday heat rises, the male drowses, and the mother walks away in the direction taken by the former leader.

The two branches of the Catarrhini are the hominoids (apes and humans) and the cercopithecoids (Old World monkeys). Cercopithecoids are the monkeys of Africa and Asia (with a small human-caused extension into Europe at Gibraltar). These Old World monkeys (Box 17.1) are the most successful living primates, whether measured by the number of individuals or species in existence, by their total biomass density, or by the diversity of environments they occupy. They display enormous variation in body size and many display strong sexual dimorphism in body size and dentition (see Table 17.1 online for physical dimensions of all cercopithecoids). There seems to be a consensus that the two subtaxa of the cercopithecoids should be classified as subfamilies, the generalized cercopithecines and the more folivorous colobines.

Box 17.1 Cercopithecoid Taxonomy

A general resemblance between the ceboids (platyrrhines) of the Americas and the cercopithecoids of the Old World led to both being called "monkeys". Molecular evidence demonstrates a profound separation between the taxa that probably occurred about 30 million years ago, which merits distinction at a high taxonomic level (see Box 8.1) and shows the Cercopithecoidea to be more closely related to the Hominoidea (apes and humans). The Ceboidea are the only superfamily in the Platyrrhini, but the Cercopithecoidea are placed in the corresponding taxon Catarrhini along with the Hominoidea.

Bilophodont molars are a key feature of Old World monkeys, taxonomically as well as functionally. In each of these teeth ridges (*lophs*) connect an anterior and a posterior pair of cusps. This trait distinguishes the cercopithecoids from ceboids and from the more closely related hominoids. Otherwise, the most obvious differences between the catarrhine taxa are hominoid specializations in terms of suspension in the body and limbs.

Cercopithecoid Behavior

Cercopithecoid Ecology

Many cercopithecoids have fully opposable thumbs in the sense that they can touch the thumb to the tips of other fingers. This makes them capable of more complex manipulations than the New World primates. Greater potential for manipulation may be adaptive for cercopithecoids because they deal with greater environmental diversity (see Table 17.2 online for ecological characteristics of all cercopithecoids).

Cercopithecoid sitting is facilitated by **ischial callosities**, pads attached to one of the bones in the pelvic girdle. Old World monkeys can sit for long periods of time. They often feed while seated on branches and rarely use the suspensory postures that are common in many New World primates. The frictional qualities of the pads allow them to use slanting branches in the sitting position (Vilensky 1978). Old World monkeys also sleep in a sitting position, but the significance of the pads for this posture is in doubt. In contrast to the versatility needed in feeding, an animal can carefully select the sleeping spot that it will use throughout an entire night.

Cercopithecoid Sociality

Troops are virtually universal among Old World monkeys. The one-male and multi-male forms are both well represented and variation between them occurs among species, among populations within some species, and in some troops over time (see Table 17.3 online for social features of all cercopithecoids). Male emigration is more common than female, so there is a strong tendency towards female philopatry in Old World monkey societies. This leads to closer and more complex relationships among females, which include allogrooming, cooperation, competition, dominance, and kinship.

Comparative mathematical treatment of social variables revealed a striking uniformity among Old World monkeys that contrasts with greater variability in social organization among genera in other primate superfamilies (Di Fiore and Rendall 1994). Since the Old World monkeys are the most ecologically diverse among equivalent primate taxa, the uniformity of social organization led the investigators to infer a significant phylogenetic influence on the social systems of this clade.

Multi-level societies (also called *modular* or *nested* systems) are a recurring feature of cercopithecoids. Grueter and colleagues have synthesized their general characteristics (Grueter et al. 2012). A modular society consists of two or more hierarchical tiers. The basic entity is a cohesive one-male unit (abbreviated OMU) in which several adult females are typically associated with the adult male. These units aggregate and disperse within one or more higher-level groupings. The second-level group is often called a *band* and the third is called a 'troop' or herd. The closest relationships occur within the OMU, which is also the basic reproductive unit. Solitary males may be present in bands and/or all-male units may be loosely attached to bands. The one-male unit differs from a one-male troop because it is profoundly affected by integration into a higher-level society. OMUs aggregate and disperse in a kind of fission–fusion, but modular societies differ from fission–fusion troops or communities in which individuals are the basic units.

Two unifying elements favoring the maintenance of modular societies are, depending on the species, male cooperation and female dispersal. Female transfers within bands and genetic relationships among leader males help to maintain inter-unit tolerance. Conspecific threat is a potentially shaping selective force in the evolution of multilevel societies in primates: males benefit from help in retaining their females. With regard to proximate cause, there is no need to infer intent to cooperate. The system works adaptively if all OMU males simply respond in the same way to the stimulus provided by an intruder (Kirkpatrick and Grueter 2010). Collective actions against bachelors may include vigilance, patrolling, and chases (Xiang et al. 2014).

Sexual skin in the adult female genital region is a recurring physical trait in Old World monkeys that strongly affects behavior (see Table 17.4 online for reproductive parameters of all cercopithecoids, and Table 17.5 online for life cycle stages of all cercopithecoids). It may change color or even swell in association with sexual receptivity. Most of the discussion about sexual skin is based on data from baboons, but the resultant hypotheses may apply to many or all of the species in which the phenomenon occurs. Baboon sexual swellings are among the largest and most colorful signals exhibited by any mammal.

Swelling of the sexual skin is also a recurring feature among the Old World monkeys (Zinner et al. 2004). Slight swellings of the genitalia around the time of ovulation occur in a variety of primates, but some Old World monkeys have large swellings that extend beyond the genitalia to surrounding tissues (Figure 17.1). These swellings result primarily from extra-cellular water retention, governed by hormonal influences. Such swellings are most likely to occur in species with multi-male troops and which lack seasonal breeding. Various explanations have been offered, but two currently competing explanations subsume most aspects of their predecessors. Explanations are based mainly on research with baboons (*Papio*) but may be broadly applicable.

The Graded Signal Hypothesis (Nunn 1999) postulates a dual function for swellings. Because they grow slowly before ovulation and decrease gradually afterward, they are

Figure 17.1 Female baboon with a sexual swelling, a recurring trait among cercopithecoids. © Glenn E. King

"probabilistic" indicators of ovulation. This allows females to attract the "best" males at the peak of swelling when conception is likely and to offer non-conceptive matings to other males in order to confuse paternity and ward off infanticide. This hypothesis explains the association between large sexual swellings and multi-male social organization. It takes account of recent theoretical developments by incorporating intersexual selection (adaptation to conflict between the sexes). It is consistent with a great deal of available information (Dixson 2012).

The Reliable Indicator Hypothesis (Pagel 1994) holds that the swellings are honest representations of a female's reproductive quality, i.e. of high fecundity and ability to raise offspring. A large swelling therefore induces males to compete for the female, giving her a chance to choose the "best male". The reliable indicator hypothesis makes the following predictions: (1) sexual swellings should occur in the context of intense female–female competition; (2) swellings should be biologically costly (otherwise they are not honest signals, according to the handicap principle); (3) high quality females (in terms of reproductive potential) have the largest swellings; (4) variance in size is greater between females than it is in one female from cycle to cycle; (5) males prefer females with larger swellings; and (6) females with larger swellings conceive more often from high quality males.

First, having larger swellings than other females indicates greater general fertility (Domb and Pagel 2001). Second, the relative size of the swelling in a particular female indicates whether her current cycle is either conceptive or non-conceptive. However, fecal estrogen levels were most strongly associated with mate guarding, which suggests that olfactory cues are as important as the visual stimulus if not more. Gesquiere et al. (2007) made comparable studies of yellow baboons, postulating that males should bias consortships toward females with a high probability of conception because mate guarding is costly. They combined data on consortships with visual scoring of sexual swellings and assays of fecal estrogen concentrations to test whether or not sexual swellings are reliable indicators of: (1) within-cycle timing of ovulation; (2) differences in conception probability among females at different maturational stages; and (3) conceptive versus non-conceptive cycles of parous females. They found that sexual swellings reflected conception probability within and among cycles. Adult males limited their consortships to the turgescent phase of cycles, and consorted more with adult females than with newly cycling adolescents. Alpha males discriminated more than did males of other ranks with respect to these factors. Male mate choice based on sexual swellings and other estrogenic cues of fertility may result in sexual selection of these female traits and enhance dominance-based reproductive skew in males. Alpha males are the least constrained in their mating behavior and can best take advantage of these cues.

A study of one wild population offered support for the third, fourth, and fifth predictions. However, Zinner et al. (2004) criticized the methodology used to reach these conclusions and argued that phylogenetic comparative tests do not support the association between large swellings and female competition that is postulated in prediction (1). To the contrary swellings are absent from unimale species where competition should be greatest. Zinner and colleagues made two other important points. They argued that large swellings are probably costly because they impose a continuous energy burden and are prone to injury and infection. In theory the adaptive benefits of the trait might be sufficient to overcome the costs, but this has yet to be demonstrated. Finally, prediction (6) is not testable with available data.

Huchard et al. (2009) tried to reconcile two hypotheses about what maximal swelling signals: female fertility within a cycle, or female quality across cycles. Published evidence favors the first hypothesis, and further indicates that larger swellings advertise higher fertility between cycles. If so, a male preference for large swellings might evolve, driving females to use swellings as quality indicators as proposed by the second hypothesis. The researchers found support for three key predictions of the female-quality hypothesis in field data from wild baboons (*Papio ursinus*): (1) inter-individual differences in swelling size are maintained across consecutive cycles; (2) females in better condition have larger swellings and higher reproductive success; and (3) males preferentially choose females with large swellings. An individual-based simulation model indicated that females producing larger swellings can achieve higher mating success even when female–female competition is low and within-female variance in the trait is high. They concluded that once sexual swellings had evolved as fertility signals, they could in certain socio-sexual systems be further selected to act as quality signals. These results reconciled the two hypotheses and suggested that mate choice for direct benefits (fertility) can lead to indirect benefits (good genes).

Most attempts to understand sexual swellings have focused only on size, and other measures have been crude. According to Higham et al. (2008, 2009), this is inconsistent with the emphasis on multiple cues in most theoretical models of mate choice. These researchers used digital photography to measure color as well as size in the sexual swellings of olive baboons. Combining these measures with hormone levels, they found that swelling size conveyed information about the timing of ovulation and the fertile period around ovulation, as well as information about female parity. Some evidence suggested that color also contains such information.

Higham et al. (2009) used their hormonal measures to assess male and female social and sexual responses. They found some significant relationships between these hormone levels and presenting behaviors. Measures of female behavior collected during the study did not appear to reveal the timing of ovulation or the fertile period. Male consortship behavior was closely tied to changes in sexual swelling size, but copulation behavior was not. However, it was linked to the timing of both ovulation and the fertile period, suggesting that males had knowledge about these timings. Swelling size may advertise the period during which males should consort with females, and then consorting males can assess the timing of the fertile period more accurately from other signals that are available only by closer inspection. An important implication is that different males may have access to different signals of ovulation at any one time. Such a system would allow females to offer different males different information simultaneously, which may resolve the **female dilemma** of how females can simultaneously assure and confuse paternity in multi-male societies.

Summary

The cercopithecoids (Old World monkeys) are one of the two main branches of the catarrhines. They are the world's most successful primates. Distinctive physical features include opposable thumbs and ischial callosities. Troops are virtually universal, including both multi-male and one-male forms. Females are typically philopatric. Major exceptions are several species organized into multi-level societies.

Sexual skin in the female genital region is a powerful signal. It may change color and/or swell to a very large size. Functional explanations for swelling are a continuing area of controversy. Much of the research is based on baboons, but the results may apply to many Old World monkeys.

Sources

Basic Bibliography

Campbell, C.A., et al. (Eds.) (2011). *Primates in perspective*. NY: OUP.

Additional References

Di Fiore, A., & Rendall, D. (1994). Evolution of social organization: a reappraisal for primates by using phylogenetic methods. *PNAS, 91*, 9941–9945.

Dixson, A.F. (2012). *Primate Sexuality*. NY: OUP.

Domb, L.G., & Pagel, M. (2001). Sexual swellings advertise female quality in wild baboons. *Nature, 410*, 204–206.

Gesquiere, L.R., et al. (2007). Mechanisms of sexual selection: sexual swellings and estrogen concentrations as fertility indicators and cues for male consort decisions in wild baboons. *H&B, 51*, 114–125.

Grueter, C.C., et al. (2012). Evolution of multilevel social systems in nonhuman primates and humans. *IJP, 33*, 1002–1037.

Higham, J.P., et al. (2008). Baboon sexual swellings: information content of size and color. *H&B, 53*, 452–462.

Higham, J.P., et al. (2009). Female reproductive signals, and male mating behavior, in the olive baboon. *H&B, 55*, 60–67.

Huchard, E., et al. (2009). Can fertility signals lead to quality signals? Insights from the evolution of primate sexual swellings. *PRSB*, DOI: 10.1098/rspb.2008.1923

Kirkpatrick, R.C., & Grueter, C.C. (2010). Snub-nosed monkeys: multilevel societies across varied environments. *EA, 19*, 98–113.

Nunn, C.L. (1999). The evolution of exaggerated sexual swellings in primates and the graded-signal hypothesis. *AB, 58*, 229–246.

Pagel, M. (1994). The evolution of conspicuous oestrous advertisement in Old World monkeys. *AB, 47*, 1333–1341.

Vilensky, J.A. (1978). The function of ischial callosities. *Primates, 19*, 363–369.

Xiang, Z.-F., et al. (2014). Snub-nosed monkeys: multilevel societies across varied environments. *AJP, 76*, 609–617.

Zinner, D.P., et al. (2004). Sexual selection and exaggerated swellings of female primates. In P.M. Kappeler & C.P. van Schaik (Eds.), *Sexual selection in primates* (pp. 71–89). NY: CUP.

18 Cercopithecoid Variation

Leaf Eaters and Cheek Pouchers

It is the dry season and the monkey eats the older, drier leaves that are now available. After a long session it sprawls on the branch to rest and allow the food to slowly digest. As its specialized digestive tract works, the monkey belches the resultant gas.

Cercopithecoids are generally divided into two groups that have often been characterized as families, but now are more often distinguished at the subfamily level (Box 18.1). The colobines are one of these branches. They are often called leaf monkeys because they display derived physical traits that facilitate leaf consumption. However, the term is problematic because these monkeys eat various other foods, and because many other primates eat significant quantities of leaves. The other clade of Old World monkeys is the cercopithecines, which are more generalized and omnivorous and have stronger terrestrial tendencies.

Box 18.1 Colobine Taxonomy

Colobinae are distinguished by digestive specializations for folivory. Some or all of the shared derived features of the skull are probably also related to folivory. These include a deep jaw, narrow incisors, high-cusped cheek teeth, and a broad space between the orbits. Colobines have relatively long legs and short thumbs. In some taxa the thumb is virtually absent.

Colobine Behavior

Colobines are primarily an Asian taxon, with about three times as many species as Africa. Asian colobines occur in the scrub deserts of western India, temperate forests of the Himalayan foothills, and swamps of Borneo. African species inhabit forests.

Common colobine features include widely spaced eyes, short snouts, a high and globular braincase (though the brain is not particularly large), lanky bodies, long legs, and long tails. They typically weigh between 5 and 25 kilograms. Their anatomy is related to arboreal behavior, especially leaping. Thumbs are reduced or absent but fingers and toes are long, which may facilitate suspensory locomotion (Struhsaker and Leland 1987).

Colobine Ecology

Colobines mostly feed on leaves and unripe fruits in the trees. Many colobines are like other primates in their preference for young leaves, which they can consume in greater quantities than other primates. They use mature leaves that most other primates cannot digest as an important fallback. The ratio of protein to fiber in the mature leaves of a habitat is the strongest predictor of colobine biomass (Oates et al. 1990). The effectiveness of the feeding strategy is demonstrated by the fact that their biomass is usually higher than that of all other primates in a community combined. These primates are equipped for their diet with sharp-cusped molar teeth, narrow incisors, and a deep mandible. Many features of the digestive tract are unique among primates and comparable to ruminants such as giraffes and kangaroos (Groves 2007).

Leaves are about half the typical diet, but the maximum rises to 90 percent for some species. Seeds and/or fruit comprise another quarter to a third. Toxins are a limiting factor, even though colobine digestion can mitigate their effects. Colobines tend to avoid ripe fruits because their specialized stomachs digest simple sugars too rapidly, causing bloating that can be fatal. Faunivory is minimal because colobines can obtain virtually all the protein they need from foliage (and the incidental consumption of insects on and in the foliage). This is more efficient than exerting energy in pursuit of prey.

Colobine behavior preserves energy as much as possible, because slow digestion of leaves constrains them to low levels of activity (like the howlers of the New World). Resting probably averages about 50 percent of the daily time budget and colobines have shorter travel distances and smaller home ranges than sympatric primates of other kinds. Maximum home range size is about 100 hectares and maximum mean daily travel about a kilometer. Most colobine troops on most days travel no more than the length of a football field. In many species a troop spends most of its time in a core area that covers only a small part of the overall range. This central area is almost exclusive to the resident troop. The digestive process makes energy conservation necessary, while the relative abundance and regular spatial distribution of food makes limited activity possible.

Colobine Sociality

The typical social unit is a one-male troop with perhaps 5 adult females and 10 immatures (size range about 5 to 15). So-called multi-male troops in many species contain no more than two adult males and seem to be temporary variants of the one-male form: longitudinal data show that many troops fluctuate between one and two males. Some units are age-graded while others seem to arise from immigration. Surveys show that the number of resident males increases as a function of the number of adult females. There is little affiliative interaction between males and a considerable amount of agonistic vocalization. Many males are solitary or members of all-male groups, and may seek extra-troop copulations or attempt takeovers.

Rates of social interaction are low compared to many other primates, including both affiliative and aggressive behaviors. Females sit together and groom each other, but other interactions are infrequent and subtle. The male tends to be spatially peripheral. He receives more grooming than he gives, and otherwise interacts with troop females only in the contexts of mating and occasional infanticide. One factor is the energy constraint imposed by the digestion process. Another may be a dispersal pattern that limits relatedness and familiarity: male emigration is ubiquitous and female emigration rates are also high in many species. Third, large food patches and the abundance of fallback foods mitigate competition.

Female colobines solicit mating with a variety of species-typical behavior patterns. Head shaking is a recurring example. Asian species are about equally divided between single-mount and multiple-mount ejaculation. The key factor seems to be the proportion of one-male troops in the species. Single-mount copulation is associated with low male–male sexual competition, which is most likely to be the case in a one-male troop.

The Defendability Index indicates that the small home ranges of colobines could potentially be protected at the borders, but this is rare. Colobine troops engage in ritualized confrontations, mostly vocal, but many seem to defend core areas rather than home ranges or else they establish intertroop dominance relations. Given the relative abundance of food, the defense of an entire territory may not be worth the cost of energy expenditure and physical risk. Additionally, conflicts may have more to do with access to mates than with food resources.

Colobine Life Cycle

Most births occur during an annual period that varies across species from two to six months. The reliable food supply probably supports this flexibility for the mothers. The infant's need for weaning foods might be the reason that there is any concentration of births at all. Another factor is food from humans. Populations that breed throughout the year generally raid crops or receive deliberate provisioning.

The functional reason for birth peaks is generally thought to be nutritional, but the proximal basis need not be the same. It may be innate in at least some species, since the pattern is retained in captivity. The IBI is generally about two years, but can vary significantly within a species. Colobines typically wean within the first year. Timing seems to vary with the quality of habitat, so the weaning schedule may depend on nutritional constraints.

Alloparenting occurs in almost all well studied colobines. Adult females other than the mother frequently hold, carry, and groom the infant. Some juvenile females and a few juvenile males also participate. Colobines differ from other Old World monkeys in that alloparenting occurs in more species, does not depend on the dominance status of the alloparents, and is rarely associated with kidnapping (removal of the infant from the mother by threat or force). The absence of kidnapping and abuse allows the low-ranking mother to feel comfortable in relinquishing her infant, and this may contribute to the frequency of the behavior.

Colobine alloparenting involves young infants, even less than a day old. This observation is supported by a growing number of colobine field studies, which counteract doubts that arose from early comparisons between captive colobines and other monkeys in the wild (McKenna 1979). Of various ultimate explanations for early alloparenting, the only one supported by colobine data (from just two species) is the Maternal Foraging Hypothesis: mothers forage more efficiently when not encumbered by infants, and this is important because lactation requires a great deal of energy. The prominence of alloparenting among colobines fits the proximal hypothesis that it is elicited by infant coat color. Infants in many species of the taxon are born with a distinctive and noticeable coat color (rare in other taxa) that seems to attract the attention of females in the troop.

Langurs

Gray langurs, also called Hanuman langurs, are the most widely distributed genus (*Semnopithecus*) of Asian colobine, ranging from Pakistan through India and from Nepal south to Sri Lanka (Figure 18.1). This is the only langur genus that displays large and stable multi-male troops (Grueter and van Schaik 2010). However, one-male troops are also common. Gray langurs live in diverse habitats, including some with extreme environments.

A species in the Himalayan Mountains (*S. schistaceus*) survives snowstorms and sub-freezing temperatures during winter nights. Heat conservation accounts for their long, thick coats and for the fact that they are the largest gray langurs. Males weigh up to 20kg. The main winter foods are cones, bark, cambium, and twigs—a clear reminder that the colobine digestive system can cope with foods other than leaves. Large troops (up to about 45 members) occupy correspondingly large home ranges.

Events during the mating season remind us that social and mating systems are not necessarily identical. Before and during the season one male drives all the others out of the troop. During this period, males move into troops as well as out of them and resident males are sometimes replaced. Mothers wean only after they become pregnant again, which results in a long IBI of 25 months.

Gray langurs in the scrub deserts of western India (*S. dussumieri*) are unusually faunivorous for colobines. When monsoons flush arthropods from the ground, the monkeys take

Figure 18.1 The gray langur is the largest and most common of the many langur species. It is also known as the Hanuman langur because it is associated with the character of that name in Hindu mythology. © Rhett A. Butler/www.mongabay.com

caterpillars and other insects for up to a quarter of their diet. These langurs are relatively sociable, devoting as much as 6 percent of their time to grooming, which may be necessary for maintenance of OMTs containing as many as 15 adult females.

Infanticide is most common in the most extreme environments. Rates of 20 to 30 percent occur in the Himalayas and in deserts. Genetic analyses support the Sexual Selection Hypothesis: males do not attack their own infants, and males who kill infants have a greater chance of siring the mother's next offspring (Borries et al. 1999). In contrast, many studies of other langurs report no infanticides and the phenomenon is rare in other Asian colobines.

Lutongs (*Trachypithecus*), living in evergreen forests east of the gray langurs, are the most specialized folivores among langurs in terms of dental structure and the size of their forestomachs. Large body size also facilitates digestion of leaves. Some lutongs sleep in caves and forage in forests that grow on almost vertical cliff faces. One possible explanation is that these monkeys depend on plant foods endemic to the setting, but nothing in their diet fits that criterion (Workman 2010).

Qihai et al. (2009) attributed the behavior to predator avoidance. They never saw predation on the monkeys, but they pointed to several cat species as potential dangers. Even if predator populations have been depleted by human activity, the lutongs might perpetuate the defensive adaptation through either genetic or cultural heritage. Alternatively, the cliffs and caves may now be a refuge from humans. The researchers rejected other hypotheses based on fieldwork in China: their subjects slept in the same places for several successive nights, which argues against parasite avoidance; they showed no clustering in a core area; and even though the caves provided more moderate temperatures, the monkeys spent more nights on the ledges than in the caves, even in rainy weather.

Langurs of the Mentawai Islands in Indonesia (*Presbytis potenziani*) are the only Old World monkeys to live in socially monogamous units throughout the species range (Fuentes 1996; Tilson and Tenaza 1976). Social monogamy seems to be obligate for the species. Each pair lives in its own range with as many as four offspring. Mentawai langurs are similar to other socially monogamous primates in monomorphy and several salient behavior patterns. Pairs are cohesive during resting, feeding, and traveling. Each pair maintains an exclusive core area with displays directed toward neighboring groups. The male emits loud calls comparable to the intertroop spacing calls of other Asian colobines. The female joins in with a three or four syllable coda to her mate's calls, making the vocal display into a duet. Both sexes engage in a jumping display that is performed only by males in other colobines. Thus, a single species of Old World monkey converges with other monogamous primates in monomorphy, joint displays, and duetting.

Odd-nosed Monkeys

Odd-nosed monkeys are Asian colobines with some distinctive physical and behavioral traits. Most obvious are their noses, which take a different form in each genus (Figure 18.2). Despite the variation, molecular evidence indicates that they actually form a clade. They all show tendencies toward multi-level social organization. The band aggregates in a common sleeping area and may travel as a unit. It usually exceeds 50 members and may have as many as 400. A band can fission into subgroups of varying size or independent OMUs. The typical OMU contains 5 to 20 individuals.

Snub-nosed monkeys (*Rhinopithecus*) are distributed from Southeast Asian tropical forests through temperate forests with some winter snow to habitats in China and Tibet where winter temperatures go below zero degrees Fahrenheit and snow cover sometimes lasts four months. This array of environments for one genus can provide a basis for testing socioecological theory.

Proboscis monkeys (*Nasalis*) are large-nosed as well as large-bodied and are strongly dimorphic in each of those traits. The nose of an adult male can be as long as 10cm and in many instances hangs over the mouth (Figure 18.2). It seems to facilitate an array of loud honking sounds that may communicate information about territory or group cohesion, or provide reassurance to infants (Kawabe and Mano 1972; Messeri and Trombi 2000). Bennett and Gombek (1993) argued instead that the nose is a sexual ornament that attracts females. If correct, this exemplifies an honest signal that conforms to the Handicap Principle. Adult males also possess a red penis and black scrotum, which may be sexual signals.

The relationship between proboscis monkeys and water is unique among the primates. They live only in coastal or riverine habitats and never go much more than one kilometer from water. They can swim as far as 20 meters under water, can swim across rivers, and sometimes leap into the water from high branches. A crocodilian is a major predator and takes adult males as well as immatures from low branches over water. The monkeys cross rivers arboreally if possible. Otherwise, they may sit and watch the water for a considerable period before entering. Monitor lizards are also potential predators.

The OMU varies in size from about 10 to 20 individuals. Female members seem to display more affiliative behavior toward one another than is typical of Asian colobines. The absence of serious aggression suggests a low level of competition, presumably because food is abundant. In one study desirable sleeping sites were the main focus of female conflicts. The tenure of the adult male in an OMU is about six to eight years and replacement seems to occur without serious aggression.

Figure 18.2 Male and female proboscis monkeys, one of the more extreme examples of primate sexual dimorphism in body size. The male displays one of the varied nasal structures that characterize the odd-nosed colobines of Asia. © WhitcombeRD via iStock

Agonistic behavior between OMUs is rare. Instead they are often near one another, especially when a number of them come together at a sleeping site to form a band. They may also join with all-male groups that have between 3 and 12 members. Despite mutual tolerance, members of an OMU will not groom or play with members of another group.

Band aggregations in the proboscis environment have possible functions other than defense against bachelor threat. One is to counter displacement from feeding sources by other species. Bands may be the best way to achieve this function because the energy-conserving way of life would not support permanent troops of the same size. Bands may also function in the coordination of large numbers of monkeys for the dangerous river crossings. Since no number of monkeys could mount an effective defense against crocodilians in the water, the relevant adaptive principle here is the dilution effect. The larger the group, the less likely it is that any particular individual will become a victim, and natural selection works on individuals.

Douc monkeys (*Pygathrix*) occur in Southeast Asia. They are half the size of the other odd-nosed genera and sexual dimorphism is less pronounced. Average male weight is about 11kg and females about 9kg. Similar to proboscis monkeys, males have a pink or red penis and a black scrotum. The female's rump changes color from white to red and the male's rump gets redder at the same time. Multiple copulations take place during estrus and copulation continues into pregnancy. Pregnant females in captivity become peripheral to the group and spend much of their time near an adult male. Reproduction takes place throughout the year, but there seem to be birth peaks.

Douc positional behavior differs from that of most other monkeys in the extent to which it involves suspensory postures and locomotion (Wright et al. 2008). The only langur that comes close is a species of *Trachypithecus*. The frequency of douc suspensory locomotion may equal or exceed that of spider monkeys. Various measures of limb proportions show that the doucs are adapted to such behavior. The suspensory capability allows a fairly large primate to exploit food among the fine branches of trees. Selection pressure for a feeding specialization may have come from intense competition with other primates. Some forested areas throughout Southeast Asia are on steep limestone cliffs, creating restricted areas that may have intensified competition.

African Colobines

An informal term for African colobines is **colobus**. At one time they were all placed in the genus *Colobus* (Groves 2007), and those expelled from that category still have *colobus* in their common names. Similarity is greater among African colobines than among Asian taxa, probably due to more contiguous geographic distribution and less ecological diversity (Oates 1994). The most obvious and important feature is greater reduction in the thumb, perhaps related to more suspensory behavior in the present or past (Struhsaker and Leland 1987). African colobines face a formidable enemy in chimpanzees, which prey on all three genera. The apes are agile in the trees as well as being large, strong, intelligent, and (to varying degrees) cooperative in their hunting.

In contrast to Asian colobines, sexual solicitation by both males and females is typical of the African forms. Sexual swellings occur in two of the genera. Multiple mounting is the only copulation pattern that has been reported, which suggests a greater amount of direct male–male competition. This is consistent with the fact that there are stronger tendencies toward multi-male troop composition in the African colobines.

Black-and-white colobus species (*Colobus*), including the largest African colobines, are spread across central Africa and occupy a wider variety of habitats than the others (Figure 18.3).

Figure 18.3 A black-and-white colobus, the most distinctive of the colobus monkeys. One-male troops are common among these species. © Matej Hudovernik via iStock

Red colobus (*Piliocolobus*) are somewhat smaller, averaging about 7.5kg. Allopatric species extend across the continent. Olive colobus (*Procolobus*), the smallest of the African colobines, are narrowly distributed in lowland rainforests of West Africa.

Some criticism of the Socioecological Model is limited to its application to folivores. This arises from an assumption that the perpetual availability of leaves in moist tropical forests obviates competition. However, some primatologists have suggested that folivore foraging patterns are not so simple and that they may be food limited in much the same way as frugivores (Snaith and Chapman 2007). Harris et al. (2009) supported this proposition by looking at the consequences of food shortage for guerezas (*Colobus guereza*). A steep decline in the two most important foods resulted in decreased reliance on these foods. The monkeys increased daily path length, number of feeding patches visited per day, size of individual feeding areas, percentage of time spent feeding, and dietary diversity. Two physiological measures provided evidence for stress: insulin production decreased in lactating females, and two of three females underwent a small but significant rise in parasite load.

Related results come from an investigation of the Ecological Constraints Principle in ursine colobus (*C. vellerosus*). This principle predicts that increased group size (equivalent to degrees of food shortage) necessitates more daily travel, which constrains group size. Teichroeb and Sicotte (2009) studied differently sized troops and found that the larger ones had increased home range size, daily path length, and percentage of time spent feeding. These changes occur because of scramble competition.

A captive study of female guerezas supported the relevance of the Socioecological Model, which predicts an egalitarian social system for this species (Grunau and Kuester 2001). The relevant findings were: (1) aggression was unritualized, bidirectional, and of low intensity; (2) communication signals did not convey dominance or subordination; and (3) allogrooming was about equally distributed and showed no kin bias.

Korstjens et al. (2007) studied sympatric species of all three colobus genera. They explored the effects of food on social systems, taking the Socioecological Model as a starting point. Weak female affiliation in all three species confirmed the predicted corollary that females did not form coalitions to defend food items against other group members. Coalitions are unnecessary because levels of female aggression are low, indicating that the distribution of food does not stimulate contest competition. Other studies rejected scramble competition because they found no connection between group size and daily path length. By the same token food distribution does not elicit cooperative intertroop aggression by females. Weakness of bonds among females facilitates female dispersal in all three species. Lack of food contestability allows for female dispersal.

Red colobus (*Piliocolobus*) bear the brunt of predation by chimpanzees (Figure 18.4). Where hunting by chimps has been studied intensively, red colobus account for about 75 percent of the apes' mammalian prey (the range across all sites is about 50 percent to 90 percent). Gombe chimps killed an average of 18 percent of the red colobus population each year, mainly immature individuals. The Gombe monkeys responded to the chimps with loud alarm calls, and adult male monkeys banded together to counterattack. The more colobus males that were involved in defense, the less likely that the hunt would succeed. Most red colobus live in fairly large multi-male troops ranging in size from 20 to 50, and troop size may be related to predation by chimpanzees. Smaller groups are found in forests where chimps are absent and there is evidence that adult sex ratios are affected by the presence or absence of chimps in the habitat. There are more males in a troop where chimps are present. Red colobus numbers do not seem to be related in a similar way to a formidable raptor (Struhsaker 2010). Another possible effect of chimp predation is a

Figure 18.4 A red colobus monkey. Infants and females are favored prey of chimpanzees, but the males in some populations fight back. © wetback via iStock

tendency toward male philopatry. Both sexes emigrate, but female dispersal seems to be much more common in most species.

Red colobus did not use the Gombe tactics in the Tai rainforest of West Africa, perhaps because they would not work in the denser canopy. There the males retreated quietly along with other troop members. Tai red colobus formed mixed-species associations with the highly vigilant Diana monkeys. Ranging patterns of troops in some populations were dictated primarily by association with the Dianas. Following experimental playbacks of chimp calls, red colobus initiated new associations with Diana troops and these associations lasted longer after playbacks of chimp calls compared with controls.

Among colobus monkeys, olives have the longest daily paths and the largest range areas per troop member (Korstjens et al. 2007; Oates 1994). An obvious explanation is that olive colobus depend on relatively high-quality foods that are scattered in clumps. However, their troops are small (2 to 12 members with a mean of 8) and their food sources dense (numerous small trees with young leaves), which should allow them to maintain the small home range typical of colobines. The ranging pattern is due to association with other monkeys, especially *Cercopithecus* species. One group associated closely with a larger group of Diana monkeys for more than three years. Keeping up with the larger monkeys shapes olive colobus activity patterns. They travel more (25 percent of scan samples) and rest less (40 percent) than other colobus monkeys. In all probability the olive colobus bear the costs of association with other species because they are a necessary defense against predators. Otherwise they depend on a cryptic profile: drab color, low rate and volume of vocalizations, and stealthy movement of small troops through dense vegetation in the middle and lower parts of the canopy. In addition to protection, Diana monkeys seem to provide olive colobus with the equivalent of a singles bar. Some extra-troop males associate with Diana monkeys and wait until female colobus show up. They attempt to copulate with these females and sometimes form new groups with them.

Cercopithecines

A monkey sits in a dense patch of foliage apart from its troop-mates. It pushes at one swollen and lumpy cheek with its fist and several small fruits work their way out. Unbothered, the monkey eats the fruits in comfort.

Cercopithecines, the other cercopithecoid subfamily, are characterized by **cheek pouches**. These are flexible areas of tissue in the lining of the cheeks that allow food to be accumulated (Figure 18.5). Unlike pouches in other mammals (e.g. squirrels) the tissue secretes a digestive enzyme. Compared to other primates, the tissue is highly innervated and strongly represented in the somatosensory area of the brain. Lambert (2005) tested functional hypotheses by comparing two distinct taxa in the same habitat. They were alike in their tendencies to: (1) use their pouches when feeding on contestable foods along with numerous conspecifics; and (2) retreat to denser vegetation after filling the pouches. Both results support the idea that cheek pouches play a role in intraspecific food competition. Retreat to a hidden place is also consistent with protection and predator evasion. In the social realm cercopithecines are distinguished by greater consistency in female philopatry and a larger role for female kinship

Figure 18.5 A female baboon, feeding on a fruit, displays the cheek pouches that are characteristic of the cercopithecines. © Glenn E. King

than in colobines or most other primates. Cercopithecines are divided into guenons and papionins, which are discussed in the next three chapters.

Summary

Colobines, mostly Asian, are physically adapted to consumption of mature leaves and unripe fruit. Slow digestion requires conservation of energy. One-male troops are typical and rates of social interaction are low. Alloparenting is common, including for very young infants. Gray langurs are widely distributed in Asia and survive in a wide variety of habitats. Other langur species display a variety of distinctive traits. Odd-nosed monkeys tend toward multi-level social organization. African colobines are preyed upon by chimpanzees in addition to the usual range of predators. Red colobus bear the brunt and display varied social adjustments. Black-and-white colobus have been the subject of important studies bearing on the relevance of the Socioecological Model to folivores. Olive colobus display several extreme responses to predator pressure. Cercopithecines are distinguished by cheek pouches and generally have a more omnivorous diet than colobines.

Sources

Basic Bibliography

Campbell, C.A., et al. (Eds.) (2011). *Primates in perspective*. NY: OUP.

Davies, G., & Oates, J. (Eds.) (1995). *Colobine monkeys*. NY: CUP.

Sterck, E.H.M. (2012). The behavioral ecology of colobine monkeys. In J.C. Mitani et al. (Eds.), *The evolution of primate societies* (pp. 65–90). Chicago: UCP.

Additional References

Bennett, E.L., & Gombek, F. (1993). *Proboscis monkeys of Borneo*. Sabah (MY): Koktas Sabah Berhad Ranau.

Borries, C., et al. (1999). DNA analyses support the hypothesis that infanticide is adaptive in langur monkeys. *PRSB*, *266*, 901–904.

Fuentes, A. (1996). Feeding and ranging in the Mentawai Island langur (*Presbytis potenziani*). *IJP*, *17*, 525–548.

Groves, C. (2007). The taxonomic diversity of the Colobinae of Africa. *Journal of Anthropological Sciences*, *85*, 7–34.

Grueter, C.C., & van Schaik, C.P. (2009). Evolutionary determinants of modular societies in colobines. *BE*, *21*, 63–71.

Grueter, C.C., et al. (2012). Evolution of multilevel social systems in nonhuman primates and humans. *IJP*, *33*, 1002–1037.

Grunau, T., Kuester, J. (2001). Dominance style in female guerezas (*Colobus guereza* Rüppell 1835). *Primates*, *42*, 301–307.

Harris, T.R. (2006). Within- and among-male variation in roaring by black and white colobus monkeys (*Colobus guereza*): What does it reveal about function? *Behaviour*, *143*, 197–218.

Harris, T.R., et al. (2006). Black and white colobus monkey (*Colobus guereza*) roars as a source of both honest and exaggerated information about body mass. *Ethology*, *112*, 911–920.

Harris, T.R., et al. (2009). Neither genetic nor observational data alone are sufficient for understanding sex-biased dispersal in a social group-living species. *Molecular Ecology*, *18*, 1777–1790.

Isbell, L.A. (2012). Re-evaluating the Ecological Constraints model with red colobus monkeys (*Procolobus rufomitratus tephrosceles*). *Behaviour*, *149*, 493–529.

Kawabe, M., & Mano, T. (1972). Ecology and behavior of the wild proboscis monkey, *Nasalis larvatus* (Wurmb) in Sabah, Malaysia. *Primates*, *13*, 213–228.

Kirkpatrick, R.C., & Grueter, C.C. (2010). Snub-nosed monkeys: multilevel societies across varied environments. *EA*, *19*, 98–113.

Korstjens, A.H., & Noe, R. (2004). Mating system of an exceptional primate, the olive colobus (*Procolobus verus*). *AJP*, *62*, 261–273.

Korstjens, A.H., et al. (2002). How adaptive or phylogenetically inert is primate social behaviour? A test with two sympatric colobines. *Behaviour*, *139*, 203–225.

Korstjens, A.H., et al. (2007). How small-scale differences in food competition lead to different social systems in three closely related sympatric colobines. In S. McGraw & 2 co-editors, *The monkeys of the Taï Forest, Ivory Coast* (pp. 72–108). NY: CUP.

Lambert, J.E. (2005). Competition, predation, and the evolutionary significance of the cercopithecine cheek pouch. *AJPA*, *126*, 183–192.

Liedigk, R., et al. (2012). Evolutionary history of the odd-nosed monkeys and the phylogenetic position of the newly described Myanmar snub-nosed monkey *Rhinopithecus strykeri*. *PLONE*, *7*, e37418.

McKenna, J.J. (1979). The evolution of allomothering behavior among colobine monkeys. *AA*, *81*, 818–840.

Messeri, P., & Trombi, M. (2000). Vocal repertoire of proboscis monkey (*Nasalis larvatus*, L.) in Sarawak. *FP*, *71*, 268–287.

Oates, J.F. (1994). The natural history of African colobines. In A.G. Davies & J.F. Oates (Eds.), *The colobine monkeys, their ecology, behavior, and evolution*. NY: CUP.

Oates, J.F., et al. (1990) Determinants of variation in tropical forest primate biomass. *Ecology*, *71*, 328–343.

Price, E.E., & Caldwell, C.A. (2007). Artificially generated cultural variation between two groups of captive colobus monkeys, *Colobus guereza kikuyuensis*. *BP*, *74*, 13–20.

Qihai, Z., et al. (2009). Sleeping site use by *Trachypithecus francoisi* at Nonggang Nature Reserve, China. *AJP*, *30*, 353–365.

Saj, T.L., et al. (2007). Controversy over the application of current socioecological theory to folivorous primates: *Colobus vellerosus* fits the predictions. *AJPA*, *113*, 994–1003.

Schel, A.M., & Zuberbuhler, K. (2012). Predator and non-predator long-distance calls in Guereza colobus monkeys. *BP*, *91*, 41–49.

Snaith, T.V., & Chapman, C.A. (2007). Primate group size and socioecological models: do folivores really play by different rules? *EA*, *16*, 94–106.

Sommer, V., & Vasey, P.L. (Eds.) (2006). *Homosexual behaviour in animals*. NY: CUP. [Chapters 1, 8, 9, 10, 14]

Stanyon, R., et al. (2012). Evolutionary molecular cytogenetics of cercopithecoid primates: past, present and future. *Cytogenetic Genome Research*, *137*, 273–284.

Struhsaker, T.T. (2010). *The Red Colobus monkeys: variation in demography, behavior, and ecology of endangered species*. NY: OUP.

Struhsaker, T.T., & Leland, L. (1987). Colobines: infanticide by adult males. In B.B. Smuts, et al. (Eds.), *Primate societies* (pp. 83–97). Chicago: UCP.

Teichroeb, J.A., & Sicotte, P. (2009). Test of the ecological constraints model on ursine colobus monkeys (*Colobus vellerosus*) in Ghana. *AJP*, *71*, 49–59.

Teichroeb, J.A., & Sicotte, P. (2012). Cost-free vigilance during feeding in folivorous primates? Examining the effect of predation risk, scramble competition, and infanticide threat on vigilance in ursine colobus monkeys (*Colobus vellerosus*). *BE&S*, 66, 453–466.

Teichroeb, J.A., et al. (2012a). Infanticide risk and male quality influence optimal group composition for *Colobus vellerosus*. *BE*, *23*, 1348–1359.

Teichroeb, J.A., et al. (2013). Factors influencing male affiliation and coalition strength in a species with male dispersal and intense male-male competition, *Colobus vellerosus*. *Behaviour*, *151*, 1045–1066.

Tilson, R.L., & Tenaza, R.R. (1976). Monogamy and duetting in an Old World monkey. *Nature*, *263*, 320–321.
Tombak, K.J., et al. (2012). Patch depletion behavior differs between sympatric folivorous primates. *Primates*, *53*, 57–64.
Wikberg, E.C., et al. (2012). Between-group variation in female dispersal, kin composition of groups, and proximity patterns in a black-and-white colobus monkey (*Colobus vellerosus*). *PLONE*, *7*, e48740.
Wikberg, E.C., et al. (2013). Individualistic female hierarchies with varying strength in a highly folivorous population of black-and-white colobus. *Behaviour*, *150*, 293–320.
Wikberg, E.C., et al. (2014). Familiarity is more important than phenotypic similarity in shaping social relationships in a facultative female dispersed primate, *Colobus vellerosus*. *BP*, *106*, 27–35.
Workman, C. (2010). Diet of the Delacour's langur (*Trachypithecus delacouri*) in Van Long Nature Reserve, Vietnam. *AJP*, *72*, 317–324.
Wright, K.A., et al. (2008). Comparisons of suspensory behaviors among *Pygathrix cinerea*, *P. nemaeus*, and *Nomascus leucogenys* in Cuc Phuong National Park, Vietnam. *IJP*, *29*, 1467–1480.
Xiang, Z-F., et al. (2014). Males collectively defend their one-male units against bachelor males in a multi-level primate society. *AJP*, *76*, 609–616.

19 A Cercopithecine Tribe

The Guenons

The dog-sized monkey sits on a branch, feeding on fruit. The yellow heart-shaped pattern around his nose marks the species identity that he shares with twenty-five troop members scattered around him. His creamy chest makes him more visible to them in the shade of the forest, while his agouti-brown back can hide him from raptors above.

Guenons are typically medium-sized arboreal monkeys found in a variety of forests throughout western and central Africa. Most of them are apparently similar to one another and until a few decades ago they were lumped into the single genus *Cercopithecus*. However, as in most other primate groups, taxonomic splits have proliferated (Box 19.1). This has created some confusion as to the meaning of the informal term "guenon". Here we use it in the broadest sense to denote the tribe Cercopithecini.

Box 19.1 Guenon Taxonomy

Guenon taxonomy has been in flux for some time now, as *Cercopithecus* has undergone a series of splits. First there was general agreement on separating the patas species from others as *Erythrocebus*, and it seems that the same has happened to the vervets. They are now *Chlorocebus*, a genus that holds an increasing number of species. The process continues. Genetic research indicates that several of the more terrestrial forest species are more closely related to patas and vervets than they are to the other guenons. This has been recognized with the suggestion that they be given their own genus, *Allochrocebus* (Fleagle 2013).

Guenon Ecology

Colorful facial markings probably serve to distinguish guenon species, individuals, and appropriate mates in shadowy habitats (Figure 19.1). There is a recurring trend toward terrestrialism, mainly to feed on terrestrial herbaceous vegetation. Niche differentiation allows up to five guenon species to coexist. Some of these differences are small, e.g. seeking different insect prey or eating different parts of the same plant.

Guenons frequently join polyspecific groups, favoring the highly vigilant Diana guenons in areas of West Africa where Dianas occur. Comparative study of one guenon species showed that, when associated with Dianas, they spent more time in exposed areas, vocalized

Figure 19.1 An arboreal guenon, illustrating one of the varied facial markings that occur among these species. © Mongabay animals_03199

more often, and displayed greater inter-individual distances. In some polyspecific groups one guenon species seems to use the other as a food guide, the wider ranging species taking advantage of the other's localized knowledge.

Guenon Sociality

The one-male troops typical of forest guenons resemble those of colobines in that the adult male tends to be peripheral and most adult interactions outside the mating season are between females. These are generally affiliative, mainly grooming. Other interactions are subtle and cohesion is maintained by mutual attention rather than interaction. Each individual monitors the others and adjusts her behavior accordingly.

Unless avoidance is easily achieved, neighboring troops are usually hostile and protect boundaries and/or resources. Adult females lead the defense, often supported by younger females. Calls keep troops apart, but confrontations occur and occasionally escalate to physical aggression. Males emit loud calls from the sidelines. They are most likely to join conflicts directly during the mating season, when opposing males are reproductive threats.

The mating system is polygynous except for influxes. Data from a few species suggest that influxes occur in some years but not in others, depending on the number of extra-troop males in the population. The length of time that an intruder stays in a troop ranges from less than a day to several months. Agonistic interactions among males reflect intense competition, and most intruders copulate with resident females. The possible outcomes for the resident male are the following: (1) he may repel or evict intruders; (2) he may coexist with intruders throughout an influx; or (3) he may leave the troop. A leader male may keep his position, but his tenure can be very much shorter.

Lacking visual signs of ovulation, proceptive females solicit males by approaching and staying near. This is presumably a strong signal, contrasting with the usual separation of the sexes. Females sometimes induce influxes with sexual solicitations, perhaps to obtain a choice among partners. Female choice is reflected in solicitation of some available males but not others. Another function may be to ensure fertilization, which could be important in a large troop where numerous females might deplete the single resident male. An important recurring pattern is non-ovulatory mating. Females sometimes copulate throughout an entire month; during pregnancy; before the mating season begins; or after it ends. This may create paternity confusion in potential invaders.

Guenon Life Cycle

In a recurring pattern, infants receive unusually extensive care from other troop members, mainly females, when they achieve enough mobility to move away from their mothers (Chism 2000). The timing may be related to subsistence. A larger infant that grabs branches and drops off the mother on a whim is a major impediment to foraging. The risk of allocare is low because the caretakers are female relatives.

The young in some guenon species seem to play less than other kinds of monkeys. In proportion to the daily activity budget, red colobus played four times as much as two guenon species (Worch 2010). The researcher offered two ecological explanations in which high levels of arthropod consumption decreased play opportunities. First, the guenons spaced themselves relatively far apart while foraging. Second, they spent relatively little time resting to facilitate digestion.

Females first give birth between four and five years of age. Males, who are significantly larger, become sexually mature at about five or six years. Most males probably become solitary when they disperse. However, all-male groups occur in several species and may have been overlooked in others because they are small (sometimes just two or three individuals) and inconspicuous. Their only intragroup vocalizations are soft and infrequent contact calls, and they may fall silent when a troop is near. On the other hand, all-male groups sometimes engage in aggressive encounters with troop males and may be the source of some influxes (as opposed to convergence of solitary males).

Guenon Blues

Blue monkeys (*Cercopithecus mitis*) have been studied more extensively than any other forest guenons, especially in the project at Kakamega that has lasted for more than 20 years. As a result, blue monkeys provide important perspectives on many aspects of forest guenon behavior. Some of the seemingly unique behaviors of blue monkeys may be truly unique, but others may be things that we have not yet been seen or understood in other forest guenons.

Dominance is a good example. Forest guenons were generally considered "egalitarian", but long-term work on blue monkeys documented a subtle dominance hierarchy (Cords 2002b). As in other guenons, aggression within the troop is rare: an average female has an agonistic interaction about once every two hours, and more than half of these are simple cases of approach–retreat without any overt aggression. Nevertheless, tabulation of a large number of these interactions displayed a linear dominance hierarchy with very few reversals in individual interactions. The hierarchy also showed great consistency over time: all but one adult and subadult females kept the same ranks for eight years.

The discovery of this hierarchy also raises a question, because rank does not seem to be related to any advantage in feeding, grooming, or reproductive success. Cords suggested several explanations. First, the advantage may have to do with longevity, a key factor in understanding primates. A few cases suggest that lower ranking females die earlier and have lower lifetime reproductive success as a result. Another possibility is that the advantages of rank are not a matter of daily life, but become apparent only in crisis situations (such as drought, perhaps). However, no such crises were observed in 21 years of research. Finally, Cords considered the possible significance of emotion: "high-ranking females are less often frustrated in achieving their immediate goals". The agonistic interactions that define the dominance hierarchy occur most often in the context of feeding on fruit. Although dominant females have no advantage in overall feeding, including fruit, they may get the most satisfying food with less effort.

Although blue monkey home ranges overlap considerably, long-term observation established the existence of borderlines related to intertroop conflict. The significance of resources is indicated by the fact that incursions are most likely to involve feeding. Major incursions (100 meters or more) involve unusually attractive resources such as a large fig tree. However, accessing resources is not necessary to trigger the proximal mechanisms that underlie defense. Crossing a boundary line is sufficient. Fights can last for as long as an hour and may become bloody. After one such conflict an adult female lay still for 40 minutes with her arms and tail spotted with blood, apparently in shock. Underscoring the fact that females have the greatest stake in territorial defense, juvenile males from the hostile troops sometimes play with each other while juvenile females confront the enemy. The aftermath of a territorial conflict seems to reflect female solidarity. There is a "frenzy" of grooming in which the females rapidly shift from one partner to another, seemingly trying to groom a variety of partners in a short time. Females are also allies against aggressive males in several contexts: feeding, infanticide, and hostile male responses to persistent solicitation from an estrus female.

The combination of territoriality and dominance in the blue monkey calls into question a basic premise of current socioeocological theory. According to the models, troops that engage in cooperative competition against other troops for resources should have egalitarian relations within the society. Blue monkeys are consistent in both territorial behavior *and* a clear-cut (though not oppressive) linear dominance hierarchy (Cords 2012b).

The consistently low level of agonism associated with blue monkey dominance raises another question in regard to socioecological theory. The models relate feeding strategies to variation in female social relationships. More specifically, increased contest competition within the troop should lead to more aggressive behavior and more pronounced dominance hierarchies. In blue monkeys rates of agonism do not vary with changing ecological conditions and availability of preferred foods. When the desirable reproductive parts of plants were less common, subordinate females avoided competition with dominants by spending more time in foraging and less time in resting. This alternative may not be available to species that are less flexible in their diets. Nevertheless, the blue monkey results suggest that socioecological models should consider the behavior of consumers as well as the distribution of their foods (Pazol and Cords 2005).

Finally, blue monkeys provide interesting perspectives on allocare. The results of a study by Forster and Cords (2005) were consistent with the Toddler Hypothesis of Chism (2000). However, they suggested that another explanation for allomothering is more important: the Practice Hypothesis, which holds that allomothering by young females prepares them for their own maternal function in later life. They noted that an infant's associations can be biased by the effect of group composition on partner availability. Controlling for this factor, Forster and Cords found that the allomothering tendency was much stronger in large juvenile females than in adult females. These juveniles were eager to care for infants from the beginning, sometimes kidnapping them. Allomothering was delayed only because the mothers did not relinquish their offspring until the toddler stage. The researchers used their data to examine two other hypotheses. The idea that allomothering is just a byproduct of maternal motivation did not fit with the fact that nulliparous females cared for infants far more than did experienced adult females. Allomothering seemed not to function mainly for social integration because despite strong female philopatry, male infants received just as much attention as females.

Guenons Exposed

Patas monkeys (*Erythrocebus patas*) and vervets (*Chlorocebus* spp.) have gone further than other guenons in adapting to terrestrial behavior and relatively open country. They live in woodland/savanna habitats throughout most of sub-Saharan Africa. Both can extract a living from savanna, but prefer the edges of woodlands where trees provide more food and refuges. Fruit constitutes no more than 10 percent of the diet. Comparison with forest guenons illuminates ecological and phylogenetic influences on behavior.

Vervets and patas display a pattern of rapid reproduction that counters difficult and unpredictable environmental conditions: high fecundity, intensive allocare, and rapid maturation. A short inter-birth interval of about 12 to 16 months is common, varying with the availability of food and water. Mothers relinquish their infants to allocare within the first few days and allomothering mainly occurs in the first three months. There is little evidence for harsh treatment by adult females that is an element of competition in some other primate species. Allocare supports early weaning and rapid independence, facilitating yearly (or at least biannual) conceptions (Chism 2000). Weaning may begin at about three months of age and is usually achieved by nine months. A female's first birth takes place at about three years unless delayed by environmental factors such as water shortage. Males display physiological aspects of sexual maturity at about that time, but may still be two or three years short of behavioral maturity and full adult size.

Patas Monkeys

Patas are the largest, the most dimorphic, and probably the most terrestrial guenons, surviving in environments that are poorer, drier, more seasonal, and less predictable. Abundant savanna grass offers little sustenance because the monkeys' unspecialized stomachs cannot digest grass stalks and their fingers are not strong enough to dig up the more digestible roots. Grass seeds are accessible and nutritious but are only available for brief periods in many areas, and are hard to collect in quantity.

Gum largely substitutes for fruit and constitutes more than a third of the diet (perhaps uniquely among catarrhines). Arthropods are also important, especially colonial ants. The monkey bites open thorns of certain acacia trees and extracts adult ants and larvae with its tongue. Ants are probably more reliable than many plant foods in highly seasonal habitats, and may provide as much as a third of daily caloric needs. Such faunivory in larger-bodied primates probably merits closer study (Isbell and Young 2007; McGraw 2014). Patas require home ranges as large as 4000 hectares, larger in proportion to troop biomass than those of any other primate. Daily path length can be as long as four kilometers. The route is often meandering, perhaps because of opportunistic foraging.

Patas depend heavily on cryptic defense. Reddish coloration blends with the brownish grasses of dry savannas. Unlike their chirping forest relatives, they rarely call while moving or foraging. Whether on the ground or in trees, patas frequently stand bipedally to scan the environment. All troop members except infants participate in vigilance and alarm calling. If crypticity fails, patas running speed may be enough to escape any predator except the cheetah (up to about 55kph).

Patas typically resemble forest guenons in that the male is peripheral while female relationships are relaxed and dominance structure is not rigid (Cords 2000). However, female relationships may be diluted in large troops that sometimes total 50 members (Chism 1999). Nakagawa (2008) reported a linear dominance hierarchy. Others attributed this to competition caused by provisioning, but Nakagawa advanced a hypothesis based on comparative study and the Socioecological Model. Nakagawa's troop spent 18 percent of their feeding time on tall trees that were widely distributed, resulting in competition in each feeding patch. Relaxed relationships existed where patas spent more than half of their feeding time on short trees that were distributed uniformly; that is, small and abundant clumps of food obviated competition.

Adult females monopolize alloparenting as long as the infant is black. They hold it, carry it, and even nurse it. Infants older than three months enter into play groups, called **crèches** because they are under more organized supervision than play groups in other primates. A single adult or juvenile female watches the infants while the mothers forage more than 100m away.

Most dispersed males spend several years in a succession of all-male groups, which seems to be unusual among guenons. Patas all-male groups stand out among anthropoids for their stability and high degree of organization. These social units may be prominent in patas because solitary individuals are at a greater disadvantage in an open environment that exposes them to predators.

Vervets

The word vervet is used here to encompass all monkeys across Africa now placed in the genus *Chlorocebus*. They are the most widespread and abundant of African monkeys other than baboons. The vervet's black face is plain compared to the markings on its forest relatives and the greenish-gold body hair is drab compared to patas coats (Figure 19.2). The face markings are not necessary for a single taxon in relatively open habitats and the body color may be better camouflage in more varied and heterogeneous environments.

Vervets span the ecological gap between patas and other guenons: some vervets live in forests, but never dense rainforests; most live in woodlands, but they may range into grasslands. They make effective use of the ground, but are not as terrestrial as was once thought. Acacia trees, common in dry African habitats, are vital to vervets. Otherwise they display enormous flexibility in feeding. They can survive on relatively restricted diets, including nutritionally poor and tough foods such as acacia pods and fibrous fruits. They can also shift to highly specialized diets when necessary, mostly fiddler crabs in one case and bamboo leaves in another (Mekonnen et al. 2010).

Multi-male troops of up to 75 members differ sharply from the small one-male troops of most guenons. However, some studies report a single alpha male with a specialized social role that might correspond to the single breeding male position of typical guenons. The one-male tendency might have a phylogenetic basis for the genus, while environment elicits various degrees of multi-male organization.

Isbell et al. (2002) reviewed and critiqued explanations for multi-male vervet troops. It is not a matter of males adjusting to the number of available females because vervets have all the characteristics of guenon one-male troops: number of females per group, length of

Figure 19.2 Vervet monkeys at ease on the ground. © Liz Leyden via iStock

breeding season, numbers of females mating per day within troops, and spatial dispersion of the troop. Cooperative defense is unlikely since the alpha male shows the strongest defense of females against outsiders and the greatest vigilance and activity against predators. Nor is home range defense likely, since the number of males in a vervet troop does not vary with the frequency or intensity of territoriality.

Isbell and colleagues proposed the Limited Dispersal Hypothesis, in which habitat configuration limits the number of troops available to a dispersing male. Vervets typically live in narrow riverine forests and each troop has just two neighbors. A potentially high cost of long-range dispersal is suggested by mortality rates among vervets forced into unfamiliar areas. Therefore, male vervets typically move into neighboring troops where they are likely to find maternal brothers and even their fathers. Kin selection predicts that relatedness will lead to tolerance on the part of residents, especially since the immigrating male is unlikely to commit infanticide because of relatedness. In an atypical habitat with swamps rather than rivers, each troop had more neighboring troops and dispersal choices, and infanticides occurred. Vervets imported to the Caribbean island of Barbados were evenly distributed and lived in one-male troops that were subject to takeovers, and infanticide seemed to be a major cause of mortality. Thus it appears that multi-male organization is a facultative trait that arises from environmental constraints on dispersal. In different conditions vervets behave more like other guenons.

In a multi-male troop males and females have separate linear dominance hierarchies. Males are dominant over females in one-to-one interactions, but a female coalition can dominate any male. Males try to form alliances with high-ranking females through proximity and grooming. An alpha male may aggressively restrict such access, thus protecting his social and breeding status.

Dominant males signal their status with the unique red, white, and blue display. The blue scrotum is shared with patas males, but the red penis and fringe of white hair are all their own. A displaying male approaches a subordinate and circles him, then lifts his tail to show the anogenital area. This is an energetically conservative way to maintain relationships (Gerald 2001). Differences in scrotal color predict eventual social status (although experiments show that this is not the only factor). In tests with paired unfamiliar males, those with scrota of similar color engage in more contact aggression with one another (Gerald et al. 2006).

Males also display the colors when they see males of another troop near a territorial border. The vigilant resident assumes a posture in which he sits with knees apart, exposing his genitalia to the intruders. However, vervet territoriality is facultative and intertroop relations vary from serious fighting to friendly intermingling. Relevant factors include habitat type, season, intensity of predation, and the configuration of resources.

The newborn vervet is an unusual burden because it is heavy in proportion to the mother's weight and it grows rapidly. The effect has been observed in captive vervets: lighter mothers are more rejecting and infants play a greater role in maintaining ventral contact. The behavior of the lighter captives is comparable to that of wild females (Fairbanks et al. 2010).

Allomothering mitigates the energetic burden. The vervet system is one in which mothers, infants, and allomothers all receive benefits that far exceed potential costs. The mother gets relief, the infant gets additional caretakers, and nulliparous siblings (the most frequent caretakers) get practice. Males and low-ranking females try to get access to high-ranking mothers through their infants. Males are seeking future mating opportunities, while the female goal is to decrease aggression from the dominants.

Fruteau et al. (2011) interpreted vervet allomothering with reference to the biomarket principle that price should fluctuate with supply. They used the following correspondences:

commodity = access to newborns; price = grooming the mother; and supply = the number of newborns in the group. Allomothers obtained access with less grooming when infants were numerous. Females handled infants longer when there were fewer infants, but only frequent grooming partners of the mother received this opportunity. These transactions were not related to kinship.

Vervet mental abilities are suggested by vocal communication. Use of different alarm calls for different dangers indicates the ability to categorize objects in a natural habitat and provide representational information about classes of environmental objects. Field experiments show that a vervet can distinguish individuals in its own and neighboring troops based on voice alone, and can classify others as belonging to particular kin groups or dominance positions. Such abilities allow vervets to be flexible and opportunistic in a highly variable environment while maintaining cohesive social structures.

Vervet social learning varies with the model. When observing the solution of a problem, vervets paid more attention to an adult female than to an adult male. The favored role of the female can be attributed to female philopatry (van de Waal et al. 2010). With regard to the solution, vervets only displayed **enhancement learning**; that is, they touched the part of the apparatus that could be used in the solution but modeling did not increase the probability that observers would solve the problem.

Summary

Guenons are medium-sized monkeys, typically arboreal, but with some terrestrial inclinations. One-male troops are the typical social organization, but mating is often affected by influxes of non-troop males. Extensive allocare is a recurring feature of the life cycle. Guenon vocalizations have been studied intensively as indicators of cognitive abilities. Two genera engage in more terrestrial activity than the others. Both have adapted to savanna conditions with a pattern of rapid reproduction. Patas typically live in one-male troops. Multi-male troops are common among vervets, but exceptions suggest that this is the result of environmental constraints.

Sources

Basic Bibliography

Campbell, C.A., et al. (Eds.) (2011) *Primates in perspective*. NY: OUP. [Chapters 14, 15, 16]

Cords, M. (2012a). The behavior, ecology, and social evolution of cercopithecine monkeys. In J.C. Mitani et al. (Eds.), *The evolution of primate societies* (pp. 91–112). Chicago: UCP.

Gautier-Hion, A., et al. (Eds.) (1988). *A primate radiation: evolutionary biology of the African guenons*. NY: CUP. [especially Fedigan and Fedigan]

Glenn, M.E., & Cords, M. (Eds.) (2002). *The guenons*. NY: Kluwer Academic/Plenum.

Mittermeier, R.A., et al. (Eds.) (2013). *Handbook of the mammals of the world, Vol. 3. Primates*. Barcelona: Lynx Edicions. [pp. 668–696]

Additional References

Bouchet, H., et al. (2013). Social complexity parallels vocal complexity: a comparison of three non-human primate species. *Frontiers in Psychology*, DOI: 10.3389/fpsyg.2013.00390.

Chism, J. (1999). Decoding patas social organization. In P. Dolhinow & A. Fuentes (Eds.), *The nonhuman primates* (pp. 86–92). Mountain View (CA): Mayfield.

Chism, J. (2000). Allocare patterns among cercopithecines. *FP*, *71*, 55–66.
Cords, M. (2000). The number of males in guenon groups. In P. Kappeler (Ed.), *Primate males* (pp. 84–96). NY: CUP.
Cords, M. (2012b). The thirty year blues: what we know and don't know about life history, group size and group fission of blue monkeys in the Kakamega Forest, Kenya. In P.M. Kappeler & D. Watts (Eds.), *Long-term studies of primates* (pp. 289–312). NY: Springer.
Fairbanks, L.A., et al. (2010). High-fiber diet promotes weight loss and affects maternal behavior in vervet monkeys. *AJP*, *72*, 234–241.
Fleagle, J.G. (2013). *Primate adaptation and evolution* (3rd ed.). NY: Academic Press.
Forster, S., & Cords, M. (2005). Socialization of infant blue monkeys (*Cercopithecus mitis stuhlmanni*): allomaternal interactions and sex differences. *Behaviour*, *142*, 869–896.
Fruteau, C., et al. (2011). Infant access and handling in sooty mangabeys and vervet monkeys. *AB*, *81*, 153–161.
Gerald, M.S. (2001). Primate colour predicts social status and aggressive outcome. *AB*, *61*, 559–566.
Gerald, M.S., et al. (2006). Artificial color treatment mediates aggression among unfamiliar vervet monkeys (*Cercopithecus aethiops*). *Animal Welfare*, *15*, 363–369.
Isbell, L.A., & Young, T.P. (2007). Interspecific and temporal variation of ant species within *Acacia drepanolobium* ant domatia, a staple food of patas monkeys (*Erythrocebus patas*) in Laikipia, Kenya. *AJP*, *69*, 1387–1398.
Isbell, L.A., et al. (1999). Rank differences in ecological behavior: a comparative study of patas monkeys (*Erythrocebus patas*) and vervets (*Cercopithecus aethiops*). *IJP*, *20*, 257–272.
Isbell, L.A., et al. (2002). Why vervets live in multi-male groups. In M. Glenn & M. Cords (Eds.), *The guenons* (pp. 173–187). NY: Kluwer.
McGraw, W.C. (2014). The 'other faunivory' revisited: insectivory in human and non-human primates and the evolution of human diet. *JHE*, *71*, 4–11.
Mekonnen, A., et al. (2010). Diet, activity patterns, and ranging ecology of the bale monkey (*Chlorocebus djamdjamensis*) in Odobullu Forest, Ethiopia. *IJP*, *31*, 339–362.
Nakagawa, N. (2008). Despotic wild patas monkeys (*Erythrocebus patas*) in Kala Maloue, Cameroon. *AJP*, *70*, 238–246.
Pazol, K., & Cords, M. (2005). Seasonal variation in feeding behavior, competition and female social relationships in a forest dwelling guenon, the blue monkey (*Cercopithecus mitis stuhlmanni*), in the Kakamega Forest, Kenya. *BE&S*, *58*, 566–577.
Rochat, M.J., et al. (2008). The evolution of social cognition: goal familiarity shapes monkeys' action understanding. *CB*, *18*, 227–232.
van de Waal, E., et al. (2010). Selective attention to philopatric models causes directed social learning in wild vervet monkeys. *PRSB*, *277*, 2105–2111.
Worch, E.A. (2010). Play behavior of red colobus monkeys in Kibale National Park, Uganda. *FP*, *81*, 163–176.

20 A Cercopithecine Tribe

The Papionins

In a troop of macaques in India a young male approaches a female in early estrus. The alpha male threatens him, leaning forward and staring with firmly rounded mouth. In a troop of macaques in Indonesia a young male approaches a female in early estrus. The alpha male sits and stretches his mouth in a yawn.

Papionins are generally larger and more sexually dimorphic than guenons. They display more variation and complexity in social organization. The taxon includes the generalized and widely successful macaques. It also includes the more narrowly distributed geladas and mandrills, which are more specialized (each in its own way) in morphology and behavior. The mangabeys, of great interest to scientists but less noticed by the general public, are also included. Baboons in the narrow sense, the genus *Papio*, get the next chapter to themselves for several reasons that will be explained there.

Box 20.1 Papionin Taxonomy

Papionin classification has been relatively stable at the genus level for some time now except for the recognition that the mangabeys, formerly considered a single genus, have been placed in two separate genera that belong to two separate clades. *Macaca* is still a single genus, though perhaps more variable than previously thought. *Mandrillus* (including mandrills and drills) is in a clade with the mangabey *Cercocebus*, diagnosed by nuchal crests and widely divergent temporal lines (especially in males), maxillary ridges in males, and a tendency toward enlarged molars. *Theropithecus* (the geladas) is characterized by the size and placement of the temporalis muscle, which optimizes occlusal forces on the molars. Details of the dentition are also highly distinctive.

Macaques

Macaques are (with one exception) Asian monkeys. Among the many species, weight varies from about 2.5kg to 18kg. Males are typically 50 percent heavier than females. The genus is composed of three evolutionary lineages. As far as possible, the generalizations below are based on fieldwork with more than one species from more than one lineage. There is also a huge quantity of information derived from captive studies, much of it from a few species. A great deal of it pertains to the rhesus macaque (*Macaca mulatta*), which is unfortunately known to many non-primatologists (even some scientists) as "the monkey".

Macaque Ecology

Macaques live in evergreen, deciduous, and coniferous forests, swamps, semideserts, and locales with severe, snowy winters. A single species may span several of these habitats. Many species spend more waking time on the ground than in the trees. Macaques in tropical forests are primarily frugivorous terminal branch feeders. Fruit, which constitutes 60 to 90 percent of the diet, is the focus of more competition and requires more than the leaves that macaques use extensively as fallback foods. Some show great flexibility in diet under varying conditions, including the use of bark and buds by northern species during the winter.

Macaque Sociality

Macaques live in multi-male troops with overlapping home ranges. The typical range in troop size is 15 to 50, but some unprovisioned troops reach 100 or more. Defense of resources seems to be a key factor. Encounters vary from hostile to peaceful, but larger troops usually supplant smaller ones. Peaceful interactions may depend in part on the presence of individuals known to each other on both sides, a likely situation since males often transfer into adjacent troops. Encounters mainly involve subadult and young adult males.

Macaque societies are organized in terms of dominance and kinship. The basic kinship unit is a matrifocal family composed of a living mother and her offspring that still reside in the same troop. Beyond this family, kinship is structured by matrilines. Each matriline constitutes a linear dominance hierarchy and all the matrilines in a troop tend to form a linear dominance hierarchy. Members of a dominant lineage outrank all members of a subordinate one. The matriline provides members with allies in dealing with unrelated troop members. Third party interventions can play a decisive role in macaque social relationships.

Grooming is extremely important to female macaques (Figure 20.1) and the genus has been prominent in studies of this behavior. Tension reduction is widely accepted as a proximal effect in many species. Studies of various macaque species show that recipients have lowered heart rates and engage in less self-directed behavior such as scratching. A former recipient is more likely to be the nearest neighbor of the former groomer during a period after grooming ends (Aureli and Yates 2010).

Performance of grooming can have a similar effect. Fecal glucocorticoids indicated that monkeys who are groomers have lower stress levels in the long run (Shutt et al. 2007). Aureli and Yates (2010) found that female black crested macaques who had recently groomed others engaged in less self-scratching and initiated fewer aggressive encounters than during control periods. They suggested that groomers are alleviating stress caused by proximity to other monkeys, noting that other macaque studies link proximity and stress.

Berman et al. (2008) used grooming in Tibetan macaques to investigate degrees of preferential treatment for relatives, which they called **kin bias intensity**. Initial observation of a troop revealed little or no kin bias in female behavior, but strong biases were apparent seven years later. The researchers considered three possible explanations: the Socioecological Model; time constraints; and risk-related responses. Time constraint simply refers to time available for various activities. Risk-related responses vary in reaction to some perceived threat (in this case, human tourists). The researchers made predictions based on each hypothesis and found that the data only supported time constraints.

Macaque reproductive success is related to latitude. Year-round breeding in tropical species makes it possible for dominant males to monopolize females as they successively come into estrus. Seasonal breeding in the larger temperate zone troops makes it impossible

Figure 20.1 A Japanese macaque, an "intolerant" species of the genus *Macaca*. © Bernhard Richter via iStock

for dominant males to control all of the simultaneously breeding females. Females can exert some choice through covert copulations with non-dominant males or visitors. Meta-analysis of 14 studies in which genetic data established paternity strongly supported the connection between high rank and reproductive benefits for males (Rodriguez-Llanes et al. 2009), with a stronger correlation in the tropical species.

Macaque Life Cycle

Adult body weight varies widely among macaque species, but patterns of growth and reproduction are also heavily dependent on food intake for both sexes. Gestation averages about 24 weeks, a week shorter in the lighter species and a week longer in the heavier species. Mothers wean their infants at some time between 6 and 12 months. Typically, a female gives birth every year or two until senescence.

Some macaque mothers may encourage or even teach certain kinds of behavior. "Dental flossing", i.e. pulling something through the teeth, apparently to dislodge food particles, has been observed in individuals of two species (Leca et al. 2010; Watanabe et al. 2007). It is characteristic of a certain group of longtailed macaques where mothers exaggerate these actions in front of their offspring, which suggests awareness of the learning process.

Most individuals reach sexual maturity at three or four years of age, before the growth of the body is completed. The first sign in females is reddening of the sexual skin. First birth usually takes place when the female is between four and six years old. Male puberty begins with testicular enlargement. During the following two years testosterone levels rise, agonistic encounters with other males increase, and dominance status rises. During this adolescent phase, a male attains full adult body mass. Sexual maturity precedes the social ability to obtain access to females.

Most males leave the natal troop during adolescence, often with other males of the same age. Additional emigrations may take place every two to five years during an individual's lifetime. Troop tenure is related to troop size and the number of available mating partners. Emigration seems to be caused by a combination of inbreeding avoidance in the old troop and female preference for novel males in the new one. Emigration by high-ranking males indicates that expulsion is not a major factor.

Solitary life and immigration both increase the mortality rate for male macaques. A new male in a troop may accept a place at the bottom of the male hierarchy and work his way up, or he may immediately challenge the most dominant residents. The latter approach is dangerous and most likely to be undertaken by males in their prime. Regardless of his place in the hierarchy, a newcomer male displays elevated rates of copulation. Female preference for novel males is presumably a key factor; the subsequent decline in a new male's sexual activity is probably due to increasing familiarity. The decline of fighting ability in older males may also be a factor. Reduced reproductive success is probably an issue of access rather than hormones, since old males have sired offspring in semi-free-ranging conditions.

Female cycles become irregular at about 20 years of age and reproduction usually ends in the middle or latter twenties. There is little evidence for post-reproductive life under natural conditions. The maximum life span for both sexes is about 35 years, which is twice the average age at death.

Macaque Cognition

Macaques in the wild and in captivity display a variety of tool using and tool modifying behaviors, and some species seem particularly prone to such behavior. Three liontailed macaques in one troop used leaves to prepare caterpillars for eating. Each monkey picked up a caterpillar with a leaf and rolled it on a surface with the leaf protecting the hand until irritating hairs were removed (Hohmann 1988). Young free-ranging liontails born in captivity used sponges of crumpled leaves to take drinking water from tree cavities (Fitch-Snyder and Carter 1993), as chimpanzees do. An adult male performed the same actions immediately after one juvenile, which suggests imitation or at least social facilitation. A confiscated adult probably learned tool use in a human setting, but others learned from him when they were placed together (Kumar et al. 2008). Captive liontails also used enrichment materials in a variety of goal-directed activities, applying leverage to objects and creating perches and ladders (Westergaard and Lindquist 1987). In a comparative experiment that allowed subjects to obtain syrup with probing tools, liontails spontaneously made and used the tools (Westergaard 1988).

Tonkean macaques seem especially manipulative, using sticks and rods to obtain food in ways that seem spontaneous and insightful (Ducoing and Thierry 2005; Ueno and Fujita 1998). In a series of experiments involving the use of a rod to obtain food, two Tonkean macaques learned the task in 6 and 8 hours while pigtailed macaques and a baboon took 8 to 14 hours. Tonkean macaques also display the ability to deceive. An experiment provided just one monkey in each trial with knowledge of hidden food (Ducoing and Thierry 2003). Higher-ranking subjects did not alter their search patterns as a result, but strongly dominated individuals did so when accompanied by a dominant. They modified their number of pauses, avoided being followed, stopped when being watched, and took wrong directions.

Captive rhesus macaques seem to display substantial cognitive ability in experiments. In categorizing unknown conspecifics according to dominance, the monkeys appear to use abstract social concepts in their awareness of relationships within the group (Bovet and Washburn 2003). They also appear to read goal-relevant gestures of humans (Hauser and Wood 2011), though this may be limited to cases with an observed and familiar goal (Rochat et al. 2008).

Despotic Macaques

Several macaque species share a set of hostile and competitive behaviors that many experts label **despotic** (Thierry 2007). Rhesus and Japanese macaques epitomize this social mode. Longtails are on the cusp, having been shifted recently to the next lower grade (Table 20.1, online). Despotic macaques constitute only 6 or 7 of about 20 species in the genus, but they occupy a greater geographic area than all other macaque species combined. This way of life has obviously been very successful.

High intensity aggression is common among despotic macaques and most conflicts are unidirectional, i.e. threatened and attacked individuals rarely respond in kind. Reconciliation is rare (rates for unrelated individuals vary from 4 to 12 percent) and so are affiliative contacts in general. A steep dominance gradient controls social life. Power determines which individuals interact with each other, affecting partners for proximity, play, and affiliation.

Much of this behavior occurs in the context of female kinship. Despotic macaques have the strongest kinship networks in the genus, showing a high preference for interaction with relatives. Kinship is the basis for coalitions that result in the steep dominance gradient.

Where several matrilines occur (in larger troops), they are ranked against one another and form a linear hierarchy. All members of a dominant lineage outrank all members of a subordinate one. Experiment shows that matriline rank ordinarily depends on the number of kin that support one another. The strength of female kin coalitions seems to mitigate male dominance, despite the males' greater size and strength. Inheritance of rank is strictly patterned within a matriline. Females support their youngest relatives in conflicts, resulting in a hierarchy that is ordered inversely according to age. This includes dominance of younger sisters over older ones.

Overthrows are a regular feature of matriline relations in captivity, at least in rhesus monkeys (Ehardt and Bernstein 1986; Oates-O'Brien et al. 2010). They also occur in other despotic species (Bernstein 1968; Hayakawa and Soltis 2011). In these events the dominant matriline loses its status due to repeated violent attacks by one or more of the other matrilines in the troop. Proposed causal factors include season, density of the captive group, sex ratio or other demographic factors, age, pregnancy, or absence of the alpha female, and the number of females or adolescent males in the dominant matriline. Oates-O'Brien et al. (2010) performed a mathematical analysis of overthrows among rhesus groups at one primate facility. The results pointed to absence of the alpha female as an important factor. Overthrows were also associated with lower densities and with smaller numbers of male juveniles and adolescents in the troops as a whole.

Gygax et al. (1997) described proximal features of an overthrow among captive longtailed macaques. The form of the assaults differed from aggressive patterns in any other context. Attackers held victims to the ground while one or several bit or chewed the hands, feet, and eyebrows. The rebelling group attacked the highest-ranking mother first, then her daughter, granddaughter, and great-granddaughter. This presumably disrupted the tendency of mothers to intervene on behalf of their daughters. In this case and a previous one the event followed assumption of alpha status by a male from the rebellious matriline.

The despotic system affects immatures in various ways. Mothers (except for those of highest rank) are very protective, restrict their young to interactions with relatives, and frequently retrieve them when they wander away. Play fighting in the most despotic species is more competitive than in others and participants show less tolerance for close physical contact in play wrestling (Thierry 2007). Interference with mating by older immatures is not tolerated by adult males.

Early dispersal is typical of despotic species. The majority of males leave their natal troops by the age of five years and may not join a new troop for months or even years. Membership in an all-male band sometimes bridges this gap, but many individuals are solitary. The low level of tolerance between males might account for this pattern. In other species males can shift from one troop to another in as little as a day, and all-male groups are rare.

There is variation within species. Japanese macaques are one of the most despotic species, but at least one population displays numerous characteristics of tolerance. These include more frequent affinitive interactions, shorter inter-individual distance, more frequent disregard for exclusion, less intense aggression, and more frequent counter-aggression (Zhang and Watanabe 2007). The despotic/tolerant distinction is useful shorthand for discussing behavioral variation, but we should not forget that it simplifies a complex gradient.

Tolerant Macaques

Tolerant or **egalitarian** social systems are characterized by more relaxed social relationships, especially greater tolerance for subordinates by dominant individuals (Riley 2010). Sulawesi macaques epitomize tolerant traits (Figure 20.2), but similar traits occur in other macaque species. High intensity aggression is rare, reconciliation is frequent, and affiliative contacts are frequent. Inter-individual distances are relatively short. If attacked, an individual usually holds its ground and either lip-smacks (a friendly signal) or screams in protest. Often the

Figure 20.2 A black crested macaque, one of the "tolerant" species of the genus. © AlanJeffery via iStock

target animal counter-attacks, which makes aggression bidirectional. Dominance gradients are not as steep as in despotic macaques and hierarchy has less influence on social contacts.

Female preference for kin is less pronounced and kin support in agonistic interactions is rare. Because kin bias is less pronounced, individuals have some freedom with regard to power networks, and dominance seems to depend more on individual characteristics than on kin ties. Coalitions and even close relationships with nonrelatives are more common. Greater tolerance among males allows for a greater number of coalitions.

The conciliatory tendency fell below 20 percent in multiple captive populations of the most despotic species (rhesus, Japanese, and longtailed macaques). The two Sulawesi species in the sample displayed tendencies around 50 percent while the relatively tolerant liontailed macaque exceeded 60 percent. The higher conciliatory tendency is correlated with higher rates of affiliative physical contact, such as mounts and clasps.

Peaceful interventions by third parties resolve some conflicts in at least two species. In Tonkean macaques, for instance, the third individual approaches one of the two in conflict, usually the aggressor, and engages in diversions such as lip-smacking, clasping, or play (Petit and Thierry 1994). The maintenance of peace within the Tonkean troop does not represent a lack of aggressiveness or fighting ability. Tonkeans at a facility in France escaped from their enclosure several times and on five occasions they entered the enclosures of rhesus and longtailed macaques and attacked them. They wounded four adults and killed several juveniles and infants.

The response of bystanders to conflict in Tonkean macaques is also noteworthy. They engage in demonstrations of affiliation with each other. These are patterned by friendship rather than kinship, females initiate affiliation sooner than males, and affiliation appears sooner after conflicts that involve physical contact. The involvement of four individuals (two antagonists and two affiliative bystanders) is called a **quadratic interaction** (DeMarco et al. 2010). Following comparable research on hamadryas baboons, this is the second study of such interactions in primates.

Egalitarian communication is an ethological reflection of despotic-tolerant variation (Thierry 2007). Among despotic macaques, as in other primates, only subordinates use the silent-bared-teeth (SBT) expression and it communicates submission and appeasement (it can reasonably be termed a grimace, at least informally). In grade 4 the SBT signals peaceful intentions on the part of either dominant or subordinate individuals. In grade 3 the same signal varies between the two meanings, depending on context. In tolerant macaques the SBT can grade into the relaxed-open-mouth (ROM) expression, sometimes called the play face. Among despotic species the ROM appears only in play. The rich repertoire of Sulawesi macaques distinguishes them from more despotic species (Thierry et al. 2000). **Clasping**, encompassing various intimate body contacts such as grasping and hugging, is more frequent in Sulawesi macaques than in despotic species. Some tolerant affiliation patterns are simply rare in other macaques, such as the *expressive run*, *turning face away*, *panting grunt*, and *affiliation call*.

More broadly, communication in the most egalitarian species lacks any status indicators. In the most despotic species subordinates express acceptance of their position with signals such as presenting. Intermediate species display signals of dominance rather than subordination. Because dominance is more contested in these species, those who would have it must claim it. Others do not voluntarily submit.

Comparison between Tonkean and rhesus macaques is suggestive with regard to control of troop movement. In the tolerant Tonkean macaques, decisions about group movement involved almost all members, while dominant and older individuals are especially prominent among the despotic rhesus in this context (Sueur and Petit 2008). The individual Tonkean macaque joins movement by others based on the actions of affiliated individuals that are not necessarily kin (Sueur et al. 2009, 2010a). Several individuals attempt different directions before the choice is made. However, as in rhesus, the final decision depends on older and dominant individuals.

Comparable factors apply to temporary troop fission (Sueur et al. 2010b). Kinship influences subgroup membership in the despotic/nepotistic rhesus. Membership decisions made by Tonkean macaques resembled **anonymous mimetism**, a system in which the individual imitates the actions of a larger group regardless of affiliations. This pattern is also to be expected from egalitarian systems. Somewhat similarly, dominance and associations had little effect on the fission of a moor macaque troop (Okamoto and Matsura 2001).

In the peaceful atmosphere of a Sulawesi troop mothers are permissive and allomaternal care is common. Lower levels of competition among females mean that there is less danger in infant handling. In larger troops an infant has contact with many females from an early age. Comparison of moor macaques with Japanese monkeys (Okamoto et al. 2000) represents other significant life history differences between despotic and tolerant species. Male Japanese macaques disperse at about five years of age and stay in non-natal troops for an average of about three years. Outsider males account for about 40 percent of matings. Male moor macaques leave their natal troops between seven and nine years of age and mating by non-troop males seems to be rare or nonexistent.

Variation occurs among tolerant macaques just as it does among the despotic species. Crested macaques are among the most tolerant species, but the long calls of males encode identity and social rank, and that rank is also reflected in the frequency of calls (Neumann et al. 2009). Tibetan macaques were considered tolerant, but continuing research suggests despotic tendencies. This is one more reminder that the despotic/tolerant dichotomy simplifies a complex gradient in a cluster of macaque behaviors.

Geladas

Geladas are a monospecific genus (*Theropithecus gelada*), sole survivors of a larger taxon represented in the fossil record. They are called *bleeding heart monkeys* because both sexes have a bare patch of red skin on the chest. Males are distinguished by a long cape or mane that extends along the back and sides.

Geladas inhabit wet, grassy plateaus in Ethiopia up to 3000m. Temperatures range from 5 to 20 degrees Centigrade. The sparse tree cover provides few resources for monkeys. Geladas do not climb trees well. They are unique among primates in their dependence on grass for food. Seeds are preferred, but the bulk of the diet is grass blades (which may exceed 90 percent). Positional behavior is correspondingly unusual. A gelada sits on its haunches and harvests all the food in a patch before moving on. If the next patch is nearby, the monkey shuffles to it rather than walking. This can account for up to 30 percent of daily travel distance.

The gelada plucks green grass blades, alternating hands, using thumbs and index fingers with the highest opposability index of all nonhuman primates. The accumulated sheaves require little biting and much chewing, so gelada dentition features small incisors and very large molar teeth. Comparative study of fibrous particles in feces determined that geladas process grass as efficiently as zebras. With the shortest and stubbiest fingers among primates, geladas use their hands like shovels to access underground plant organs for fallback foods.

Foraging time varies from 50 to 85 percent of the day, the highest of any primate, largely due to the low quality diet. Despite their highly specialized diet, geladas can be flexible.

During an outbreak of locusts, the monkeys fed on them in large quantities (Fashing et al. 2010). They seemed somewhat frightened at first, but quickly turned to pursuit of the insects by leaping into the air to catch them or pouncing on them when they landed.

Geladas sleep on cliffs formed by gorges that run through their plateaus (Figure 20.3). Dominant individuals spend less time in a relaxed posture than others, which may represent a higher level of alertness and readiness to respond to nocturnal threats (Noser et al. 2003). When geladas ascend to forage on the plateau, they rarely go more than one kilometer from the edge of the cliff.

Geladas live in modular societies that include stable all-male groups. The latter consist mainly of subadult males, but usually contain some older individuals. A band is composed of OMUs that share a common range and often forage together, usually with one or more all-male groups. Some bands include more than 300 individuals. A *community* is composed of two to four bands with extensively overlapping ranges. OMUs from the same community can form a *herd*, usually a temporary foraging group.

Females are the heart of the OMU. The gelada male is peripheral and the group can persist despite his death (or experimental removal). He tries to maintain grooming relationships with all females in his unit, but females usually prefer each other and respond to the male in a perfunctory manner. A female in a close relationship with the male probably has no relatives in the unit. Females in alliance with relatives have a lifetime reproductive advantage because they harass subordinates, who thus require more cycles to conceive. In larger OMUs females are divided into sets of two or three individuals who groom each other almost exclusively and support each other in disputes. Genealogy of a captive group indicates that females in a set are close relatives, sisters or mothers and daughters. Coalition females have higher rank and are likely to attack and harass others.

After a few years in an all-male group, a typical young male tries to begin his reproductive career. One option is to become a follower in an OMU and establish relationships with juvenile females and low-ranking adult females. Though initially submissive to the leader, the follower begins to support females against him. Finally, the follower leaves with female associates to form his own OMU.

The alternative, used by older and stronger males, is to take over an established OMU. Ritualized attempts may be practice for the challenger or a test of the resident male (Snyder-Macklin 2010). Both of them groom the females intensely, as the leader intermittently tries to chase the other away. If the challenger persists, a serious fight occurs. However, this seems futile in that female decisions determine the final outcome. If a majority of the females accept the grooming solicitations of the newcomer, the incumbent loses even if he has dominated. Low-ranking females in large groups are particularly willing to shift allegiance, and challengers are especially attracted to large groups. Takeover attempts attract all-male groups and may lead to a series of challenges.

Males who attempt a takeover rather than become followers may acquire more females, but they are more likely to fail, and they do not last long even if they succeed. Large harems attract challengers. In the end the two reproductive strategies are about equally successful over a lifetime. These appear to be two evolutionarily stable strategies (ESS) because neither has a selective advantage over the other.

The male's long and silky mane is probably an honest signal of his general condition. Whether due to such signals, mate guarding or female choice, the mating system seems to be effective for the male. His testes are relatively small, showing that there is relatively little sperm competition. Uniquely, geladas have sexual skin on the chest as well as the perineal area. This may be the evolutionary result of the unique feeding style. Spending much of the

Figure 20.3 A male gelada. The terrain behind him illustrates the cliffs that provide shelter for the species. © RosaFurtado via iStock

day sitting or shuffling, a female displays her perineum for relatively little time. Male geladas also have a patch of red skin on the chest and changes in color seem to convey male quality (Bergman et al. 2009). Leaders have redder chests than other males and leaders with larger units had redder chests than other leaders. This study used digital means to quantify redness, an example of the diverse technology now used in field research.

A gelada study supported the Social Complexity Hypothesis, which holds that social cognition is costly. In playback experiments geladas were either unable or unmotivated to keep track of the individual identities of other males in their society—even males with whom they had a large degree of social overlap (Bergman 2010). As noted above, changes in chest color can provide important social information independent of individual recognition (Bergman et al. 2009).

Mandrills

Monkeys of the genus *Mandrillus* occupy thick rainforests in western Africa. The mandrill (*M. sphinx*) has especially prominent ridges on each side of the nasal bones, as well as a beard, crest of hair, and mane (illustrated in the Introduction). Southern and northern populations display profound differences in DNA, indicating a long phylogenetic divergence (Telfer et al. 2003), which may correspond to significant behavioral differences. Most of our detailed information about mandrill social behavior and life history comes from members of the southern population that have been enclosed (semi-free-ranging) at the research station in Gabon, which must be considered with caveats, as discussed in Chapter 1.

The mandrill is the largest and most dimorphic of all monkeys: at over 45 kilograms males are more than twice the size of females. Male facial skin is alternately purple, blue, and red. The ischial pads and the surrounding skin are lilac-tinged, which may facilitate group contact while moving through dense vegetation (Nowak 1999). This skin brightens when the monkeys are excited. Males have the highest canine crowns of all papionins and the most pronounced sexual dimorphism in canine size.

The highly omnivorous mandrills (Hoshino 1985; Norris 1988) have been called **forest floor gleaners** because they get so much food from the ground, including seeds and tough-shelled fallen fruits, mushrooms, and invertebrates. Mandrills use bodily strength along with craniodental specializations to perform strenuous activities such as nut cracking, log rolling, and bark ripping.

Wild mandrills form unique foraging groups that can include more than 800 members (Abernethy et al. 2002; Hoshino et al. 1984). Such a group is called a **horde**. Despite its size, the horde is a relatively stable and cohesive unit. It may spread out into foraging subgroups, but daily movement of the total group is coordinated. The horde in the southern population may include only females and young on a permanent basis. Adult males are solitary for most of the year and only join the females for a mating season of about three months. This is a more restricted breeding pattern, and a much more restricted social pattern, than in almost any other anthropoid. Mandrills are the only papionins that do not breed year-round. Observers of the northern population saw multi-male groups in all but one month of the year and a one-male group that seemed to persist throughout the year. These results suggest that the populations differ significantly in some genetic influences on behavior. This interpretation is especially salient, given the evidence for a lengthy phylogenetic separation of the two mandrill populations.

During the mating season, males engage in intense and continuous competition that involves marking trees with a sternal gland, threats, fighting, and "marathon grunting" (Jolly

2007). This competition accounts for the huge size of male bodies and canines, as well as specialized fat deposits that carry the male through a mating season when little feeding is possible. The signals sent by the multi-colored skin inform rivals of a male's readiness to contend.

Researchers used the southern colony to test hypotheses about mating. First, they found that males choose partners based on reproductive quality, which the males demonstrated by mate-guarding higher-ranking females more than lower-ranking ones, and parous females more than nulliparous ones (Setchell and Wickings 2006). One cue for males may be the facial color of females, which varies from black to pink and is stronger in older and multiparous females (Setchell et al. 2006).

Male perception and choice were indicated by a pattern of mate guarding more on peri-ovulatory days than on other receptive days, and during conceptive cycles more than non-conceptive cycles (Setchell et al. 2005). Mate guarding was a good predictor of paternity, but consistently overestimated the reproductive success of the alpha male. This was apparently due to sneak copulations by males from age four years up.

Setchell and Wickings (2004) tested the Reliable Indicator Hypothesis for large sexual swellings. They found no significant relationship between swelling size and reproductive history, and the alpha male did not prefer females with larger swellings. Mate guarding by any male was not related to swellings. The only hint of support for the hypothesis was the fact that females with larger swellings were more likely to have a sperm plug after copulation.

Female mandrills make choices based on the brightness of red coloration in male faces (Setchell 2005). Peri-ovulatory females spend more time closer to bright males, are more responsible for being close to them, avoid fewer approaches from them, avoid their perineal inspections less often, solicit sex from them more often, and groom only the brightest male. These behavior patterns fulfill criteria intended to evaluate the effect of mate choice on communication. Bright color is sexually dimorphic, varies among males, is a basis for preference in the context of reproduction, and influences male reproductive success (as indicated by mating success). Even though (or because) males are nearly three times their size, females can resist unwanted advances. Females can run faster than adult males and climb into branches that are too small for the males. Mate-guarding males are often left at the bottom of a tree when a female jumps from that tree to another.

Mandrills have one of the longest adolescent periods among cercopithecine monkeys (Charpentier et al. 2004). The testicles descend between three and four years of age and some males begin having sneak copulations soon after. Testicular volume increases markedly at about 5.5 years and secondary sexual traits begin to develop about a year later (Setchell and Dixson 2002). However, full adulthood in terms of both size and secondary sexual characteristics does not arrive until nine or ten years. Presumably this has to do with the enormous body size and elaborate secondary characteristics of male mandrills. Development into a sexually contending adult male seems to be socially conditioned. There is evidence that the presence of older breeding males inhibits the development of secondary sexual characteristics, such as red coloration on the face, rump, and genitalia. Variation in color, along with facial and gestural signals, mediates conflict among adults. Intense aggressive encounters are more likely to occur between males of similar color, which shows that they are serious competitors for dominance. Males that show arrested development of such characteristics may be delaying their entry into the dominance competition. Red color is associated with androgen level, but the lower level of androgens may be the result rather than the cause of lower status. This seems to be a case of reproductive suppression.

Laidre (2006) investigated social cognition in the Gabon colony and determined that the mandrills engaged in two forms of manipulation, varying with the circumstance. Food-associated calls occurred at a high rate when the monkeys were provisioned at a feeding cage. When food was placed elsewhere in the enclosure, discoverers suppressed their calls. However, later arrivals called loudly. More arrivals attracted by the calls seemed to create a diversion that allowed latecomers to gain access to the food. Social cognition is sometimes linked to mind reading, the attribution of knowledge or beliefs to other individuals. In this case it appears that there was no connection. The nonvocal discoverers of food behaved in ways inconsistent with mind reading, such as openly moving around with unconcealed food.

Laidre (2011) reported a gesture that seems to be cultural because it was not mentioned in 19 other mandrill studies. The gesture consisted of covering the eyes with one or both hands for periods up to 30 minutes, often raising an elbow in the air. Seven individuals out of 23, representing all age–sex classes, performed the behavior. It began spontaneously with a three-year-old female, spread to others, and has persisted through changes in personnel. Laidre speculated that the gesture functions to discourage interruptions, a kind of "do not disturb" sign.

The Mangabey Issue

Mangabeys are medium-sized (5 to 13kg), frugivorous, long-limbed, long-tailed, and hollow-cheeked (Fleagle and McGraw 2002). Similarities to guenons may be convergences or retention of primitive traits, or both. The monkeys loosely called mangabeys were once taxonomically lumped, but are now divided into two genera. Despite their superficial similarity to each other, one genus (*Cercocebus*) forms a clade with the mandrills while the other (*Lophocebus*) is more closely related to *Papio* (discussed in the next chapter).

Cercocebus occurs in a patchy distribution across equatorial Africa, mainly in rainforest. Like the related mandrills, they commonly feed and travel on the ground, specializing in fallen fruit, fungi, and invertebrates. They often eat hard seeds and nuts (Range et al. 2007), crushing them with expanded premolars, a feeding process that they display throughout ontogeny (McGraw et al. 2014). However, they have displayed ecological flexibility by traveling through non-forest habitats to expand a home range (Wieczkowski 2005) and tracking changes in forest composition with dietary shifts (Wieczkowski and Kinnaird 2008). They handle predator risk with a high level of vigilance, alarm calling before other primates in the vicinity. Even the predator-sensitive Diana monkeys, though highly arboreal, feed in the lower levels of the forest when associating with *Cercocebus* (McGraw and Bshary 2002).

Cercocebus live in large multi-male troops varying from about 20 to more than 100 members. While troops have year-round resident males, it appears that other males live in all-male groups except during the breeding season. These mangabeys differ from most other papionins in having well-defined breeding seasons.

Attacks on mothers and infants by intruding males have been observed in wild *Cercocebus*, undermining the view that infanticide is unlikely in multi-male troops (Fruteau et al. 2010). However, mothers often fought back and troop males who were potential fathers guarded and supported both mothers and infants. During 136 attacks on a troop, only one of seven infants was killed. Mothers use the calls of males to distinguish between those that were in the troop during the period of conception and those that were not (Range 2005).

Free-ranging mangabeys demonstrated spatial memory to be a key factor in foraging (Janmaat et al. 2006), apparently relocating previously used fruit trees and distinguishing between those that have and have not carried fruit in the immediate past. The monkeys traveled toward unseen fruiting trees with a frequency greater than chance and did so at a rapid rate. The researchers considered the alternative explanation of sensory cues and concluded that various circumstances made this unlikely.

Cercocebus demonstrated unusual facility with the **reverse contingency task**, in which the subject chooses between two quantities of food and receives the array that was *not* chosen (Albiach-Serrano et al. 2007). This requires the subject to distinguish between the two quantities and to understand the connection between choosing the poorer reward and obtaining the better one. It also requires self-control to inhibit the natural first reaction. The *Cercocebus* version presented the subjects with raisins in two arrays with the varying proportions. Three of four subjects solved the problem at levels well above chance and performed well on retests seven to ten months after the original experiment (Albiach-Serrano et al. 2007). Before this experiment only apes and one macaque species had demonstrated the ability to perform the task without needing procedural modifications.

Lophocebus are highly arboreal inhabitants of continuous evergreen forest. They feed in the canopy on figs and other soft fruits wherever possible, although they use hard-shelled fruits and seeds as fallback resources. Similar facial shape in the two mangabey genera may be the result of convergent masticatory adaptations for the exploitation of hard-shelled fruits.

Summary

Papionins are generally larger and more dimorphic than guenons. Multi-male troops with complex social organization are typical. Macaques are among the most successful primates, covering a wide geographic area in Asia and adapting to a wide variety of environments. Female kinship is a vital aspect of social organization and grooming a vital expression of relationships. Several macaque species display tool-using proclivities. Macaque species can be classified along a continuum from despotic to tolerant, with implications for many aspects of social life. Mangabeys are actually two unrelated genera with very different ways of life.

Geladas live in modular societies, but OMU females are largely independent of the male. Mandrills, the most dimorphic monkeys, are forest floor gleaners. They occur in two highly distinct populations, each of which displays unusual social and mating patterns. Coloration seems to be very important in sexual interaction. Male maturation is apparently affected by competition and reproductive suppression. *Cercocebus* "mangabeys" are closely related to mandrills and are similar to them ecologically. *Lophocebus* are more like their relatives, the *Papio* baboons.

Sources

Basic Bibliography

Campbell, C.A., et al. (Eds.) (2011). *Primates in perspective*. NY: OUP. [Chapters 14, 15, 16]

Cords, M. (2012). The behavior, ecology, and social evolution of cercopithecine monkeys. In J.C. Mitani et al. (Eds.), *The evolution of primate societies* (pp. 91–112). Chicago: UCP.

Additional References

Abernethy, K.A., et al. (2002). Hordes of mandrills (*Mandrillus sphinx*): extreme group size and seasonal male presence. *JZL*, 258, 131–137.

Albiach-Serrano, A., et al. (2007). Mangabeys (*Cercocebus torquatus lunulatus*) solve the reverse contingency task without a modified procedure. *AC*, *10*, 387–396.

Aureli, F., & Yates, K. (2010). Distress prevention by grooming others in crested black macaques. *BL*, 6, 27–29.

Bergman, T.J. (2010). Experimental evidence for limited vocal recognition in a wild primate: implications for the social complexity hypothesis. *PRSB*, *277*, 3045–3053.

Bergman, T., et al. (2009). Chest color and social status in male geladas (*Theropithecus gelada*). *IJP*, *30*, 791–806.

Berman, C.H., et al. (2008). Variation in kin bias over time in a group of Tibetan macaques at Huangshan. *Behaviour*, *145*, 863–896.

Bernstein, I.S. (1968). Spontaneous reorganization of a pigtail monkey group. *Proc. 2nd Int. Cong. Primat. 1*, 48–51.

Bouchet, H., et al. (2013). Social complexity parallels vocal complexity: a comparison of three non-human primate species. *Frontiers in Psychology*, DOI: 10.3389/fpsyg.2013.00390.

Bovet, D., & Washburn, D.A. (2003). Rhesus macaques (*Macaca mulatta*) categorize unknown conspecifics according to their dominance relations. *JCP*, *117*, 400–405.

Charpentier, M., et al. (2004). Changes in social interactions during adolescence in male mandrills (*Mandrillus sphinx*). *AJP*, *63*, 63–73.

Charpentier, M., et al. (2005). Constraints on control: factors influencing reproductive success in male mandrills (*Mandrillus sphinx*). *BE*, *16*, 614–623.

De Marco, A., et al. (2010). Conflicts induce affiliative interactions among bystanders in a tolerant species of macaque (*Macaca tonkeana*). *AB*, *80*, 197–203.

Ducoing, A.M., & Thierry, B. (2003). Withholding information in semifree-ranging Tonkean macaques (*Macaca tonkeana*). *JCP*, *117*, 67–75.

Ducoing, A.M., & Thierry, B. (2005). Tool-use learning in Tonkean macaques (*Macaca tonkeana*). *AC*, *8*, 103–113.

Ehardt, C.L., & Bernstein, I.S. (1986). Matrilineal overthrows in rhesus monkey groups. *IJP*, *7*, 157–181.

Fairbanks, L.A., et al. (2010). High-fiber diet promotes weight loss and affects maternal behavior in vervet monkeys. *AJP*, *72*, 234–241.

Fashing, P.J., et al. (2010). Behavior of geladas and other endemic wildlife during a desert locust outbreak at Guassa, Ethiopia. *Primates*, *51*, 193–197.

Fitch-Snyder, H., & Carter, J. (1993). Tool use to acquire drinking water by free-ranging lion-tailed macaques (*Macaca silenus*). *Laboratory Primate Newsletter*, *32*, 1–2.

Fleagle, J.F., & McGraw, W.S. (2002). Skeletal and dental morphology of African papionins: unmasking a cryptic clade. *JHE*, *42*, 267–292.

Fruteau, C., et al. (2010). Infanticide risk and infant defence in multi-male free-ranging sooty mangabeys, *Cercocebus atys*. *BP*, *83*, 113–118.

Fruteau, C., et al. (2011). Infant access and handling in sooty mangabeys and vervet monkeys. *AB*, *81*, 153–161.

Gerald, M.S. (2001). Primate colour predicts social status and aggressive outcome. *AB*, *61*, 559–566.

Gerald, M.S., et al. (2006). An experimental examination of female responses to infant face coloration in rhesus macaques. *BP*, *73*, 253–256.

Gygax, L., et al. (1997). A matrilineal overthrow with destructive aggression in *Macaca fascicularis*. *Primates*, *2*, 149–158.

Hauser, M.D., & Wood, J.N. (2011). Replication of 'Rhesus monkeys correctly read the goal-relevant gestures of a human agent.' *PRSB*, *278*, 158–159.

Hayakawa, S., & Soltis, J. (2011). Troop takeover and reproductive success of wild male Japanese macaques on Yakushima island (*Macaca fuscata yakui*). *IJZ*, *Volume 2011*, Article ID 308469, 8 pages.

Hohmann, G. (1988). A simple case of tool use in wild lion-tailed macaques (*Macaca silenus*). *Primates*, *29*, 565–567.

Hoshino, J. (1985). Feeding ecology of mandrills (*Mandrillus sphinx*) in Campo Animal Reserve, Cameroon. *Primates*, *26*, 248–273.

Hoshino, J., et al. (1984). Preliminary report on the grouping of mandrills (*Mandrillus sphinx*) in Cameroon. *Primates*, *25*, 295–307.

Huffman, M.A. (2010). Consort intrusion and female mate choice in Japanese macaques (*Macaca fuscata*). *Ethology*, *75*, 221–234.

Isbell, L.A., & Young, T.P. (2007). Interspecific and temporal variation of ant species within *Acacia drepanolobium* ant domatia, a staple food of patas monkeys (*Erythrocebus patas*) in Laikipia, Kenya. *AJP*, 69, 1387–1398.

Janmaat, K.R.L., et al. (2006). Evidence for a spatial memory of fruiting states of rainforest trees in wild mangabeys. *AB*, *72*, 797–807.

Jolly, C.J. (2007). Baboons, mandrills, and mangabeys. In C.J. Campbell et al. (Eds.), *Primates in perspective*, 1st ed. (pp. 240–251). NY: OUP.

Kumar, S., et al. (2008). Observations on tool use in captive lion-tailed macaque (*Macaca silenus*). *Current Science*, *94*, 925–928.

Laidre, M.E. (2006). Manipulation without mind-reading: information suppression and leakage during food discovery by mandrills (*Mandrillus sphinx*). *Behaviour 143*, 365–392.

Laidre, M.E. (2011). Meaningful gesture in monkeys? Investigating whether mandrills create social culture. *PLoS ONE*, 6, e14610.

Leca, J.B., et al. (2010). The first case of dental flossing by a Japanese macaque (*Macaca fuscata*): implications for the determinants of behavioral innovation and the constraints on social transmission. *Primates*, *51*, 13–22.

McGraw, W.S., & Bshary, R. (2002). Association of terrestrial mangabeys (*Cercocebus atys*) with arboreal monkeys. *IJP*, *23*, 311–325.

McGraw, W.S., et al. (2014). Dietary variation and food hardness in sooty mangabeys (*Cercocebus atys*): implications for fallback foods and dental adaptation. *AJPA*, *154*, 413–423.

Neumann, C., et al. (2009). Loud calls in male crested macaques, *Macaca nigra*: a signal of dominance in a tolerant species. *AB*, *79*, 187–193.

Norris, J. (1988). Diet and feeding behavior of semi-free ranging mandrills in an enclosed gabonais forest. *Primates*, *29*, 449–463.

Noser, R., et al. (2003). Sleep and social status in captive gelada baboons (*Theropithecus gelada*). *BBR*, *147*, 9–15.

Nowak, R.M. (1999). *Walker's primates of the world*. Baltimore: Johns Hopkins University Press.

Oates-O'Brien, R., et al. (2010). Predictors of matrilineal overthrows in large captive breeding groups of rhesus macaques (*Macaca mulatta*). *Journal of the American Association for Laboratory Animal Science*, *49*, 196–201.

Okamoto, K., & Matsura, S. (2001). Group fission in moor macaques (*Macaca maurus*). *IJP*, *22*, 481–493.

Okamoto, K., et al. (2000). Life history and demography of wild moor macaques (*Macaca maurus*). *AJP*, *52*, 1–11.

Petit, O., & Thierry, B. (1994). Aggressive and peaceful interventions in conflicts in Tonkean macaques. *AB*, *48*, 1427–1436.

Range, F. (2005). Female sooty mangabeys (*Cercocebus torquatus atys*) respond differently to males depending on the male's residence status – preliminary data. *AJP*, *65*, 327–333.

Range, F., et al. (2007). The structure of social relationships among sooty mangabeys in Tai. In R. Noë, et al. (Eds.), *Monkeys of the Taï Forest*. NY: CUP.

Riley, E.P. (2010). The endemic seven: four decades of research on the Sulawesi macaques. *EA*, *19*, 22–36.

Rochat, M.J., et al. (2008). The evolution of social cognition: goal familiarity shapes monkeys' action understanding. *CB*, *18*, 227–232.

Rodriguez-Llanes, J.M., et al. (2009). Reproductive benefits of high social status in male macaques (*Macaca*). *AB*, *78*, 643–649.

Setchell, J.M. (2005). Do female mandrills prefer brightly colored males? *IJP*, *26*, 715–735.
Setchell, J.M., & Dixson, A.F. (2002). Developmental variables and dominance rank in adolescent male mandrills (*Mandrillus sphinx*). *AJP*, *56*, 9–25.
Setchell, J.M., & Wickings, E.J. (2004). Sexual swelling in mandrills (*Mandrillus sphinx*): a test of the reliable indicator hypothesis. *BE*, *15*, 438–445.
Setchell, J.M., & Wickings, E.J. (2006). Mate choice in male mandrills (*Mandrillus sphinx*). *Ethology*, *112*, 91–99.
Setchell, J.M., et al. (2005). Mateguarding and paternity in mandrills: factors influencing alpha male monopoly. *AB*, *70*, 1105–1120.
Setchell, J.M., et al. (2006). Signal content of red facial coloration in female mandrills (*Mandrillus sphinx*). *PRSB*, *273*, 2395–2400.
Shutt, K., et al. (2007). Grooming in Barbary macaques: better to give than to receive? *BL*, *3*, 231–233.
Sueur, C., & Petit, O. (2008). Organization of group members at departure is driven by social structure in *Macaca*. *IJP*, *29*, 1085–1098.
Sueur, C., et al. (2009). Selective mimetism at departure in collective movements of *Macaca tonkeana*: an experimental and theoretical approach. *AB*, *78*, 1087–1095.
Sueur, C., et al. (2010a). Sequence of quorums during collective decision making in macaques. *BE&S*, *64*, 1875–1885.
Sueur, C., et al. (2010b). Short-term group fission processes in macaques: a social networking approach. *JEB*, *213*, 1338–1346.
Telfer, P.T., et al. (2003). Molecular evidence for deep phylogenetic divergence in *Mandrillus sphinx*. *Molecular Ecology*, *12*, 2019–2024.
Thierry, B. (2007). Unity in diversity: lessons from macaque societies. *EA*, *16*, 224–238.
Thierry, B., et al. (2000). The influence of phylogeny on the social behaviour of macaques. *Ethology*, 713–728.
Ueno, Y., & Fujita, K. (1998). Spontaneous tool use by a Tonkean macaque (*Macaca tonkeana*). *FP*, *69*, 318–324.
Watanabe, K., et al. (2007). Long-tailed macaques use human hair as dental floss. *AJP*, *69*, 940–944.
Westergaard, G.C. (1988). Lion-tailed macaques (*Macaca silenus*) manufacture and use tools. *JCP*, *102*, 152–159.
Westergaard, G.C., & Lindquist, T. (1987). Manipulation of objects in a captive group of lion-tailed macaques (*Macaca silenus*). *AJP*, *12*, 231–234.
Wieczkowski, J. (2005). Examination of increased annual range of a Tana mangabey (*Cercocebus galeritus*) group. *AJPA*, *128*, 381–388.
Wieczkowski, J., & Kinnaird, M. (2008). Shifting forest composition and primate diets: a 13-year comparison of the Tana River mangabey and its habitat. *AJP*, *70*, 339–348.
Zhang, P., & Watanabe, K. (2007). Extra-large cluster formation by Japanese macaques (*Macaca fuscata*) on Shodoshima Island, central Japan, and related factors. *AJP*, *69*, 1119–1130.

21 Genus *Papio*

The Real Baboons

The land cruiser's tire bumps over a rock and startles a mother baboon. Infant clinging to her abdomen, she disappears over a low hill. Moments later an adult male, twice the mother's size, lopes over the hill and sits down next to the vehicle. He stares at the primatologist for several minutes and then slowly walks away. The message has been sent.

The term **baboon** has been applied to several papionin genera that are not closely related. This informal category is based on traits that are shared due to parallel or convergent evolution: large bodies, extreme sexual dimorphism, long muzzles, very long male canines (twice the length of female canines), and largely terrestrial habits. Baboons in this broad sense also share some behavioral patterns due to functional convergence. The **yawn threat** that displays the large canines is a ritualized signal that may have evolved from yawning as a displacement behavior in agonistic situations (Figure 21.1). Male–female relationships, though varied across taxa, have probably been shaped by extreme sexual dimorphism. For a female, a male twice her size presents a strong threat for food competition, infanticide, or unwanted copulation (Figure 21.2). By the same token an affiliated male provides a strong ally.

Common Baboons

Found throughout most of Africa, *Papio* is one of the most widely distributed primate genera. The term **common baboon** encompasses numerous and geographically extensive *Papio* subtaxa that typically live in troops (Box 21.1). Though often called "savanna baboons" because many populations forage in grasslands, common baboons (Figures 21.1 and 21.2) actually span diverse habitats from evergreen forests to deserts and from seashores to mountains.

Box 21.1 Baboon Taxonomy

There is little doubt about the classification of the "baboon" genera. A recent cladistic analysis of craniodental data using both quantitative and qualitative traits (Gilbert 2013) confirms the distinct status of the genus *Papio* as well as a phylogenetic relationship with the mangabey *Lophocebus*. A key shared derived trait is the extensive development of the maxillary fossae.

While there is little doubt about the populations that belong in the genus *Papio*, there has been a great deal of controversy about how many species are included. One possibility is that all *Papio*, from the deserts of Arabia to the cold southern coast of South Africa, belong in a single species. According to the rules of nomenclature, that species must be called *P. hamadryas*. Others would reserve that classification for the northeastern populations of the genus. The remainder might constitute anywhere from four to ten species. The guinea baboons of West Africa seem quite distinct and the common ("savanna") populations were divided into olive, yellow, and chacma species. As in other primate taxa, there has been a trend toward elevating subspecies and newly described populations to species status.

Figure 21.1 Yawn threat by a male chacma baboon. © Curt Busse

Figure 21.2 A male and female in close proximity demonstrate the degree of sexual dimorphism that is typical of common baboons and similar monkeys. © Glenn E. King

Common Baboon Ecology

Baboons are the most successful primates in colonizing hot African plains. Lacking special mechanisms to protect the brain (Hill 2006), they compensate with behavioral thermoregulation. They rest in shade as much as possible and their midday drinking sessions facilitate evaporative cooling (Figure 21.3). Baboons can subsist on almost any plant food but are often selective, consuming just one part of each plant (buds, for example). The use a precision grip with thumb and forefinger to manipulate small items like grass blades and seeds. Rapid retrieval of such items can provide a substantial amount of food in a short time. Baboons also have the strength to dig up underground food sources that are not accessible to sympatric primates such as vervets. These include *corms* (nutrient storage organs of some plants), roots, and bulbs, found by observing clues like small shoots. There is a tradeoff: processing these nutritious resources increases total feeding time (which decreases when baboons can get fruit) (Hill and Dunbar 2002).

Baboons differ from other monkeys in the extent to which they exploit vertebrates, especially mammals. They kill small antelopes, infants of larger antelopes, hares, and vervets. They often pluck prey from hiding places, but they also chase more mobile individuals. Adult males do most of the hunting and usually take kills away from others.

A leopard is about twice the size of a male baboon and matches or exceeds the monkey in dental weaponry. On the other hand, several adult male baboons can injure or kill a leopard and these are the usual odds in a daytime confrontation. Cooperative male defense is reported more often for common baboons than for other primates. A single male can intimidate less formidable predators, such as cheetahs. Regarding ultimate cause, males may be defending

Figure 21.3 Baboons rest in shade during midday. © Glenn E. King

their offspring and/or their mating partners. Benefits outweigh costs in most cases because male baboons are large, strong, and have dangerous canine teeth. However, adult males flee from lions along with the rest of the troop. Baboons signal danger with a unique call, the *wahoo*. More technically called the *two-phase bark*, it sounds like a short bark followed by a sharp inhalation.

Common Baboon Sociality

A typical troop of 40 to 80 individuals includes 2 to 5 adult males and 10 to 12 adult females. Females with infants occupy central positions, probably in part because they are seeking maximum safety for their offspring, and partly because they are surrounded by other baboons who are interested in the babies. Pregnant females are on the periphery, which can be explained in proximate terms by a need to concentrate on feeding, and by avoidance of even ordinary aggression from other troop members.

Adult females form stable linear hierarchies, including the ranking of matrilines in larger troops (much like the closely related macaques). Females spend hours of each day in grooming, perhaps because alliances are so important. Subordinates woo dominant females by grooming them. Females without effective allies may transfer from one troop to another, which raises the question of whether alliances are the *cause* of female philopatry rather than an effect (Barton 2000).

Silk et al. (2006a) investigated the basis for female–female relationships and found that the strongest bonds connected a female with her mother and sisters; beyond that, bond strength decreased as relatedness decreased. Ties between sisters became stronger after their mother's death, while relationships with maternal aunts weakened. The aunt–niece connection is apparently a byproduct of their association with the linking relative. If close maternal kin are not available, females seek alternatives. Bonds formed with paternal sisters are ordinarily intermediate in strength between ties to maternal sisters and to nonkin.

A strong attraction to similar-aged females may be explained by the fact that such peers are often paternal cousins. The durability of social bonds followed the same patterns as their strength (Silk et al. 2006b). Fourteen dyads maintained close relationships for at least five years: mother/daughter (5 pairs), maternal sisters (5), paternal kin (3), and possible paternal sisters (1).

There were also 13 pairs of nonkin who maintained bonds for at least three years. Females of adjacent ranks were attracted to each other. One explanation is that such associations generate less tension than subordinate attempts to affiliate with higher-ranking individuals; however, a relevant study found no evidence for lower stress in these pairs (Castles et al. 1999).

Silk et al. (2004) tested Biomarket Theory with female coalitions. Third-party interventions occurred in no more than 4 percent of all disputes and there was no evidence for a biomarket. Females did not reciprocate agonistic support and neither did they exchange it for grooming. They supported close maternal kin more often than others, and they did negotiate alliances with nonrelatives. Long-term investment in kin was clear. However, it remained possible that market forces affected short-term relationships (Silk et al. 2006b).

Biomarket Theory predicts that females will primarily balance the amount of grooming they trade within single bouts, particularly when all partners offer similar value. If some partners can provide other benefits, like reduced aggression, females may exchange grooming for those benefits. Then, grooming will not be evenly balanced within bouts. The pattern of grooming in a study troop did not fit the predictions (Frank and Silk 2009a). Two-thirds of all grooming bouts were completely one-sided and females did not consistently provide more grooming to higher-ranking partners. Grooming was more evenly balanced across multiple bouts than within single bouts, suggesting that females are not constrained to complete exchanges within single transactions.

Distributed leadership, also called democratic or egalitarian, is typical of baboon troops (Norton 1986; Rowell 1972). This means that decision-making is shared by the whole group or by all members of a subgroup. One or more individuals, called indicators, set off in a particular direction. When a sufficient number of deciders follow, the troop moves in that direction. Adult males are the most frequent indicators and adult females (or some segment of them, usually the oldest and/or most dominant) are the deciders.

Special friendship seems ubiquitous among common baboons. This is an adult male–female relationship signified by elevated rates of proximity and grooming and by the male's attention to the female's infant. Lactating females seek male friends to protect them and their infants against harassment by other females. Harassment can be indirect and yet have drastic consequences. An investigation of rough handling found that it was a significant predictor of infant mortality by age three months (Kleindorfer and Wasser 2004). Rarity of male infanticide renders it unlikely as a cause of special friendship, but defense against predators might be of some significance.

In many cases a male friend protects genetic investment in his own offspring (Nguyen et al. 2009; Smuts 1985). DNA showed that almost half of all male friends in a population were the genetic fathers (Figure 21.4). Mating effort can explain friendship behavior in non-fathers. The female recipient of such care may be predisposed to future matings with the male, and so might other females who observed the male's behavior. Females' grooming of

Figure 21.4 A male baboon holds an infant, possibly his own offspring. © Curt Busse

male friends has been explained in terms of the Biomarket Hypothesis: males trade protection for grooming (Nguyen et al. 2009).

The linear hierarchies of adult males are subject to frequent change. Some fights are about food and others arise when a male protects an infant from infanticide, but the main cause is competition for reproductive success (Hausfater 1975; Strum 1987). Nevertheless, the sneak copulation and other alternative strategies are available.

A male coalition, consisting of two and occasionally three individuals, can play a major role in mating effort. In one study, coalitions succeeded in causing a consort turnover in 35 of 55 attempts. Alberts et al. (2003) found that coalitions were the main cause of **queue jumping** (deviations from Priority-of-Access). Partners are likely to be older than prime and of middle or lower rank (Bercovitch 1988). Younger, stronger males do not need help and do not benefit from sharing; the oldest and weakest males cannot find partners (Noe and Sluijter 1995). To avoid coalitions dominant males often consort early in the day in the sleeping refuge (trees or cliff), where it is more difficult for coalition males to coordinate with one another.

Partner selection seems to be based mainly on fighting ability (Bercovitch 1988). Having an affiliative relationship may be a factor, but it can be explained as cementing an existing alliance (Noe and Sluijter 1995). Recurrent partners rarely spend extra time in affiliative behavior and spatial proximity can be explained by the common focus of coalition. It may be that coalition behavior is each partner's opportunistic response to a situation, and only appears cooperative. This interpretation avoids anthropomorphism, but seems to be undercut by the occurrence of preliminary recruiting behaviors. Signals exchanged by prospective partners include head flagging, staccato grunting, and mounting. Applying Biomarket Theory, Noe (1992) argued that cooperation is established by a form of bargaining through repeated interactions of the same individuals. The choice of a partner is a matter of "shopping" for the best available product.

Packer (1977) argued influentially that coalitions exemplify reciprocal altruism. In his study the initiating male obtained the female in all of six successful episodes. However, more extensive data showed that the supporting male was just as likely to get the female, which indicates that the behavior is self-interested (Bercovitch 1988; Noe and Sluijter 1995). Cost is an important factor. Comparative evidence indicated that the behavior required little time (usually less than 30 minutes) and energy (mostly threatening with occasional short bursts of running and fighting) and that wounds were rare and usually slight. In short, joining a coalition is a low-cost behavior with the prospect of a substantial benefit.

Troop fission can take place in several different ways. Van Horn et al. (2007) found that kinship and friendship influenced group choice by some females; however, the relative importance of these factors varied across fission events. Age peers other than paternal kin had no effect on group choice, and the effect of maternal kin did not differ from average kinship to all group-mates. Most females in an offshoot group were subordinate to fewer females after fission, but females often preferred to remain with social superiors who were close maternal kin. Researchers inferred the following association priorities: (1) close maternal kin if numerous enough to influence fitness; (2) close paternal kin; (3) social bonds with nonkin.

Both sexes enhance the primate stare threat with a **brow raise** that displays eyelids of a lighter color than the rest of the integument. **Grunting** is a deep rhythmic sound that conveys various meanings, from contact maintenance to friendliness to cooperation. Male relationships, usually tense, can be mitigated by greetings (Smuts and Watanabe 1990). The minimum **male greeting ritual** involves one male approaching another while sending signals of affiliation or appeasement, such as lip-smacking and crouching, followed by presentation of the rump. The other individual then touches or grasps the rump and sometimes mounts. Participants may embrace and in the most intense cases one or both will grasp the genitals of the other, which is unique to baboons.

Explanations for male greeting include appeasement of dominant individuals in tense situations, assessment of aggressive tendencies in male rivals, reconciliation after conflict, and communication of tolerance and cooperativeness. The risk of allowing genital handling is a high cost that indicates an honest signal. As such, it implies more than ordinary appeasement. More likely it is a way of communicating mutual tolerance and willingness to cooperate or at least to coordinate activities. More specific functions may vary from one baboon species to another.

Common Baboon Life Cycle

The baboon infant (Altmann 1980) is a gangly kilogram of pink skin and black hair, about 7 percent of its mother's size. It clings continuously for two or three weeks. Other troop members show great interest by approaching and trying to touch the infant (Figure 21.5). Adult males greet infants with friendly behaviors such as grunting and lip-smacking, embrace them, and sometimes carry them. They tolerate older infants crawling over their legs, bodies, and heads.

Frank and Silk (2009b) tested Biomarket Theory on grooming between new mothers and females who attempt to handle their infants. Females provided a greater share of grooming within bouts when mothers had young infants, and they groomed mothers longer when they

Figure 21.5 Growing up in a baboon troop entails a frequently changing constellation of relationships. Here, a young baboon interacts with another while being groomed by an adult female. © Glenn E. King

handled the infants. However, females did not compete for access to infants with increased grooming as availability declined. A change resulted from mothers decreasing their grooming rather than non-mothers increasing their proportion. These results suggest that females do trade across currencies, but market pricing does not always explain the pattern of exchange.

By six months the infant rides on its mother's back and spends a third of its waking time away from her. Its coat and skin have turned to adult colors, and adult responses have changed. The mother still grooms her offspring regularly and huddles with it in the sleeping place, but her other attentions are greatly reduced. The infant must maintain proximity and solicit help. Other females have little or no interest and sometimes punish young ones who irritate them by nipping them or pushing them flat on the ground. Mothers retaliate against other females for such treatment, but they punish their own infants for similar reasons. Males protect brown infants but do not carry them or give them unconditional tolerance.

Weaning takes place between 10 and 15 months of age. The infant's cries may represent conflict of interests with the mother or they may be an honest signal of need (Dirks et al. 2010). Lines in the tooth enamel of growing individuals indicate stress. In baboons they located the highest frequencies of stress at about six months of age, coinciding with inferred reduction of nursing; and at 11 months, inferred to coincide with cessation of nursing. These data suggest that infant cries are honest signals.

Older infants and juveniles form playgroups and often range far from their mothers. Juveniles sometimes forage together and seem to share information on food (Altmann 1999). Juveniles in groups are often attracted to a dominant male. An individual often has a particular male friend (or friends), past or present associates of that juvenile's mother. Threatened juveniles run to adult males rather than their mothers.

Sex differences in behavior become prominent at about three years. Females spend increasing time grooming with adult females. Adolescent females, supported by kin, challenge unrelated females for dominance. Eventually some challenge their own relatives, including their mothers. Occasional female emigration is probably related to two factors (Anderson 1987). First, the advantage of living with maternal relatives is diluted when a troop is small or frequently divides into small subgroups. Second, absence of predation reduces risks in the transfer process. On the other hand, predation may stimulate transfer. One troop received ten adult female immigrants after a leopard attacked a nearby troop (Rasmussen 1981).

Among male juveniles play fighting increases and dominance relations become apparent. A male enters adolescence and stops playing at about the age that females reach menarche. He begins serious competition by dominating older juveniles and challenging adult females. Growth continues until the male reaches adult size, twice that of females, at about eight years. He is sexually mature and may start his adult reproductive career.

The typical male emigrates between eight and ten years of age, apparently of his own volition (Smith 1992). Often a young male becomes increasingly isolated from female relatives. He may be attracted to novel females in other troops (Packer 1979) or he may respond to the proportion of adult males in the natal troop (Alberts and Altmann 1995). Alternative strategies are responses to variance in competitive ability. Analysis of males in a troop disclosed that only high-quality males responded to the absolute number of females and only low-quality males were sensitive to excess male number.

Responses to immigrants vary (Smith 1992). Resident males are often hostile and the newcomer must gain entrance by fighting or bluffing. However, there are instances of little or no aggression and some males are even friendly toward immigrants. It may be that there is less impetus to fight if the female-to-male ratio is high. A less obvious factor is the age structure: older males are less likely to fight a newcomer.

A male makes several transfers in his life and the entry process changes with age and experience. Some older males can quickly establish consortships and social alliances. As males age, they take one of two life paths. Some change troops more frequently, perhaps because declining abilities and rank leave no reason to stay. Others enter a troop during middle age and stay for a long period, sometimes until death. They have probably fathered many offspring in the troop and benefit genetically from protecting them (Rasmussen 1981).

As many as a third of the males in some populations stay in their natal troops and become reproductively active (Smith 1992). Proximate causes may be achievement of high rank and success in consortships. In one study, however, sons of low-ranking females were more likely to mate within the natal troop (Alberts and Altmann 1995). This is apparently because high-ranking females have more daughters, which limits the sexual opportunities of their sons due to inbreeding avoidance. The same study indicated that mating within the natal troop is not a successful strategy: though males avoided close relatives, they produced fewer viable offspring than emigrants.

Silk et al. (2010) studied fitness benefits from female relationships and found that variation in breeding lifespan accounts for approximately 50 percent of the variation in lifetime fitness. Dominance rank and the quality of close social bonds have independent effects on longevity. High-ranking females live longer than lower-ranking females. Females who form stronger and more stable social bonds with other females live longer than females with less viable relationships.

Longevity associated with female relationships can extend to their offspring (Silk et al. 2009). This effect is independent of maternal dominance rank and the number of kin. In particular, females who formed stronger bonds with their mothers and adult daughters experienced higher offspring survival rates. For females lacking mothers or adult daughters, offspring survival was closely linked to bonds with maternal sisters.

Common Baboon Cognition

Mental traits contribute to the success of baboons as opportunistic omnivores. They find food almost anywhere because they are inquisitive. They can extract foods that are well hidden or well protected because they are intelligent and manipulative. These capabilities make it possible for some baboon populations to depend completely on underground food sources during part of the year.

Schmitt and Fischer (2009) found evidence for inferential reasoning by exclusion. They tested baboons in object choice experiments with differing information about food located in one of two cups. All seven subjects located the reward when they had previously seen it. With partial information (only one cup was opened), four baboons apparently inferred the location of the reward by reasoning. The other three may have adopted the alternative strategy of avoiding the empty cup.

Sapolsky and Share (2004) described a possible example of culture. When half of the males in a troop succumbed to tuberculosis, circumstances dictated that the more aggressive males died and atypically unaggressive individuals survived. A decade later, troop males were still unaggressive. Due to dispersal, none of the males who survived the epidemic remained in the troop. Thus, new males joining the troop must have adopted the unique social pattern. Features of this male culture included high rates of grooming and affiliation with females, a relaxed dominance hierarchy, and physiological measures suggesting less stress among low-ranking males. Models explaining transmission of this culture centered on the treatment of transfer males by resident females.

Hamadryas Baboons

Hamadryas baboons are somewhat smaller than other *Papio*, but size dimorphism is comparable. Males have pink faces, grayish-brown hair, and a silvery gray mane or cape on the shoulders. The following description emphasizes Ethiopian hamadryas. Human presence and absence of predators have pervasively affected Arabian populations.

Hamadryas Ecology

Hamadryas typically live in hot, dry thornscrub with annual rainfall of less than 650mm at altitudes up to 3000m. Trees are small and scarce in most localities. Foraging hamadryas spread out across a front of more than 200m, which avoids most competition for food. Some populations depend on underground food and water at certain times of the year, so digging is crucial. Distinctive features of hamadryas life are mainly expressed in their modular social organization and related aspects of the life cycle.

Hamadryas Sociality

The basic unit of everyday life is an OMU that typically contains one or two females, though there may be more (Figure 21.6). Spatial cohesion is demonstrated by the fact that almost all grooming (98 percent) takes place within the unit. The females have been accumulated by the male, rather than having a prior relationship with one another. A female who strays or approaches another male is likely to be threatened and chased. Many females display scars from neck bites, a semi-stereotypical punishment behavior (probably at least partly ritualized, despite the wounds) that distinguishes hamadryas herding. It elaborates on behavior sometimes displayed by common baboons during inter-troop encounters. After such an episode the hamadryas female follows the male closely and grooms him vigorously. Though coercion is a factor in the relationship, females also receive benefits. The leader protects them from predators and infanticide, and shields them during storms. During travel a male will wait for a female who is slowed by age, injury, or carrying a dead infant. Seventy percent of these relationships last for more than three years.

A male follower accompanies about a third of the OMUs. When young he is allowed to sit near the adult females who sometimes groom him. Gradually the senior male allows him to play a role in leading the group, but a fully adult follower no longer has contact with the females. This is acceptable to him because he will eventually acquire one or more of the group's females. He may inherit some or all of the adults, but more likely he will depart with a prepubertal female. He treats her as a juvenile and she becomes a sexual partner only after she is mature. In either case a young male follower develops bonds with females that may facilitate later mating relationships.

Hamadryas social organization is modular. The band is equated with the common baboon troop because of its spatial and functional unity as well as composition and size. It is a multi-male group that combines several OMUs along with bachelor males. Most bands probably have 50 to 100 members, though some contain 200 or more. The larger bands probably result from unusual resources in certain areas, such as palm fruits that are available throughout the year. A band has its own area on the sleeping cliff. In the morning friendly bands often travel

Figure 21.6 A male hamadryas baboon displays his canine teeth (in what is probably an ordinary yawn rather than a threat) as a female grooms him. © mrtom-uk via iStock

for a short distance together, though they always split and go in different directions for the rest of the day. Eventually each band fragments into its OMU components, but it often assembles at a waterhole around midday.

An occasional function of the band is to confront alien conspecifics on the sleeping cliff. Such disputes are accompanied by *baahu* calls that seem homologous with the *wahoo* of common baboons. Fights on the cliffs are probably about females rather than sleeping space, since they start after strangers have already settled on a ledge. Some males try to abduct females during the early turmoil, before clear battle lines form. Conflicts between males from different bands often end when both parties flee, and sometimes both bands abandon the cliff at the end of a dispute. A plausible explanation for withdrawals is that loss of females is the paramount concern and flight is preferable to any significant risk of such a loss.

In some populations OMUs form **clans** of about 20 members within the band. Where they exist, clans seem to be the more typical foraging groups than OMUs and perhaps the more important social units. Clan members stay together when traveling within a larger group and when they are on the sleeping cliff. They groom each other more often than they do others in the same band. Males of a clan engage in cooperative predator defense, forming a shield against hyenas while the others retreat. Observation of a captive population indicates that both male and female clan members have advantages over other hamadryas in contest competition for clumped food (Colmenares et al. 2006).

The clan is male-philopatric, an unusual pattern among Old World monkeys. Adult females in one study transferred between clans 19 times while only one adult male transfer was seen. A follower male is readily tolerated because he belongs to the same clan as the OMU leader and is likely to be a relative. In a captive colony maternal brothers had the most cohesive relationship among males. Two anecdotes suggest the strength of clan ties. First, a juvenile male stayed with his father when his mother was abducted into another clan. Second, an infant of an abducted female returned to his natal clan at the age of three years.

Romero and Castellanos (2010) applied new statistical techniques to data on agonistic interactions, showing that leader males are generally dominant over followers and that dominance ability does not correlate with age. The values of different adult males were very close, suggesting that dominance relationships are not rigid. Male dominance values were positively correlated with the number of adult females per OMU but not with feeding priority. Greeting interactions occurred more frequently between individuals with similar dominance values, including a ritual called *notifying* that is essentially the same as the male greeting in common baboons. Fraser and Plowman (2007) tested six hypothesized functions of notification and accepted three: alliance, peacekeeping, and submission.

OMU females are strongly oriented toward the leader male and the main fights among them concern grooming access to the male. Relations between females vary, but a female typically engages in social interaction with other females as much as with the leader male. The frequency of female–female interactions varies with the number of females in the OMU rather than the availability of the leader. Females risk punishment from leaders by visiting females in other OMUs, possibly relatives. Cycling females are closer to the male than those who are pregnant or lactating. Swollen females are closest and receive the most grooming. Females maintained kinship bonds in a captive colony and mother–daughter dyads did the most grooming.

The OMU is the reproductive unit. A leader never copulates with outside females. Effective mate guarding is implied by testes size indicating a minor role for sperm competition (1.3 percent of body weight, versus 2 to 3.5 percent in common baboons). Hamadryas females emphasize the strategy of **paternity concentration**, behavior that gives the male a high level of paternal certainty. The female benefits from greater protection for herself and her infants. Two aspects of female sexuality enhance paternity concentration. One is the fact that females rarely make sexual presentations. These are frequent among common baboons except during consortships, so hamadryas females are in effect in permanent consortship. The other factor is the relatively low frequency of copulation calls, which limits the possibility of eliciting male competition.

Despite the emphasis on paternity concentration, hamadryas females may practice paternity confusion by engaging in surreptitious copulations with subadult and juvenile males (about 20 percent of the copulations in one study). This may confuse paternity for males who could potentially take over the OMU and commit infanticide. The effectiveness of the ploy is enhanced by the fact that even juveniles are capable of ejaculation among hamadryas.

A leader sometimes observes copulations with follower males without interfering, or he punishes the female but not the follower. Since the follower is probably a clan relative, tolerance may be favored by inclusive fitness. These events could be stages in the process of a leader voluntarily relinquishing his females.

Hamadryas Life Cycle

Trivers and Willard (1973) predicted that when parental condition has differential effects on the fitness of male and female offspring, parents who are in good condition should bias investment toward the sex that benefits most. Silk and Strum (2010) suggested that mixed results from tests of this model might have resulted from the proxy measures used for parental condition (e.g. dominance rank). They made visual assessments of maternal body condition and found that maternal condition at conception had no consistent relationship with birth sex ratios. Mothers in poorer condition at the time of conception resumed cycling significantly later, but the sex of their infants had no effect on resumption of cycling. Thus, maternal condition seems to influence female reproductive ability but there is no facultative adjustment of the birth sex ratio.

Hamadryas infants are in danger when an OMU is taken over. A captive study suggested that situation-dependent swellings are a defensive mechanism (Zinner and Deschner 2000). Following takeovers, five of six lactating females immediately developed sexual swellings and copulated (though none conceived in the first few cycles) and four of five dependent infants survived.

Hamadryas life histories diverge from those of common baboons at puberty with enlargement of the testes soon after they descend. This is probably to take advantage of occasional sneak copulations, which can be adaptive as long as the probability of conception is above zero and stealth minimizes risk. The low energy expenditure in copulation balances the low probability of benefit. In a captive population, males under four years of age fathered at least 2.5 percent of offspring (Zinner et al. 2006). Despite early growth, hamadryas testes end up smaller than those of common baboons in accord with successful mate guarding.

At about six years of age a male withdraws from his peers, and then joins either an all-male group or an OMU as a follower. He also develops an interest in mothering small baboons: pulling them close, hugging them, and carrying them on his abdomen or back. Followers do this twice as often as subadult females. A young male's first subjects are mainly male infants, but between six and eight years his interest shifts to females. This is a prelude to adopting a young female as the first member of his OMU when he is between 8 and 11 years old (Sigg et al. 1982). The male either lures her from his own OMU or abducts her from another (or from any OMU if he is not a follower). She is his only partner for as long as 30 months, until he acquires an adult female. At that point the younger female sometimes leaves.

A male maintains his females for only a few years in his prime and then a younger male displaces him. The challenge can come from the same clan or others. Twenty percent of adult males are bachelors, waiting to take over a harem. In some cases an older follower usurps most or all of the group rather than leaving with just one female. Injured leaders are likely to be raided by other leaders as well as bachelors. Followers performed most of the takeovers at one site; at another site nearby more than half of the takeovers were by other leaders. A deposed male may stay with his OMU or at least with his band. Some fight for the band even if they have no females left, perhaps because they are supporting relatives.

Over seven years of a study, almost all changes in OMU membership resulted from takeovers rather than voluntary transfers (Swedell et al. 2011). Two-thirds of the takeovers occurred within bands and one third across bands. Of 22 cases providing relevant data, 77 percent occurred within clans. These results indicate that hamadryas dispersal is not analogous to sex-biased dispersal in other taxa, because: (1) females do not disperse voluntarily; (2) females are most often rearranged within bands, which does not promote gene flow; (3) females undergo social dispersal but not usually locational dispersal; and (4) while male hamadryas are far more philopatric than females, they have been observed to disperse. It thus appears that the ancestral baboon pattern of female philopatry and male dispersal has evolved into a system in which neither sex is motivated to disperse.

After staying with her first male for up to 30 months, a female will probably start moving around. In one band 27 females changed affiliation over 5.5 years. Twenty of 26 immature females stayed in their own clan. Ultimately, 6 stayed in the same clan, 8 went to other clans, and 18 went to another band. A young female hamadryas is shielded from competition with other females during the time that she is the sole mate of a young male. As a result, she can start her reproductive career earlier than females in other baboon species.

Chronic year-round heat stress could explain the female strategy of paternal concentration. The energy budget of hamadryas females might not sustain the additional sexual activity required to establish paternal confusion (Swedell and Saunders 2006). I suggest that unusual stress on reproduction may also be related to the rarity of weaning conflict in hamadryas. Whatever the conflict is really about (e.g. nursing or activity patterns), environmental stresses could result in closer concordance of mother and infant concerns.

Summary

Common baboons are widespread because of omnivory and effective predator defense. Females, often in ranked matrilines, are the core of the multi-male troop. Large sexual swellings convey reproductive information. Most males change troops several times in pursuit of mating opportunities. Typical male strategies include coalitions and special friendships. Distinctive communication signals include the raised-brow threat, wahoo bark, and male–male greeting. Chacmas lack male coalitions. Special friendships provide a defense against high rates of infanticide. Hamadryas baboons live in modular societies. An OMU male controls females with threats and neck bites and females are oriented toward him. A male often starts his own OMU by kidnapping a female juvenile.

Sources

Basic Bibliography

Alberts, S.C., & Altmann, J. (2012). The Amboseli Baboon Research Project. In P.M. Kappeler & D.P. Watts (Eds.), *Long-term field studies of primates* (pp. 261–288). NY: Springer.

Altmann, S.A., & Altmann, J. (1970). *Baboon ecology: African field research.* Chicago: UCP.

Cheney, D.L., & Seyfarth, R.M. (2007). *Baboon metaphysics: the evolution of a social mind.* Chicago: UCP.

Swedell, L. (2011). African papionins. In C.A. Campbell & 4 co-editors, *Primates in perspective* (pp. 241–277). NY: OUP.

Swedell, L., & Leigh, S.R. (Eds. 2006). *Reproduction and fitness in baboons*. Chicago: UCP.

Additional References

Alberts, S.C., and Altmann, J. (1995). Balancing costs and opportunities: dispersal in male baboons. *AN*, *145*, 279–306.

Alberts, S.C., et al. (2003). Queuing and queue-jumping: long-term patterns of reproductive skew in male savannah baboons, *Papio cynocephalus*. *AB*, *65*, 821–840.

Altmann, J. (1980). *Baboon mothers and infants*. Cambridge MA: HUP.

Altmann, S.A. (1999). *Foraging for survival, yearling baboons in Africa.* Chicago: UCP.

Anderson, C.M. (1987). Female transfer in baboons. *AJPA*, *73*, 241–250.

Barton, R.A. (2000). Socioecology of baboons: the interaction of male and female strategies. In P. Kappeler (Ed.), *Primate males* (pp. 167–203). NY: CUP.

Bercovitch, F.B. (1988). Coalitions, cooperation, and reproductive tactics among adult male baboons. *AB*, 36, 1198–1209.

Bergman, T., & Kitchen, D.M. (2009b). Comparing responses to novel objects in wild baboons (*Papio ursinus*) and geladas (*Theropithecus gelada*). *AC*, *12*, 63–73.

Castles, D.L., et al. (1999). Social anxiety, relationships and self-directed behaviour among wild female olive baboons. *AB*, 58, 1207–1215.

Colmenares, F., et al. (2006). One-male units and clans in a colony of hamadryas baboons (*Papio hamadryas hamadryas*): effect of male number and clan cohesion on feeding success. *AJP*, 68, 21–37.

Dirks, W., et al. (2010). The relationship of accentuated lines in enamel to weaning stress in juvenile baboons (*Papio hamadryas anubis*). *FP*, *81*, 207–223.

Frank, R., & Silk, J.B. (2009a). Grooming exchange between mothers and non-mothers: the price of natal attraction in wild baboons (*Papio anubis*). *Behaviour*, *136*, 889–906.

Frank, R., & Silk, J.B. (2009b). Impatient traders or contingent reciprocators? Evidence for the extended time course of grooming exchanges in baboons. *Behaviour*, *136*, 1123–1135.

Fraser, O., & Plowman, A.B. (2007). Function of notification in *Papio hamadryas*. *IJP*, *28*, 1439–1448.

Gilbert, C.C. (2013). Cladistic analysis of extant and fossil African papionins using craniodental data. *JHE*, *64*, 399–433.

Hausfater, G. (1975). Dominance and reproduction in baboons (*Papio cynocephalus*): a quantitative analysis. *Contributions to primatology*, Vol. 7. Basel (Switzerland): Karger.

Hill, R.A. (2006). Thermal constraints on activity scheduling and habitat choice in baboons. *AJPA*, *129*, 242–249.

Hill, R.A., & Dunbar, R.I.M. (2002). Climatic determinants of diet and foraging behaviour in baboons. *Evol. Ecol.* 16: 579–593.

Jolly, C.J. (2007). Baboons, mandrills, and mangabeys. In C.J. Campbell & 4 co-editors, *Primates in perspective*, 1st ed. (pp. 240–251). NY: OUP.

Kleindorfer, S., & Wasser, S.K. (2004). Infant handling and mortality in yellow baboons (*Papio cynocephalus*). *BE&S*, 56, 328–337.

Nguyen, N., et al. (2009). "Friendships" between new mothers and adult males: adaptive benefits and determinants in wild baboons (*Papio cynocephalus*). *BE&S*, *63*, 1331–1344.

Noe, R. (1992). Alliance formation among male baboons: shopping for profitable partners. In A.H. Harcourt, & F.B.M. De Waal (Eds.), *Coalitions and alliances in humans and other animals* (pp. 284–321). NY: OUP.

Noe, R., & Sluijter, A.A. (1995). Which adult male savanna baboons form coalitions? *IJP*, *16*, 77–105.

Norton, G.W. (1986). Leadership: decision processes of group movement in yellow baboons. In J. Else & P.C. Lee (Eds.), *Primate ecology and conservation* (pp. 145–156). NY: CUP.

Packer, C. (1977). Reciprocal altruism in *Papio anubis*. *Nature*, *5593*, 441–443.

Packer, C. (1979). Inter-troop transfer and inbreeding avoidance in *Papio anubis*. *AB*, *27*, 1–36.

Rasmussen, D.R. (1981). Communities of baboon troops (*Papio cynocephalus*) in Mikumi National Park, Tanzania. *FP*, *36*, 232–242.

Romero, T., & Castellanos, M.A. (2010). Dominance relationships among male hamadryas baboons (*Papio hamadryas*). *Journal of Ethology*, *28*, 113–121.

Rowell, T.E. (1966). Forest living baboons in Uganda. *JZL*, *149*, 344–365.

Rowell, T. (1972). *The social behavior of monkeys*. New York: Penguin.

Sapolsky, R.M., & Share, L.J. (2004). A pacific culture among wild baboons: its emergence and transmission. *PLOS Biology*, *DOI:10.1371/journal.pbio.0020106*

Schmitt, V., & Fischer, J. (2009). Inferential reasoning and modality dependent discrimination learning in olive baboons (*Papio hamadryas anubis*). *JCP*, *123*, 316–325.

Sigg, H, et al. (1982). Life history of hamadryas baboons: physical development, infant mortality, reproductive parameters and family relationships. *Primates*, *23*, 473–487.

Silk, J.B., & Strum, S.C. (2010). Maternal condition does not influence birth sex ratios in anubis baboons (*Papio anubis*). *PLONE*, 5: 312750.

Silk, J.B., et al. (2004). Patterns of coalition formation by adult female baboons in Amboseli, Kenya. *AB*, *67*, 573–582.

Silk, J.B., et al. (2006a). Social relationships among adult female baboons (*Papio cynocephalus*) I. Variation in the strength of social bonds. *BE&S*, *61*, 183–195.
Silk, J.B., et al. (2006b). Social relationships among adult female baboons (*Papio cynocephalus*) II. variation in the quality and stability of social bonds. *BE&S*, *61*, 197–204.
Silk, J.B., et al. (2009). The benefits of social capital: close social bonds among female baboons enhance offspring survival. *PRSB*, *276*, 3099–3104.
Silk, J.B., et al. (2010). Strong and consistent social bonds enhance the longevity of female baboons. *CB*, *20*, 1359–1361.
Smith, E.O. (1992). Dispersal in sub-Saharan baboons. *FP*, *59*, 177–185.
Smuts, B.B. (1985). *Sex and friendship in baboons*. NY: Aldine.
Smuts, B.B., & Watanabe, J.M. (1990). Social relationships and ritualized greetings in adult male baboons (*Papio cynocephalus anubis*). *IJP*, *11*, 147–172.
Strum, S.C. (1987). *Almost human*. New York: W.W. Norton.
Swedell, L., & Saunders, J. (2006). Infant mortality, paternity certainty, and female reproductive strategies in hamadryas baboons. In L. Swedell & S.R. Leigh (Eds.), *Reproduction and fitness in baboons* (pp. 19–51). NY: Springer.
Swedell, L., et al. (2011). Female "dispersal" in hamadryas baboons: transfer among social units in a multilevel society. *AJPA*, *145*, 360–370.
Trivers, R.L., & Willard, D.E. (1973). Natural selection of parental ability to vary the sex ratio of offspring. *Science*, *179*, 90–92.
Van Horn, R.C., et al. (2007). Divided destinies: group choice by female savannah baboons during group fission. *BE&S*, *61*, 1823–1837.
Zinner, D., & Deschner, T. (2000). Sexual swellings in female hamadryas baboons after male take-over. 'Deceptive' swellings as a possible female counterstrategy against infanticide. *AJP*, *52*, 157–168.
Zinner, D., et al. (2006). Early sexual maturity in male hamadryas baboons (*Papio hamadryas hamadryas*) and its reproductive implications. *AJPA*, *129*, 584–590.

22 The Hominoid Superfamily

Apes Small and Large

Wearing the purple gown of a mandarin, a Chinese painter sits among tall trees on the cool mountainside. He watches a slender ape hang comfortably from a branch by its long arms and then swing easily to a cluster of fruit. The painter dips his delicate brush into the thin ink and begins to render the gibbon on paper with strokes as graceful as the movements of the ape.

Apes differ from monkeys in physical features associated with suspension (see Table 22.1 online for physical dimensions of all apes). Mobile shoulders and long arms extend their reach in a wide range of directions. Their elbows and wrists are flexible and grasping is facilitated by long hands and fingers and a robust hallux. The short, stiff lower back supports erect posture and the flexible movement of shoulders and arms. Hanging by one or both hands (usually to feed), or grasping with three appendages, provides greater stability than above-branch activity. Thus all but the largest apes can feed among small branches (see Table 22.2 online for ecological characteristics of all apes). Apes have wide faces, broad in the nasal region and palate. Their flat molar teeth, with five cusps on the lower molars, are probably adapted to the consumption of fruit.

Every stage of life is longer for apes. Only the smallest apes and largest monkeys overlap in this respect. The minimum inter-birth interval for almost all ape species is two years, which is the maximum for almost all other primates (see Table 22.4 online for reproductive parameters of all apes). Ape gestation lasts for at least seven months and weaning is rarely completed before the age of two years. The ape juvenile period is twice as long as that of monkeys, which allows for a great deal of protected learning. Apes reach puberty at about eight years, but full physical and social maturity takes longer (see Table 22.5 online for life cycle stages of all apes).

Hominoids constitute a superfamily equivalent to the cercopithecoids. Hominoids differ in that they are divided at the family rather than the subfamily level. The two ape families are informally called the gibbons and the great apes. The families are considered separately in this chapter. The great apes, our closest living relatives, will be examined more closely over the following four chapters.

Gibbon Behavior

The main gibbon habitat is lowland evergreen rainforest in the Malay Peninsula and Indonesia. They can also be found in semi-deciduous rainforests further north and in favorable pockets throughout Southeast Asia and southern China. Gibbons are the smallest apes, most of them about the size of a large domestic cat (Bartlett 1999). Four genera are currently recognized (Box 22.1).

Box 22.1 Gibbon Taxonomy

There seems to be a consensus that the gibbons are a clade and should all be assigned to the family Hylobatidae, which is characterized by short snouts and globular skulls, large orbits with protruding rims, minimal sexual dimorphism, and the longest arms in proportion to body size of any living primate (Fleagle 2013; Tuttle 2014). Gibbons are further distinguished from other apes by their smaller size and by ischial callosities comparable to those of cercopithecoids. However, these are not indicative of a close phylogenetic relationship because the callosities develop postnatally in gibbons and prenatally in monkeys. Four genera are currently recognized: *Hylobates* (common gibbons), *Nomascus* (crested gibbons), *Hoolock* (hoolocks), and *Symphalangus* (siamangs). They are distinguished by chromosome numbers, by the absence, presence, and size of laryngeal air sacs, and by the distribution of body mass among the limbs (Tuttle 2014; Zihlman et al. 2011).

An increasing body of genetic evidence also supports the monophyly of the Hylobatidae and of the four commonly recognized genera (Carbone et al. 2014; Meyer et al. 2012; Thinh et al. 2010). These data indicate that the gibbon clade separated from the other apes between 15 and 18 million years ago, and that the lineages leading to the extant genera underwent a rapid radiation at some time between 10 and 5 million years ago. This may have been precipitated by major geographic changes in Southeast Asia that resulted in cycles of habitat compression and expansion. Interbreeding among the lineages has complicated genetic analysis of gibbon phylogeny.

Gibbon Ecology

Gibbons display two kinds of brachiation (Bertram 2004). The **pendular** form, swinging from hand to hand, depends on continuous contact with the superstrate and is relatively slow. **Ricochetal** brachiation involves aerial phases between contacts, sometimes covering more than 9 meters. This faster and more efficient form entails a novel kind of energy transfer that requires unique anatomy and maneuvers (Figure 22.1).

Combining brachiation with climbing and bipedal running along the tops of branches, gibbons travel rapidly through densely canopied forests without leaving the trees. During rare movement on the ground, they use a bipedal gait in which the long arms are held up and out of the way. Feeding on large arthropods is probably the main reason for coming to the ground.

The average diet consists of about 60 percent fruit, 30 percent leaves, and 10 percent arthropods and flowers. Gibbons eat more fruit than sympatric monkeys and orangutans, and more than most other primates that are labeled frugivores. Gibbons have been labeled "ripe-fruit specialists", which leads to two important points. One is the need for precise terminology and definitions. Gibbons are specialists in that the fruits they eat display similar chemistry and morphology. However, they are generalists in that they exploit many different plant genera. A related lesson concerns comparative generalization. Marshall et al. (2009) concluded from their fieldwork that gibbons are relatively unspecialized, conflicting with an earlier study at another site that characterized gibbons as the most specialized primates in the region. The longer study by Marshall and colleagues clarified two points: (1) the breadth of gibbon diet varies on a seasonal basis, which requires at least a year of research; and

Figure 22.1 A young gibbon demonstrates the aerial phase of ricochetal brachiation. © 4FR via iStock

(2) gibbons become more generalized feeders when food is scarce, a situation that may appear only after several years of fieldwork.

Gibbons strongly prefer the succulent and sugary pulp of ripe fruits, but figs sometimes constitute almost half of the fruit that they eat. Variation may be related to the availability of figs in various habitats. Unripe fruits and liana products are other significant fallback foods. Fallback foods, more than staples, differentiate gibbons from sympatric primates.

There are several mutually compatible explanations for gibbon dietary flexibility. One is that relatively simple gut morphology does not constrain them to a narrow diet. Another is that rapid locomotion allows them to visit many small food patches. Finally, gibbons may be able to process a wider variety of fruits than avian competitors because of their manual dexterity. The apes may also have an advantage in the form of a brain that can store superior knowledge of resource locations.

Several gibbon foraging patterns probably lessen competition. First, they do most of their foraging in the upper canopy. Second, they use small terminal branches that are unavailable to other primates and most other mammals because they combine relatively small size with suspensory and quadrumanous capabilities. Third, they use scattered small trees that are valuable only to small groups of small animals. Gibbon day ranges are 1.5 times longer than those of sympatric langurs and 2.5 times longer than those of orangutans. Fourth, they omit the typical primate midday rest and end their daily activities 1.5 to 3 hours before sunset. Foraging at midday while other primates are resting, gibbons descend into the lower canopy that other primates are likely to use when foraging. In doing so the gibbons also escape much of the midday heat. Early retirement avoids competition with monkeys that are in their second foraging session of the day (Ahsan 2001).

Predation on gibbons tests the Socioecological Hypothesis that small primate groups occur only where predator pressure is low. This supposedly applies to gibbons, due to their relatively large size, high canopy habits, and speed and agility. However, evidence for predation comes from gibbon remains and from a high rate of disappearance at a site where poaching was minimal (Morino 2010; Uhde and Sommer 2002). Gibbon speed and agility are countered by the fact that raptors and snakes strike suddenly and the significance of size is diminished by raptor concentration on the young. Anti-predator behaviors, some costly, provide further evidence. Gibbons frequently scan the canopy and the ground, respond to alarm calls and mobbing noises of other species, and engage in mobbing. Reluctance to come to the ground, even to retrieve an infant, suggests the significance of terrestrial predators. Raptors flying low over the trees cause young gibbons to drop from the tree crown or crouch against the trunk. Adult gibbons distinguish between dangerous and harmless animals, as in the case of a group that mobbed a python but ignored a poisonous snake (Nettelbeck 1998b).

Early retirement may function to avoid predators that are active at dusk (Reichard 1998), a hypothesis consistent with rapid movement to sleeping places (in 5 to 15 minutes). Predator defense is also suggested by the choice of emergent trees and by the fact that individuals (other than mother and infant) never sleep in the same spot and usually occupy separate trees. This may minimize a concentration of odor that might attract predators. Local factors may also influence these choices. One species used trees without vines because of biting ants that inhabited chambers within thick vines (Whitten 1982) and/or because human hunters used the vines to climb the trees (Tenaza and Tilson 1985). Once sleeping places are selected, the apes usually rest in a sitting posture on ischial callosities like those of Old World monkeys.

Early retirement seems to challenge the Optimal Foraging Theory, which predicts that animals will use all available time to maximize energy intake. Gibbons appear to be time minimizers rather than energy maximizers, active for only as long as it takes to fulfill their caloric needs. However, this is not quite accurate either: gibbon foraging time does not decline when food is abundant. A likely explanation lies in territoriality. The gibbon food supply depends on exclusive use of limited resources in a small area, so time must be devoted to defense.

Home range, daily travel, and territoriality are closely related in gibbons. The average home range of a pair-group is about 40 hectares. Day travel is lengthy for such small ranges, about 1 to 1.5 kilometers. The combination of small home range and long day range makes it possible for gibbons to regularly monitor their space and a typical group defends about 70 to 80 percent of its range. The location of food sources is probably the most important influence on range use, but sleeping trees and patrol of territorial boundaries may also be significant. The territorial hypothesis explains the proportioning of time by gibbons, but it does not seem to explain the sequence in the daily activity pattern. Most territorial activity takes place in the morning.

Gibbon Sociality

In gibbons as in marmosets and nocturnal strepsirrhines, increasing breadth and depth of research has added complexity to a stereotype of obligate monogamy. Some groups, up to 10 percent of some populations, are socially and sexually polygamous. Typically such groups are polyandrous, but polygyny is common in several species, and some groups contain multiple males (Malone and Fuentes 2009). These results come from longer field projects and broader comparisons among gibbon genera, species, and social units (see Table 22.3 online for social features of all apes).

The great majority of social units observed in the great majority of gibbon species contain only one adult of each sex, so the pair seems to be the basic gibbon social adaptation. It is a facultative trait with a strong central tendency. One explanation is a variant of the Socioecological Model called the Female Dispersion Hypothesis. This interpretation has female gibbons spreading out across a homogeneous habitat and each defending her own territory; males join females and the result is long-term pairing because each male can defend only one mate against other males. The last point dovetails with the Mate Guarding Hypothesis, which puts the emphasis on the greater defendability of a single female as a vehicle for increased paternal certainty. Mate guarding occurs in many primates, but the Mate Guarding Hypothesis is a specific explanation for long-term pairing.

Brockelman (2009) noted that these analyses equated territorial behavior with mate defense. He argued that these are two different phenomena in gibbons. Territorial conflicts are frequent, noisy group events at borders; mate competition involves deep penetration of the home range by a lone outsider with little or no vocalization, resulting in potentially fatal fighting between two males. The two functions may be in direct conflict with one another: extra-group copulations have been observed when males left mates to defend borders. Brockelman argued that territoriality is parental investment in the form of resource defense. Combined with aggression or avoidance between females, it can result in two-adult social units. Brockelman criticized the practice of designating mates as a resource for males. He pointed out that competition for physical resources depends on the overall density of a population while mate competition depends on the relative frequencies of the two sexes. If the focus is on physical resources, a male should aggressively defend a territory against all intruders. The sex-specific responses of gibbons are more consistent with mate guarding to raise paternal certainty. Since monomorphy makes it difficult for males to restrict females, the male's only option is to defend against intruding males.

Some analyses used the Defendability Model of Mitani and Rodman (1979) to conclude that a male gibbon *could* defend more than one mate. These interpretations explained pairing as an efficient way to defend against sexually selected infanticide. As in some other cases, the infanticide-defense explanation has been applied on the basis of circumstantial evidence, such as a long IBI and observation of post-immigration copulations. There is only one probable case of infanticide by an immigrant male. A comparable explanation puts the focus on predators. Male gibbons are vigilant against predators and active in mobbing them.

According to Bartlett (2011), food availability may be a limiting factor. Pairing rather than solitary foraging may be favored by sharing information about resource location or the avoidance of competition with each other (associated with a small shared territory). This is consistent with at least one version of Sexual Selection Theory. The idea that females defend resources against each other while males defend against male sexual competitors accounts for the sex-specific responses to outsiders.

Apart from the issue of pairing, small group size and territoriality are probably closely related to one another. Ecological constraints play a key role in limiting group size. Some major factors that affect group size and territorial defense are the high diversity of foods, relatively small food patches, and a high level of foraging efficiency. Given the small size of the group, it would be difficult for the male(s) to defend the boundaries of a large area.

Extra adult males in polyandrous groups are in some cases mature or nearly mature offspring that are delaying emigration. Others seem to be nonbreeding older males, perhaps former breeders defeated by immigrants. There are some cases of sexual polygyny. In a few cases, two females in a polygynous group were seen to carry infants that, by virtue of age similarity, could not have been produced by just one of those females. However, no researcher seems to have demonstrated that the resident male sired both of the infants in such cases. One or both might have resulted from extra-group copulation. Social polygyny probably occurs occasionally in gibbons, as it does in strepsirrhines that typically live in pairs, but there is no conclusive evidence for sexual polygyny.

Although the relationship of paired gibbons is often described as a *bond*, there is little evidence for an emotional connection that merits the term. Rates of social contact are typically low, vocalizations are at a minimum, with contact calls apparently absent, and the only facial expression used in communication is an open-mouth threat (Leighton 1987). Gibbon facial muscles are simple in comparison to those of primates that live in large groups (Burrows et al. 2011). Adults groom with each other during rest periods, but a study of one species indicated that grooming is primarily hygienic (Reichard and Sommer 1994). Groomers gave little attention to surfaces that are easily cleaned by self-grooming and concentrated on upper body areas that were likely to be infested by parasites and other foreign matter as a result of the vertical posture assumed in most activities.

The issue of bonding illustrates the problem of inferring proximal mechanisms from overt behavior. As is said of many callitrichids, the association between male and female could be based solely on site attachment on the part of each individual and intrasexual aggression against intruders. On the other hand, the low level of affiliative behavior has also been attributed to the strong cohesion of the group and the concomitant familiarity of all members with the daily routine (Chivers 2005).

Limited data suggest that males groom females more than the reverse. This has been taken to mean that males contribute more to the presumed bond, possibly because they must overcome female resistance to feeding competition (Palombit 1996). Travel also seems to give females priority. A family moves in single file, with the female leading the majority of travel bouts and the male usually bringing up the rear (Barelli et al. 2008a). One interpretation relates this pattern to the nutritional needs of reproducing females. Another suggests that the male provides defense for the group. Territorial behavior is often a joint activity and may contribute to a male–female bond.

Perhaps the best evidence for a bond is the stability of relationships that last eight years or more. An alternative explanation, however, is that desertion (usually by males) seems to be an ineffective reproductive strategy. Finding an amenable new partner is difficult, and success is sometimes followed by a long delay before the first birth. Some deserting males return to their original partners. High frequencies of mate switching, desertion, and extra-group copulations do occur, but they seem to be associated with poor living conditions that produce high levels of disease and death (Chivers 2001; Savini et al. 2009).

Songs are ubiquitous among gibbons and distinguish them from other hominoids and from most other primates. A song is a series of notes, usually of more than one type, that form a recognizable sequence. These loud songs typically last 10 to 30 minutes, but some approach 2 hours. Gibbons usually sing in the morning, either around dawn or two to three hours later (preference varying with species). A male starts by combining notes into one or several short phrases that often become more complex as the song goes on. Complexity refers to the number of notes in a phrase, the number of distinct types of notes, and the degree of frequency modulation.

The duet is a song in which a paired male and female coordinate their vocalizations. The most conspicuous contribution of the female is the **great call**, typically a highly rhythmic series of long notes in which the tempo and/or the peak frequency increases. It is comparable to loud calls of males in many other primate species, and can be heard almost a mile away. The great call seems to have a genetic core, but it is flexible in several ways (Cheyne et al. 2007).

A male usually stops vocalizing when a great call begins and he provides a **coda**, a special reply phrase, at the end. The female also produces short phrases comparable to those of the male. A pair may sing together spontaneously every day or once in several days.

Haimoff (1986) compared the acoustical features of gibbon duets with those of other socially monogamous primates: spectral tarsiers, indris, and Mentawai langurs. He concluded that their vocal similarities represent functional convergence resulting from similarities in both social organization and ecological niche. The similarities among gibbon songs are strong enough to suggest homology (Geissmann 2002), but variation lends itself to phylogenetic reconstruction.

Numerous functions have been proposed for gibbon duets, but most revolve around two themes: (1) territoriality and/or intergroup spacing; and (2) creating and/or reinforcing the pair relationship (Geissmann and Orgeldinger 2000). These categories are not mutually exclusive. Furthermore, male and female songs might have different functions. Cowlishaw (1992) tested hypotheses with comparative data and concluded that female song is strongly associated with territorial defense while male solo songs are related to the estimated density of unmated males (floaters), which indicates a mate defense function. Playback experiments with several species supported the mate defense explanation by showing that the calls of strangers agitate resident gibbons much more than do those of neighbors. Increasing diversity in the comparative evidence suggests functional variation of songs and duets from one taxon to another (Geissmann 2002).

A female gibbon's genitals swell slightly and change color around the time of ovulation. Research on several *Hylobates* species indicates that copulation is more common when a female is swollen and that ovulation is probable at maximal swelling (Barelli et al. 2008b). The pattern seems to fit the Graded Signal Hypothesis (Cheyne et al. 2007), which was based on monkeys in multi-male troops. This might seem odd for pair-living species. However, there are polyandrous groups (almost half the groups at one site) and even without polyandry the pattern is consistent with extra-pair copulations.

Both male and female gibbons engage in extra-pair copulations (EPCs), which may approach 10 percent of all copulations in some populations. The male behavior presumably represents the basic mammalian strategy of multiple fertilization attempts. Females may be trying to confuse paternity, a plausible idea if takeovers are really common and result in a significant number of infanticides. Another reason for female extra-group copulation is to attract and test a potential new partner when the current one is perceived as inadequate in some way. EPCs may be more common than is now known. They often happen during the confusion of intergroup encounters and can easily be overlooked. Other instances may be inhibited because one of the potential participants is not habituated to the observer.

Gibbon Life Cycle

Virtually hairless in front, a neonate clings to its mother's chest and belly for warmth and the two are in constant contact for two or three months (Ahsan 2004). The mother carries her infant ventrally until weaning, and she is the main caretaker. Adult males provide little or no help, but juveniles sometimes groom, carry, and play with infants. The infant begins to brachiate when four or five months old, and begins bipedal walking a little later.

Weaning is typically completed by the age of two years. The mother's refusal to accept a frightened offspring into her lap may be another sign that the infant stage is coming to an end.

Infants and juveniles utter a few brief phrases during the songs of their parents, which is probably more of a learning process than an effective contribution. Another learning process is indicated by the fact that immatures try to mob harmless animals that are ignored by adults (Uhde and Sommer 2002). Some typical behaviors seem to require less learning. A male and two females from different caged groups of juveniles were released together. Two of them quickly formed a pair that excluded the third, and the paired individuals displayed behaviors much like those of normally reared wild gibbons (Paluck et al. 1970). This old experiment bears on recent discussions about "monogamy" in gibbons. It suggests an innate tendency toward pairing, which fits with the evidence that the majority of natural social units contain only two adults. Occasional additions presumably result from flexible responses to environmental and demographic factors.

Food transfer to juveniles occurs in three gibbon genera (Fan and Jiang 2009; Fox 1984; Nettelbeck 1998a; Schessler and Nash 1977). It is more common in captivity than in the wild, but the main initiators in both situations are juveniles. The juvenile usually begs by reaching toward the food and sometimes grabs it. Adults usually show some resistance but often acquiesce. Begging is not motivated by hunger, because it occurs when food is abundant. It does not arise from easier access because it occurs when the same food is readily available from the environment. It is not a matter of gaining preferred foods, because these may be a small percentage of the foods taken. It is not likely to be the basis for learning edible foods because the main practitioners are juveniles rather than weanlings, and because the takers may discard the food immediately without any inspection. Learning preferred food more likely comes about because adults eating preferred foods emit soft vocalizations that often attract young individuals. One plausible explanation for gibbon food sharing is that begging is a form of play that arises mainly in captive conditions where food and spare time are both abundant. Another is that food transfer develops social bonds between the young and their elders and maintains them until dispersal.

Adolescence begins at about age five and research with captive groups indicates that social development continues until about seven (Burns et al. 2011). Both sexes reach adult body size and the subadult behavioral stage when they are around eight years old. A subadult male may receive aggression from his father and become spatially peripheralized. After the birth of an infant he stays closer to his mother than to his father.

The issue of a female's first birth illustrates the distinction between field and captivity. Geissmann (1991) concluded from a review of captive studies that first births took place between the ages of six and eight years. A long-term field study by Reichard and Barelli (2008) established a mean age of 11 years for first birth. Similarly, the captive inter-birth interval was usually stated as three to four years but the interval was four to five years in a field study (Palombit 1995). The IBI of more than three years means that the loss of even one infant diminishes the reproductive success of a female gibbon more than it does that of a monkey. Actual lifetime reproductive success may fall far short of the ten offspring that had been postulated. Variance in reproductive success seems to be much greater than expected for a basically monogamous species. Studies of birds suggest that this leads to more intense reproductive competition and greater selectivity in mating, which fits with evidence for desertion and mate-switching (Palombit 1995). On the other hand, as noted earlier, desertion and switching are probably inhibited because establishment of a new pair can be a long and difficult process.

Some studies (e.g. Burns et al. 2011) report that maturing individuals are evicted from their natal groups by intrasexual aggression, presumably a defense against replacement by the maturing individual. The adult is not necessarily the parent, since turnovers in personnel are common and infanticide seems to be rare. However, there are also cases of tolerance and lengthy delay of dispersal. In one study females dispersed at age eight, while the average departure age for males was ten. Brockelman et al. (1998) suggested that extra subadults help the parents by acting as social partners for infants and by assisting in territorial defense. Young males in several species disperse over relatively short distances, perhaps into a neighboring territory or one beyond that (data for female dispersal are inadequate).

Captive research addressed the role of observational learning in the development of sexual motivation and capability. Mootnick and Nadler (1997) collected data on 31 gibbons of several species and three genera that had been raised from infancy apart from their mothers. Twenty-four (75 percent) became sexually competent as adults, with no observation of experienced animals. As in rhesus monkey experiments, the companionship of a peer was important for relatively normal development. Also as in rhesus, males were more adversely affected than females by a long period of isolation after maternal separation. The researchers attributed this to greater complexity of the male role in copulation. Sexual competence in both sexes seemed to develop through trial-and-error learning with a socially compatible conspecific, reinforced by sexual stimulation. In other words, an innate motivational system responded to external stimuli to initiate a learning process (though it is possible that some of the motor patterns had innate components). Bearing on another issue concerning innate and learned components of behavior, females in this sample generally became adequate parents without observing parental behavior in others.

A similar survey addressed inbreeding avoidance by comparing inbred and outbred captive gibbons (Mootnick et al. 2005). Unlike other primates, contact during early rearing did not inhibit mating in these gibbons. Mortality rates for the inbred and outbred cohorts did not differ, indicating that inbreeding had no deleterious effects. Gibbons may differ from other primates in this regard, perhaps because dispersal is limited. Gibbons may not emigrate far when the environment is saturated with conspecifics, and it may be adaptive under any circumstances to remain near the natal range in order to use accumulated knowledge of the resources. A population that has survived the initial effects of inbreeding may become increasingly viable as the exposed deleterious alleles are eliminated by natural selection. Inbreeding may actually be adaptive in a stable saturated environment because it maintains adaptive genotypes. Such circumstances are due in part to human activity, but fragmentation of the forests may have begun with drying trends tens of millions of years ago.

New pairs are formed in several different ways. According to some, it is more common for an individual in an established group to be replaced by a same-sex intruder, or by a maturing offspring of the same sex. The new individual may expel the predecessor or fill a vacancy. Some cases involving offspring result in incest, but these relationships are usually transient and/or reproductively ineffective. When a young adult is excluded from the natal group, he or she may receive parental help in acquiring a new territory. Daughters are more likely to receive such aid, while sons travel farther away. Having a territory, the young individual can attract a mate.

The longevity record for gibbons is 60 years, the second highest for nonhuman primates, exceeded only by a chimpanzee (Geissmann et al. 2009). Hominoids in general have longer life histories than other primates, but longevity is correlated with body size and gibbons are much smaller than other hominoids. From this viewpoint gibbon longevity is an unusual feature that requires further research.

Gibbon Cognition

Limited research on gibbon cognition may be attributed to poor performance and lack of motivation in the few experiments done (Cunningham et al. 2006). Informal observations suggest that gibbon problem-solving ability is impaired by temperament. They are easily frustrated and react emotionally to initial failures. Interpretation of experimental results is also confounded by variables such as sample size, sex, individual experience, and taxonomic status.

A rake test implicated sex difference (Cunningham et al. 2011). Fourteen of 22 gibbons spontaneously used a rake to obtain a reward. The sex difference appeared in the effect of prior experience. Interaction with the rake before testing significantly reduced latencies to solution for females, but not for males. The researchers suggested that the underlying proximate mechanisms are related to the natural energetic needs of reproductive females. These females have a greater need to explore potential resources, but the exploratory response is inhibited by maternal investment. The ability to learn about a new object quickly may help females to strike a balance between conflicting needs.

One of the few broadly comparative experiments included ten gibbons of four species (including one hybrid) representing three of the four genera, males and females, some born in the wild and others in captivity, ranging in age from 7 to 30 years (Fedor et al. 2008). The investigators performed several Piagetian tests of object permanence (the understanding that objects continue to exist when out of sight). During invisible displacements, an experimenter hid the target object in her hand before putting it into a box. There was great variation across individuals, but some gibbons performed better than chance on all but one test.

A field study investigated goal orientation in the travel paths of white-handed gibbons (Asensio et al. 2011), using the change-point test (CPT) developed by Byrne and colleagues for baboons (Noser and Byrne 2014). The gibbon researchers found that locations of significant travel direction changes were mainly related to preferred fruit sources. They inferred that the movements were planned, because consecutive change-points were beyond the gibbons' sight. The apes apparently grasped the connections among preferred food sources, implicating a cognitive map based on effective spatial memory. This is consistent with an experiment in which captive gibbons used spatial memory to recall where hidden items could be found and what those items were (Scheumann and Call 2006).

Fieldwork on gibbons produced the first evidence for referential communication in a free-ranging ape. Their songs in response to predators (represented by experimental models) differed consistently from their ordinary duets (Clarke et al. 2006). The apes produced the songs only in response to terrestrial predators, not raptors. Males who were absent from the group hurried back and began to sing a version of the predator song before seeing the threat. Neighboring groups responded similarly. Predator songs were introduced by more use of a particular kind of note, differed in the number of some notes, and lasted longer. The first female great call was delayed in predator songs and the males replied sooner.

Great Apes

The great apes are orangutans, gorillas, common chimpanzees, and bonobos (sometimes called pygmy chimpanzees). They are more closely related to each other than any of them are to the gibbons (Box 22.2). The term *great ape* refers to size: adults weigh at least 30kg and the smallest females weigh three times more than adult male siamangs. Great ape faces are

Box 22.2 Great Ape Taxonomy

Compared to gibbons, great apes are much larger and have more robust canine teeth, broader premolars, a very broad ilium, and a robust fibula (Fleagle 2013). There is a consensus that all belong in the superfamily Hominoidea. Previously, based on obvious anatomical similarities not shared with humans (e.g. long arms), the three genera were assigned to the family Pongidae. This paralleled the family Hominidae containing the genus *Homo* and its closest (fossil) relatives. Molecular evidence then showed that African apes (gorillas and chimpanzees) are more closely related to humans than to orangutans. Moreover, some evidence indicated that chimpanzees are more closely related to humans than either is to gorillas. There is a virtual consensus that these changed perspectives should be reflected in the classification. The taxon Pongidae has been discarded and all great ape genera have been sunk into Hominidae along with *Homo*. Groves (2001) divided the Hominidae into the subfamily Ponginae, containing only *Pongo* (the orangutans), and the Homininae for the other genera: *Gorilla*, *Pan*, and *Homo*.

more **prognathous** (protruding) than those of gibbons and the canine teeth are blunter. Male great apes have larger bodies and canine teeth than females.

Great Ape Ecology

Most weaned great apes construct nests almost daily, which distinguishes them from all the other anthropoids (Fruth and Hohmann 1996; Russon et al. 2007; Stewart et al. 2007; Stewart and Pruetz 2013; van Casteren et al. 2012). An individual pulls, bends, and breaks vegetation into a circular structure around its body, weaving parts of it, in a process that takes less than 10 minutes. The common features of this process, even in captivity, suggest innate components derived from the common ancestor of great apes; however, there is no question that effective performance is learned.

The typical nest is built for sleeping in a tree at night. There are several functional explanations and all may be correct for various circumstances. Because great apes are broader and heavier than monkeys, they may be less stable when sleeping in trees; nests probably make falls less likely. Greater stability might also allow sleeping in peripheral branches that are less accessible to tree-climbing cats. Nests might also provide camouflage against tree-climbing predators.

Nest comfort is an issue for both proximate and ultimate reasons. Some apes make extra effort to increase comfort, such as lining the nest with leafy twigs. Comfort seeking may be an important motivation for nest building. Greater comfort probably enhances sleep, which can result in greater energy for daytime activities as well as neural maintenance and memory consolidation. The nest provides a comfortable way for the large-bodied ape to sleep in a recumbent position, which may be necessary because large-bodied mammals lose muscle tone during REM sleep. Thermoregulation via insulation is another nest comfort with a potentially important function.

Great apes use more complex food processing techniques than monkeys (Byrne 2004), which enhance the flexibility and effectiveness of foraging. Two essential components are

precision handling and **bimanual role differentiation**. The former entails a grip in which the thumb opposes the fingertips. The latter means that the hands perform different but complementary actions (such as holding a food object at a favorable angle with one hand while manipulating it with the other).

Great apes use gestures, especially with the hands, with a degree of freedom that makes communication very flexible. They choose particular gestures for particular contexts and follow up with others if a first signal does not work. They also routinely adjust communication to the attentional state of the intended recipient, using visual gestures only if the recipient is paying attention, and moving to a position where they can be seen if necessary. They use some gestures in more than one context and they use a significant percentage in combinations (though most of these are simple repetitions). Great apes learn gestures, as evidenced by large differences among individuals and differences among subgroups in a community or captive population. Idiosyncratic gestures arise that are understood by other group members.

Great Ape Life Cycle

A slower pace of reproduction and life history differentiates great apes from monkeys and gibbons. Average gestation lasts at least a month longer than in gibbons and weaning takes place at least two years later. An infant sleeps in the mother's nest for several years. The infant begins to make its own night nests between the ages of three and five, an important milestone in development. A juvenile associates almost constantly with its mother for several years after weaning.

Lewis (2005) asserted that great apes are unique in using almost the entire play repertoire that is scattered across other primate species. Play is further elaborated by communication. It is the main functional context for the flexible gestures that distinguish great apes from other primates. The number of gestures used by an individual peaks during the juvenile stage, when play is most prominent (Call and Tomasello 2007).

A male great ape reaches puberty by ten years of age, but adult males limit adolescent copulation. The timing of a female's first birth seems to be deeply rooted in her physiology, since captivity makes little difference in all but one species (Robson and Wood 2008). The length of gestation and the outcome of the longer fetal growth period follow the rule that larger primates have larger babies. Female maximum lifespan in the wild is 40 to 50 years, resulting from low adult mortality.

Age-related deficits and deaths are more common than in other primates, which alters individual behavior and probably affects social relationships. Aging is apparent from signs such as slow movement, worn teeth, and sagging skin.

Great Ape Cognition

At least two schools of thought hold that great apes surpass other nonhuman primates in cognitive ability. According to one of these, the apes can form **secondary representations**. These are derived from mental images that reflect external situations (representations), but they go further. Secondary representations can be separated from reality in order to model hypothetical situations, such as those in the past or future (Suddendorf and Whiten 2001, 2003). A.E. Russon attributes **metacognition** to great apes. This entails a strong form of representation in which mental images can be called to mind in the absence of sensorimotor cues.

Captive apes use a variety of tools with flexibility and innovation. They pole-vault or construct bridges to get across gaps, and climb on stools or ladders to raise themselves higher. If a single object does not fulfill the latter function, they can stack boxes in order to reach a goal. There are many other examples of great apes using a variety of tools for a variety of functions. Tool use is more significant than object manipulation because it involves conceptually relating an object to another object (Tomasello and Call 1997). Mental representations of the world are apparent when an animal finds new and creative ways to use an object without trial and error (i.e. demonstrates insight or foresight). Suddendorf and Whiten (2001) viewed tool use as exemplifying secondary representation and reasoning in the means–ends sense. The tool user keeps an ultimate goal in mind while achieving necessary sub-goals.

Problem solving entails a number of mental functions. Tomasello and Herrmann (2010) summarized some of the most important features of great ape cognition that have been suggested by various lines of research. The apes understand the permanence of physical objects, as shown by their responses when objects are hidden from them. They can place objects into categories and engage in rudimentary quantification. They have some understanding of causal relationships in events involving these objects.

Theory of mind, often abbreviated as TOM, refers to an individual's ability to understand the perceptions and cognition of another individual. The word *theory* in this context pertains to an idea held in the mind of a subject rather than an outside observer. Researchers have a theory (or hypothesis) that the apes have a theory (of mind). The terminology is a source of confusion, but the phrase theory of mind is deeply entrenched in the literature. There is evidence that some form of TOM exists in all great apes. In gestural communication, for example, adjustments in relation to the attentional state of the receiver indicate that the signaler understands something about how the other individual perceives and interprets the communication (Call and Tomasello 2007). Great apes can infer the perceptions and goals of other individuals, which implies an understanding that the others are intentional actors. They may understand the basis for decision-making in others (Tomasello and Herman 2010).

Gaze following can be an indicator of TOM, if there is evidence that the subject tries to take the visual perspective of the other. All great ape genera can perform in this manner (Bräuer et al. 2005). Experimental subjects followed the gaze of a human experimenter to distant locations and around barriers. Sometimes they looked back at the experimenter when finding no target in empty space, apparently double-checking on where the human was looking. When the experimenter gazed at a point behind a barrier, the apes changed position to look at that point. This behavior goes beyond looking in the same direction as others because this is a useful cue. The apes conformed to a perspective-taking model in which the observer *understands* that when others look in a certain direction it is because *they* are seeing something interesting.

Summary

Gibbons live in Asian tropical forests and use two forms of brachiation: pendular and ricochetal. Most gibbons are ripe fruit specialists. They avoid food competition through midday foraging and early retirement to sleeping places. Gibbons are monomorphic and tend to be socially monogamous, though group composition varies and there may be some sexual polygamy. Gibbons sing complex songs, often duets, which may be associated with sex or bonding or defense against conspecifics or predators.

Young gibbons seem to learn singing, but pairing may be an innate tendency. Juvenile food begging may be for play or bonding. Parents may evict dispersing young or help them to establish new territories. Parents sometimes tolerate delayed dispersal, possibly because young adults help with infants or defense. Significant cognitive performances come mostly from captives, but there is evidence for referential signaling in the wild.

Great apes are large, sexually dimorphic, and largely quadrumanous. They use complex food processing techniques, make nests, and use tools. Voluntary gestural communication facilitates social groups and networks larger than those of gibbons. Relationships with non-kin are prominent. All life cycle stages are long. Immatures spend years closely associated with their mothers. The play repertoire is large and diverse. First pregnancy typically occurs at 10 years and a male reaches full physical and social maturity at about 13. Diverse evidence indicates higher cognitive levels than other primates in both physical and social domains. Some researchers consider ape–human cognitive comparisons to be anthropomorphic, but others accept mental images as part of great ape cognition.

Sources

Basic Bibliography

Bartlett, T.Q. (2011). The Hylobatidae. In C.A. Campbell et al. (Eds.), *Primates in perspective*, 2nd ed. (pp. 300–312). NY: OUP.

Chivers, D.J. (2005). Gibbons. In J. Caldecott & L. Miles (Eds.), *World atlas of great apes and their conservation* (pp. 205–214). Berkeley: UCP.

Gursky-Doyen, S., & Supriatna, J. (Eds.) (2010). *Indonesian primates*. NY: Springer. [Chapters 5 through 9]

Lappan, S., & Whittaker, D.J. (Eds.) (2009). *The gibbons, new perspectives on small ape socioecology and population biology*. NY: Springer. [Chapters 8 through 17]

Sodaro, V., & Sodaro C. (Eds.) (2001). *The apes: challenges for the 21st century*. Chicago: Brookfield Zoo.

Additional References

Ahsan, M.F. (2001). Socio-ecology of the hoolock gibbon (*Hylobates hoolock*) in two forests of Bangladesh. In *The apes: challenges for the 21st century* (pp. 286–299). Brookfield IL: Brookfield Zoo.

Ahsan, M.F. (2004). Infant behaviour and development in hoolock gibbon (*Bunopithecus hoolock hoolock*). *Zoos' Print Journal*, *19*, 1435–1436.

Asensio, N., et al. (2011). Gibbon travel paths are goal oriented. *AC*, *14*, 395–405.

Badraun, J.C., et al. (1998). Hand modulation of vocalization in siamangs *Hylobates syndactylus*. *International Zoo Yearbook*, *36*, 84–89.

Barelli, C., et al. (2008a). Female white-handed gibbons (*Hylobates lar*) lead group movements and have priority of access to food resources. *Behaviour*, *145*, 965–981.

Barelli, C., et al. (2008b). Mating patterns and sexual swellings in pair-living and multimale groups of wild white-handed gibbons, *Hylobates lar*. *AB*, *75*, 991–1001.

Barelli, C., et al. (2011). Is grooming used as a commodity in wild white-handed gibbons, *Hylobates lar*? *AB*, *82*, 801–809.

Barelli, C., et al. (2013). Extra-pair paternity confirmed in wild white-handed gibbons. *AJP*, *75*, 1185–1195.

Bartlett, T.Q. (2011). The Hylobatidae. In C.J. Campbell, et al. (Eds.), *Primates in perspective* (2nd ed.) (pp. 300–312). NY: OUP.

Bertram, J.E.A. (2004). New perspectives on brachiation mechanics. *YPA*, *47*, 100–117.

Bräuer, J., Call, J., & Tomasello, M. (2005). All great ape species follow gaze to distant locations and around barriers. *Journal of Comparative Psychology*, *119*, 145–154.
Brockelman, W.Y. (2009). Ecology and social system of gibbons. In S. Lappan & D.J. Whittaker (Eds.), *The gibbons* (pp. 211–239). NY: Springer.
Brockelman, W.Y., et al. (1998). Dispersal, pair formation and social structure in gibbons (*Hylobates lar*). *BE&S*, *42*, 329–339.
Burns, B.L., et al. (2011). Social dynamics modify behavioural development in captive white-cheeked (*Nomascus leucogenys*) and silvery (*Hylobates moloch*) gibbons. *Primates*, *52*, 271–277.
Burrows, A.M., et al. (2011). Evolution of the muscles of facial expression in a monogamous ape: evaluating the relative influences of ecological and phylogenetic factors in hylobatids. *AR*, *294*, 645–663.
Byrne, R.W. (2004). The manual skills behind hominid tool use. In A.E. Russon & D.R. Begun (Eds.), *Evolutionary origins of great ape intelligence* (pp. 31–44). New York: Cambridge University Press.
Call, J., & Tomasello, M. (Eds.) (2007). *The gestural communication of apes and monkeys*. Manhaw, NJ: LEA.
Carbone, Lucia, et al. (2014). Gibbon genome and the fast karyotype evolution of small apes. *Nature*, *513*, 195–201.
Cheyne, S.M., Chivers, D.J., & Sugardjito, J. (2007). Covariation in the great calls of rehabilitant and wild gibbons (*Hylobates albibarbis*). *The Raffles Bulletin of Zoology*, *55*, 201–207.
Chivers, D.J. (2001). The swinging singing simians: fighting for food and family in Far East forests. In V. Sodaro & C. Sodaro (Eds.), *The apes: challenges for the 21st century* (pp. 1–27). Chicago: Brookfield Zoo.
Clarke, E.A., et al. (2006). The syntax and meaning of wild gibbon songs. *PLONE*, *1*, e73.
Clemens, Z., et al. (2008). Observations on paternal care in a captive family of white-handed gibbons (*Hylobates lar*). *Gibbon Journal*, *4*, 46–50.
Cowlishaw, G. (1992). Song function in gibbons. *Behaviour*, *121*, 131–153.
Cunningham, C.L., et al. (2006). Object manipulation to obtain a food reward in hoolock gibbons, *Bunopithecus hoolock*. *AB*, *71*, 621–629.
Cunningham, C.L., et al. (2011). A sex difference in effect of prior experience on object-mediated problem-solving in gibbons. *AC*, *14*, 599–605.
Fan, P.-F., and Jiang, X.-L. (2009). Predation on giant flying squirrels (*Petaurista philippensis*) by black crested gibbons (*Nomascus concolor jingdongensis*) at Mt. Wuliang, Yunnan, China. *Primates*, *50*, 45–49.
Fan, P., et al. (2010). Social structure and group dynamics of the cao vit gibbon (*Nomascus nasutus*) in Bangliang, Jingxi, China. *FP*, *81*, 245–253.
Fedor, A., et al. (2008). Object permanence tests on gibbons (*Hylobatidae*). *JCP*, *122*, 403–417.
Fleagle, J.G. (2013). *Primate adaptation and evolution* (3rd ed.). New York: Elsevier Science.
Fruth, B., & Hohmann, G. (1996). Nest building behavior in the great apes: the great leap forward? In W.C. McGrew, L.F. Marchant, & T. Nishida (Eds.), *Great ape societies* (pp. 225–240). NY: CUP.
Geissmann, T. (1991). A reassessment of age of sexual maturity in gibbons (*Hylobates* spp.). *AJP*, *23*, 11–22.
Geissmann, T. (2002). Duet-splitting and the evolution of gibbon songs. *Biological Reviews*, *77*, 57–76.
Geissmann, T. (2008a). Gibbon paintings in China, Japan, and Korea: historical distribution, production rate and context. *Gibbon Journal*, *4*, 1–38.
Geissmann, T. (2008b). Inter-group conflict in captive siamangs (*Symphalangus syndactylus*). *Gibbon Journal*, *4*, 51–55.
Geissmann, T., & Anzenberger, G. (2009). Hormonal correlates of the ovarian cycle in the yellow-cheeked crested gibbon (*Nomascus gabriellae*), and a review of ovarian cycles in gibbons (Hylobatidae). *Gibbon Journal*, *5*, 61–73.
Geissmann, T., & Orgeldinger, M. (2000). The relationship between duet songs and pair bonds in siamangs, *Hylobates syndactylus*. *AB*, *60*, 805–809.
Geissmann, T., et al. (2009). Longevity in gibbons (*Hylobatidae*). *Gibbon Journal*, *5*, 81–92.

Groves, C. (2000). *Primate taxonomy*. Washington DC: Smithsonian Institution Press.
Haimoff, E.H. (1986). Convergence in the duetting of monogamous Old World primates. *JHE, 15*, 51–59.
Heschel, A., & Fuchsbichler, C. (2009). Siamangs (*Hylobates syndactylus*) recognize their mirror image. *IJCP, 22*, 221–233.
Islam, M.A., & Feeroz, M.M. (1992). Ecology of hoolock gibbon of Bangladesh. *Primates, 33*, 451–464.
Lappan, S., & Morino, L. (2014). Mating in the presence of a sexual competitor: audience effects may promote male social tolerance in polyandrous siamang (*Symphalangus syndactylus*) groups. *Behaviour, 151*, 1067–1089.
Leighton, D.R. (1987). Gibbons: territoriality and monogamy. In B.B. Smuts et al. (Eds.), *Primate societies* (pp. 135–145). Chicago: UCP.
Lewis, K. (2005). Social play in the great apes. In A.D. Pellegrini & P.K. Smith (Eds.), *The nature of play: great apes and humans*. New York: Guilford Press.
Liebal, K. (2007). Gestures in siamangs (*Symphalangus syndactylus*). In J. Call & M. Tomasello (Eds.), *The gestural communication of apes and monkeys* (pp. 131–158). Mahwah NJ: Lawrence Erlbaum Associates.
Malone, N., & Fuentes, A. (2009). Ecology and evolution of hylobatid communities. In S. Lappan & D.J. Whittaker (Eds.), *The gibbons* (pp. 241–264). NY: Springer.
Marshall, A.J., Cannon, C.H., & Leighton, M. (2009). Competition and niche overlap between gibbons (*Hylobates albibarbis*) and other frugivorous vertebrates in Gunung Palung National Park, West Kalimantan, Indonesia. In S. Lappan & D.J. Whittaker (Eds.), *The gibbons* (pp. 161–188). NY: Springer.
Meyer, T.J., et al. (2012). An alu-based phylogeny of gibbons (Hylobatidae). *Molecular Biology and Evolution, 29*, 3441–3450.
Mitani, J.C., & Rodman, P.S. (1979). Territoriality: the relation of ranging pattern and home range size to defendability, with an analysis of territoriality among primate species. *BE&S, 5*, 241–251.
Mootnick, A.R., & Nadler, R.D. (1997). Sexual behavior of maternally separated gibbons (*Hylobates*). *Developmental Psychobiology, 31*, 149–161.
Mootnick, A.R., et al. (2005). Familiarity during immaturity: implications for the captive propagation of gibbons. *IJP, 26*, 1417–1433.
Morino, L. (2010). Clouded leopard predation on a wild juvenile siamang. *FP, 82*, 362–368.
Nettelbeck, A.R. (1998a). Observations on food sharing in wild lar gibbons (*Hylobates lar*). *FP, 69*, 386–391.
Nettelbeck, A.R. (1998b). Encounters between lar gibbons (*Hylobates lar*) and binturongs (*Arctictis binturong*). *FP, 69*, 392–396.
Noser, R., & Byrne, R.W. (2014). Change point analysis of travel routes reveals novel insights into foraging strategies and cognitive maps of wild baboons. *AJP, 76*, 399–409.
Palombit, R.A. (1995). Longitudinal patterns of reproduction in wild female siamang (*Hylobates syndactylus*) and white-handed gibbons (*Hylobates lar*). *IJP, 16*, 739–760.
Palombit, R.A. (1996). Pair bonds in monogamous apes: a comparison of the siamang *Hylobates syndactylus* and the white-handed gibbon *Hylobates lar*. *Behaviour, 133*, 321–356.
Palombit, R.A. (1997). Inter- and intraspecific variation in the diets of sympatric siamang (*Hylobates syndactylus*) and lar gibbons (*Hylobates lar*). *FP, 68*, 321–337.
Paluck, R.J., et al. (1970). Formation and development of a group of juvenile *Hylobates lar*. *Primates, 11*, 185–194.
Rafacza, M.L., et al. (2012). Hormonal correlates of paternal care differences in the Hylobatidae. *AJP, 74*, 247–260.
Rafacza, M.L., et al. (2013). Hormonal and behavioral patterns of reproduction in female hylobatids. *Animal Reproduction Science, 137*, 103–112.
Reichard, U. (1998). Sleeping sites, sleeping places, and presleep behavior of gibbons (*Hylobates lar*). *AJP, 46*, 35–62.

Reichard, U.H., & Barelli, C. (2008). Life history and reproductive strategies of Khao Yai *Hylobates lar*: implications for social Evolution in Apes. *IJP*, *29*, 823–844.

Reichard, U., & Sommer, V. (1994). Grooming site preferences in wild white-handed gibbons (*Hylobates lar*). *Primates*, *35*, 369–374.

Robson, S.L., & Wood, B. (2008). Hominin life history: reconstruction and evolution. *Journal of Anatomy*, *212*, 394–425.

Russon, A.E., Handayani, P., Kuncoro, P. & Ferisa, A. (2007). Orangutan leaf-carrying for nest building: unravelling cultural processes. AC, *10*, 189–202.

Savini, T., et al. (2009). Varying ecological quality influences the development of polyandry in white-handed gibbons (*Hylobates lar*). *Biotropica*, *41*, 503–513.

Schessler, T., & Nash, L.T. (1977). Food sharing among captive gibbons (*Hylobates lar*). *Primates*, *18*, 677–689.

Scheumann, M., & Call, J. (2006). Sumatran orangutans (*Pongo abelii*) and a yellow-cheeked crested gibbon (*Nomascus gabriellae*) know what is where. *IJP*, *27*, 575–602.

Stewart, F.A., & Pruetz, J.D. (2013). Do chimpanzee nests serve an anti-predatory function? *AJP*, *75*, 593–604.

Stewart, F.A., et al. (2007). Do chimpanzees build comfortable nests? *AJP*, 69, 930–939.

Suddendorf, T., & Collier-Baker, E. (2009). The evolution of primate visual self-recognition: evidence of absence in lesser apes. *PRSB*, *276*, 1671–1677.

Suddendorf, T., & Whiten, A. (2001). Mental evolution and development: evidence for secondary representation in children, great apes, and other animals. *Psychological Bulletin*, *127*, 629–650.

Suddendorf, T., & Whiten, A. (2003). Reinterpreting the mentality of apes. In J. Fitness and K. Sterelny (Eds.), *From mating to mentality: evaluating evolutionary psychology* (pp. 173–196). New York: Psychology Press.

Tenaza, R., & Tilson, R.L. (1985). Human predation and Kloss's gibbon (*Hylobates klossii*) sleeping trees in Siberut Island, Indonesia. *American Journal of Primatology*, *8*, 299–308.

Thinh, V.N., et al. (2010). Mitochondrial evidence for multiple radiations in the evolutionary history of small apes. *BMCEB*, *10*, *74*.

Tilson, R.L. (1979). Behaviour of hoolock gibbon (*Hylobates hoolock*) during different seasons in Assam, India. *Journal of the Bombay Natural History Society*, *76*, 1–16.

Tomasello, M., & Call, J. (1997). *Primate cognition*. New York: Oxford University Press.

Tomasello, M., & Herrmann, E. (2010). Ape and human cognition: what's the difference? *Current Directions in Psychological Research*, *19*, 3–8.

Tuttle, R.H. (2014). *Apes and human evolution*. Cambridge: Harvard University Press.

Uhde, N.L., & Sommer, V. (2002). Antipredatory behavior in gibbons (*Hylobates lar*, Khao Yai/Thailand). In L.E. Miller (Ed.), *Eat or be eaten* (pp. 268–291). NY: CUP.

Ujhelyi, M., Merker, B., Buk, P., and Geissmann, T. (2000). Observations on the behavior of gibbons (*Hylobates leucogenys*, *H. gabriellae*, and *H. lar*) in the presence of mirrors. *JCP*, *114*, 253–262.

van Casteren, A., et al. (2012). Nest-building orangutans demonstrate engineering know-how to produce safe, comfortable beds. *PNAS*, *109*, 6873–6877.

Whitten, A.J. (1982). The ecology of singing in Kloss gibbons (*Hylobates klossii*) on Siberut Island, Indonesia. *IJP*, *3*, 33–51.

Zihlman, A.L., Mootnick, A.R., and Underwood, C.E. (2011). Anatomical contributions to hylobatid taxonomy and adaptation. *IJP*, *32*, 865–877.

23 Great Apes of Asia

Orangutans

A vaguely orange shape looms in the trees. The guide gets a questioning look from one of the tourists and says, "Yes, that is the one many still call the Old Man of the Forest. But I think now it must be the Old Person of the Forest."

Orangutans are the great apes of Asia. Their ancestors occupied almost every suitable habitat throughout Asia, but orangutans now survive only in pockets on the islands of Sumatra and Borneo. They are the most sexually dimorphic living primates except for mandrills. Averaging about 80kg, males are more than twice the size of females. Males have cheek pads that may function to increase apparent size and/or modify vocalizations (Figure 23.1). There are at least two species (Box 23.1).

Box 23.1 Orangutan Taxonomy

After a long period in which orangutans were considered a single species, genetic studies demonstrated two species that are at least roughly associated with two major islands of Indonesia: *Pongo pygmaeus* (Borneo) and *Pongo abelii* (Sumatra). There may be some relevant behavior patterns, as suggested in the text. Many taxonomists support the division of *Pongo pygmaeus* into three subspecies: *P.p. pygmaeus*, *P.p. morio*, and *P.p. wurmbii*.

Orangutan Behavior

Orangutan Ecology

The densest orangutan populations occur in rainforests with annual precipitation of about 2000mm to 4500mm, especially peat and freshwater swamps and alluvial forests in river valleys. Orangutans also occupy dry lowland and hill forests, but at low densities. They are the most arboreal of the great apes and the heaviest forest canopy animals in the world. They have the longest arms, hands, and feet of the great apes, and flexible limb joints give them a wider range of motion (Figure 23.1). Each leg can be extended at an angle of over 90 degrees.

All age-sex categories use the same basic positional behaviors because they all follow the same arboreal pathways (Manduell et al. 2012; Thorpe et al. 2009). About 75 percent of orangutan locomotion occurs with the body in an orthograde position, either in quadrumanous

Figure 23.1 Orangutan feeding in a suspensory position. © Rhett A. Butler/www.mongabay.com

climbing or arm-swinging. Movements are usually slow and careful. Adult females are the most cautious, perhaps because they are usually carrying an infant or guiding an older offspring along the safest paths. However, orangutans are capable of rapid movement, such as leaping and tumbling, especially when escaping from agonistic situations.

Orangutans use unique strategies to control and exploit compliant branches and trunks (Thorpe et al. 2007; van Casteren et al. 2013). The apes sway them to bridge gaps and allow transfer from one to another with less energy expenditure than jumping. They often use pronograde suspension in the most compliant branches to reach essential fruits and leaves, which are most abundant in the terminal zones of trees. They hang under a branch (facing either toward or away from it), bearing weight with three limbs, and reaching for food with the fourth (Myatt and Thorpe 2011). Orangutans seem almost as comfortable eating while upside down as they do when right side up.

The fruit supply in Southeast Asia varies more in time and space than in any other rainforests. **Mast fruitings** take place every two to ten years, apparently driven by climatic events associated with the El Niño phenomenon. During these irregular events the majority of trees in a forest (sometimes almost 90 percent) bear fruit simultaneously, but a period of scarcity usually follows. There is also more variation between and within annual cycles in Southeast Asian forests than elsewhere in the world. Orangutans experience prolonged periods of negative energy balance on a regular basis (Harrison et al. 2010).

Wild orangutans mitigate the effect of food fluctuations by feeding on twice as many plant genera as other great apes. They also have the ability, unusual among primates, to store excess food as fat. This accounts for a greater tendency toward obesity in captivity compared to other apes. It has been hypothesized that the large size of orangutans as a species is needed to store a sufficient amount of fat. A study of captive hybrid subjects suggests that the genus has adapted with a metabolism that uses a low level of energy (Pontzer et al. 2010). Daily energy expenditure (DEE) relative to body mass was lower than that of any of the mammals in a large sample, other than sloths. It is important to note that these captive orangutans engaged in an activity routine comparable to that of congenerics in the wild.

Sumatran forests are more consistently productive than those of Borneo (Wich et al. 2011). Fruit is never less than half of the diet for Sumatran orangutans while the proportion may fall to 10 percent in Borneo. A five-year study of urinary metabolites in the Bornean apes (Vogel et al. 2011) found that protein intake was at a level inadequate for humans (one tenth the intake of mountain gorillas). Bornean orangutans seem to have adapted to consumption of tougher foods as fallbacks with more robust mandibles and perhaps thicker tooth enamel (Vogel et al. 2014).

Orangutans use two alternative subsistence strategies. They *search-and-find*, moving and feeding continuously when there is a regular supply of fruit. This strategy is most effective in peat swamps, and in forests where figs are abundant. They *sit-and-wait* during periods of low fruit availability, resting as much as possible to conserve energy and allow for digestion of low quality fallback foods.

Strangling figs are an important fallback for orangutans wherever they are available. More common fallbacks include herbaceous and woody vegetation, seeds, unripe or acidic fruit, and leaves. Orangutans have simple stomachs, but their large bodies help with leaf digestion and can tolerate a certain amount of toxic material. Use of less desirable foods diminishes scramble competition with sympatric primates, especially the highly frugivorous gibbons. In contest competition size gives orangutans dominance over other primates.

Insects and honey are readily available to these highly arboreal apes and they sometimes endure mass attacks by stinging or biting species to obtain these foods. Orangutans occasionally catch and eat slow lorises and perhaps a few other vertebrates (Hardus et al. 2012). Loris captures occurred during periods of fruit scarcity, indicating some nutritional significance. Females spend more time than males in obtaining insects, and females figure prominently in the few known cases of vertebrate predation, which suggests that extra protein is related to reproductive needs.

Orangutan foraging uses both skill and strength. They get social insects by smashing their nests and use their strong hands skillfully to penetrate mechanically defended foods such as seeds and palm hearts. Orangutans obtain some insects, as well as honey, with sticks or vegetation wielded as tools. Similarly, some use tools to pry out palm hearts and other edible portions of plants. One orangutan covered its head with large leaves while raiding a hive for honey.

Like nocturnal strepsirrhines, orangutans solve recurring subsistence problems with solitary foraging. The proportion of time spent in all social interactions, including mating, can be as little as 1.3 percent. Some other anthropoids also have low rates of social interaction (colobines, for example), but they maintain daily proximity to one another while orangutans do not.

Like other highly arboreal primates, such as New World monkeys, Bornean orangutans sometimes come to the ground. Observers may have inhibited such behavior in the past, but it has been documented by a camera-trap study (Ancrenaz et al. 2013; Loken et al. 2013). All age–sex classes have been recorded. Various resources may provide a general explanation, but terrestriality may be especially important in male dispersal.

Orangutan nests seem uniquely sophisticated (van Casteren et al. 2012). These apes choose strong, rigid tree branches for the structural parts of the nests that support their weight and weaker, more flexible branches for the nest's linings. They break the structural branches halfway across, leaving them attached, while they completely sever the branches used for lining, which suggests that they understand the mechanical properties of wood. Orangutans sometimes add one or more final touches to enhance comfort, such as a blanket, roof, or pillow. The pillow, added to most night nests, is made from small twigs that are bitten lightly as if to blunt the sharp ends. Comfort also seems to be a factor in the choice of stable trees with stilt roots or buttresses (Cheyne et al. 2013). Closer study of captive orangutans demonstrated an association between sleep quality and more complex and comfortable beds (Samson and Shumaker 2013).

Orangutan Sociality

Orangutan social interactions are more limited than in any other anthropoids. An individual may spend days or weeks out of contact with conspecifics. This separation probably results from intense competition for food, which usually takes the scramble form since individuals rarely come together. However, contest competition may appear when females aggregate in a large fig tree. Adult females have individual home ranges that are fairly stable over time. Daughters tend to establish ranges overlapping with those of their mothers, resulting in recurring associations, and other related females may interact frequently. In others words there is a loose female philopatry that results in female clusters defined by social interaction and kinship (van Noordwijk et al. 2012). Maternally-related Bornean females met more often than unrelated females with similar range overlaps. They associated longer and displayed some feeding tolerance, though affiliative behavior was minimal. Related females allowed and even facilitated play between their offspring, while they sometimes interfered when offspring played with non-relatives.

Social interaction between adult orangutans may be no more than a glance (Kaplan and Rogers 2002; van Schaik and Aureli 2000). Glances, however, can be significant. Videotape analysis shows that orangutans pay attention to both humans and other orangutans by looking at them sideways while the head is turned at an angle away from the subject. They often maintain visual contact with repeated glances toward an individual or object of interest. Such glances are often one of the few indicators of a social relationship in wild orangutans. This work suggests that orangutan attention structure is based on age rather than hierarchy.

Male home ranges are at least three times larger than those of females. Some males were called *wanderers*, supposedly having no fixed association with place. However, male home ranges are so large that most of them extend beyond the boundaries of any feasible study area. So-called wanderers may have fairly well defined ranges that are not apparent to researchers. The longer that research goes on at a particular site, the more likely it is that supposed transients will reappear. The wanderer supposition illustrates the need for long-term studies, and provides another caution against making inferences without direct evidence.

Flanged males are those with full growth of facial flanges and other adult male adornments (Figure 23.2). They rarely associate with other orangutans except when mating, and they usually avoid each other. Meetings are likely to result in serious fights that involve grappling and biting.

A flanged male usually tolerates unflanged males within his range and in proximity to him, sometimes as close as 20 meters, even when in consort. Genetic analysis at one site dispelled the obvious hypothesis that this is due to kinship. Such associations are based on nothing more than tolerance; there is no affiliative behavior involved. Expulsion of unflanged

Figure 23.2 Adult male orangutan face with flanges. © Rhett A. Butler/www.mongabay.com

males, who are smaller and more agile, may not be worth the energetic cost. Unflanged males are generally tolerant of one another and often travel together. They sometimes display affiliative behavior, such as play and isosexual interactions. They may associate with females at the same time with little or no conflict between them.

Orangutans have more calls and other sound signals (30 or more) than any other great ape except the bonobo. This may seem anomalous in light of orangutans' limited sociality. The repertoire might be a relict of a time when orangutans were more social (phylogenetic inertia), but the nature of the signals suggests that they are functional in the present. Many encounters between the apes are agonistic and more than half of the signals are associated with behavioral categories such as fighting, intimidation, annoyance, confrontation, and fear. Only a third of orangutan signals convey affiliative relations between adults.

The long call, uttered only by flanged males, may exceed three minutes in length and can be heard for more than a kilometer in dense forest (Spillman et al. 2010). It is produced with full inflation of the laryngeal sac, acting something like a bagpipe. One possible function is to warn away other males. The dominant flanged male in a region emits loud calls at the highest rate and males that heed them can avoid altercations. The calls may occur spontaneously, in response to another male's long call, or in response to a disturbance such as a falling tree. Females can distinguish the individual and the circumstances. Those with dependent offspring move away from spontaneous calls, but ignore the ones elicited by disturbances. Females without offspring may be attracted.

Around the time of ovulation a female seeks out the dominant male and solicits sex. Proceptive behavior is important because there is no swelling or other visual indicator of fertility. Matings with flanged males take place in the context of consortships that last for several days or more. The male closely guards the female and reacts aggressively to the presence or calls of other males. Unflanged males are likely to be chased away. A cost of consort is an adjustment of the male's activity budget to accommodate the longer day range of the female. Another is the risk of conflict with another flanged male, resulting in significant energy expenditure and possible injury. Given the costs of mating, flanged males tend to be somewhat selective about their partners. Like other primate males, they readily accept parous females but sometimes reject nulliparous ones. Matings are more common during high fruit availability. Measurement of estrogen levels suggests that conception is less likely when food is scarce.

Females are less proceptive toward subordinate flanged males, and ordinarily have no interest in unflanged males. Both kinds of males sometimes engage in physical coercion that is commonly called **forced mating** (the term *rape* has anthropocentric connotations). Female orangutans usually resist strenuously, screaming and trying to escape, but they are successful in only about 10 percent of cases. This degree of sexual coercion is virtually unique among primates and is probably possible because the dispersed nature of orangutan life leaves a female with few allies. Females are never wounded during these encounters, which indicates that they function as potential reproductive acts on the part of the males (rather than being a pathology, as some interpret it).

The fact that females are not injured during forced mating allows an alternative explanation: these copulations are not really coercive, but rather a kind of wrestling match in which the female can evaluate the health and vigor of a prospective mate. This is the Mate Assessment Hypothesis. Even if the female's proximate motivation is entirely negative, the same ultimate explanation might apply. The successful male, if he also becomes a father, has imparted good genes to the female's offspring. Supposedly this argument fails because females provide no resistance to the dominant males that are their preferred partners.

However, the dominant flanged males have already proven themselves; from the female viewpoint, only non-dominants need testing. The strongest argument against the Mate Assessment Hypothesis is probably that females scream during an assault, a behavior that will summon the dominant male (if close enough) to drive the other away. If the function of these interactions is to test male quality for mating, the proximate response of females should be to struggle without vocalizing.

Female orangutans have at least two other possible mechanisms of choice. One is to test a male's aggressive tendencies in the context of food transfer. A female screams loudly when a male refuses to tolerate her taking some of his food or tries to take food from her, which brings the association to an early conclusion. The foods involved are equally available to both sexes, so the food itself is not the issue. A second choice mechanism is early termination of a sexual interaction. This is effective if ejaculation by male orangutans takes a relatively long time. Forced matings typically end after about eight minutes while voluntary copulations may last as long as half an hour.

If forced copulation is undesirable for females (rather than a female choice mechanism), they may need protection. To secure a protector a female should raise his certainty of paternity, which orangutan females do by showing a strong preference for one particular dominant male. Research at one site showed that females who associated with a flanged male received less sexual harassment from unflanged males (Fox 2002). They seemed to travel rapidly toward the dominant male's loud calls when they were pursued.

Females may also prefer the dominant male for protection against infanticide. This is rendered dubious by the fact that no male infanticides or attempts have been observed at any orangutan research site during about four decades and 80,000 hours of study (Beaudrot et al. 2009). The counter-argument is that infanticide is rarely observed in any primate species, but is indicated by female counter-strategies (Knott et al. 2010). A possible example in orangutans is that females willingly copulate with subordinate males during periods of rank instability, which can be interpreted as creating paternity confusion. Observation of infanticides (if they exist) in orangutans is said to be especially unlikely because the ape's low reproductive rate means the infant cohort is very small at any given time. By the same token, a single infanticide could have an enormous impact on reproductive success. The orangutan's slow life history makes this male strategy seem likely.

Why do both forms of adult male orangutan exist? Large-bodied males have an advantage in direct mating competition. However, the large body is a metabolic challenge in itself and raises the energetic cost of travel. The reproductive strategy of flanged males entails aggressiveness against peers that increases the risks of injury and death. The high level of testosterone required for the flanged state may in itself shorten life. Unflanged males may have some reproductive success and at less cost. The concept of the Evolutionarily Stable Strategy may apply. If all males are unflanged, then one who becomes flanged has a great reproductive advantage. When others become flanged, there are eventually so many that the advantage of each one is greatly diminished. Then it becomes more effective to stay in the unflanged state. Eventually the number of flanged males will diminish and it will again be to the advantage of males to assume that state. The evolutionary result is a balance between the two approaches to reproductive success.

Voluntary copulation among orangutans has several unusual features, beginning with a lengthy period of diversified foreplay. Partners may sit next to each other, touching casually. A male (especially a subadult) may put his arm around the female and touch her face with his lips. Males also lick and fondle the female genitalia. The female similarly stimulates the male genitals, aids intromission, and performs her own pelvic thrusts during intercourse.

Orangutans are the only nonhuman primates in which copulation is always carried out face to face. Couples assume various positions, with one or the other partner reclining, sitting, or hanging from a tree limb, but the partners are always able to look at each other. There is time for it, because orangutan copulations are the longest among nonhuman primates. The average is about ten minutes, but the act may last as long as half an hour. Females occasionally scream near or at the end, suggesting that they may experience orgasm.

Orangutan Life Cycle

Orangutan life history is unusually slow, characterized by a large body, late maturation, low reproductive rate, and long life. A female gives birth in a nest large enough for her to fully stretch out. The nest is high in the canopy and the birth takes place at night, both of which may be defenses against predators. The height of the nest protects against terrestrial carnivores while darkness provides a shield against raptors that might snatch the newborn.

The infant is in constant contact with its mother during the first four months of its life, but contact time has been reduced to about 25 percent by the age of one year (Munn and Fernandez 2006). The infant rides its mother ventrally for nearly a year and may still be riding at the age of two. In the latter part of this phase a mother usually carries her infant cradled on the hip, after which the infant shifts to her back (Figure 23.3).

The infant develops climbing and swinging skills by the age of two years, but sometimes moves while holding the hand of another individual (Munn and Fernandez 2006). For some time after the riding phase the infant is likely to follow its mother closely, grasping the hair on her rump. The three-year-old has achieved a high level of locomotor competence, though it still needs help in crossing large gaps in the canopy (van Noordwijk and van Schaik 2005). It begins to spend some of its time in a tree separate from the one occupied by its mother. Infant mortality rates are low, presumably because of a completely arboreal existence in the close company of large mothers.

Infants show great interest in their mothers' nest building from an early age. They begin to practice at about six months and may make numerous attempts in a single day. By the age of three years an orangutan can construct real nests, as opposed to the earlier ones made in play, and use them for brief rests (van Noordwijk and van Schaik 2005). At this age the youngster makes population-specific adult nesting sounds as it works. At some research sites infants can make rain hats without help by one or two years. The infant plays with its mother, and with peers when they are available.

Weaning begins between ages five and six, sometimes precipitating conflict. Weaning is typically completed when the youngster is seven years old, and the mother occasionally shows intolerance with regard to other kinds of food. Around this time the mother stops playing with her offspring, daytime proximity begins a rapid decline, and the two sleep in separate nests. The youngster now has an energy budget like that of an adult.

When an offspring is between ages seven and nine, the mother bears another infant. Proximity between her and the juvenile gradually decreases to adult levels. By the age of 10 or 11, the youngster is an adolescent that has added ranging competence to its other skills and is ready to transition into adult society (van Noordwijk and van Schaik 2005). The long period of dependence may be caused by late acquisition of ranging competence or by prolonged need for protection.

Adolescent female orangutans are particularly sociable. If not with their mothers, they associate with each other or with adolescent males. These young females feed, play, or just sit together. When resources are scarce temporary feeding aggregations sometimes form at

Figure 23.3 Female orangutan with an infant riding dorsally. © Rhett A. Butler/www.mongabay.com

fruiting trees, and juveniles in these groups are likely to play with one another. Adolescent males sometimes associate with their mothers and sometimes travel with each other, but eventually they isolate themselves.

Females reach sexual maturity at age 10 or 11 years and begin to pursue dominant males just as adult females do. A female typically gives birth for the first time at about age 12. Data from a thousand captive female orangutans indicates that the age of 10 years is pivotal. More than half (57 percent) of younger females died within 30 days of having their first infant, while there were no deaths among females whose first birth occurred when they were older than 10. Early parity seems to result in complications during and after parturition (Cocks 2007b), suggesting that adolescent sterility is especially important for orangutans.

The orangutan IBI of seven to nine years, the longest of any mammal, may result from repeated suppression of ovarian function by environmental conditions (Knott and Kahlenberg 2007). Hormone levels fall during low fruit availability, decreasing the probability of conception at a time when the potential mother would have trouble getting the energy needed for pregnancy. The inter-birth interval is shorter in captive orangutans with reliable food supplies (Cocks 2007b). It is not clear whether depression of ovarian function was favored by natural selection or is an unavoidable consequence of external conditions.

The long IBI may be adaptive for orangutans. Captive females with shorter IBIs (less than four years) have a significantly lower survival rate, which may result from long-term physiological stress (Cocks 2007b). An alternative functional explanation links the long IBI of wild orangutans to solitary foraging and the long maturation period (van Noordwijk and van Schaik 2005). One component of this hypothesis is the young orangutan's need to learn a variety of skills before it can function on its own. On the other hand, the mother cannot have her foraging impaired by a dependent juvenile while she is carrying a young infant. By weaning the juvenile close to the age of complete independence, she increases its chances for survival when she turns her attention to the new offspring.

In male maturation the pivotal issue is when or whether to make the irreversible transition from unflanged to flanged. Males can remain in this stage of development until at least the age of 20 years, but an undeveloped male can become flanged in less than a year with a rush of testicular steroids and growth hormones. Unflanged status is a species-specific morphological stage for adult males that avoids both the stresses of becoming flanged and the costs of being fully flanged (Maggioncalda et al. 2002). It is not an alternative path, since all males eventually develop fully adult characteristics; instead, all male orangutans have a long and highly variable maturation period (Delgado and van Schaik 2000). Unflanged males have reproductive opportunities because the levels of male hormones in their bodies are enough to cause spermatogenesis, though not enough to trigger maturation of secondary sexual characteristics.

Maturation seems to be connected with the reproductive prospects of full adulthood. Captive studies show that the presence of a single fully mature male does not necessarily prevent the transition. It may be that the number of fully flanged males in a natural habitat determines whether or not the transition will be to the advantage of an unflanged male. The proximate stimulus could be the density of calling males. It is not chronic stress due to subordination that delays maturation. A captive study of hormone levels showed that developing males, rather than unflanged ones, experience elevated levels of stress.

Male vulnerability is an important feature of orangutan life history. Birth ratios display a slight bias in favor of males, but this is drastically reversed among adults. Males are occasionally killed by predators and perhaps more often by each other, but the data are not sufficient to demonstrate the most important causes of male mortality.

Orangutan Cognition

Wild orangutans use a variety of tools. Those used for subsistence probably have the greatest functional significance, but several more are used for self-care. Orangutan technological capabilities have been confirmed in captive experiments. Orangutans learn quickly from other animals and from humans. Rehabilitants imitate many human activities, including hammering nails, sawing logs, brushing teeth, shooting blowguns, and washing laundry. At one camp they created leaf bowls to hold a mixture of milk and bananas for slow consumption. One implication of this last example is that not all innovations stem from practical needs. Some orangutan tools may be important for subsistence, but the leaf bowl seems to be a product of pleasure seeking.

Foraging behavior other than tool use suggests cognitive capabilities. Orangutans have a prodigious memory for the locations of food sources, including those that are cryptic and/or temporary, and seem to use a mental map to take the most direct routes to these sources. The apes are also aware of seasonal variations in availability. When not using such stored information, an orangutan typically forages in a zigzag pattern (at least in Borneo). However, it readily breaks off when it perceives clues to irregular food sources. These often come in the form of animal behavior. The apes follow the flight of frugivorous birds (they seem to recognize the hornbill in particular as a fruit indicator) or the noise made by birds and mammals at a fruiting site. These indications of intelligence are borne out by numerous tests on orangutans in captivity (Delgado and van Schaik 2000). By human standards they are among the most intelligent of animals, matching and sometimes surpassing the vaunted chimpanzee.

Orangutans differ from other primates with regard to inequity aversion. Tested in the same conditions as chimpanzees, orangutans showed no inequity aversion (Brosnan et al. 2011). The researchers see cooperativeness as the key. An animal that engages in a significant amount of cooperation benefits from attention to equity in the distribution of the returns. Chimpanzees are cooperative and orangutans are not. Similarly, among New World primates, squirrel monkeys differ from capuchins in that they are gregarious but not cooperative.

Orangutans differ sharply from African apes in some other tests of mental patterns. In one experiment, for example, the subjects viewed a human demonstrator in two conditions, either freely choosing to use a tool or having to use it. Only orangutans (and enculturated chimpanzees) showed a preference for the freely chosen tool-using performance. This indicates an understanding of another individual's intentions (Buttleman et al. 2008).

Summary

Orangutans are highly arboreal despite their large size. Mast fruiting affects their food supply. They search during abundance and rest during scarcity. Insectivory is important to females. Both sexes use tools that provide evidence for cognition and culture. Adults are often alone, but maintain contact with calls. Females sometimes form social clusters. Forced copulations occur and dominant males may provide protection from sexual harassment. Orangutans have low infant mortality and very slow life history. The long IBI facilitates juvenile learning, but may be the result of sporadic ovarian suppression caused by irregular food supply. Flanged males grow to double the female size and develop secondary sexual characteristics. Only flanged males can become dominant, but undeveloped males can quickly become flanged. Western orangutans live in richer habitats, are more gregarious, and have longer life histories.

In the poorer habitats of eastern orangutans, negative energy balance seems to impair female reproduction. Male aggression and forced mating are more common. Long calls function in spacing males rather than attracting females.

Sources

Basic Bibliography

Knott, C.D., & Kahlenberg, S. (2011). Orangutans. In C. J. Campbell et al. (Eds.), *Primates in perspective* (pp. 290–305). NY: OUP.
Russon, A.E., & Begun, D.R. (Eds.) (2004). *The evolution of great ape intelligence.* NY: CUP.
Wich, S.A., et al. (Eds.) (2009). *Orangutans.* NY: OUP.

Additional References

Ancrenaz, M., et al. (2013). Coming down from the trees: is terrestrial activity in Bornean orangutans natural or disturbance driven? *Scientific Reports, 13,* Epub.
Beaudrot, L.H., et al. (2009). Why male orangutans do not kill infants. *BE&S, 63,* 1549–1562.
Bräuer, J., et al. (2005). All great ape species follow gaze to distant locations and around barriers. *JCP, 119,* 145–154.
Brosnan, S.F., et al. (2011). Responses to inequity in orangutans. *FP, 82,* 56–70.
Buttelmann, D., et al. (2008). Behavioral cues that great apes use to forage for hidden food. *AC, 11,* 117–128.
Byrne, R.W. (2004). The manual skills behind hominid tool use (pp. 31–44). In A.E. Russon & D.R. Begun (Eds.) *Evolutionary origins of great ape intelligence.* NY: CUP.
Call, J., & Tomasello, M. (2008). Does the chimpanzee have a theory of mind? 30 years later. *Trends in Cognitive Science, 12,* 187–192.
Cheyne, S.M., et al. (2013). How orangutans choose where to sleep. *Asian Primates Journal, 3,* 13–17.
Cocks, L. (2007a). Factors affecting mortality, fertility, and well-being in relation to species differences in captive orangutans. *IJP, 28,* 421–428.
Cocks, L. (2007b). Factors Influencing the well-being and longevity of captive female orangutans. *IJP, 28,* 429–440.
Delgado, R.A., & van Schaik, C.P. (2000). The behavioral ecology and conservation of the orangutan (*Pongo pygmaeus*). *EA, 9,* 201–218.
Fox, E.A. (2002). Female tactics to reduce sexual harassment in the Sumatran orangutan (*Pongo pygmaeus abelii*). *BE&S, 52,* 93–101.
Fruth, B., & Hohmann, G. (1996). Nest building behavior in the great apes: the great leap forward? In W.C. McGrew et al. (Eds.), *Great ape societies* (pp. 225–240). NY: CUP.
Hardus, M.E., et al. (2012). Behavioral, ecological, and evolutionary aspects of meat-eating by Sumatran orangutans (*Pongo abelii*). *IJP, 33,* 287–304.
Harrison, M.E., et al. (2010). Orangutan energetics and the influence of fruit availability in the nonmasting peat-swamp forest of Sabanau, Indonesian Borneo. *IJP, 31,* 585–607.
Kaplan, G.T., & Rogers, L.J. (2000). *Orangutans: their evolution, behavior, and future.* Cambridge, MA: Perseus.
Kaplan, G., & Rogers, L.J. (2002). Patterns of gazing in orangutans (*Pongo pygmaeus*). *International Journal of Primatology, 23,* 501–526.
Knott, C.D., and Kahlenberg, S. (2007). Orangutans in perspective: forced copulations and female mating resistance. In C.J. Campbell, et al. (Eds.), *Primates in perspective* (pp. 290–305). NY: OUP.
Knott, C.D., et al. (2010). Female reproductive strategies in orangutans, evidence for female choice and counterstrategies to infanticide in a species with frequent sexual coercion. *PRSB, 277,* 105–113.

Lewis, K.P. (2005). Social play in the great apes. In P.K. Smith & A.D. Pellegrini (Eds.), *The nature of play: great apes and humans* (pp. 27–53). NY: Guilford.

Loken, B., et al. (2013). Terrestriality in the Bornean orangutan (*Pongo pygmaeus morio*) and implications for their ecology and conservation. *AJP*, *75*, 1129–1138.

Maggioncalda, A.N., et al. (2002). Male orangutan subadulthood: a new twist on the relationship between chronic stress and developmental arrest. *AJPA*, *118*, 25–32.

Manduell, K.L., et al. (2012). Forest structure and support availability influence orangutan locomotion in Sumatra and Borneo. *AJP*, *74*, 1128–1142.

Munn, C., & Fernandez, M. (2006). Infant development. In C. Sodaro (Ed.), *Orangutan husbandry manual* (pp. 135–148). Chicago: Brookfield Zoo, http://www.czs.org/Centers-of-Excellence/Center-for-Animal-Welfare/Animal-Husbandry/Orangutan-Husbandry-Manual (last accessed 8.9.15)

Myatt, J.P., & Thorpe, S.K.S. (2011). Postural strategies employed by orangutans (*Pongo abelii*) during feeding in the terminal branch niche. *AJPA*, *146*, 73–82.

Pontzer, H., et al. (2010). Metabolic adaptation for low energy throughput in orangutans. *PNAS*, *107*, 14048–14052.

Prasetyo, D., et al. (2012). Nest structures in Bornean orangutan. *Jurnal Biologi Indonesia*, 8, 217–227.

Robson, S.L., & Wood, B. (2008). Hominin life history: reconstruction and evolution. *Journal of Anatomy*, *212*, 394–425.

Samson, D.R., & Shumaker, R.W. (2013). Documenting orang-utan sleep architecture: sleeping platform complexity increases sleep quality in captive *Pongo*. *Behaviour*, *150*, 845–861.

Spillman, B., et al. (2010). Acoustic properties of long calls given by flanged male orang-utans (*Pongo pygmaeus wurmbii*) reflect both individual identity and context. *Ethology*, *116*, 385–395.

Stewart, F.A., & Pruetz, J.D. (2013). Do chimpanzee nests serve an anti-predatory function? *AJP*, *75*, 593–604.

Stewart, F.A., et al. (2007). Do chimpanzees build comfortable nests? *AJP*, 69, 930–939.

Suddendorf, T., & Whiten, A. (2001). Mental evolution and development: evidence for secondary representation in children, great apes, and other animals. *Psychological Bulletin*, *127*, 629–650.

Thorpe, S.K.S., et al. (2007). Orangutans utilise compliant branches to lower the energetic cost of locomotion. *BL*, 3, 253–256.

Thorpe, S.K.S., et al. (2009). Orangutans employ unique strategies to control branch flexibility. *PNAS*, *106*, 12646–12651.

Tomasello, M., & Herrmann, E. (2010). Ape and human cognition: what's the difference? *Current Directions in Psychological Research*, *19*, 3–8.

van Casteren, A., et al. (2012). Nest-building orangutans demonstrate engineering know-how to produce safe, comfortable beds. *PNAS*, *109*, 6873–6877.

van Casteren, A., et al. (2013). Factors affecting the compliance and sway properties of tree branches used by the Sumatran orangutan (*Pongo abelii*). *PLONE*, 8, e67877.

van Noordwijk, M.A, & van Schaik, C.P. (2005) Development of ecological competence in Sumatran orangutans. *AJPA*, *127*, 79–94.

van Noordwijk, M.A., et al. (2012). Female philopatry and its social benefits among Bornean orangutans. *BE&S*, 66, 823–834.

van Schaik, C.P., & Aureli, F. (2000). The natural history of social relationships in primates (pp. 307–333). In F. Aureli and F.B.M. de Waal (Eds.), *Conflict management*. Berkeley: UCP.

Vogel, E.R., et al. (2011). Bornean orangutans on the brink of protein bankruptcy. *BL*, 8, 333–336.

Vogel, E.R., et al. (2014). Food mechanical properties, feeding ecology, and the mandibular morphology of wild orangutans. *JHE*, *75*, 110–124.

Wich, S.A., et al. (2011). Forest fruit production is higher on Sumatra than on Borneo. *PLONE*, 6, e21278.

24 Great Apes of Africa
Gorillas

She sits on a stump in a familiar dim clearing, surrounded by rainforest vegetation, listening to approaching sounds of crashing and crunching. A black giant looms out of the forest, turning its head warily from side to side. Then the warm brown eyes focus on the researcher with mild interest and she is equally calm. They know each other well.

Gorillas are the largest primates and among the most sexually dimorphic. Males average about 180kg and females about 90kg. Pioneering studies focused on a small population at unusually high altitudes in the far eastern section of the genus distribution (Fossey 1983; Schaller 1963). These mountain gorillas differ from other populations in physical and behavioral characteristics. The relatively numerous gorillas in the lowlands of central and western Africa are more typical of the genus. At least two species are now recognized (Box 24.1).

Box 24.1 Taxonomy of African Apes and Gorillas

African apes differ from orangutans in having long, narrow heads, large brow ridges, flat foreheads, and dark hair. The laryngeal air sacs that enhance vocalization are internal rather than forming external pouches as in orangutans. Between chimpanzees and gorillas, the wrist anatomy varies. The columnar structure of the gorilla wrist supports terrestrial movement, while the flexible *Pan* wrist is more suitable to arboreal locomotion (Kivell and Schmitt 2009).

Once considered a single species, gorillas have been substantially subdivided in recent years (Groves 2001; Grubb et al., 2003; Sarmiento and Oates 2000). There is virtual consensus about two taxa: the western species (*Gorilla gorilla*) and the eastern species (*Gorilla beringei*).

There is wide agreement about subspecies. The western gorilla includes the western lowland gorilla (*Gorilla gorilla gorilla*) and the smaller population of Cross River gorillas (*Gorilla gorilla diehli*). Eastern gorillas are divided into mountain gorillas (*Gorilla beringei beringei*) and Grauer's gorillas (*Gorilla beringei graueri*). Higher taxonomic status may be accorded to the Virunga and Bwindi populations of mountain gorillas.

Research on western gorillas illustrates some of the difficulties in fieldwork, and one unusual opportunity. These gorillas travel farther than other gorillas, leave less obvious trails in the dense ground vegetation of their forest habitats, and are harder to habituate because of hunting by humans. However, they also frequent marshy clearings called **bais** where primatologists can observe them in the open from raised platforms.

Gorilla Behavior

Gorilla Ecology

Dependence on terrestrial herbaceous vegetation (THV) leads gorillas to favor open canopy forest with dense ground cover. However, they survive in a wide range of environments from forest–savanna edges to swampy forests to mountains up to 3800m. Most gorilla movement is on the ground, though females and young are capable climbers. Gorillas are called **knuckle walkers**, a misnomer for a unique locomotor pattern in which weight on the forearms is supported by the middle phalanges of the fingers (Richmond and Strait 2000). Knuckle walking combines effective ground locomotion with forelimb specializations for climbing and suspension.

Gorillas prefer ripe fruit and most of them can eat fruit regularly throughout most or all of the year. Swamplands provide succulent fruits, along with aquatic plants that may supply minerals (Magliocca and Gautier-Hion 2002). Nevertheless, environment and large size constrain gorillas to varying degrees of herbivory. All gorillas eat substantial quantities of herbaceous and woody vegetation, enough to make them the most herbivorous great apes (Figure 24.1). They are eclectic but selective, consuming a wide variety of plant foods but making predictable choices within each category. Fallback foods include leaves, pith, and bark. Whole stems are usually fibrous, but the inner cores are succulent (Rothman et al. 2006). Gorillas can eat a considerable amount of low quality food when necessary. Lacking digestive specializations like those of leaf-eating monkeys, the apes depend on body mass for tolerance of secondary compounds.

Across the genus, consumption of plant foods that may have medicinal value includes about 170 species belonging to about 60 families. Among these are several *Cola* species (Huffman and Cousins 2005). Sometimes called kola nuts, the seeds contain caffeine and theobromine and are chewed by African people as a stimulant. Gorillas also consume three plants containing hallucinogens, but these chemicals also have medicinal properties for the cardiovascular system, and as an antiseptic and a cough remedy.

Gorillas use size and strength to obtain insects, breaking termite mounds into small pieces with their hands, for example, and using their massive jaws to consume ant nests whole (Cipolletta et al. 2007). Contest competition sometimes occurs at termite mounds, resulting in more displacements than at other food sources. Perhaps due to the low protein content of their fruit, western gorillas eat more invertebrates than congeners do. Around 25 to 30 percent of fecal samples contain insect remains (Deblauwe et al. 2003). Quantities seem small, but the analysis probably underestimates termite consumption because feces only yield remains of the least common caste, the soldiers. Some researchers concluded that the nutritional *quality* of insects consumed by the gorillas is low and suggested medicinal or other functions (Deblauwe and Janssens 2008).

Foraging gorillas spend only minutes in any one tree, moving slowly between fruit sources and feeding on ground vegetation as they travel. Because staple foods are abundant, the apes rarely travel far. Day range is seldom more than one kilometer and annual home range varies from $15km^2$ to $30km^2$. Annual range encompasses occasional visits to distant sites to exploit unusual food sources (such as bamboo or heavily fruiting trees). Monthly home range is probably more closely linked to foraging effort and therefore to environmental conditions.

Figure 24.1 Female gorilla feeding herself and nursing an infant. © Rhett A. Butler/ www.mongabay.com

Western gorillas have larger day ranges and home ranges than other populations because of seasonal and spatial irregularities in the supply of important foods. They travel farther when more fruit and termites are available.

Gorillas display low "site fidelity", meaning that they are prone to shift home range across months and years. One group's annual range averaged about 15.5km², but it traveled over more than 40km² during 92 months (a good example of the need for long-term field research). Home ranges overlap extensively, but aggressive confrontations between groups are rare. Social factors, including group fission and encounters with lone outsiders, can influence a group's ranging behavior.

Gorillas display diverse and distinctive sleeping habits (Mehlman and Doran 2002; Sanz et al. 2007; Tutin et al. 1995). They build about a third of their night nests in trees, typically at low levels. Sleeping on the ground, with or without a nest, is preferred, possibly because it is cooler. Gorillas are also likely to sleep near or on the ground where herbaceous undergrowth is plentiful. They build nests high in swamp forests and are more likely to make nests when the weather is wet or cold. Ground nesting is more common in the dry season, on precipitous slopes, and in light gaps and clearings. Gorillas nest in trees in the wet season and in primary forest where saplings are the preferred understory (DeVere et al. 2011).

Gorillas in a single population may build as many as six different kinds of nests, arboreal and terrestrial. Preference for particular materials may affect various aspects of the process, including height and reuse (Iwata and Ando 2007). Some plants preferred for the nest foundation fit the Comfort Hypothesis, as evidenced by comparable human uses. Gorilla nests are large, up to 2 meters in diameter, but can be constructed in minutes (Stanford 2008).

Some gorillas differ from other great apes in reusing specific nest sites, though not the nests themselves (Sunderland-Groves et al. 2009). One explanation is that it places the apes close to preferred fruit trees during fruiting season (Iwata and Ando 2007). Age–sex zoning at sleeping places may reflect danger from leopards (Weiche and Anderson 2007). Adult male group leaders frequently sleep on the ground while most others are in trees.

Gorillas in a zoo study (Lukas et al. 2003) transported materials to their nesting spots and engaged in a variety of manipulations, including fluffing and tucking. Juveniles built nests when still sleeping with their mothers, indicating proficiency in captive-born individuals. Thus gorilla nesting may have innate components (Mehlman and Doran 2002). The zoo gorillas built higher nests in colder weather and spent more time on the bare floor in warmer weather, supporting thermoregulatory interpretations of wild gorilla behavior. Vertical zoning appeared in another captive group. Weiche and Anderson (2007) noted the value of night observations without disturbing the subjects, which is difficult to do in the field.

Gorilla Sociality

Gorillas live in stable cohesive social units. A typical group of about ten includes four or five adults and four or five immatures of various ages. It may contain one fully adult male or several. The leader is a **silverback**, an adult male at a stage in life where the hair on his back is silvery gray (Figure 24.2). This may function as a signal of maturity to other group members and might also enhance his apparent size, which could intimidate predators and (more importantly) male rivals. The distinctive back color can also serve as a focal point for followers. Observers have noted anecdotally that females persistently respond to the leader's timing, direction, and speed.

Orientation toward the leader seems to be related to his role as a protector. Group members congregate near him during rest periods. Females stay closer when they have

Figure 24.2 A silverback male gorilla. © Rhett A. Butler/www.mongabay.com

infants, and juveniles as they become more independent of their mothers. After the death of a silverback, members of his group were twice as likely to nest in trees as before. When a new adult male joined that group, the females returned to their previous sleeping habits (Yamagiwa and Kahekwa 2001). Without such a timely replacement, a group usually disintegrates and the females join other males (Caillaud et al. 2008).

Bradley et al. (2008) investigated a suggestion of fission–fusion by using DNA from hair and feces to trace individual movements. They could connect the evidence with individuals because gorillas are unique among apes in leaving feces in their nests. Results showed that tracking could easily conflate separate groups due to the frequency of overlapping home ranges and intergroup encounters. Nest counts were misleading because a western gorilla sometimes makes more than one nest in a night. This practice also led to the erroneous idea that many groups contained more than one silverback, because silverback dung (distinguished by its size) was found in more than one nest at a site. DNA analysis showed that the same individual made those nests.

DNA research established genetic relationships among adults in western groups and demonstrated that kinship may be important in their social life (Bradley et al. 2007). Average relatedness within groups was higher than expected based on the assumption of random dispersal. Forty percent of adult females had an adult female relative in the group. Females seemed to avoid related silverbacks when changing groups, presumably an incest avoidance mechanism based on familiarity. Female mobility is a key factor in gorilla social organization. The typical female leaves her natal group to join another silverback and she may make one or more additional transfers. Mobility is facilitated by the rarity of strong ties among females.

Solitary males sometimes follow a social unit for several days, presumably seeking to detach a female. Most solitaries are silverbacks, mature and ready for leadership. Encounters with other groups also provide females with an opportunity to transfer to another male. A leader usually responds to another leader or a lone male agonistically, with avoidance, flight, or aggression. Leaders sometimes try to herd their females away from such threats (Robbins 2009; Robbins and Sawyer 2007).

In an aggressive confrontation two silverbacks usually exchange displays, hooting and beating their chests. **Chest beating** (Figure 24.3) is a unique signal in which cupped hands are slapped against the chest in rapid alternation (contrary to popular stereotype, gorillas do not use their fists). Ritualized threats occasionally escalate to serious fighting and a few silverbacks die from injuries. Familiarity between leader males may sometimes mitigate interactions.

Disputes between females tend toward ambiguous outcomes. Relationships among females are highly variable and individualistic. This is probably due to the availability of essential foods; minimal food competition provides little incentive to establish clear dominance relationships or to form coalitions. Some sharp disputes do occur in connection with patchy resources such as fruit trees. The agonistic behavior of female gorillas parallels that of monkey species in similar circumstances (Stokes 2004), and is consistent with the Socioecological Model.

Kinship mitigates aggression between females. Some females in the same group are related as mother and daughter or as sisters, because some females do not emigrate and because related females sometimes transfer together. The behavior of related gorillas fits sociobiological expectations: affiliative interactions, especially reciprocal grooming and resting in proximity, occur at higher rates than among unrelated individuals. Females thought to be paternal relatives are intermediate in such behaviors. Kin are more tolerant of one another and may help each other (Bradley et al. 2007), although they do not form coalitions. Aggression between females is less frequent in western gorillas than in mountain gorillas (Stokes 2004).

Relationships seem to shift with reproductive status and there is evidence for sociosexual competition in a group of zoo gorillas (Stoinski et al. 2009). All females in that group engaged in mating outside their conceptive periods. Regardless of reproductive condition, they were more likely to engage in sexual behavior on days when other females were sexually active.

Gorilla anatomy is consistent with sociosexual behavior. Extreme sexual dimorphism, small testes, and a short estrus period with no external signs signal a single breeder system with intense competition for mates and no sperm competition. Estrus lasts for 1 or 2 days out of every 28 and coincides closely with ovulation. Females, who solicit the majority of copulations, indicate estrus with proceptive behavior. Genetic studies show that, in contrast to many other primates, females breed only with resident males (Stoinski et al. 2009; Doran-Sheehy et al. 2009). However, females can obtain new mating opportunities by transferring.

Male aggression toward females is common, but usually consists of displays rather than physical attacks and females sometimes ignore threatening males (Robbins 2009; Stokes 2004). The mild aggression may be an attempt to intimidate as a means of mate retention, or it may be a form of courtship related to female mate choice. A silverback leader intervenes in some disputes between females, bringing them to an end. This may be another mate retention tactic. Since the leader always sides with the less dominant or losing individual, he makes it less likely that those females will transfer.

Female mate choice is correlated with male body size and musculature (Caillaud et al. 2008), which is consistent with the Good Genes Hypothesis. More particular to gorillas, the

Figure 24.3 Young gorilla beating its chest. © Rhett A. Butler/www.mongabay.com

size of a male's external crest was linked to the number of females in his group. This fibrous and adipose structure may reflect the health of the male (extra-large crests develop in captive males). The researchers also found that the number of male agonistic displays during dyadic encounters was positively correlated with the total number of females present. Sexually receptive female gorillas preferred more aggressive males in a captive experiment that prevented direct coercion with a doorway between cages that was only large enough for a female (Nadler and Collins 1984).

On average about 12 percent of the groups in a population are not the typical social units. Most members of these groups are males ranging from juveniles to young adults, but there is almost always at least one silverback. These units are often called bachelor groups; however, since immature females occasionally join, the neutral term *non-breeding group* seems preferable (Levrero et al. 2006). Some non-breeding groups last for years, although membership changes frequently. The young males benefit from protection and from the opportunity to develop social relationships (Levrero et al. 2003). Silverbacks who care for the young individuals may be enhancing their inclusive fitness (Enciso et al. 1999).

Western home ranges overlap extensively and social units meet often, perhaps four times more frequently than among mountain gorillas (Doran-Sheehy et al. 2004). Some encounters involve an exchange of displays, and intergroup dominance is implied by the many encounters that end with one group leaving the area. Most such meetings involve large fruiting trees or groves containing uncommon food species (Bermejo 2004; Tutin 1996). Indirect evidence from one site suggests that peaceful mingling occurs and just 3 of 43 meetings resulted in a fight (but one fight was fatal).

Some western groups may have ongoing social relationships. Sometimes social groups track each other (Doran and McNeilage 2001) and genetic research suggests that such behavior may represent a community based on kinship ties (Bradley et al. 2004). Twelve of the 14 silverbacks in the study area were related to one or more others as siblings, half-siblings, or father and son. These connections formed the basis for two clusters, containing three and four groups. The structure implies a form of philopatry in which male dispersal is limited and some males maintain friendly or tolerant relations with one another.

Meetings at bais are likely to be peaceful, but this can be attributed to the fact that the gorillas are eating as they enter and do not stop until shortly before they leave (Parnell 2005). The rich resources provide little reason for competition and the open nature of the environment allows silverbacks to monitor their females with ease. Aggressive responses did occur in 30 percent of shared occupations, but none involved physical contact. Juveniles from different groups played with each other, which may establish relationships that affect adult interactions (Parnell 2005). Sometimes immatures transfer from one group to another. Comparison of several sites indicates that about 60 percent of encounters are neutral and 30 to 40 percent involve some level of agonism (Magliocca and Gautier-Hion 2004).

Gorilla Life Cycle

Compared to other great apes, gorillas fit the general correlation between smaller relative brain size and faster life history. However, the rapidity of gorilla life is contrary to theoretical expectation based on body size. This may be related to their distinctive foraging adaptation. Using an abundant supply of high quality THV as their staple food, many gorillas lead a relatively easy life and can afford faster growth and reproduction. Growth and reproductive rates of wild and captive gorillas are very similar, indicating that wild populations are near their full potential. Most maturation data come from mountain gorillas, but available information from other populations suggests comparable rates.

Within the genus life history is slower for western gorillas, at least at certain stages (Magliocca et al. 1999; Nowell and Fletcher 2007; Weber and Vedder 1983). Weaning, a female's first birth, and male adulthood all take place one to two years later in western gorillas (Groves and Meder 2001; Robbins 2011). Breuer et al. (2009) take this as support for the Risk Aversion Hypothesis of Janson and van Schaik (1993): where food supply is poor or uncertain, extended immaturity reduces the risk of starvation due to conspecific competition. Western gorilla ecology fits the conditions because these apes depend heavily on ripe fruit, which is a seasonal resource. Breuer and colleagues also note that more arboreal movement, larger home ranges, and greater complexity in foraging (all applicable to western gorillas) can reduce the amount of energy available to an individual for physical maturation. The slow maturation of male western gorillas may be a factor in the small number of multi-male groups. If a male's tenure as group leader ends before his son becomes a silverback, then there is no chance for an age-graded multi-male group to develop.

Observed gorilla births have been brief and easy for both primiparous and multiparous females (Beck 1984; Nadler 1974; Stewart 1977). The newborn weighs about 2kg. Males and females grow at the same rate throughout the infant and juvenile stages. About one-fourth of infants die in the first year in the majority of populations, mostly from predation and disease.

Regarding the controversy about learning maternal behavior, one would expect the need to be especially strong in great apes because they are so dependent on learning in general. However, many captive female gorillas that have never observed maternal care become competent mothers with their first birth. Naïve females are able to clean the baby, help it to cling, and nurse it (Beck 1984; Nadler 1974). Cases of neglect and abuse may be attributable to aspects of captivity rather than lack of knowledge. Some agitated and neglectful mothers, for example, became competent when restored to the company of an adult male (the species-typical protector).

Rarity of infanticide among wild western gorillas seems to be reflected in the social preferences of captive females. During the first six months after parturition, mothers preferred to associate with other females and juveniles rather than with the silverback. This suggests that the silverback's protector role is limited, which is consistent with sparse evidence for infanticide (Stoinski et al. 2003). Close association with a silverback appears to be a facultative behavior that shifts to a very low level in captivity where the female perceives no danger from infanticide or predation.

A juvenile sleeps in its mother's night nest until the next infant is born. Loss of maternal support is somewhat compensated by adult male attention. During the third year, as proximity to the mother decreases, proximity to the silverback increases. Juveniles and silverbacks groom each other, though the young usually do more. The silverback intervenes in disputes between immatures and supports the younger in most cases. Some silverbacks increase affiliative contact with immatures who have lost their mothers (through death or emigration), grooming them more and nesting with them.

Maestripieri and Ross (2004) addressed general hypotheses of play in relation to gorillas. Considering motor training in the context of gorilla sexual dimorphism and the male protector role, they predicted that male infants would display a higher frequency of play than females and would prefer to play with other males. Given weak bonds among adult female gorillas and their greater attraction to males, the Social Relationship Hypothesis predicted that female infants would prefer to play with males. All predictions were confirmed, supporting the view that play patterns are functionally related to the adult life of gorillas. These results were consistent with the findings of earlier captive studies (Brown 1988; Meder 1990). Wild females prefer to play in smaller groups (Robbins et al. 2004) and seem to spend less time in developing social relationships, perhaps because they are more likely to leave the natal group (Ferriss 2005).

Puberty at about age six marks the beginning of a two-year period during which male and female developmental trajectories diverge. Young females experience their first labial swellings, estrus behavior, and completed copulations. However, this is also the time of adolescent sterility. Females reach adult size after age eight and usually give birth for the first time at ten years.

Most females leave the natal group during this period. Primary and secondary transfers are ubiquitous and some females transfer five or six times (Robbins et al. 2004; Stokes et al. 2003). Some transfers are made by choice and some result from a leader's death. Female western gorillas are more likely to emigrate from larger groups and more likely to join smaller groups, which seems to be related to male reproductive strategies. Although infanticide appears to be rare in western gorillas, females faced with the death of their silverback seem to follow strategies that minimize the probability of infanticide. Male protection and male quality both seem to be important in making choices.

As in other highly dimorphic primates, male gorillas mature more slowly than females. A male is still immature until the age of 12. He is called a *blackback* because his coat has yet to undergo the color change of adulthood. Blackbacks sometimes sire offspring in the same group as a silverback, and they always have the option of emigrating and attempting to attract females of their own.

The silvered back signals adulthood, but a male is socially a *young silverback* until he reaches full physical growth at about age 15. There is no suppression (or adaptive delay) of maturation as in orangutans. The majority of males disperse before full adulthood. Although many leaders display increasing aggression toward maturing males (even juveniles in some cases), there is no clear case of expulsion. Final departure seems voluntary, following months of gradually increasing time away and distance traveled. Most males become solitary for years, until they can obtain their own females. Some join non-breeding groups. The male emigration rate is high, perhaps twice that of mountain gorillas (Robbins et al. 2004), which seems to imply greater pressure on group leaders from outsiders seeking mates. Females probably die between their mid-twenties and early thirties while males are likely to die by their late twenties (Robbins and Robbins 2004). This is in accord with other very dimorphic animals, in which males succumb to the wearing effects of competition.

Gorilla Cognition

Adult males at a bai engage in social displays that might be considered tool modification in that they produce large plumes of water (Parnell and Buchanan-Smith 2001). The body splash is a run or leap into water as deep as 1.5 meters. The other style is a slap with one or both hands. Two-thirds of these displays are agonistic and the rest apparently play. Serious displays seem to be intended as intimidation, since silverbacks typically perform them toward other silverbacks. Females never display like this and are rarely targeted. The water displays resemble the charges and ground-slaps commonly used by gorillas, which suggest that they may be modifications of innate action patterns. These unusual variants occur in places where visibility extends to about 500 meters. The gorillas may have invented them with an awareness of the circumstances.

One adult female at a bai used a stick to test water depth and to steady herself as she tried to cross a pool of water. Another adult female used the detached trunk of a shrub as a stabilizer during food processing and then as a bridge to cross a deep section of the swamp. Most great ape tool use relates to food, but these examples demonstrate that features of the habitat can stimulate such behavior (Breuer et al. 2005).

Such signs of cognitive skills are rare in wild gorillas compared to orangutans and chimpanzees. However, the apparent gap is narrowed by behavior in captivity. Captive gorillas display more creativity in object manipulation, tool use and tool modification. Parker and Mitchell (1999) compiled a list of observed performances, which can be exemplified by one versatile group (Fontaine et al. 1995). Several members often used sticks to reach food or objects of curiosity. They shaped long pieces from logs by breaking or chewing, and discarded unsuitable pieces. One female used thin sticks or twigs, chewed to a point, to clean her infant's nostrils and navel. She also used coconut fibers to clean the baby. Another female once dipped coconut fibers into a small hole containing water and squeezed the water into her mouth. All but one of the females in this group tried to use logs as ladders.

Diverse examples continue to enter the literature. Individuals in three separate groups at two zoos, for instance, engaged in chaff blowing to separate chaff from a handful of grain. This was apparently more than play because the technique seemed to increase food processing efficiency and food intake (Pika and Tomasello 2001). The technique must have been invented independently at least three times. It should be noted that captives are all western gorillas so there may be innate variation at the population or species level.

Experiments have elaborated on gorilla abilities. In one case they carried branches to an apparatus to extract food, then broke them and removed projections. Statistical analysis showed a correlation between the length of sticks and the depth of holes. The gorillas often began with the longest stick and modified it to the right length by breaking it into two or more pieces (Pouydebat et al. 2005).

The following example of object manipulation by a captive female may represent more complex cognition. After losing her infant two weeks earlier, she pulled up weeds and formed them into a ball. Three times she placed the ball on her neck, tipped her shoulder forward, and caught the ball in her hands, replicating a way in which she handled her lost infant. She also carried the ball in the crook of her arm, just as she had carried her infant. This behavior is comparable to that of wild apes and monkeys with dead infants, which suggests the significance of the ball for the gorilla. If this was an artificial substitute for the infant, it may have been based on a capacity for pretending.

Most of the behaviors cited above are evidence for innovation. Several examples demonstrate social learning and some imply culture. The fact that 19 different individuals used the splash displays in the bai suggests that this might be an example of culture. Variations in insect consumption might also be considered cultural. Among five sites studied, some insect species were exploited in one or more places, and little used or ignored in others. These variations seemed not to be due to availability (Deblauwe et al. 2003).

Genty et al. (2009) studied gorilla gestures as possible examples of culture. Comparison of three captive groups and one in the wild yielded a catalog of 102 gesture types, but only six of them appeared to be unique traditions within single social groups. Thus gorillas resemble other great apes in providing this kind of evidence for culture, but their repertoire is much smaller. Genty and colleagues considered it puzzling that gorillas do not regularly extend their gesture repertoire. They tentatively concluded that gorillas do not perceive any function for inventing new gestures, even in novel situations. If they are correct, this attitude seems to be comparable to the apes' apparently indifferent reactions to the mirror response tests (see below). Both patterns could arguably arise from a species personality.

Genty et al. (2009) considered the Ontogenetic Ritualization Hypothesis and concluded that it does not explain most gorilla gestures because they are derived from species-typical displays rather than ordinary functional actions, but are used flexibly and with awareness of the intended receiver's attentional state. These researchers did not find a clear distinction between innate and learned gestures. They concluded that gorilla gestural communication is based on a species-typical repertoire that is enlarged and elaborated by behavioral flexibility and learning. Similar reasoning can be applied to the water displays in the bais.

Hand clapping illustrates the mixed features of gorilla gestures. As described by Kalan and Rainey (2009), it has a highly stereotyped form that suggests genetic influence: in every case an individual, always a female, brought the hands together in front of the body twice in rapid succession, and repeated this action after an interval of one to two minutes. Earlier observations on another population recorded basically similar behavior with a little more variability in the number of claps in a bout (Fay 1989). The motivational source is a basic one, fear or anxiety in the face of an intruder (usually a human observer), and the function is to alert other individuals (especially the silverback). The fact that this behavior occurs only in western gorillas is consistent with a species-specific pattern. On the other hand, the behavior seems to be far from universal among western gorillas, which suggests independent innovation in various groups.

Gorillas have been a puzzle with regard to mirror self-recognition. They seem to demonstrate MSR in some tests, but often do not (Ledbetter and Basen 1982). The work on this problem exemplifies continuing refinement in behavioral experimentation. Of 23 gorillas studied by Swartz et al. (1999), six displayed self-directed behavior guided by a mirror. This was a relatively poor showing, but seemed to indicate that some gorillas are capable of self-recognition. The failures might have been caused by a natural tendency to gaze aversion or an inhibition response to human observers. Additional tests used a video camera instead of direct observation and an angled mirror device to avoid direct eye contact between gorilla and image. With these modifications gorillas still did not provide strong evidence for self-recognition, even after four years of exposure to mirrors (Shillito et al. 1999).

Nicholson and Gould (1995) started by training a gorilla to find an object that was only visible with a mirror. After the training the ape significantly increased mirror gazing along with face-directed behaviors and positive responses to the mark test. Posada and Colell (2007) used a subject who was well adjusted to captivity, and began by habituating him to the observers and to the mirror as an object. Showing no aversion to eye contact, he was interested in the mirror and responded positively to the mark test. The percentage of successes across all the experiments may be lower than in orangutans and chimpanzees, but the research shows that gorilla performance can be enhanced with better conditions and preparation.

In another experiment a male gorilla was marked with odorless and tasteless dye without anesthesia. Observers scored his videotaped behavior without knowing which trials used a mirror. Results showed that the gorilla touched the marked area more when he was in front of the mirror (Allen and Schwartz 2008).

Summary

Gorillas and chimpanzees are closely related. Both use the so-called knuckle-walking locomotor pattern. Gorillas are the largest primates and are highly sexually dimorphic. They prefer fruit but will eat THV and other vegetation. A silverback male leads and protects a group and is usually the only breeding male. Solitary males try to obtain females from groups and females may willingly transfer to a single male or an established group. Retention is a major concern of a leader. Kinship affects some female relationships, but females are generally not close. The leader is their main adult orientation. Gorilla life history pace is rapid for their size, but slower among western gorillas. Infanticide is much more common among mountain gorillas than western gorillas. Evidence is slowly accumulating that gorilla cognitive skills are greater than has long been assumed.

Sources

Basic Bibliography

Caldecott, J., & Miles, L. (Eds.) (2005). *World atlas of great apes and their conservation*. Berkeley: UCP.
Campbell, C.A., et al. (Eds.) (2011). *Primates in perspective*. NY: OUP. [Chapters 19, 20]
Harcourt, A.H., & Stewart, K.J. (2007). *Gorilla societies*. Chicago: UCP.
Kappeler, P.M., & Watts, D.P. (Eds.) (2012). *Long-term field studies of primates*. New York: Springer. [Chapter 17]
Robbins, M., et al. (Eds.) (2005). *Mountain gorillas*. NY: CUP.
Taylor, A.B., & Goldsmith, M.L. (Eds.) (2008). *Gorilla biology*. NY: CUP.
Watts, D.P. (2012). The apes: taxonomy, biogeography, life histories, and behavioral ecology. In J.C. Mitani et al. (Eds.), *The evolution of primate societies* (pp. 113–142). Chicago: UCP.

Additional References

Allen, M., & Schwartz, B.L. (2008). Mirror self-recognition in a gorilla (*Gorilla gorilla gorilla*). *Journal of Integrative Biosciences*, *5*, 19–24.
Beck, B.J. (1984). The birth of a lowland gorilla in captivity. *Primates*, *25*, 378–383.
Bermejo, M. (2004). Home-range use and intergroup encounters in western gorillas (*Gorilla g. gorilla*) at Lossi forest, North Congo. *AJP*, *64*, 223–232.
Bradley, B.J., et al. (2004). Dispersed male networks in western gorillas. *CB*, *14*, 510–513.
Bradley, B.J., et al. (2007). Potential for female kin associations in wild western gorillas despite female dispersal. *PRSB*, *274*, 2179–2185.
Bradley, B.J., et al. (2008). Genetic identification of elusive animals: re-evaluating tracking and nesting data for wild western gorillas. *JZL*, *275*, 333–340.
Breuer, T., et al. (2005). First observation of tool use in wild gorillas. *Plos Biology*, 3, e380.
Breuer, T., et al. (2009). Physical maturation, life-history classes and age estimates of free-ranging western gorillas. *AJP*, *71*, 106–119.
Brown, S.G. (1988). Play behaviour in lowland gorillas. *Primates*, *29*, 219–228.
Caillaud, D., et al. (2008). Influence of male morphology on male mating status and behavior during inter-unit encounters in western lowland gorillas. *AJPA*, *135*, 379–388.
Cipolletta, C., et al. (2007). Termite feeding by *Gorilla gorilla gorilla* at Bai Hokou, Central African Republic. *IJP*, *28*, 457–476.
Cousins, D., & Huffman, M.A. (2002). Medicinal properties in the diet of gorillas. *African Study Monographs*, *23*, 65–89.
Deblauwe, I., & Janssens, G.P.J. (2008). New insights in insect prey choice by chimpanzees and gorillas in southeast Cameroon: the role of nutritional value. *AJPA*, *135*, 42–55.
Deblauwe, I., et al. (2003). Insectivory by *Gorilla gorilla gorilla* in southeast Cameroon. *IJP*, *24*, 493–502.
DeVere, R.A., et al. (2011). Nest site ecology of the Cross River gorilla at the Kagwene Gorilla Sanctuary, Cameroon, with special reference to anthropogenic influence. *AJP*, *73*, 253–261.

Doran, D.M., & McNeilage, A. (2001). Subspecific variation in gorilla behavior: the influence of ecological and social factors. In M.M. Robbins et al. (Eds.), *Mountain gorillas: three decades of research at Karisoke* (pp. 123–149). NY: CUP.

Doran-Sheehy, D.M., et al. (2004). Impact of ecological and social factors on ranging in western gorillas. *AJP*, *64*, 207–222.

Doran-Sheehy, D.M., et al. (2009). The strategic use of sex in wild female western gorillas. *American Journal of Primatology*, *71*, 1011–1020.

Enciso, A.E., et al. (1999). Social interactions between captive adult male and infant lowland gorillas. *Zoo Biology*, *18*, 53–62.

Fay, J.M. (1989). Hand-clapping in western lowland gorillas (*Gorilla gorilla gorilla*). *Mammalia*, *53*, 457–458.

Fontaine, B., et al. (1994). Observations of spontaneous tool making and tool use in a captive group of Western Lowland Gorillas (*Gorilla gorilla gorilla*). *FP*, *65*, 219–223.

Fossey, D. (1983). *Gorillas in the mist*. NY: Houghton Mifflin Harcourt.

Genty, E., et al. (2009). Gestural communication of the gorilla (*Gorilla gorilla*): repertoire, intentionality, and possible origins. *AC*, *12*, 527–546.

Groves, C. (2001). *Primate taxonomy*. Washington DC: Smithsonian Institution Press.

Groves, C., & Meder, A. (2001). A model of gorilla life history. *Australian Primatology*, *15*, *1*, 2–15.

Grubb, P., et al. (2003). Assessment of the diversity of African primates. *IJP*, *24*, 1301–1357.

Huffman, M.A., & Cousins, D. (2005). Potential medicative value in the diet of gorillas. In GRASP *World great ape atlas*. Berkeley: University of California Press.

Iwata, Y., & Ando, C. (2007). Bed and bed-site reuse by western lowland gorillas (*Gorilla g. gorilla*) in Moukalaba-Doudou National Park, Gabon. *Primates*, *48*, 77–80.

Janson, C.H., et al. (1993). Ecological risk aversion in juvenile primates. In M.E. Pereira & L.A. Fairbanks (Eds.), *Juvenile primates: life history, development, and behavior* (pp. 57–74). Chicago: UCP.

Kalan, A., & Rainey, H. (2009). Hand-clapping as a communicative gesture by wild female swamp gorillas. *Primates*, *50*, 273–275.

Kivell, T.L., & Schmitt, D. (2009). Independent evolution of knuckle-walking in African apes shows that humans did not evolve from a knuckle-walking ancestor. *PNAS*, *106*, 14241–14246.

Ledbetter, D.H., & Basen, J.A. (1982). Failure to demonstrate self-recognition in gorillas. *AJP*, *2*, 307–310.

Levrero, F., et al. (2006). Living in nonbreeding groups: an alternative strategy for maturing gorillas. *AJP*, *68*, 275–291.

Lukas, K.E., et al. (2003). Nest building in captive *Gorilla gorilla gorilla*. *IJP*, *24*, 103–124.

Maestripieri, D., & Ross, S.R. (2004). Sex differences in play among western lowland gorilla (*Gorilla gorilla gorilla*) infants: implications for adult behavior and social structure. *AJPA*, *123*, 52–61.

Magliocca, F., & Gautier-Hion, A. (2002). Mineral content as a basis for food selection by western lowland gorillas in a forest clearing. *AJP*, *57*, 67–77.

Magliocca, F., & Gautier-Hion, A., (2004). Inter-group encounters in western lowland gorillas at a forest clearing. *FP*, *75*, 379–382.

Magliocca, F., et al. (1999). Population structure and group composition of western lowland gorillas in north-western Republic of Congo. *AJP*, *48*, 1–14.

Meder, A. (1990). Sex differences in the behaviour of immature captive lowland gorillas. *Primates*, *31*, 51–63.

Mehlman, P.T., & Doran, D.M. (2002). Influencing western gorilla nest construction at Mondika Research Center. *IJP*, *23*, 1257–1285.

Nadler, R.D. (1974). Periparturitional behavior of a primiparous lowland gorilla. *Primates*, *15*, 55–73.

Nadler, R.D. (1989). Sexual initiation in wild mountain gorillas. *IJP*, *10*, 81–92.

Nadler, R.D., & Collins, D.C. (1984). Research on reproductive biology of gorillas. *Zoo Biology*, *3*, 13–25.

Nadler, R.D., et al. (1984). Research on reproductive biology of gorillas. *Zoo Biology*, *3*, 13–25.

Nakamichi, M., et al. (2004). Baby-transfer and other interactions between its mother and grandmother in a captive social group of lowland gorillas. *Primates*, *45*, 73–77.
Nicholson, I.S., & Gould, J.E. (1995). Mirror mediated object discrimination and self-directed behavior in a female gorilla. *Primates*, *36*, 515–521.
Nowell, A.A., & Fletcher, A.W. (2007). Food transfers in immature wild western lowland gorillas (*Gorilla gorilla gorilla*). *Primates*, *47*, 294–299.
Parker, S.T., & Mitchell, R.W. (1999). The mentalities of gorillas and orangutans in phlyogenetic perspective. In S.T. Parker et al. (Eds.), *The mentalities of gorillas and orangutans, comparative perspectives* (pp. 397–411). NY: CUP.
Parnell, R.J. (2005). Forest clearings: a window into the world of gorillas (Box 7.2). In J.O. Caldecott & L. Miles (Eds.), *World atlas of great apes and their conservation* (pp. 113–114). Berkeley: UCP.
Parnell, R.J., & Buchanan-Smith, H.M. (2001). An unusual social display by gorillas. *Nature*, *412*, 294.
Pika, S., & Tomasello, M. (2001). "Separating the wheat from the chaff": a novel food processing technique in captive gorillas (*Gorilla g. gorilla*). *Primates 42*, 167–170.
Posada, S., & Colell, M. (2007). Another gorilla (*Gorilla gorilla gorilla*) recognizes himself in a mirror. *AJP*, *69*, 576–583.
Pouydebat, E., et al. (2005). Use and manufacture of tools to extract food by captive *Gorilla gorilla gorilla*: experimental approach. *FP*, *76*, 180–183.
Richmond, B.G., & Strait, D.S. (2000). Evidence that humans evolved from a knuckle-walking ancestor. *Nature*, *404*, 382–385.
Robbins, M.M. (2009). Male aggression towards females in mountain gorillas: courtship or coercion? In M.N. Muller and R. Wrangham (Eds), *Sexual coercion in primates: an evolutionary perspective on male aggression against females* (pp. 112–127). Cambridge: Harvard University Press.
Robbins, M.M., & Robbins, A.M. (2004). Simulation of the population dynamics and social structure of the mountain gorillas in the Virunga Volcanoes. *AJP*, *63*, 201–223.
Robbins, M.M. & Sawyer, S.C. (2007). Intergroup encounters in mountain gorillas of Bwindi Impenetrable National Park, Uganda. *Behaviour*, *144*, 1497–1519.
Robbins, M.M., et al. (2004). Social structure and life-history patterns in western gorillas (*Gorilla gorilla gorilla*). *AJP*, *64*, 145–159.
Robbins, A.M., et al. (2011). Lifetime reproductive success of female mountain gorillas. *AJPA*, *146*, 582–593.
Rothman, J.M., et al. (2006). Decaying wood is a sodium source for mountain gorillas. *BL*, *2*, 321–324.
Sanz, C., et al. (2007). Distinguishing between the nests of sympatric chimpanzees and gorillas in the Goualougo Triangle, Republic of Congo. *Journal of Applied Ecology*, *44*, 263–272.
Sarmiento, E.E., & Oates, J.F. (2000). Cross River gorillas: a distinct subspecies, *Gorilla gorilla diehli* Matschie, 1904. *American Museum Novitates*, *3304*, 1–55.
Schaller, G.B. (1963). *The mountain gorilla*. Chicago: UCP.
Shillito, D.J., et al. (1999). Factors affecting mirror behaviour in western lowland gorillas, *Gorilla gorilla*. *AB*, *57*, 999–1004.
Stanford, C.B. (2008). *Apes of the Impenetrable Forest*. Prentice Hall.
Stewart, K.J. (1977). The birth of a wild mountain gorilla (*Gorilla gorilla beringei*). *Primates*, *18*, 965–976.
Stoinski, T.S., et al. (2003). Proximity patterns of female western lowland gorillas (*Gorilla gorilla gorilla*) during the six months after parturition. *AJP*, *61*, 61–72.
Stoinski, T.S., et al. (2009). Proximate factors influencing dispersal decisions in male mountain gorillas, *Gorilla beringei beringei*. *AB*, *77*, 1155–1164.
Stokes, E. J. (2004). Within-group social relationships among females and adult males in wild western lowland gorillas. *AJP*, *64*, 233–246.
Stokes, E.J., et al. (2003). Female dispersal and reproductive success in wild western lowland gorillas (*Gorilla gorilla gorilla*). *BE&S*, *54*, 329–339.

Sunderland-Groves, J.L., et al. (2009). Nesting behavior of *Gorilla gorilla diehli* at Kagwene Mountain, Cameroon. *IJP*, *30*, 253–266.

Swartz, K.B., et al. (1999). Comparative aspects of mirror self-recognition in great apes. In S.T. Parker & 2 co-editors, *The mentalities of gorillas and orangutans* (pp. 283–294). NY: CUP.

Tutin, C.E.G. (1996). Ranging and social structure of lowland gorillas in the Lope Reserve, Gabon. In W.C. McGrew (Ed.), *Great ape societies* (pp. 58–70). NY: CUP.

Tutin, C.E.G., et al. (1995). Nest building by lowland gorillas in the Lopé Reserve, Gabon. *IJP*, *16*, 53–76.

Weber, A.W., & Vedder, A. (1983). Population dynamics of the Virunga gorillas: 1959–1978. *Biological Conservation* *26*, 341–366.

Weiche, I., & Anderson, J. (2007). Influence of social and environmental factors on nesting behaviour in captive gorillas. *FP*, *78*, 154–165.

Wittiger, L., & Sunderland-Groves, J.L. (2007). Tool use during display behavior in wild Cross River gorillas. *AJP*, 69, 1307–1311.

Yamagiwa, J. & Kahekwa, J. (2001). Dispersal patterns, group structure and reproductive parameters of eastern lowland gorillas at Kahuzi in the absence of infanticide. In M. Robbins, P. Sicotte, & K.J. Stewart (Eds.), *Mountain gorillas* (pp. 89–122). New York: Cambridge University Press.

25 Great Apes of Africa

Common Chimpanzees

A party of male common chimpanzees moves through the forest on the ground, silently, in single file. The observer follows with mounting excitement, but falls behind in denser undergrowth. Frustrated, he wonders if this will turn out to be a hunt or a boundary patrol.

Common chimpanzees are one or more species of the genus *Pan* (Box 25.1). The term serves to distinguish them from one species that has been called the "pygmy" chimpanzee, also widely known as the bonobo (see Chapter 26). Common chimpanzees are heavier than bonobos, ranging from about 35 to 65 kilograms.

> **Box 25.1 Common Chimpanzee Taxonomy**
>
> Common chimpanzees are distinguished from gorillas by molecular evidence and by physical traits. The latter include broader incisors and cheek teeth, shallower faces and mandibles, and less development of cranial crests. Common chimpanzee limbs are less robust and more similar in length. They have narrower hands and feet with thinner and more curved digits.
>
> Common chimpanzees are generally recognized as a single species, *Pan troglodytes*, based on molecular and morphological data. There is also agreement about four subspecies. These are the Eastern (*P.t. schweinfurthii*), the Central (*P.t. troglodytes*), the Western (*P.t. verus*), and the Nigerian (*P.t. ellioti*, formerly called *P.t. vellerosus*). Some see a possible species distinction between the Western common chimpanzees and the rest (Watts 2012). They are separated by a greater molecular and morphological gap than any of the other subgroups.

Common Chimpanzee Behavior

Common Chimpanzee Ecology

Common chimpanzees are knuckle walkers with a locomotor pattern comparable to that of their close relatives, the gorillas. However, according to Kivell and Schmitt (2009), the anatomical details differ. The chimpanzee wrist is more flexible and suited for arboreal behavior despite their terrestrial tendencies.

Common chimpanzees display longer daily ground travel than other apes, about 5km (Figure 25.1). However, they are anatomically specialized for vertical climbing (Figure 25.2). This facilitates feeding in multiple trees every day, but raises the question of how natural selection can maintain arboreal specializations despite the cost in terrestrial locomotion. Pontzer and Wrangham (2004) rejected the idea that arboreal efficiency minimizes total energy expenditure because common chimpanzees use ten times more energy per day in

Figure 25.1 Female chimpanzee with infant shows the knuckle-walking position of the hands. © Eric C. Matthews

Figure 25.2 Young chimpanzee displays vertical climbing ability. © Eric C. Matthews

terrestrial travel than in vertical climbing, a deficit that arboreal efficiency is unlikely to overcome. Access to food seemed unlikely because less arboreal species (e.g. baboons) readily obtain food in trees, and because most common chimpanzee fallback foods are terrestrial. Predation risk was ruled out because the apes survive their terrestrial limitations even though predation risk is greatest on the ground.

The researchers concluded that safety is the most important selective factor in maintaining common chimpanzee arboreal anatomy. Surveys of skeletal material reveal postcranial trauma, such as healed fractures, in roughly one-third of common chimpanzee individuals. This is about twice the frequencies found in bonobos and gorillas (Jurmain 1997; see also Carter et al. 2008) and presumably would reach an unacceptable level without the specializations. Hoffman (2012) supports the findings but notes that some of the injuries, especially to the arms, might have been sustained in fights.

Common chimpanzees use trees seasonally. In one study they spent less time in trees during months when fruit was plentiful, which seems to suggest that food distribution is the key factor. However, thermoregulation was indicated by a positive correlation between time on the ground and maximum daily temperature (Takemoto 2004). A "common sense" explanation might be that the ground is cooler in the tropical heat. However, the researcher concluded that his subjects rested in trees during cool periods because the higher ambient temperatures conserved their energy.

Fruit is the target of much foraging effort even when scarce. Some figs are so nutritious as to be virtually a staple food for some common chimpanzees. One particular species, for example, is probably responsible for extremely high population density at the Ngogo research site in Uganda. Other populations have no such abundance and for them a few pulp-fruit species may be critical (Basabose 2002). However, common chimpanzees can subsist on a diversified diet that may include more than 300 different items (Hunt and McGrew 2002) and they can use either major fallback strategy: eating a wider variety of foods or lower quality foods.

Common chimpanzees in several locales consume tree gum. About 30g of one analyzed species provide little energy, but fulfill daily requirements for several minerals (Ushida et al. 2006). Common chimpanzees elsewhere eat algae, sometimes for an hour or more, and may enter water to obtain them (Devos et al. 2002). Pursuit of such ephemeral foods requires wide ranging and good memory (McGrew, in Lonsdorf et al. 2010).

Common chimpanzee feeding varies through the day, often shifting to leaves or other non-fruit foods. They eat leaves from saplings when the proportion of nonstructural carbohydrates is greatest (Carlson et al. 2013). One proximate explanation for the shift is taste fatigue. Another is that the apes actually track the profitability of resources (Ganzhorn and Wright 1994), which suggests a cognitive mechanism. A third hypothesis considers a social cause: common chimpanzees start the day by renewing relationships when they congregate at fruit trees and diversify the diet later (Mitani 2009). There are also two functional explanations: (1) fibrous leaves are eaten before sleep to facilitate digestion; (2) the more protinaceous food is consumed after energy needs are met with fruit. Most of these hypotheses are not mutually exclusive.

Lodwick et al. (2004) connected the length of the foraging day to social factors. They found that low-ranking males took longer to get enough food because they were delayed by deference to others. Estrus females took longer because sexual activity used time and energy. Chemurot et al. (2012) quantified the weight of food consumed. Adult females consumed slightly less per day than adult males even though they spent more time feeding, possibly because non-estrus females are distracted by offspring and/or wary of competing males.

Common chimpanzees disperse seeds, some from hard-shelled fruits that few other animals can access. A cultivation experiment demonstrated that seeds excreted by common chimpanzees had higher germination rates than those from naturally fallen fruits (Takasaki and Uehara 1984). Nishida proposed that common chimpanzees have a coevolutionary relationship with some fruit tree species.

Animal foods constitute up to 10 percent of the diet. Analysis at one site suggested that termites provided a higher concentration of some amino acids than did plants in the common chimpanzee diet (Hladik 1977). Common chimpanzees differ from other great apes in eating social insects more frequently and/or consuming a greater biomass of them (DeBlauwe and Jannsens 2008; McGrew 2001) and using an array of tools for the purpose (McGrew, in Lonsdorf et al. 2010). Such tool use seems to be more frequent in areas where protein is less available from leaves or monkey hunts (Raubenheimer and Rothman 2013). Though widespread, termite eating is not universal among common chimpanzees, and use of tools for the purpose is less common (Stewart and Piel 2014).

The most common tools for insectivory are flexible probes of vegetation used to obtain ants, termites, bees, and honey from arboreal and underground nests (Figure 25.3). Driver ants (also called army ants) are more problematic because of painful bites. The apes obtain them with two kinds of "dipping"—inserting a long stick into the ants' temporary nest or a shorter tool into a moving column. An investigatory probe is a stick or other tool used to touch a strange object and then sniffed. A larger variant is used at some sites to extract small mammals from holes in the ground or trees. The leaf-sponge is another kind of tool that is widely used. This is a hand-held mass of crushed leaves used to get water from tree-holes.

Mammal hunting has been observed at all long-term research sites (Figure 25.4). Common chimpanzees favor ungulates and monkeys, with a strong bias toward red colobus. Hunting

Figure 25.3 Captive chimpanzee using a termiting tool in the Oregon Zoo. © Eric C. Matthews

Figure 25.4 Chimpanzee feeding on a carcass. © Curt Busse

behavior varies with the availability of these monkeys and tends to be infrequent where they are absent. The typical hunting success rate is 50 percent, exceeding that of several large African carnivores. However, the apes make far fewer attempts and can afford to wait for relatively favorable conditions. Even where hunting is frequent, observers only see about seven events per month. Common chimpanzees favor smaller brained prey (Schultz and Dunbar 2006), which may indicate less flexibility in escape or defense behavior.

One factor in hunting success is the number of participants, which exemplifies the importance of demography in explaining primate behavioral variation (Mitani 2006). In an unusually large community of about 150 members, hunting parties with up to 14 males swamped red colobus defenses and achieved a success rate of about 80 percent. Forests with denser canopies offer monkeys more escape routes, but this difficulty can be overcome by a large number of hunters and perhaps by increased cooperation. Hunting frequency is a somewhat different matter. It was low at one site when the community contained ten fully adult males (Newton-Fisher et al. 2002). Seasonality and availability of red colobus affect frequency.

Research at two sites demonstrated that common chimpanzee hunting is associated with food abundance (Mitani and Watts 2001; Watts and Mitani 2002). Hunting is uncertain and plentiful foods of other kinds compensate for low or negative payoff (Gilby and Wrangham 2007). Even successful hunting is time-consuming and energetically costly (Mitani 2006). Common chimpanzees hunt when abundance gives them the opportunity, because meat is a high quality food that is highly prized.

Meat provides no more than 3 percent of the food supply to a common chimpanzee population, which raises the question of how or whether individuals benefit. The Meat Scrap Hypothesis postulated that small quantities of meat are valuable for micronutrients, such as zinc and vitamin B-12, that are rare in plants (Gilby et al. 2008) and that participants in group hunting can gain significantly more micronutrients than when hunting alone (Tennie et al. 2009).

McGrew (in Lonsdorf 2010) suggested that an individual would do better to exploit insects, which can provide the same nutrients as meat. This may apply only to females, because males hunt more often and more successfully. Based on observations and reasonable assumptions, the average adult male common chimpanzee can obtain 25 to 70 kilograms of meat in a year (Nishida 2012). These quantities suggest significant protein and calorie gains for at least some individuals. An isotope study confirmed that some adult males obtain a substantial amount of protein from meat. Isotope signals from hair and bone collagen allowed the researchers to build a record of meat-eating throughout the hunting season, compensating for constraints on direct observation (Fahy et al. 2013).

Common chimpanzee hunting has been explained in terms of sharing, such as the exchange of meat for sex. Some field reports seem to support the hypothesis (Gomes and Boesch 2009), but direct observations of such exchanges are rare (Mitani and Watts 2001). Some long-term studies concluded that the presence of estrus females does not predict hunting and may even be associated with decreased hunting (Gilby et al. 2006). Male-to-male meat exchanges are more common than male-to-female and may be an important reason for hunting.

The Coercion Hypothesis holds that sharing arises from harassment rather than social exchange. Gilby (2006) found that: (1) a meat possessor's feeding rate decreased with the number of supplicants; (2) harassment increased the likelihood of sharing; and (3) harassment decreased after sharing (see also Stevens 2004). According to his data, there was no causal connection between mating and male–female sharing, and male–male sharing was not associated with grooming relationships. Males were more likely to share with females who were frequent grooming partners, but this can be explained by increased harassment from those emboldened partners. Whatever may be the case for meat, social functions in sharing are indicated by the fact that chimpanzees will beg for a share of fruit when the same food is readily available for gathering (Nakamura and Itoh 2001; Slocombe and Newton-Fisher 2005).

Common Chimpanzee Sociality

Common chimpanzee communities display high levels of temporal variation in spatial cohesion, party size, and party composition (Figure 25.5). Using data from 14 populations, Lehmann et al. (2007) constructed a mathematical model to represent climate, time budget, and limits on community size. The model accurately predicted two things: variation among habitats; and an average size of 77. The number may seem high, but it fits with probable values for times before reduction due to human activities (a reminder that primatology does not deal with pristine populations). Even in the present, the Ngogo community has had as many as 165 members. One important inference from the model was that cost of travel rather than feeding is the limiting factor. Another was that fission–fusion makes it possible for common chimpanzees to absorb travel costs while living in a large community. Without high F–F the limit would be 15 to 20 members, close to the size of gorilla groups. The researchers suggested that large communities are advantageous in reproduction and/or territorial defense. We might add hunting, because the high success rate at Ngogo is the result of large hunting parties drawn from the unusually large community.

Research at Budongo showed neutral or negative relationships between party size and various measures of food abundance, and a weak relationship with food dispersion (Reynolds 2005; see also Sakura 1994). As food abundance increased, the common chimpanzees used more patches instead of increasing party size at the richest patches. The researchers concluded that: (1) both abundance and dispersion should be considered in connection with common chimpanzee grouping patterns; and (2) the impact of food declines as abundance increases.

Itoh and Nishida (2007) noted the importance of methodology in this discussion. They distinguished between the *face-to-face party*, measured by scan sampling, and the *nomadic party*, including all individuals counted during a specified time period. The size of nomadic parties varied with fruit availability (measured by fruiting plant density and mean potential patch size). Face-to-face party size was relatively stable and displayed little or no correlation with fruit availability.

Larger parties are related to a positive correlation between the number of males and the number of estrus females in the party (Hashimoto et al. 2001). Other possible influences are hunting frequency, territorial activity, and predation risk (Mitani and Watts 2005; Sakura 1994). Party size seems to adjust to several factors that may vary from one locality to another.

Male–male relations are prominent in common chimpanzee society (Figure 25.6). Newton-Fisher (1999) studied small parties and found that the strength of association within dyads made a significant contribution to variance in party size. He attributed the personal associations of common chimpanzees to tactical decisions, demonstrated by shifts over time in response to different alpha males. Mitani (2009) found stability in dyadic male relationships. Twenty-eight subjects formed bonds averaging seven years in duration, some lasting the entire ten years of the study. Other research showed that male–male bonds are more stable than those between females, and the fitness benefit was demonstrated by long-term genetic and behavioral data: coalitionary aggression enabled males to father more offspring than other males of similar dominance status (Gilby et al. 2012b).

Philopatry suggests that kinship is important to males. DNA studies have demonstrated a tendency for the average relationship between males to be significantly stronger than those between females (Inoue et al. 2008; Morin et al. 1994a). However, demography shows that a male is unlikely to have close maternal relatives in his community throughout his life (Mitani 2009). Consequently, males bond with non-kin of the same age and dominance classes.

Figure 25.5 A chimpanzee party traveling on the ground. © Eric C. Matthews

Figure 25.6 One adult male chimpanzee grooms another. © Eric C. Matthews

Choruses of pant hooting represented male bonds (Fedurek et al. 2014; Mitani and Gros-Louis 1998). A male was more likely to join in if the originator was a long-term associate. Pant hooting also reflected short-term relationships. Two males were more likely to chorus together on days when they engaged in some mutual activity, such as grooming or a coalition. Both grooming and choruses predicted feeding proximity, while pant hooting was the better predictor of coalitions and joint non-vocal displays. Chorusing might play a causal role in affiliation.

Male common chimpanzees gather in larger groups, where grooming is a frequent behavior (Figure 25.7). Functions of such groups include hunting and territorial behavior. Male-male grooming networks are wider than those of male–female or female–female combinations, which suggests that male relationships provide community stability (Sakamaki 2009).

Dominance and mating success are correlated when there are fewer males in a stable hierarchy. High-ranking males have greater mating success when females are most likely to conceive, and this results in higher rates of paternity (Constable et al. 2001; Vigilant et al. 2001). An alpha male may have the most offspring in a community during his tenure (Inoue et al. 2008), but monopolization of females is impossible when there is a large male cohort (Newton-Fisher et al. 2010).

Males are more aggressive than females and most of their aggression is directed at other males. Intense reinforcement of dominance may be needed because of the fission–fusion process, which precludes continuous monitoring of other males' relationships (Muller and Wrangham 2004a). Violence within the community can cause injury and death, and even alpha males may be victims (Kaburu et al. 2013). Serious aggression against highly dominant males emphasizes the importance of alliances and larger coalitions. Some cases display the flexibility of males in changing sides. All adult males dominate all females and females are sometimes victims of male violence.

Figure 25.7 Adult male chimpanzees form a grooming chain. © Curt Busse

Female–female relationships, though less violent, are also characterized by dominance. Rank affects reproductive success by influencing body mass, foraging strategies, and association patterns (Murray et al. 2007). Dominant females travel together and support each other in food competition, which allows them to occupy prime feeding sites (Williams et al. 2002). Each has a high quality core area (Kahlenberg et al. 2008). Competition is usually moderate because female reproductive success depends on long-term access to resources (Pusey et al. 2008), in contrast to male competition for immediate mating opportunities.

Within the large community at Ngogo females formed associations called *cliques* (Wakefield 2008, 2013). Members engaged in affiliative interactions more often than expected by chance and displayed minimal food competition. Though less gregarious than males, females spent more than 60 percent of their time in association with one or more other females.

Four types of mating occur among common chimpanzees (Tutin 1979). *Opportunistic mating* is responsible for the majority of conceptions. This means that sperm competition is an important factor and accounts for the relatively large testes of chimps. *Possessive mating* is often successful for alpha males. An alpha may accomplish this alone or with the help of allies (Watts 1998). *Consortship*, which varies greatly in frequency from one site to another, can bring some success to low-ranking males (Constable et al. 2001). Extra-group mating accounts for less than 10 percent of conceptions and may be virtually absent from some communities (Newton-Fisher et al. 2010).

Male mate choice follows several criteria. First, males can apparently judge fecundity by the state of a female's sexual swelling (Figure 25.8). During females' most fertile periods,

Figure 25.8 Female chimpanzee with a sexual swelling. © Curt Busse

males are most likely to associate with them, sexually solicit them, interrupt copulations with other males, and otherwise compete with other males (Emery Thompson and Wrangham 2008). Males groom estrus females significantly more often than they do anestrus females and initiate these grooming interactions more often than do the estrus females (Slater et al. 2008). Koyama et al. (2011) found two patterns consistent with the biomarket: (1) males groomed swollen females less as the number of those females in the group increased; and (2) dominant males groomed less than lower ranking males.

Common chimpanzee males prefer older females (based on copulatory approaches, male party sizes, average rank of mating partners, and rate of male mating aggression) (Muller et al. 2006). Similarly, Proctor et al. (2011) found that captive male preferences reflected age, parity, and fertility status. Thus, male interest seems to be directed toward experienced females who have a history of parity and are currently fertile.

Territoriality in the form of boundary defense is ubiquitous among common chimpanzees, manifested in boundary patrols (Figure 25.9). This behavior does not occur in other great apes and is rare among other primates (similar behavior occurs in spider monkeys, which have a similar fission–fusion social system). Every 10 to 20 days a party of 5 to 15 common chimpanzees, all or mostly male, moves quickly and quietly to part of the boundary and looks for signs of outsiders. If an encounter takes place, which happens in fewer than half of the patrols, the party may withdraw or approach. An approach may end with a display or an attack. According to mathematical analysis based on playback experiments, attack is indicated when one side outnumbers the other by a ratio of at least 1.5 (Wilson and Wrangham 2003). This supports the Imbalance of Power Hypothesis, which proposed that attacks only take place when one side has a substantial advantage (Wrangham 1999). Discovery of a lone outsider usually results in a prolonged attack that may be fatal.

Figure 25.9 Chimpanzee patrol. © Curt Busse

Common chimpanzee territoriality must be adaptive because it occurs despite high energetic cost. Patrollers cover long distances and spend more time traveling and less time feeding than during control periods (Amsler 2010). Mitani and Watts (2005) favored the Range Expansion Hypothesis, emphasizing that successful aggression leads to control of more land and resources. They doubted that males are trying to recruit females, because secondary transfer by female common chimpanzees rarely happens under any circumstances. Participants seem to be aware of what they are doing and its potential consequences. When traveling at the periphery they remain silent, or, if in a relatively large group, they advertise their strength by vocalizing loudly (Wilson et al. 2007, 2012).

Sobolewski and colleagues tested the Challenge Hypothesis in relation to male common chimpanzee territoriality and relations with estrus females, assuming both to involve fitness-enhancing behavior. Testosterone concentrations in urine were significantly higher than baseline levels when males accompanied parous estrus females, replicating an earlier study with a smaller sample (Sobolewski et al. 2013). They also found that urinary testosterone increased during territorial behavior, but decreased during hunting (another highly active endeavor, but not aggressive) (Sobolewski et al. 2012).

The last point leads to an old debate about the motivational connection between aggression and hunting in any mammals that hunt. The connection seemed to be supported by the fact that Gombe common chimpanzees were more likely to patrol borders on days when they hunted red colobus monkeys. However, Gilby et al. (2013a) demonstrated that the connection is ecological rather than motivational. Distance traveled on a patrol increases the probability of an encounter with the monkeys, as does movement in the woodlands of the community periphery. Woodlands also make hunting more attractive because it is more likely to be successful than in forest.

Common Chimpanzee Life Cycle

Intergroup infanticide has been reported by four of five long-term studies and the number of victims is comparable to the number of adults killed (Sherrow and Amsler 2007). Usually the number of attackers, most or all male, overwhelms the mother. Typically the infant is killed quickly in contrast to the prolonged attacks on adults, but the difference may be due only to the infant's greater vulnerability. The Sexual Selection Hypothesis does not apply because the killers rarely have an opportunity to mate with the mothers, who typically flee back to their own communities. Inter-community infanticides are more like aggression against adults of other communities. The number of attackers are consistent with Imbalance of Power, and since the attacks always take place during border patrols, they are consistent with Range Expansion. Elimination of future competitors could also be a function (Muller and Mitani 2005).

Adult females sometimes kill infants in their own communities, probably in connection with resource competition (Townsend et al. 2007). The mother may be an established community member, but somewhat weak and isolated. Infanticides by females might be more common if not for the fact that recent immigrants tend to stay near adult males, who may intervene.

In the transition from infant to juvenile, social play changes with regard to complexity, asymmetry, innovativeness, and choice of playmates (Cordoni and Pelagi 2011). It becomes more symmetrical because competitive elements increase the probability of real fighting. Older partners play more roughly when the mother of the younger one is present, and both participants increase the use of play signals (Flack et al. 2004). It seems that the enhanced communication represents the function (and perhaps the intention) of forestalling termination of play by the mother.

Rotational play (pirouetting) is common (Nishida and Inaba 2009). The individual spins around in a quadrupedal posture while moving forward in a straight line. Infants do this in any context, but juveniles and adolescents do it mainly during travel. Development of general physical ability may be the main function, but there are sex differences. Males tend to pirouette faster than females and are more likely to continue the behavior in adolescence. Perhaps this is preparation for balanced movement during serious fighting in adulthood.

Ontogeny of manual skill is an important aspect of common chimpanzee development. Corp and Byrne (2002) traced this process in relation to extraction of fruit from an inedible matrix. Infants obtained portions of the fruit from their mothers and processed them with more varied techniques than adults. The infants could process whole fruits by the age of two years, but did not master the adult techniques until age four. This entailed maturation of manual abilities, including precision gripping, bimanual role differentiation, and differentiation of digital roles. Social factors included synchronous feeding, close attention to the mother, and food sharing.

Juvenile males tend to stray from their mothers to be near adult males. As adolescents they spend much less time with their mothers and continue trying to associate with adult males, but the adults tend to reject them as potential sexual competitors. The young males become somewhat peripheral to society as they undergo the process of trying to gain entry to an adult male grooming network. There are no all-male groups for them to join.

Juvenile and adolescent males sometimes threaten older and apparently stronger individuals such as mature females. Females usually ignore juveniles, but often respond agonistically to adolescents. Two explanations for the male behavior are exploratory aggression and rank improvement. Males stop sooner when females ignore them and persist when females retaliate. This is more consistent with rank improvement, a developmental pattern also found in some monkeys.

Assertive behavior can be dangerous. Coalitions of older males occasionally attack young adults who seem to threaten them in some way. One individual who refused to signal submission was beaten, and left the community for three months (Nishida et al. 1995). Another was killed after a rapid rise in the hierarchy that left him with no strong allies and poor social integration (Watts 2004). A third attack may have resulted from anticipation of sexual competition: it occurred when a single female became fully tumescent during a time when few females were cycling (Fawcett and Muhumuza 2000).

A male common chimpanzee attains fully adult musculature and social maturity at about 16 years of age. He now faces the task of moving up in the hierarchy, which includes maneuvering as an individual and in concert with others. Some individuals opt out of the competition; for those that are strong and healthy, this may be a matter of personality.

Juvenile females usually stay close to their mothers at the age when sons begin following adult males. Females copulate frequently as they approach puberty. Initially they interact with immature males in a manner that is sometimes playful. Later, they show a more serious sexual interest in adult males (who usually reject them). A female typically reaches full sexual maturity at about 13 years of age.

Typically the majority of females transfer from the natal community to another, where they may encounter serious aggression from resident females (Pusey et al. 2008). These attacks include coalitionary aggression, which is otherwise rare in female common chimpanzees (Kahlenberg et al. 2008). A new female immigrant tries to establish relationships with males for protection. They may defend her, presumably perceiving a potential sexual partner rather than a food competitor. Eventually she befriends some resident females, but food competition remains an issue.

A wild common chimpanzee at birth has a life expectancy of less than 15 years. The mean adult lifespan after sexual maturity is about 15 years (Hill et al. 2001). These survival rates are lower than those of common chimpanzees in zoos or captive colonies, which indicates the level of danger in natural environments. Female mortality is lower than that of males, as expected from sexual selection theory. However, considering both lifespan and reproductive data, the average production of daughters per female is 0.8. This suggests that reproductive success is low for most females despite their longer lifespan, and the low rate for female offspring places populations in a precarious position.

Common Chimpanzee Cognition

Captive studies based on controlled experiments that limit choices are extremely prominent in common chimpanzee cognitive research. In contrast, wild chimpanzees confront unique ecological and social challenges that probably make unusual demands on their mental abilities (Mitani et al., in Lonsdorf et al. 2010). A probable error based on captive studies is the claim that primates cannot understand causality concerning unobservable objects (Vonk and Subiaul 2009). Boesch (2013) pointed out that wild chimpanzees find and extract underground resources (Boesch et al. 2009; Sanz and Morgan 2009). Lifelong ecological deprivation probably accounts for the failure of captives in many tests. Morimura and Mori (in Lonsdorf et al. 2010) compared problem-solving abilities in captive-born and wild-born common chimpanzees. The latter consistently outperformed the former in using material for bedding and tools.

Evidence for common chimpanzee mental maps comes from a variety of contexts. First the chimpanzees travel efficiently from one food source to another within their own territories. Second, Ngogo chimpanzees engage in nonrandom movements to preferred fruit trees in the territories of neighboring communities (apparently to attack neighbors feeding there). Third, apes at two sites form parties that search for red colobus monkeys with nonrandom movements that indicate mental maps of previous successful hunts (Mitani et al. 2010).

The fluid nature of fission–fusion organization suggests a need for unusual cognitive capabilities for constant reassessment of social relationships (Mitani et al. 2010). Studies at Ngogo demonstrated that preferred grooming partners of males vary in number and identity due to fluctuations in party size. Beyond keeping track of partner identities, coalitionary behavior demands mental scorekeeping. The chimpanzees exchange meat and grooming for support; that is, they track interactions using currencies (goods and services) that may be similar or different. Furthermore, the long lifespan means that social calculations must be carried out over extended periods.

Crockford et al. (2013) questioned the cognitive basis for such relationships, disputing the idea of so-called mental bookkeeping to keep track of exchanges. They postulated that male–male relationships in common chimpanzees are facilitated by oxytocin, the hormone involved in parental and sexual bonding in mammals. Urine samples showed higher levels after grooming with a bonded partner than with others. Oxytocin may facilitate bonds because it acts directly on the neural reward and social memory systems of the brain.

Ecological and social challenges confronting these apes imply high adaptive value for memory (Mitani et al. 2010). Observations and experiments indicate that chimpanzees have better short-term memory than humans (Inouye and Matsuzawa 2007) and that they recall events and aspects of the environment for longer periods (Horner 2010). Sayers and Menzel (2012) performed a naturalistic experiment in which subjects watched foods being hidden in a large wooded area. After a delay, they were able to direct naïve humans to the food locations (using gestures and vocalizations). Furthermore, their directions were ordered in a way that closely matched the value of the different foods (i.e. the ratio of energy to handling time for each food).

Common chimpanzees exceed other apes in variety and complexity of technology. The strongest cognitive implications arise from the degree of sequential tool use in exploiting honey in trees and underground resources that include honey, social insects, and tubers (Boesch 2013). Some common chimpanzees use three tools to obtain honey (Boesch et al. 2009; Sanz and Morgan 2009): a *pounder* to break the hive covering, an *enlarger* to open the honey chambers inside, and an *extractor* to achieve the final goal. The user must be aware of the different tool types and their functions, the order in which they must be used, and the reward that will only be achieved at the end of the sequence.

The same tool may be applied repeatedly, sometimes creating a long delay in the sequence. One individual used a pounder more than 280 times before breaking the hive covering and moving on to the next tool. This seems to require patience as well as a cognitive grasp of the process. A further complication is that a tool may not work and an alternative must be chosen. When replacing a pounder, for instance, a chimpanzee finds one that works by trial and error, but only tries branches that could function (Sanz and Morgan 2009). Anticipation of a future event keeps the apes working at the task. Similarly, chimpanzees carry termiting tools to above ground nests and use sticks to probe for underground nests (Boesch 2013).

The spontaneity of captive common chimpanzees in solving collaborative problems suggests that the apes understand the potential role of partners in cooperation (Melis et al. 2010). This leads to an important distinction between motivation and cognition with regard to hunting. Though chimps hunting in groups may each be acting selfishly, they could still understand that cooperation increases the chances of success. Melis et al. (2010) also considered the cognitive distinction between *social tools* and collaborative partnerships. The former denotes perception of others as no more than devices to achieve selfish ends. The latter denotes perception of shared intentions and a shared (not just common) goal. Shared intentions can be identified by communication that partners use to try to influence each other.

Numerous experiments have been directed toward the Theory of Mind Hypothesis in common chimpanzees, moving Tomasello and Call (in Lonsdorf et al. 2010) from skepticism to support. They noted the continuing doubt of other researchers, based on the concept of **behavior reading** (taking cues from behavior without understanding the underlying mental states) as an alternative to **mind reading**. Tomasello and Call answered this objection with a new experimental paradigm in which they devised diverse experiments aimed at testing for a particular psychological state. They argued that the behavior reading hypothesis could be sustained only by creating different behavior reading explanations for all their results, most of which would have no empirical basis. Furthermore, they noted that most of the relevant ape studies are modeled on studies of human children; thus children would also be excluded from having a Theory of Mind and perhaps from having any social cognitive skills at all.

Schmelz et al. (2013) built on experimental evidence that common chimpanzees can predict the behavior of other individuals based on what the others can see, hear, know, or infer. They examined the possibility that a chimpanzee can act on the assumption that another has preferences similar to its own. Subjects chose a certain box with a food reward when choosing first or without competition. They avoided the same box if another individual had chosen first. Work like this provides evidence that common chimpanzees understand other individuals as mentally representing their goals and using their perceptions of the world to devise strategies for achieving those goals. In short, these apes understand that others have a perception-goal psychology like their own (Tomasello and Call, in Lonsdorf 2010).

Crockford et al. (2012) criticized the food paradigm used in experiments because they saw it as conflating perception of food with the intention to get the food. As a remedy, they offered a field experiment based on a common chimpanzee alarm call (the *hoo* that elicits cryptic behavior). They chose an alarm vocalization because it is costly to the producer (in the sense of risk) and therefore not likely to occur without a prospective benefit. The experimental danger was a model of a venomous snake. The researchers found that the vocalization was used significantly more often when other members of a party were ignorant of the apparent danger. They explained this as based on attribution of a mental state (they also interpreted their results as support for volitional control of vocalization and prosocial motivation in common chimpanzees).

Common Chimpanzee Culture

Common chimpanzees have played the key role in cultural primatology (Whiten, in Lonsdorf et al. 2010). Early research established the possibility of culture, and Goodall (1973) compiled a list of relevant traits from five field sites. She noted that the cultural interpretation was speculative, but expressed hope that it would inspire further research. Thus an important body of knowledge began with anecdotal evidence and informed speculation. More data and more sophisticated methods produced a list of 39 probable cultural variants among common chimpanzees, a list that has been continuously augmented.

The cultural interpretation requires systematic evidence that the traits in question are socially learned. Observations of young common chimpanzees watching adult behavior are suggestive, but do not substitute for experimental control (Byrne 2007). Menzel (1972) pioneered such experiments by showing that rapid habituation to novel objects could become a group tradition transmitted to new members. Recent work showed that different techniques for achieving a desired outcome could become distinct traditions after being seeded in separate captive groups (Horner et al. 2010; Whiten, in Lonsdorf et al. 2010). Further experiment demonstrated that such traditions could be transmitted from one group to another, which shows how regional traditions can become established among wild chimpanzees. This research also demonstrated the importance of social status and personality in the origin of traditions. Females of high rank, competence, and social tolerance played crucial roles. Similarly, in the wild, particular males play a special role in hunting leadership and such males may be responsible for the presence, weakness, or absence, of group hunting traditions in different populations of wild chimpanzees.

Nakamura (2010) argued against over-emphasis on traditions involving complex cognition, such as tool technology, and noted that many common chimpanzee cultural patterns are relatively simple. One example is *leaf clipping*, biting that produces an audible sound. This is a courtship display in one population, an expression of frustration in another, and a preliminary to buttress-drumming in another. The form of behavior is similar while the function (or meaning) varies. Nakamura also pointed out that some cultural variants are quantitative rather than qualitative; for example, females in some populations have denser grooming networks than those in others. Despite continuing skepticism, the evidence for common chimpanzee culture is becoming more abundant and more sophisticated (McGrew, in Lonsdorf et al. 2010; Whiten, in Lonsdorf et al. 2010).

Summary

Common chimpanzees live in fission–fusion communities that range in size from roughly 15 to 150. Parties within them are highly flexible. Common chimpanzees hunt relatively large mammals with greater frequency than other primates. They display relatively complex technology and culture. Male bonds are strong and males are consistently dominant over females. Common chimpanzees provide strong evidence for complex cognitive abilities such as Theory of Mind. Research in western Africa suggests some major regional differences in common chimpanzee behavior. Some primatologists use the behavior of savanna common chimpanzees as a model for the adaptations of early hominins.

Sources

Basic Bibliography

Boesch, C. (2012). *Wild cultures*. NY: CUP.
Boesch, C., & Boesch-Acherman, H. (2000). *The chimpanzees of the Tai Forest*. NY: OUP.
Goodall, J. (1986). *The chimpanzees of Gombe*. Cambridge MA: Belknap (HUP).
Kappeler, P.M., & Watts, D.P. (Eds.) (2011). *Long-term field studies of primates*. NY: Springer. [Chapters 14, 15, 16]
Lonsdorf, E.V. et al. (Eds.) (2010). *The mind of the chimpanzee*. Chicago: UCP.
Nishida, T. (2012). *Chimpanzees of the lakeshore*. NY: CUP.
Reynolds, V. (2005). *The chimpanzees of the Budongo Forest*. NY: OUP.
Stumpf, R.M. (2011). Chimpanzees and bonobos. In C.J. Campbell et al. (Eds.), *Primates in perspective, 2nd ed.* (pp. 340–356). NY: OUP.

Additional References

Amsler, S.J. (2010). Energetic costs of territorial boundary patrols by wild chimpanzees. *AJP, 72*, 93–103.
Basabose, A.K. (2002). Diet composition of chimpanzees inhabiting the montane forest of Kahuzi, Democratic Republic of Congo. *AJP*, 58, 1–21.
Boesch, C. (2013). Ecology and cognition of tool use in chimpanzees. In C.M. Sanz, et al. (Eds.), *Tool Use in Animals—Cognition and ecology* (pp. 21–46). NY: CUP.
Boesch, C., et al. (2009). Complex tool sets for honey extraction among chimpanzees in Loango National Park, Gabon. *JHE*, 56, 560–569.
Byrne, R.W. (2007). Culture in great apes: using intricate complexity in feeding skills to trace the evolutionary origin of human technical prowess. *PTRSB, 362*, 577–585.
Carlson, B.A., et al. (2013). Diurnal variation in nutrients and chimpanzee foraging behavior. *AJP, 75*, 342–349.
Carter, M.L., et al. (2008). Skeletal pathology in *Pan troglodytes schweinfurthii* in Kibale National Park, Uganda. *AJPA, 135*, 389–403.
Chemurot, M., et al. (2012). Amount of plant foods eaten and sexual differences in feeding among wild chimpanzees (*Pan troglodytes*) of Kanyawara Community. *ISRN Zoology*, doi.org/10.5402/2012/120250.
Constable, J.L., et al. (2001). Noninvasive paternity assignment in Gombe chimpanzees. *Molecular Ecology, 10*, 1279–1300.
Cordoni, G., & Pelagi, E. (2011). Ontogenetic trajectories of chimpanzee social play: similarities with humans. *PLONE*, DOI: 10.1371/ journal.pone. 0027344.
Corp, N., & Byrne, R.W. (2002). Leaf processing by wild chimpanzees: physically defended leaves reveal complex manual skills. *Ethology, 108*, 673–696.

Crockford, C., et al. (2012). Wild chimpanzees inform ignorant group members of danger. *CB*, *22*, 142–146.

Crockford, C., et al. (2013). Urinary oxytocin and social bonding in related and unrelated chimpanzees. *PRSB*, *280*, DOI: 10.1098/rspb.2012.2765.

DeBlauwe, I., & Janssens, G.P. (2008). New insights in insect prey choice by chimpanzees and gorillas in southeast Cameroon: the role of nutritional value. *AJPA*, *135*, 42–55.

Devos, C., et al. (2002). New record of algae feeding and scooping by *Pan t. troglodytes* at Lokoue Bai in Odzala National Park, Republic of Congo. *Pan Africa News*, 9, 19–21.

Emery Thompson, M., & Wrangham, R.W. (2008). Male mating interest varies with female fecundity in chimpanzees. *IJP*, *29*, 885–905.

Fahy, G.E., et al. (2013). Meat-eating and hunting specialization in adult male chimpanzees, *PNAS*, *110*, 5829–5833.

Fawcett, K., & Muhumuza, G. (2000). Death of a wild chimpanzee community member: possible outcome of intense sexual competition. *AJP*, *51*, 243–247.

Fedurek, P., et al. (2014). Social and ecological correlates of long-distance pant hoot calls in male chimpanzees. *BE&S*, 68, 1345–1355.

Flack, J.C., et al. (2004). Play signaling and the perception of social rules by juvenile chimpanzees (*Pan troglodytes*). *JCP*, *118*, 149–159.

Ganzhorn, J.U., & Wright, P.C. (1994). Temporal patterns in primate leaf eating: the possible role of leaf chemistry. *FP 63*, 203–208.

Gilby, I.C. (2006). Meat sharing among the Gombe chimpanzees: harassment and reciprocal exchange. *AB*, *71*, 953–963.

Gilby, I.C., & Wrangham, R.W. (2007). Risk-prone hunting by chimpanzees (*Pan troglodytes schweinfurthii*) increases during periods of high diet quality. *BE&S*, *61*, 1771–1779.

Gilby, I.C., et al. (2006). Ecological and social influences on the hunting behaviour of wild chimpanzees, *Pan troglodytes schweinfurthii*. *AB*, *72*, 169–180.

Gilby, I.C., et al. (2008). Economic profitability of social predation among wild chimpanzees. *AB*, *75*, 351–360.

Gilby, I.C., et al. (2013a). Ecology rather than psychology explains co-occurrence of predation and border patrols in male chimpanzees. *AB*, *86*, 61–74.

Gilby, I.C., et al. (2013b). Fitness benefits of coalitionary aggression in male chimpanzees. *BE&S*, *67*, 373–381

Gomes, C. M., & Boesch, C. (2009). Wild chimpanzees exchange meat for sex on a long-term basis. *PLONE* 4, e5116.

Goodall, J. (1973). Cultural elements in a chimpanzee community. In E. Menzel (Ed.), *Precultural primate behaviour* (pp. 144–184). Basel: S. Karger.

Hashimoto, C., et al. (2001). What factors affect the size of chimpanzee parties in the Kalinzu Forest, Uganda? *IJP*, *22*, 947–959.

Hill, K., et al. (2001). Mortality rates among wild chimpanzees. *JHE*, *40*, 437–450.

Hladik, C.M. (1977). Chimpanzees of Gabon and chimpanzees of Gombe: some comparative data on the diet. In T.H. Clutton-Brock (Ed.), *Primate ecology* (pp. 481–501). NY: Academic Press.

Hoffman, S. (2012). Ape fracture patterns show higher incidence in more arboreal species. *Discussions@ studentpulse.com*, 8, 1–21.

Horner, V. (2010). The cultural mind of chimpanzees: how social tolerance can shape the transmission of culture. In E.V. Lonsdorf et al. (Eds.), *The mind of the chimpanzee*. Chicago: Chicago: UCP.

Horner, V., et al. (2010). Prestige affects cultural learning in chimpanzees. *PLONE*, 5, e10625.

Hunt, K.D., & McGrew, W.C. (2002). Chimpanzees in dry habitats at Mount Assirik, Senegal and at the Semliki-Toro Wildlife Reserve, Uganda. In C. Boesch et al. (Eds.), *Behavioural diversity in chimpanzees and bonobos* (pp. 35–51). NY: CUP.

Inoue, E., et al. (2008). Relatedness in wild chimpanzees: influence of paternity, male philopatry, and demographic factors. *AJPA*, *137*, 256–262.

Inoue, S., & Matsuzawa, T. (2007). Working memory of numerals in chimpanzees. *CB*, *17*, R1004–R1005.
Itoh, N., & Nishida, T. (2007). Chimpanzee grouping patterns and food availability in Mahale Mountains National Park, Tanzania. *Primates*, *48*, 87–96.
Jurmain, R. (1997). Skeletal evidence of trauma in African apes, with special reference to the Gombe chimpanzees. *Primates*, *38*, 1–14.
Kaburu, S.S., et al. (2013). Death of the alpha: within-community lethal violence among chimpanzees of the Mahale Mountains National Park. *AJP*, *75*, 789–797.
Kahlenberg, S.M., et al. (2008). Immigration costs for female chimpanzees and male protection as an immigrant counterstrategy to intrasexual aggression. *AB*, *76*, 1497–1509.
Kivell, T.L., & Schmitt, D. (2009). Independent evolution of knuckle-walking in African apes shows that humans did not evolve from a knuckle-walking ancestor. *PNAS*, *106*, 14241–14246.
Koyama, N.F., et al. (2012). Supply and demand predict male grooming to swollen females in captive chimpanzees (*Pan troglodytes*). *AB*, *84*, 1419–1425.
Lehmann, J., et al. (2007). Fission–Fusion social systems as a strategy for coping with ecological constraints: a primate case. *Evolutionary Ecology*, *21*, 613–634.
Lodwick, J.L., et al. (2004). From nest to nest—influence of ecology and reproduction on the active period of adult Gombe chimpanzees. *AJP*, *64*, 249–260.
McGrew, W.C. (2001). The other faunivory: primate insectivory and early human diet. In C.B. Stanford & H.T. Bunn (Eds.), *Meat-eating and human evolution* (pp. 160–178). NY: OUP.
Melis, A.P., et al. (2010). Collaboration and helping in chimpanzees. In E. Lonsdorf et al. (Eds.), *The mind of the chimpanzee* (pp. 265–281). Chicago: UCP.
Menzel, E.W. (1972). Spontaneous invention of ladders in a group of young chimpanzees. *FP*, *17*, 87–106.
Mitani, J.C. (2006). Demographic influences on the behavior of chimpanzees. *Primates*, *47*, 6–13.
Mitani, J.C. (2009). Male chimpanzees form enduring and equitable social bonds. *AB*, *77*, 633–640.
Mitani, J.C., & Gros-Louis, J. (1998). Chorusing and call convergence in chimpanzees: tests of three hypotheses. *Behaviour*, *135*, 1041–1064.
Mitani, J.C., & Watts, D. (2001). Why do chimpanzees hunt and share meat? *AB*, *61*, 915–924.
Mitani, J.C., & Watts, D. (2005). Correlates of territorial boundary patrol behavior in wild chimpanzees. *AB*, *70*, 1079–1086.
Mitani, J.C., et al. (2010). Chimpanzee minds in nature. In E. Lonsdorf et al. (Eds.), *The mind of the chimpanzee* (pp. 181–191). Chicago: UCP.
Morin, P.A., et al. (1994a). Kin selection, social structure, gene flow and the evolution of chimpanzees. *Science*, *265*, 1193–1201.
Morin, P.A., et al. (1994b). Paternity exclusion in a community of wild chimpanzees using hypervariable simple sequence repeats. *Molecular Ecology*, *3*, 469–478.
Muller, M., & Mitani, J.C. (2005). Conflict and cooperation in wild chimpanzees. In P.J.B. Slater, et al. (Eds.), *Advances in the study of behavior* (pp. 275–331). NY: Elsevier.
Muller, M.N.,& Wrangham, R.W. (2004a). Dominance, aggression and testosterone in wild chimpanzees: a test of the challenge hypothesis. *AB*, *67*, 113–123.
Muller, M.N., & Wrangham, R.W. (2004b). Dominance, cortisol and stress in wild chimpanzees (*Pan troglodytes schweinfurthii*). *BE&S*, *55*, 332–340.
Muller, M.N., et al. (2006). Male chimpanzees prefer mating with old females. *CB*, *16*, 2234–2238.
Murray, C.M., et al. (2007). A new case of intra-group infanticide in the chimpanzees of Gombe National Park. *IJP*, *28*, 23–37.
Nakamura, M. (2010). Ubiquity of culture and possible social inheritance of sociality among wild chimpanzees. In E. Lonsdorf et al. (Eds.), *The mind of the chimpanzee* (pp. 156–167). Chicago: UCP.
Nakamura, M., & Itoh, N. (2001). Sharing of wild fruits among male chimpanzees: two cases from Mahale, Tanzania. *Pan Africa News*, *8*, 28–31.
Newton-Fisher, N.E. (1999). Association by male chimpanzees: a social tactic? *Behaviour*, *136*, 705–730.

Newton-Fisher, N.E., et al. (2002). Hunting of mammalian prey by Budongo Forest chimpanzees. *FP*, *73*, 281–283.
Newton-Fisher, N.E., et al. (2010). Paternity and social rank in Budongo forest chimpanzees. *AJPA*, *142*, 417–428.
Nishida, T., (1995). A within-group gang attack on a young adult male chimpanzee: ostracism of an ill-mannered member? *Primates*, *36*, 207–211.
Nishida, T., & Inaba, A. (2009). Pirouettes: the rotational play of wild chimpanzees. *Primates*, *50*, 333–341.
Pontzer, H., & Wrangham, R.W. (2004). Climbing and the daily energy cost of locomotion in wild chimpanzees: implications for hominoid locomotor evolution. *JHE*, *46*, 315–333.
Proctor, D.P., et al. (2011). Male chimpanzees' grooming rates vary by female age, parity, and fertility status. *AJP*, *73*, 1–8.
Pusey, A.E., et al. (2008). Severe aggression among female *Pan troglodytes schweinfurthii* at Gombe National Park, Tanzania. *IJP*, *29*, 949–973.
Raubenheimer, D., & Rothman, J.M. (2013). The nutritional ecology of entomophagy in humans and other primates. *ARo Entomology*, 58, 141–160.
Sakamaki, T. (2009). Group unity of chimpanzees elucidated by comparison of sex differences in short-range interactions in Mahale Mountains National Park, Tanzania. *Primates*, *50*, 321–332.
Sakura, O. (1994). Factors affecting party size and composition of chimpanzees (*Pan troglodytes verus*) Bossou, Guinea. *IJP*, *15*, 167–183.
Sanz, C., & Morgan, D. (2009). Flexible and persistent tool-using strategies in honey-gathering by wild chimpanzees. *IJP*, *30*, 411–427.
Sayers, K., & Menzel, C.R. (2012). Memory and foraging theory: chimpanzee utilization of optimality heuristics in the rank-order recovery of hidden foods. *AB*, *84*, 795–803.
Schmelz, M., et al. (2013). Chimpanzees predict that a competitor's preference will match their own. *BL*, *9*, DOI: 10.1098/rsbl.2012.0829.
Sherrow, H.M., & Amsler, S.J. (2007). New intercommunity infanticides by the chimpanzees of Ngogo, Kibale National Park, Uganda. *IJP*, *28*, 9–22.
Shultz, S., & Dunbar, R. (2006). Chimpanzee and felid diet composition is influenced by prey brain size. *BL*, *2*, 505–508.
Slater, K., et al. (2008). The influence of oestrous swellings on the grooming behaviour of chimpanzees of the Budongo Forest, Uganda. *Behaviour*, *145*, 1235–1246.
Slocombe, K.E., & Newton-Fisher, N.E. (2005). Fruit sharing between wild adult chimpanzees (*Pan troglodytes schweinfurthii*): a socially significant event? *AJP*, 65, 385–391.
Sobolewski, M.E., et al. (2012). Territoriality, tolerance and testosterone in wild chimpanzees. *AB*, *84*, 1469–1474.
Sobolewski, M.E., et al. (2013). Female parity, male aggression and testosterone. *Primates*, *54*, 81–88.
Stevens, J.R. (2004). The selfish nature of generosity: harassment and food sharing in primates. *PRSB*, *271*, 451–456.
Stewart, F.A., & Piel, A.K. (2014). Termite fishing by wild chimpanzees: new data from Ugalla, western Tanzania. *Primates*, *55*, 35–40.
Takasaki, H., & Uehara, S. (1984). Seed dispersal by chimpanzees: Supplementary Note 1. *African Study Monographs*, 5, 91–92.
Takemoto, H. (2004). Seasonal change in terrestriality of chimpanzees in relation to microclimate in the tropical forest. *AJPA*, *124*, 81–92.
Tennie, C., et al. (2009). The meat-scrap hypothesis: small quantities of meat may promote cooperative hunting in wild chimpanzees (*Pan troglodytes*). *BE&S*, 63, 421–431.
Townsend, S.W., et al. (2007). Female-led infanticide in wild chimpanzees. CB, *17*, R355–R356.
Tutin, C.E.G. (1979). Mating patterns and reproductive strategies in a community of wild chimpanzees (*Pan troglodytes schweinfurthii*). *BE&S*, 6, 29–38.
Ushida, K., et al. (2006). Nutritional significance of the selective ingestion of *Albizia zygia* gum exudate by wild chimpanzees in Bossou, Guinea. *AJP*, 68, 143–151.

Vigilant, L., et al. (2001). Paternity and relatedness in wild chimpanzee communities. *PNAS*, *98*, 12890–12895.

Vonk, J., & Subiaul, F. (2009). Do chimpanzees know what others can and cannot do? Reasoning about "capability". *AC*, *12*, 267–286.

Wakefield, M. (2008). Grouping patterns and competition among female *Pan troglodytes schweinfurthii* at Ngogo, Kibale National Park, Uganda. *IJP*, *29*, 907–929.

Wakefield, M.L. (2013). Social dynamics among females and their influence on social structure in an East African chimpanzee community. *AB*, *85*, 1303–1313.

Watts, D.P. (2004). Intracommunity coalitionary killing of an adult male chimpanzee at Ngogo, Kibale National Park, Uganda. *IJP*, *25*, 507–521.

Watts, D.P. (2012). Long-term research on chimpanzee behavioral ecology in Kibale National Park, Uganda. In P.M. Kappeler & D. Watts (Eds.), *Long-term field studies of primates* (pp. 313–228). New York: Springer.

Watts, D.P. & Mitani, J.C. (2002). Hunting behavior of chimpanzees at Ngogo, Kibale National Park, Uganda. *IJP*, *23*, DOI:10.1023/ A:1013270606320.

Williams, J.M., et al. (2002). Female competition and male territorial behaviour influence female chimpanzees' ranging patterns. *AB*, *63*, 347–360.

Wilson, M.L. & Wrangham, R.W. (2003). Intergroup relations in chimpanzees. *ARA*, *32*, 363–392.

Wilson, M.L., et al. (2007). Chimpanzees *Pan troglodytes* modify grouping and vocal behaviour in response to location-specific risk. *Behaviour*, *144*: 1621–1653.

Wilson, M.L., et al. (2012). Ecological and social factors affect the occurrence and outcomes of intergroup encounters in chimpanzees. *AB*, *83*, 277–291.

Wrangham, R.W. (1999). The evolution of coalitionary killing. *YPA*, *42*, 1–30.

26 Bonobos

Lightweight Chimpanzees

A male bonobo swings toward a cluster of attractive fruits where three females are feeding. They threaten him away. Another male, the adult son of one of the females, approaches and is allowed to feed.

Bonobos are the closest relatives of chimpanzees and resemble them closely in physical features (Box 26.1). Though sometimes called *pygmy chimpanzees*, bonobos simply have a more slender build than other *Pan* species rather than being a miniature version. They overlap common chimpanzees in height and weight because of their long limbs and muscular thighs. Bonobos are scattered across the Democratic Republic of the Congo in various kinds of forest. Some exploit secondary forest and savanna in mosaic habitats. Despite resembling common chimpanzees in many features, bonobos display some drastic differences in behavior.

Box 26.1 Bonobo Taxonomy

Currently bonobos are all included in the single species *Pan paniscus* in the same genus as common chimpanzees. Bonobos differ from chimpanzees in having more gracile skulls, less robust limbs, and longer hands and feet. The face is black with red lips and prominent side-whiskers. The skull is round and head hair is long and parted in the middle. Despite sexual dimorphism in weight, there is virtually none in the limb skeleton or dentition. They are knuckle walkers like common chimpanzees, but it has not been determined that they knuckle walk in the same way.

Bonobo Ecology

Some bonobo locomotion in dense forest is unusual. In **quadrumanous scrambling** (Doran and Hunt 1994) an individual negotiates foliage and terminal branches using all four appendages opportunistically, without any particular gait. Even more distinctive is leaping and diving (about 10 percent of bouts at one site), which involves launching into the air either feet-first or head-first. Bonobos demonstrated their power in experiments by vertically jumping twice as high as humans (Scholz et al. 2006). However, these apes mostly travel on the ground. An individual may cover about two kilometers during a typical foraging day.

The typical bonobo diet is about half fruit and a quarter THV. Comparisons suggest that THV is more important to bonobos than to chimpanzees as a source of protein, and that they use THV in all seasons while it is only a fallback for chimpanzees. Bonobos at one site exploited seeds heavily during a stressful season, gathering them from the ground instead of THV (Georgiev et al. 2011). Seed abundance seemed to ameliorate competition and allow female associations to continue unimpaired. Preference for fruit and potential reliance on seeds suggest that bonobo feeding ecology may be more like that of common chimpanzees than has been recognized (Georgiev et al. 2011; Hohmann et al. 2010). Some bonobos wade into water to obtain algae (Figure 26.1).

Despite an increasing number of reports, it remains clear that faunivory is less significant for bonobos than for chimpanzees (Hohmann and Fruth 2008). Bonobos hunt less varied prey, do so less frequently, and are more individualistic and opportunistic. Levels of carbon and nitrogen isotopes in hair samples from nests indicated that plants provide the needed protein and that fauna play a marginal role in the diet (Oelze et al. 2011). THV and other protinaceous plants (perhaps seeds) consumed by the bonobos may lessen the need for animal foods. Surbeck and Hohmann (2008) considered and rejected several explanations for less faunivory including insufficient data and human interference. They also rejected the idea that hunting was diminished due to selection against aggression in the evolution of female gregariousness because, in contrast to chimpanzees, female bonobos actively participate in hunting. Compared to chimpanzees and other great apes, insectivory is surprisingly absent from the bonobo diet. Research at one site demonstrated limited insectivory despite the presence of insect taxa favored by common chimpanzees and the kinds of vegetation needed to make insect probes (McGrew et al., 2007).

Bonobo Sociality

The absolute size of bonobo parties falls into the same range as that of common chimpanzees (Furuichi 2009), but this exemplifies the need for care in choosing and interpreting measures

Figure 26.1 A bonobo wades in water with a stick, possibly to gather algae. © USO via iStock

of behavior. Considered as ratios, chimpanzee parties are 10 to 20 percent of community size while bonobo parties are 25 to 50 percent. Bonobos range in large parties when feeding on preferred fruits. They often form even larger nest groups, sleeping close together, and assemble in the same mixed party the following morning. Despite larger size, bonobo parties are more stable, which may explain the smaller variety of greeting behaviors in bonobos and the lower frequency of mutual grooming. Evaluation during every hour of the day at Wamba disclosed that mean size remained the same, averaging about 8.5 between two studies (Mulavwa et al. 2008, 2010).

Seeking an explanation of party sizes, comparative studies indicate that THV and fruit patches are too variable to account for a systematic difference between the *Pan* species (Furuichi 2009). Another possibility is the bonobo feed-as-you-go foraging pattern that means slower travel between fruit patches. Females, especially those with infants, are better able to keep up with a party and augment its size. However, estimated daily travel distances are not drastically different according to extant data: about 2.6 kilometers for mixed parties of bonobos and 2 to 4 kilometers for female chimpanzees.

Chapman et al. (1994) concluded that species differences in competition levels accounted for differences in social cohesion. White (1996) hypothesized that the use of large arboreal fruit patches reduces competition and favors larger parties to defend the resource. Comparisons of Lomako bonobos with chimpanzees at the forested Kibale site support this explanation. Although food abundance and mean party size were correlated in both cases, focal sampling showed that parties were smaller for more months at Kibale than at Lomako.

Higher levels of female cohesiveness can partly explain bonobo party size in proximal terms (Furuichi 2009, 2011). The female attendance ratio is greater than that of males in parties of any size. This affects composition as well as size. Most bonobo parties contain adults of both sexes with females outnumbering males, and all-female parties are also common. Another factor in composition is that a community contains social subgroups of individuals who tend to share specific parts of the community's range and who are likely to form parties with each other.

Bonobos have been called *female bonded*, but this requires two qualifications. First, there is evidence that female–male relationships are at least as important. Second, the proximal basis for female association is not necessarily a personal or emotional one. Females certainly display affinity for one another, as party composition shows (Hohmann and Fruth 2002). They can forage close together in a relaxed way with few aggressive interactions or signals of dominance and submission. Female–female grooming is common in parties at rest. Unrelated females in a wild community or a captive group form coalitions to control access to desirable food and resist domination by males. This contrasts sharply with the emphasis on male–male relationships in chimpanzees. Kin selection is not a factor, since females disperse.

Mother–son bonds and other female–male relationships also distinguish bonobos from chimpanzees (Furuichi 2009). Female–male grooming was more frequent than female–female at Wamba. When mothers and sons were excluded from the analysis, results were mixed: unrelated females displayed the strongest proximity among adults, but grooming was more frequent between unrelated males and females (Furuichi and Ihobe 1994). The tendency toward large parties seems to be based on female aggregation, supplemented by female bonds with adult sons.

Female dyads at Lomako rarely lasted beyond one field season while some male–female relationships spanned several years (Hohmann et al. 1999). Hohmann and Fruth (2002) compared observed statistical values and found that only the male–female associations occurred more often than would be expected by chance. Association of female bonobos in groups is important, but dyadic male–female relationships appear to be at the core of at least some bonobo societies.

Stevens et al. (2006) compared five captive bonobo groups and found that neither grooming nor proximity set female–female relationships apart from other dyads. Females were distinguished only by the formation of coalitions. Mothers groomed and supported their sons more than they did unrelated males. Female bonobos seem to form coalitions opportunistically, rather than based on interpersonal bonds, with tolerance as the most essential feature. This may account for the unusual behavior called *genito-genital rubbing*, usually shortened to GG rubbing, abbreviated here as GGR. This involves two individuals (in any of various positions) rubbing their sexual skins together, or one individual thrusting against the other. The bonobo clitoris may be a proximal factor because it is larger than in other apes and is located more ventrally than in chimpanzees.

The apparent function of GGR (inferred from context and by correlation with other behaviors) is expression or reinforcement of social bonds (Zamma and Fujita 2004). These contacts occur more frequently after aggression or when one individual is monopolizing food, which suggests specific functions of tension relief and reconciliation. Low-ranking females initiate GGR more often than dominants. However, one study found that GGR was more frequent when food was abundant and decreased when food was rare (White and Lanjouw 1992). The behavior may vary with food quality, distribution, accessibility, type of competition, and extent of monopolization (Hohmann and Fruth 2000).

Bonobos display a strong preference for sociosexual behavior in consolation as opposed to reconciliation (Palagi et al. 2004). Aggression and consolation have figured prominently in recent studies of bonobo social behavior (Clay and de Waal 2013; Furuichi 2011). Hohmann et al. (2009) tested the Tension Regulation Hypothesis for genital contact. They looked at the relationship between such behavior and stress, measured by cortisol levels, with equivocal results.

Bonobos have been characterized as *female-dominant*: female aggression against males is common while the reverse is rare, and females sometimes displace males from preferred feeding positions. White and Wood (2007) distinguished between social dominance and feeding priority. As in lemurs, feeding priority may result from either female dominance or male deference. Another important point is that coalition success is not the same thing as individual dominance. These researchers studied wild bonobos and found that: (1) all adult and subadult males were dominant over females in dyadic encounters; (2) male aggression toward females was common; and (3) male submission to female aggression was rare. Even so, females had feeding priority in small food patches. They entered before males more often than expected and fed more than males at the beginning of a session. As in lemurs, this could represent male deference. Unlike lemurs, however, such deference cannot be explained in terms of nutritional stress on females.

An alternative explanation is the threat of female coalitions, though apparently White and Wood did not observe any. Coalitions are more likely to occur among a larger number of females, and bonobo feeding parties are larger than other kinds of parties. Coalitions are common among captive bonobos where (in the absence of fission–fusion) the same potential partners are always together.

Females occupied the highest ranks in six captive groups but could not evoke submissive responses from males in all contexts. Every group contained high-ranking males who could dominate some females (Stevens et al. 2007). Paoli et al. (2006a) constructed a dominance matrix for a large captive group that included both males and females. During one study period, the alpha female dominated three of four males; during a second period the alpha male dominated all females. Paoli and colleagues found that hierarchies were more linear when groups were smaller. It seems clear that individual dominance ranks among male and female bonobos are intertwined to a greater degree than in many other primates, including chimpanzees.

Dominance interactions between male bonobos are rarely intense, but they do result in clear status for at least the alpha and omega males. Male status is also related to that of the mother. Furuichi (2011) attributed the relatively peaceful nature of bonobo society to the preeminence of females. However, there is experimental evidence that proximal factors in males also play a role. Male bonobos and chimpanzees were placed in a situation where a dominant individual would obtain more food. Testosterone increased in the chimps while cortisol increased in the bonobos, indicating that the chimps perceived the event as a dominance contest while the bonobos viewed it as a source of stress (Wobber et al. 2010).

Peering, a distinctive bonobo pattern, is prolonged gazing at another animal from a distance of 30cm or less (Idani, 1995). It has been interpreted as begging, an initiation of affinitive interaction, or a precopulatory behavior. These interpretations are doubtful, because peering is rarely followed by food transfer, sex, or any other social interaction (Stevens et al. 2005). Peering is mostly directed toward older females, which suggests that it acknowledges female status (Johnson et al. 1999). The behavior was closely linked to the agonistic hierarchy in one captive group (Vervaecke et al. 2000), but in only one of five captive groups after controlling for sex and age (Stevens et al. 2005). The latter research group looked at the behavior of recipients *before* peering took place and found that bonobos most often peered at others who had an interesting resource or commodity (e.g. food or grooming). The apparent connection with dominance arises from the fact that dominant individuals often possess attractive resources (Stevens et al. 2005). A possible alternative is that dominant individuals are intrinsically interesting.

Female bonobos play a special role in food sharing that can be attributed to either dominance or priority. Fruth and Hohmann (2002) recorded 20 females and 10 males who were in possession of large fruits. Females had them in 80 percent of all episodes and in 88 percent of the episodes in which the food was begged. Female supplicants were successful more often than males. Meat ownership was similarly biased toward females.

Fruth and Hohmann used their data to test explanations for food sharing. Kin selection could not account for the female-biased sharing because few if any females in a community are related. Reciprocity was eliminated because very few recipients reciprocated. Buy-off was consistent with the data in that most sharing was passive, but it was contradicted by the fact that threatened owners gave less rather than more. Mutualism was the most likely explanation because both parties benefited: owners with supplicants kept a food item longer, especially when the latter prevented others from joining the group around the owner.

Bonobos are more vocal than other great apes. They use numerous calls for both short and long distances, including synchronized choruses. The most common long-distance vocalization is the *high-hoot*. Hooting occurs most often when the apes arrive at feeding sites in the morning and at prospective nesting sites in the afternoon. In addition to their calls bonobos engage in frequent exchanges of looks and gestures; they also break branches and pound on tree buttresses (Savage-Rumbaugh et al. 2005).

Bonobos produce a variety of complex calls that seem to convey information about food quality (Clay and Zuberbühler 2009, 2011). Grunting (the only food call of chimpanzees) is associated with low quality foods. Even so, it is acoustically more complex than the same call in chimps. Bonobos also display four other acoustically different call types, which they emit in complex sequences. Some calls are associated with specific food preferences, while others are more general and some are also used in contexts other than food. Barks, the vocalizations with the highest pitch, are responses to the highest quality foods. Regarding proximate causation, high pitch might point to a high level of arousal, although this conclusion is debatable because the concept of arousal has proven difficult to define. Alternatively, high pitch might make the calls more audible to other bonobos. A study with captive animals indicated that males might signal in order to attract potential mates, while females might benefit by attracting coalition partners (Van Krunkelsven et al. 1996).

Sex is an everyday event for all age and sex classes of bonobos. Many female characteristics promote a high level of sexuality. The sexual cycle is 40 days, longer than in chimpanzees, and swelling occurs throughout most of the cycle. Maximum swelling and postconception swelling last longer than in chimpanzees. Ovulation is less reliably tied to swelling and often occurs after detumescence. Copulation rates between successive births are higher for bonobos because of the quick resumption of cycling and longer tumescence.

Males initiate almost all heterosexual mating attempts. Opportunistic mating is the most common form while consortship and possessive mating are rare. Males display little competition for females, and extra-community mating is frequent. Because the sexual cycle places few limits on female bonobos, there are usually a number of females available at any given time. This makes it more difficult for any male to monopolize mating opportunities. The female determines whether or not copulation occurs. Female cohesiveness and dominance may give females greater mate choice (Hohmann and Fruth 2002). Females may tend to choose dominant males; there is evidence that these males have more mating success (Surbeck et al. 2011). The presence of mothers enhances the mating success of sons, which reduces the proportion of matings by the highest-ranking male. This sexual system may contribute to female dominance over males (Surbeck 2012a, 2012b).

Bonobos in wild communities display varied responses to other groups, but are generally tolerant, possibly due to low levels of feeding and mating competition. Parties from different communities interact often, generating excitement but little serious aggression. Sometimes groups feed or rest together. Excited males tend to avoid close contact with outsiders, but females may temporarily join the alien group and copulate with its males. Encounters often begin with hostility and end peacefully. However, about a third of the encounters at Lomako involved physical aggression and more males augmented party size on days following such encounters (Hohmann and Fruth 2002).

Bonobo Life Cycle

The average inter-birth interval is shorter than in chimpanzees. Female bonobos often resume cycling just a year after giving birth. This may be related to greater food availability in bonobo habitats (Wrangham 2002). Bonobos are born with side-whiskers that distinguish them from chimpanzees. At Wamba (Kano 1992) the infant clings continuously underneath the mother until about three months old, when some gradual separation begins. At six months the infant starts to totter around its mother, but she will bar any further movement with her hand. At about this time the infant becomes interested in solid food and begins to beg by extending a hand toward food items in the mother's hand or mouth. Ventral carriage remains typical until about two years, although the infant occasionally rides on its mother's back.

Adult males share food and nesting spaces with infants. Sometimes an unrelated adult male embraces an infant or carries it on his back. These responses are consistent with the sexual system that gives males little indication of which infants they may have fathered. The pattern may parallel that of some macaques: paternalistic behavior is associated with promiscuous sexuality, single-mount ejaculation, and an even sex ratio in the community.

Male infants display penile erections as early as six months old and begin sexual behavior within the first year. After his mother engages in GGR, a male infant will cling to the other female and insert his penis. In an experimental feeding test with wild-born sanctuary orphans (two to four years old), infant bonobos displayed various sociosexual behaviors with both opposite-sex and same-sex partners that were never seen in the chimpanzee infants (Woods and Hare 2011). All subjects were orphaned while still in close contact with their mothers, so the behavior is unlikely to be the result of learning. Bonobo sexuality seems to be species-typical. However, the experiment leaves the function of the behavior somewhat uncertain. Since sociosexual behavior occurred more often during food tests, it may have resulted from positive arousal due to the food, or stress due to potential competition.

A Wamba juvenile of two years or more typically rides on its mother's back, but a mother may take her juvenile ventrally up to age four in crisis situations (such as an intraspecific fight or a predator scare). As the juvenile ventures farther away from her, the mother signals when travel is about to begin. After vocalizing, and then walking a short distance, she stands with one foot slightly lifted and the sole facing backward. During the juvenile phase a mother plays with her offspring less and grooms it more.

The juvenile begs for food by touching its mother's mouth while sitting on her back or in her lap. Though she moves her head and looks away, she allows the youngster to persist and finally gives in. Examples of this behavior given by Kano (1992) involved sugarcane, a provisioned food. Occasionally something other than tolerated theft occurs. A mother strips the skin from some sugarcane, breaks off a piece, and holds it in her mouth until her juvenile takes it. Juveniles also beg from other adults, but with little success, perhaps as part of establishing social relationships. Weaning conflict is minimal. In contrast to chimpanzee studies, maternal rejection and juvenile tantrums have not been observed. Bonobo mothers are generally indulgent and do not threaten or punish their offspring.

Wamba juveniles often use branches in social play. When one runs away with a branch, the playmate takes it as a signal to give chase and try to grab the branch. When the possessor finally loses it, he becomes the pursuer. In contrast to other hominoids juvenile bonobos rarely engage in sex play with each other. In any case sexual interactions are rare for juvenile female bonobos, whose external genitalia have been characterized as extremely small (Kano 1992). The ratio of male to female sexual episodes is about seven to one. Various sexual interactions with adults occur regularly, often at the invitation of the adults.

Adult tolerance for juveniles declines as adolescence approaches. Adults chase and threaten adolescents and sometimes bite them. Mothers rarely intervene, so adolescents are usually found on the periphery of any party they are in. Dominance seems not to be fixed among adolescents. Aggressive interactions are not violent and often turn into play.

Throughout adolescence (7 to 14 years) and into adulthood, Wamba males almost always traveled with their mothers and engaged in mutual grooming with them. Occasionally the male begged food from his mother while the reverse occurred much less often. A mother often intervened when her son was attacked, either with a warning vocalization or a physical response. Sons rarely reciprocated, possibly because senior females are seldom attacked.

A Wamba female who is approaching adolescence becomes less social than the males and stays on the periphery of her group, often alone in a tree. This may be a refuge from decreasing adult tolerance, but it may also be preparation for emigration, which is typically done by adult females but sometimes by adolescents.

An adult female's relationships within a community begin with immigration. In contrast to chimpanzees, an incoming female establishes bonds with one or more senior females who are already in residence. Thus immigrant females can form alliances against males if necessary and occupy the central position in parties (Furuichi 2009).

Adult bonobos play more often and more roughly than adult chimpanzees. The most common combination is two females, which seems to be related to a dual role of play in social solidarity and social assessment (Palagi 2006). Female dominance over males (or equality) depends on cooperation, which requires relationship maintenance, especially because there is fluidity in the fission–fusion system. Dominance among females depends on continuing social assessment, which is also affected by the fluidity of social grouping. The tendency to rougher play accounts for the greater frequency and intensity of play signals, especially the *full play face* with the upper tooth row bared. Bared teeth make the play face resemble the silent bared teeth face (SBT), which indicates appeasement.

Bonobo Cognition

Hermann et al. (2010b) compared rescued bonobos with chimpanzees on six scales of cognitive skills. Bonobos scored higher in two measures that the experimenters took to represent Theory of Mind: following an experimenter's gaze to a target and understanding what an experimenter intended to do. The difference between the *Pan* species is consistent with the closer social ties and more peaceful behavior of bonobos in the wild.

Rilling et al. (2011) used a variety of imaging techniques to investigate neurological mechanisms underlying bonobo society. Bonobos have more gray matter than chimpanzees in brain regions that are involved with perceiving distress in the self and in other individuals. They also have a larger pathway between two brain regions that implies greater top-down control of aggressive impulses and bottom-up biases against harming others. The researchers hypothesized that this neural system supports increased empathic sensitivity in bonobos along with tension-dissipating behaviors such as sex and play. The implicated brain structures include several that are traditionally considered part of the limbic system or the so-called emotional brain (including the amygdala and hypothalamus). The key pathway described by the research links the amygdala with the prefrontal cortex.

These findings are consistent with evidence that some bonobo success in social tasks owes as much to emotional tendencies as to cognitive ability. This is illustrated by an experiment with both *Pan* species. Paired individuals were given the opportunity to retrieve food by means of cooperation (Hare et al. 2007). When the food was difficult to monopolize, the two species displayed equal cooperation. When the food was highly monopolizable, the bonobos were more successful. The experimenters attributed this to a higher level of social tolerance that did not succumb to the temptation of trying to monopolize the food. The result supported the Emotional Reactivity Hypothesis that predicted more cooperation in bonobos because of their general tolerance level. It did not fulfill the Hunting Hypothesis that predicted better cooperation in chimpanzees because their behavioral repertoire includes cooperative hunting.

Another aspect of temperament may distinguish the *Pan* species from each other. In experiments where the opportunity to gain a better food reward entailed risk, bonobos preferred a fixed option while chimps chose risk with a chance of a better return. This fulfilled a prediction based on chimp exploitation of riskier food sources in nature (Heilbronner et al. 2008; but cf. Rosati and Hare 2010). Hare (2009) reviewed other experimental comparisons and found the results to be consistent with the natural ecologies of the two species. Bonobos in general live in richer environments than chimpanzees and the steadier food supply can explain why bonobos place a lower value on future food payoffs and avoid risky foraging decisions. Lower levels of food competition can explain the bonobo capability for more flexible cooperation in solving joint problems that require high levels of social tolerance.

Hermann et al. (2010b) concluded from their experiments that the level of tool-using behavior is lower in bonobos than in chimpanzees. Gruber et al. (2010) disputed this, based on observations of five groups of captive and rescued bonobos and comparison with chimpanzees in the literature. They concluded that tool-using performances by the two species take place in a variety of contexts, are about equal in their diversity, and can reach similar levels of complexity. The most obvious difference between the species emerged at a bonobo sanctuary where all adults engaged in tool-based play. The researchers see this as support for the more general hypothesis that neotony has been a feature of bonobo evolution.

Meat-eating and hunting provide possible *cultural patterns* in that striking differences across several research sites seem unrelated to availability or any other ecological reason (Fruth and Hohmann 2002; Surbeck and Hohmann 2008). Duikers are the most common prey at two sites but not taken at a third, and flying squirrels are prey at only one site. Meat sharing occurs at two or more sites but is absent from another (Hohmann and Fruth 2008). However, the absence of sharing may be due to exploitation of smaller prey (Ihobe 1992).

Savage-Rumbaugh et al. (1996) suggested that Wamba bonobos manipulated vegetation to communicate about the direction of group travel. They found numerous instances of flattened and broken vegetation and several times saw male bonobos creating these patches with apparent intent. The apes broke off branches, carried them to another location, dropped them there, and then continued to travel. The arrangements pointed in travel directions along bonobo trails, seemingly to communicate with following community members. The majority occurred where trails were harder to follow. Beyond culture, this behavior seems to have implications for spatial and social cognition.

Summary

The *Pan* species, the smallest great apes, display moderate sexual dimorphism. They live in fission–fusion societies and share food in complex ways. Their societies are multi-male and male-philopatric. Their behavior indicates high levels of cognitive function. Females develop large sexual swellings. Bonobos are characterized by more frequent bonding among unrelated females, more complex male–female dominance patterns, less mating competition, and social use of sexuality.

Sources

Basic Bibliography

Boesch, C., et al. (Eds.) (2002). *Behavioural diversity in chimpanzees and bonobos*. NY: CUP.

Caldecott, J., & Miles, L. (Eds.) (2005). *World atlas of the great apes and their conservation*. Berkeley: UCP.

Furuichi, T., & Thompson, J. (Eds.) (2009). *The bonobos: behavior, ecology and conservation*. NY: Springer.

Furuichi, T., et al. (2011). Long-term studies on wild bonobos at Wamba, Luo Scientific Reserve, D. R. Congo. In P.M. Kappeler & D.P. Watts (Eds.), *Long-term field studies of primates* (pp. 413–433). NY: Springer.

Kano, T. (1992). *The last ape, pygmy chimpanzee behavior and ecology*. Stanford CA: Stanford University Press.
Stumpf, R. (2011). Chimpanzees and bonobos: inter- and intraspecies diversity. In C.A. Campbell et al. (Eds.), *Primates in perspective*, 2nd ed. (pp. 340–356). NY: OUP.
Susman, R., (Ed.) (1984). *The pygmy chimpanzee*. NY: Plenum.

Additional References

Chapman, C.A., et al. (1994). Party size in chimpanzees and bonobos: a reevaluation of theory based on two similarly forested sites. In R.W. Wrangham, et al. (Eds). *Chimpanzee cultures* (pp. 41–58). Cambridge MA: HUP.
Clay, Z., & de Waal, F.B.M. (2013). Bonobos respond to distress in others: consolation across the age spectrum. *PLONE*, *8*, e55206.
Clay, Z., & Zuberbuehler, K. (2009). Food associated calling sequences in captive bonobos. *AB*, *77*, 1387–1396.
Clay, Z., & Zuberbuehler, K. (2011). The structure of bonobo copulation calls during reproductive and non-reproductive sex. *Ethology*, *117*, 1158–1169.
Clay, Z., & Zuberbuehler, K. (2012). Communication during sex among female bonobos. *SR*, *2*, 291.
Clay, Z., et al. (2011). Female bonobos use copulation calls as social signals. *BL*, *7*, 513–516.
de Waal, F.B. (1988). The communicative repertoire of captive bonobos (*Pan paniscus*) compared to that of chimpanzees. *Behaviour*, *106*, 183–251.
Doran, D.M., & Hunt, K.D. (1994). Comparative locomotor behavior of chimpanzees and bonobos: species and habitat differences. In R.W. Wrangham et al. (Eds). *Chimpanzee cultures* (pp. 93–108). Cambridge MA: HUP.
Fruth, B., & Hohmann, G. (2002). How bonobos handle hunts and harvests: why share food? In C. Boesch et al. (Eds.), *Behavioural diversity in chimpanzees and bonobos* (pp. 231–243). NY: CUP.
Furuichi, T. (2009). Factors underlying party size differences between chimpanzees and bonobos. *Primates*, *50*, 197–209.
Furuichi, T. (2011). Female contributions to the peaceful nature of bonobo society. *EA*, *20*, 131–142.
Furuichi, T., & Ihobe, H. (1994). Variation in male relationships in bonobos and chimpanzees. *Behaviour*, *130*, 211–228.
Georgiev, A.V., et al. (2011). Seed predation by bonobos (*Pan paniscus*). *Primates*, *52*, 309–314.
Gruber, T., et al. (2010). A comparison of bonobo and chimpanzee tool use: evidence for a female bias in the *Pan* lineage. *AB*, *80*, 1023–1033.
Hare, B. (2009). What is the effect of affect on bonobo and chimpanzee problem solving? In A. Berthoz & Y. Christen (Eds.), *The neurobiology of the umwelt: how living beings perceive the world* (pp. 89–102). NY: Springer.
Hare, B., et al. (2007). Tolerance allows bonobos to outperform chimpanzees on a cooperative task. *CB*, *17*, 619–623.
Heilbronner, S., et al. (2008). A fruit in the hand or two in the bush? Divergent risk preferences in chimpanzees and bonobos. *BL*, *23*, *4*, 246–249.
Hermann, E., et al. (2010a). Bonobos and chimpanzees infer the target of an actor's attention. *AB*, *83*, 345–353.
Hermann, E., et al. (2010b). Differences in the cognitive skills of bonobos and chimpanzees. *PLONE*, *5*, e12438.
Hohmann, G. & Fruth, B. (2000). Use and function of genital contacts among female bonobos. *AB*, *60*, 107–120.
Hohmann, G., & Fruth, B. (2002). Dynamics in social organization of bonobos (*Pan paniscus*). In C. Boesch et al. (Eds.), *Behavioural diversity in chimpanzees and bonobos* (pp. 138–150). NY: CUP.
Hohmann, G., & Fruth, B. (2008). New records on prey capture and meat eating by bonobos at Lui Kotale. *FP*, *79*, 103–110.

Hohmann, G., et al. (1999). Social bonds and genetic ties: kinship, association and affiliation in a community of bonobos *(Pan paniscus)*. *Behaviour, 136*, 1219–1235.
Hohmann, G., et al. (2009). The relationship between socio-sexual behavior and salivary cortisol in bonobos: tests of the tension regulation hypothesis. *AJP, 71*, 223–232.
Hohmann, G., et al. (2010). Plant foods consumed by Pan: exploring the variation of nutritional ecology across Africa. *AJPA, 141*, 476–485.
Horn, A.D. (1979). The taxonomic status of the bonobo chimpanzee. *AJPA, 51*, 273–282.
Idani, G. (1995). Function of peering behavior among bonobos (*Pan paniscus*) at Wamba, Zaire. *Primates, 36*, 377–383.
Ihobe, H. (1992). Observations of meat-eating behavior of wild bonobos (*Pan paniscus*) at Wamba, Republic of Zaire. *Primates, 33*, 247–250.
Johnson, C., et al. (1999). Peering in mature, captive bonobos (*Pan paniscus*). *Primates, 40*, 397–407.
McGrew, W.C., et al. (2007). Prospects for bonobo insectivory: Lui Kotal, Democratic Republic of Congo. *IJP, 28*, 1237–1252.
Mulavwa, M., et al. (2008). Seasonal changes in fruit production and party size of bonobos at Wamba. In T. Furuichi & J. Thompson (Eds.), *The bonobos: behavior, ecology, and conservation* (pp. 121–134). NY: Springer.
Mulavwa, M.N., et al. (2010). Nest groups of wild bonobos at Wamba: selection of vegetation and tree species and relationships between nest group size and party size. *AJP, 72*, 575–586.
Oelze, V.M., et al. (2011). Exploring the contribution and significance of animal protein in the diet of bonobos by stable isotope ratio analysis of hair. *PNAS, 108*, 9792–9797.
Palagi, E. (2006). Social play in bonobos (*Pan paniscus*) and chimpanzees (*Pan troglodytes*): implications for natural social systems and interindividual relationships. *AJPA, 129*, 418–426.
Palagi, E., et al. (2004). Reconciliation and consolation in captive bonobos (*Pan paniscus*). *AJP, 62*, 15–30.
Paoli, T., et al. (2006a). Reevaluation of dominance hierarchy in bonobos (*Pan paniscus*). *AJPA, 130*, 116–122.
Paoli, T., et al. (2006b). Perineal swelling, intermenstrual cycle, and female sexual behavior in bonobos (*Pan paniscus*). *AJP, 68*, 333–347.
Reichert, K.E., et al. (2002). What females tell males about their reproductive status: are morphological and behavioral cues reliable signals of ovulation in bonobos (*Pan paniscus*)? *Ethology, 108*, 583–600.
Rilling, J.K., et al. (2012). Differences between chimpanzees and bonobos in neural systems supporting social cognition. *Social Cognitive and Affective Neuroscience, 7*, 369–379.
Rosati, A., & Hare, B. (2010). Chimpanzee and bonobos distinguish between risk and ambiguity. *PRSB, 7*, 15–18.
Savage-Rumbaugh, S., et al. (1996). Language perceived. In W. McGrew et al. (Eds.), *Great ape societies* (pp. 1173–1184). NY: CUP.
Savage-Rumbaugh, S., et al. (2005). Culture prefigures cognition in *Pan/Homo* bonobos. *Theoria, 54*, 311–328.
Scholz, M.N., et al. (2006). Vertical jumping performance of bonobo (*Pan paniscus*) suggests superior muscle properties. *PRSB, 273*, 2177–2184.
Stevens, J.M.G., et al. (2005). Peering is not a formal indicator of subordination in bonobos (*Pan paniscus*). *AJP, 65*, 255–267.
Stevens, J.M.G., et al. (2006). Social structures in *Pan paniscus*: testing the female bonding hypothesis. *Primates, 47*, 210–217.
Stevens, J.M.G., et al. (2007). Sex differences in the steepness of dominance hierarchies in captive bonobo groups. *IJP, 28*, 1417–1430.
Surbeck, M., & Hohmann, G. (2008). Primate hunting by bonobo at LuiKotale, Salonga National Park. *CB, 18*, R906–R907.
Surbeck, M., & Hohmann, G. (2013). Intersexual dominance relationships and the influence of leverage on the outcome of conflicts in wild bonobos (*Pan paniscus*). *BE&S, 67*, 1767–1780.

Surbeck, M., et al. (2011). Mothers matter! Maternal support, dominance status and mating success in male bonobos (*Pan paniscus*). *PRSB*, *278*, 590–598.

Surbeck, M., et al. (2012a). Mate competition, testosterone and intersexual relationships in bonobos, *Pan paniscus*. *AB*, *83*, 659–669.

Surbeck, M., et al. (2012b). Social correlates of variation in urinary cortisol in wild male bonobos (*Pan paniscus*). *H&B*, *62*, 27–35.

Van Krunkelsven, E., et al. (1996). Food calling by captive bonobos (*Pan paniscus*): an experiment. *IJP*, *17*, 207–217.

Vervaecke, H., et al. (2000). Dominance and its behavioral measures in a captive group of bonobos (*Pan paniscus*). *IJP*, *21*, 47–68.

White, F. (1996). Comparative socio-ecology of *Pan paniscus*. In W.C. McGrew & 2 co-editors, *Great ape societies* (pp. 29–41). NY: CUP.

White, F.J., & Chapman, C.A. (1994). Contrasting chimpanzees and bonobos: nearest neighbor distances and choices. *FP*, *63*, 181–191.

White, F.J., & Lanjouw, A. (1992). Feeding competition in pygmy chimpanzees: variation in social cohesion. In T. Nishida et al. (Eds.), *Topics in primatology*, *Vol. 1* (pp. 67–79). Tokyo: Tokyo University Press.

White, F.J., & Wood, K.D. (2007). Female feeding priority in bonobos, *Pan paniscus*, and the question of female dominance. *AJP*, *69*, 1–14.

Wobber, V., et al. (2010). Differential reactivity of steroid hormones in chimpanzees and bonobos when anticipating food competition. *PNAS*, *107*, 12457–12462.

Woods, V. & Hare, B. (2011). Bonobo but not chimpanzee infants use socio-sexual contact. *Primates*, *52*, 111–116.

Wrangham, R.W. (2002). The cost of sexual attraction. In C. Boesch et al. (Eds.), *Behavioural diversity in chimpanzees and bonobos* (pp. 231–243). NY: CUP.

Wrangham, R.W., et al. (2010). Evidence for delayed development of social behavior and cognition in bonobos relative to chimpanzees. *CB*, *20*, 226–230.

Zamma, K., & Fujita, S. (2004). Genito-genital rubbing among the chimpanzees of Mahale and Bossou. *Pan Africa News*, *11*, 5–8.

27 Human Origins

The Last Common Ancestor

A long-armed primate rises from her nest in a tree. She swings from branch to trunk and climbs down to the ground as her neighbors do the same. She walks slowly on two feet to a nearby tree. Plucking low hanging fruits, she cradles the accumulation in her long arm. Subordinate females look on but none contest her.

The term *human* is ambiguous and laden with controversial connotations. We will limit it to the species *Homo sapiens* and refer to the clade that produced the human species by the zoological term **hominins** (Box 27.1). The term encompasses living humans and all fossil forms that are more closely related to us than to any of the apes.

Box 27.1 Hominin Taxonomy

Our own taxonomic group has undergone drastic changes over the last few decades. A slowly diminishing number of popular books and articles continue to use the term *hominid* for the clade that distinguishes us from all other living primates, including the genus *Pan*. Hominid now properly denotes the taxon that encompasses our own evolutionary lineage along with those of the living great apes: family Hominidae. A further split placed us in the subfamily Homininae to join us with the African apes, separated from Ponginae. Ponginae contains only one living genus, *Pongo* (orangutans). We are now distinguished from all other living primates in the tribe Hominini. Genus *Homo* still identifies us along with our recent ancestors and some close collateral lines (such as *H. erectus* and *H. neandertalensis*). We continue to flatter ourselves and our most recent ancestors with the species name *H. sapiens*.

We will consider the relationship between humans and other primates from several points of view: (1) the use of living primates to illuminate the behavior of early hominins; (2) the persistence of primate behavioral traits in recent humans; and (3) the effects of recent humans on other primates. This chapter introduces the concept of primate models and applies it to the basal ancestor of the hominins, often called the **Last Common Ancestor** or **LCA**.

Primate Models for Hominin Evolution

There are two main reasons for using primate ecology and behavior to reconstruct hominin life. One is that gaps in the fossil and archeological records leave us with no direct evidence for behavior. The other is that we want to know about behaviors that leave no record, such as social relationships. In cases where we have both prehistoric evidence and living models, they complement one another.

The word *model* has been used widely but loosely. Moore (1996) listed 12 synonyms, ranging from *hypothesis* (an idea amenable to testing) to *just-so story* (a fairy tale, or someone else's model). We will briefly survey several kinds of models that primatologists have advocated.

Referential Models

A **referential model** links one real phenomenon to another one that is more difficult to study. The real phenomenon in each model discussed here is the behavior of extant nonhuman primates; the less amenable referent is the behavior of human ancestors. These **analogical models** (Stanford 2012) may use a particular taxon for reconstruction, based on phylogenetic closeness to the hominin lineage or ecological similarity to a phase of hominin evolution. This has been called the "species model" (Foley 1987) because a single species is often used, but the same approach can be applied to multiple species that are connected by phylogeny or analogy (a genus, for example). Referential models can do the following: (1) show what is possible for a particular kind of animal; (2) suggest important new ideas; and (3) provide detailed scenarios from which testable predictions can be derived. Primatologists commonly use referential models today.

Some critics dismiss this approach because it supposedly assumes that each species has typical behaviors shaped by one or a few variables, whereas real primate behavior is extremely flexible. The criticism is valid as far as it goes, but it ignores two facts. First, some behaviors are consistent across a taxon (e.g. chimpanzees live in fission–fusion communities). Second, an analogical model can take into account the range of behavioral variation in a taxon (e.g. chimpanzee social cohesion increases from forest to savanna).

Conceptual Models

Conceptual models, advocated by Tooby and DeVore (1987), are theories rather than real phenomena, consisting of analytically specified relationships within sets of concepts or variables. The **strategic modeling** that produces these constructs is based on: (1) the gene as the unit of selection; and (2) the idea of animals as strategists promoting their inclusive fitness. This approach is based on behavioral ecology, sociobiology, and evolutionary theory. Tooby and DeVore listed 25 principles of strategic modeling, which seem to restate the basic ideas of the theories just mentioned.

Dunbar (2001) advocated two mathematical approaches. One seeks relationships between traits that are responsible for distribution across the primates. I call this the **correlational model**. Plotting one variable against another (such as group size and body mass) across all of the primates is the basis for a statistical description of the relationship by means of a regression equation. A hominin species can then be interpolated in order to predict (or, more accurately, retrodict) the unknown characteristic. Of course, this works only if one of the variables is already known for the fossil species. Thus the approach is largely limited to behaviors that are correlated with physical traits such as body mass.

This limitation is overcome (at least in part) by the **systems model**, which uses multivariate regression equations to construct functional models of behavioral ecology. The multivariate approach is more likely than others to reflect underlying biological complexities. A systems model represents a population or some other group rather than a whole species, which is also closer to biological reality.

These mathematical models can be powerful predictors. However, the concomitant is that they require much more data than other models (Dunbar 2001). Furthermore, they leave gaps. The traits in question must be widespread to facilitate mathematical treatment, but some important traits are unusual (male–female friendships in baboons, for example). The systems approach is supposedly realistic because it operates at the community or population level. However, it omits the realities of individual behavior and proximal mechanisms.

Another problem is that conceptual models tend to be very speculative and thus more likely to be refuted by new evidence. Tooby and DeVore, for example, made the theory-based assumption that social animals must emigrate from their natal groups in order to avoid inbreeding depression. Subsequent fieldwork demonstrated that neither sex consistently emigrated in Barbary macaques, and other examples have since been discovered.

An approach that includes both referential and conceptual models is likely to be the most effective because they can increase understanding through reciprocal refinement. Stanford (2012) pointed out that they overlap, noting that the extensive analysis by Tooby and DeVore ended up with an early hominin much like a chimpanzee. Much the same can be said of the effort by Strum and Mitchell (1987), which reconstructed a hominin similar to a baboon in its behavior.

Functional Models

Functional modeling is based on comparable adaptations. This refers to: (1) analogies between taxa based on similar environmental factors; or (2) linkage between behavioral patterns such that changes in one affect the other. Functional models can address several different issues: (1) *probable adaptation*: what would a human ancestor have had to do to survive in a certain set of conditions? (2) *capability*: could ancestral hominins have performed a postulated behavior and did they have alternatives? and (3) *limits*: what behaviors were impossible for human ancestors, no matter the need?

Another kind of functional model has been called **attribute analysis** (Foley 1987). It starts with the goal of reconstructing a particular trait in a human ancestor. An analogous trait in an extant primate is then found and studied in order to gain a better understanding of the ancient trait. In other words, known functional explanations for an extant trait provide the model.

Primate Models for the LCA

The genus *Pan* contains the closest living relatives of humans. Thus, living *Pan* and *Homo* have a common ancestor. For humans this is the Last Common Ancestor (abbreviated LCA), the last ancestor shared with other living primates. Molecular evidence from living primates indicates that the hominin lineage emerged between eight and four million years ago. The main fossil candidates for LCA date to within this span. The LCA itself is more or less a theoretical construct. The species had to exist, but there is no fossil form that we can name as the LCA with any certainty. The few fossil candidates are fragmentary and to some extent contradictory in their traits. Living primates can be used to augment the direct evidence.

As in the study of physical evolution, the phylogenetic perspective reasons from common descent, the idea that closely-related species are likely to have some homologous behaviors (shared ancestral patterns). More precisely, a phylogenetic model postulates that key components of behavioral traits are under specific genetic influences that descended from the LCA.

Chimpanzee Model

The chimpanzee referential model is predominant in reconstructions of early hominin behavior. It is generally agreed that chimpanzees are our closest living relatives, which makes them the starting point for any phylogenetic approach. Traits that they share with living humans are strong candidates for origin in the LCA. For example, human smiling and the chimpanzee play face may be derived from the same expression in the LCA.

Functional considerations are also important because humans and chimpanzees, despite their close relationship, have undergone divergent evolution. Where they differ, it is often not obvious which of them (if either) represents the ancestral condition. To the extent that chimpanzees are more similar to the probable LCA, they constitute the more likely model. The LCA and chimpanzees are comparable with regard to environment, body size, and encephalization. These features suggest inferences about such matters as energy need, predator pressure, and cognitive abilities.

Sayers and Lovejoy (2008) criticized the chimpanzee model on two counts: (1) a quadrupedal animal is a dubious model for a biped; (2) other animals are similar to chimps with regard to behaviors of anthropological interest, such as tool use and social hunting. The second point seems to belie the first, since any plausible alternative animals are inevitably quadrupedal (or saltatory or winged). Furthermore, locomotion is not critical in the comparison of every behavioral pattern. Ecologists have compared bison and certain kangaroos, despite drastically different locomotor systems, because both are large, plains-living herbivores (Bolen 1998).

The idea that animals other than chimpanzees should be used in hominin reconstructions is nothing new. They can be used to supplement and complement inferences from chimpanzees. With regard to behaviors of specifically anthropological interest, such as tool use, the best models other than chimpanzees are likely to be other primates (most notably, orangutans).

Bonobo Model

The bonobo has been suggested as a better model for hominin evolution than the common chimpanzee, largely based on its anatomy and locomotion (Zihlman 1996). The bonobo seems to be the more generalized species in having a smaller body (in weight), smaller facial and canine dimensions, and less sexual dimorphism in the dentition. Longer and heavier legs like those of the bonobo, suggesting both strength and a lower center of gravity, could have made the transition to bipedalism easier.

Even if the LCA most closely resembled the bonobo physically, this does not mean that it was like the bonobo in all aspects of behavior. The implications of increased terrestriality, for example, are better modeled by common chimpanzees. Nevertheless, the bonobo may give us some significant insights. Most importantly, the bonobo gives us a broader perspective on the possible social life of the LCA (de Waal 2001; Zihlman 1996), suggesting less aggression, more female influence, and greater flexibility in sexuality. The LCA may have displayed a mixture of traits comparable to both *Pan* species.

Larger Phylogenetic Samples

Many critics of *Pan* models simply want to subsume them in larger primate samples with more inferential power. The *Pan* species offer unique insights for human ancestry, but they also present serious problems. Each species, and the generic lineage that they share, has its own evolutionary history and distinctive characteristics. Similarities between chimpanzees and humans may have a homologous basis, but they may be homoplasies. If chimpanzees and humans differ, there is no guarantee that either trait is the ancestral one. This means we should expand our sample whenever possible (King 1972).

The first option is to include gorillas in an African ape phylogenetic model (Cameron and Groves 2004; Ghiglieri 1987, 1989; Wrangham and Pilbeam 2001). This is still a very limited sample and gorillas differ from chimpanzees in many ways, which limits possible generalizations. The next step is to include orangutans to form a great ape phylogenetic model. For example, tool use in chimpanzees and humans may have had separate origins and the dearth of tools in wild gorillas did little to resolve the issue. However, the fact that tool-using capability exists in all the great apes (albeit rudimentary in gorillas) strengthens the homology interpretation and the inference that such behavior existed in the LCA.

Several writers have constructed their models from surveys of the catarrhines (Cachel 2006; Zihlman and Bolter 2004). It is also possible to generalize across the anthropoids, including New World primates as well as the catarrhines. Finally, Strier (2001) advocated that we consider the primate order as a whole. This approach articulates with conceptual models like those of Dunbar. Such broader perspectives are especially important where the great apes all differ from one another, as they do in social organization. These perspectives can also give us a better understanding of traits shared between apes and many other primates, such as the influence of kinship. The most fundamental features of sociality, for example, probably come from the ancestors of all primates and persisted through our LCA. Cohesive communities are more characteristic of the anthropoids and dominance is especially prominent among catarrhines. These larger samples can provide broadly documented homologies that are likely to have been present in the LCA. The price is that these broad perspectives yield fewer relevant traits and less detail.

Behavior of the Last Common Ancestor

Now we apply various models to the reconstruction of the LCA, in conjunction with direct evidence for environment and morphology.

LCA Ecology

From the LCA to about two million years ago most hominin environments were mosaics (Figure 27.1) that mixed woodlands and a variety of more open areas (Bobe 2011; Cerling et al. 2011). The LCA (and the early hominins that followed) had access to both woodlands and more open areas (Potts 2012; White et al. 2009). Evidence associated with the fossils (such as faunas and pollens) indicates that they preferred the woodland components of their environments. In addition, almost every hominin site provides evidence for access to substantial bodies of water, whether lakes, rivers, or floodplains (Aronson et al. 2008; Reynolds et al. 2011).

The general primate heritage tells us that the LCA was almost certainly arboreal and probably, like all living apes, engaged in suspensory behavior. However, variation among

Figure 27.1 A mosaic habitat in the Okavango Swamp area of Botswana. © Curt Busse

extant apes has inspired different positional models with different implications for the origin of **bipedalism**, the regular use of two-legged locomotion. The behavior of some chimpanzees suggested the assumption of upright postures in trees for feeding. In the Bwindi Forest they displayed a high rate of bipedal posture (0.73 bouts per observation hour), almost entirely during arboreal feeding (Stanford 2006). This behavior may be more common at other sites than is generally known, since the special observing conditions at Bwindi made it more apparent (researchers were at eye-level with chimpanzees in treetops across narrow ravines). Bipedal posture, possibly advantageous for foraging in fruit trees, may have originated before bipedal locomotion.

Other hypotheses also locate the origin of bipedalism in trees. The Vertical Climbing Hypothesis, probably the most widely favored, sees the key positional mode as climbing up and down on supports such as vines and tree trunks (Crompton et al. 2008; Richmond et al. 2001). In African apes this is kinematically more similar to human bipedalism than occasional bipedal locomotion on the ground; hip extension, for example, is greater in vertical climbing than in ape bipedalism. Vertical climbing in the LCA would therefore have facilitated the evolution of hominin bipedalism on the ground.

Orangutan arboreal bipedalism is even more like that of humans on the ground (Crompton et al. 2008). While African apes keep their hindlimbs flexed during these bouts, orangs extend their legs like humans. Hand-assisted bipedalism allows the Asian apes to move among supports that are too small and flexible to access in any other way. They react to branch flexibility by increasing knee and hip extension like humans running on a springy track (all other primates react in the opposite way) (Thorpe et al. 2007). If the postural bipedalism of chimpanzees in the trees facilitates feeding success, it seems likely that the orangutan's locomotor capability is even more advantageous for that function.

Another model of arboreal bipedalism in the LCA is based on the gibbon—or, more precisely, half of the gibbon. This version of the LCA had a generalized hominoid upper body, but a long-legged gibbon-like lower body that allowed it to stand bipedally and engage in short bursts of bipedal running (Richmond et al. 2001; Tuttle 1981). This gives us an LCA that weighed about 9 to 14 kilograms and climbed vertically on tree trunks and vines. It stood bipedally in foraging and for display. However, both ape and hominin fossils indicate that the LCA was considerably bigger than a gibbon. Furthermore, gibbon running is mechanically different from that of humans and gibbons almost never walk bipedally (Crompton et al. 2008).

Positional behavior of African apes suggests a trend toward terrestriality in the LCA. Though well suited to suspensory behavior, more than 85 percent of their locomotion is ground travel between feeding and resting sites (Doran 1996). At least one primate model for terrestrial bipedalism is particularly relevant to woodland habitats like those of the LCA. Observing chimpanzees in such habitats, Hunt (1994, 1996) noted that they adopted bipedal postures most frequently when feeding on the small fruits of small trees in woodland or open-forest settings (Figure 27.2). The apes either reached up from the ground, or while standing on a branch and stabilizing themselves by grasping another branch. Both stances allowed them to reach deeper into the trees, but the terrestrial stance also allowed them to use both hands in harvesting the fruits. This model articulates with that of Stanford (2006) based on Bwindi chimpanzees feeding in trees. Together they suggest that bipedalism originated as a postural feeding adaptation that began in the trees and persisted on the ground. Another terrestrial hypothesis, less in favor for the LCA, emphasizes carrying (Figure 27.3).

Figure 27.2 Baboons, like chimpanzees, sometimes use erect posture to obtain low-hanging fruits. Increased importance for early hominins might have led to selection for bipedal anatomy. © Glenn E. King

Figure 27.3 Baboons and other primates sometimes use a bipedal gait to carry objects over short distances. Here again, increased importance for early hominins might have led to selection for bipedal anatomy. © Glenn E. King

The LCA must have been omnivorous like most other primates. Gut anatomy and function are much the same throughout the living hominoids, including humans (Milton 1999). This indicates a homologous basis for subsistence behavior. The LCA probably preferred ripe fruit and other high quality plant parts, but was flexible enough to feed on any suitable plants that were available in their various habitats. Rather than insisting on particular subsets of food, they could have adjusted to regional and local variations.

The LCA may have eaten terrestrial herbaceous vegetation, as African apes do. Some THV is hard to digest, but herbaceous plants are good fallback foods because they are evenly distributed and generally available throughout the year (Goldsmith 2003), producing new leaves and shoots continuously (Tutin 2003). Herbaceous plants are mostly low in energy, but they provide protein. THV is generally abundant, although biomass does vary significantly from one place to another (Ganas et al. 2009; White et al. 1995).

Lucas et al. (2008) suggested that LCA dentition resembled that of chimpanzees. Wrangham considered the chimpanzee to be the best model for LCA feeding because skull morphology of the extant apes is like that of relevant fossils and body mass matches the best fossil estimates. Bonobos are less likely because of derived features such as small molars and pedomorphic cranial anatomy. However, the wear pattern of relevant fossil teeth most closely resembles that of the orangutan among living apes, which suggests a frugivorous diet with few or no abrasive foods (Haile-Selassie et al. 2009). On the other hand, the large size of the cheek teeth and the thickness of molar enamel indicate some hard foods (White et al. 2009). The percentage of C4 foods eaten by individuals varied from about 10 percent to 25 percent.

From the phylogenetic viewpoint it seems almost certain that the LCA consumed arthropods and possibly some meat. All great ape species, like most other primates, eat insects and some vertebrates. Meat-eating has been documented in both *Pan* species and orangutans in the wild, and gorillas readily eat meat in captivity.

Controversy revolves around how the LCA or early hominins might have obtained meat. In its simplest form this argument pits hunting against scavenging. The *Pan* species both give us evidence for hunting. Chimpanzees also engage in some scavenging—as do most other mammalian predators—but hunting is overwhelmingly predominant in both forest and woodland habitats. It seems likely that the LCA also hunted more than they scavenged. Llorente et al. (2012) described hunting by rehabilitant chimpanzees as similar to that of wild chimps despite the rehabilitants' lack of experience. They concluded that this is species-typical behavior, which implies a genetic basis that might be derived from the LCA.

Paleontological evidence indicates that the LCA faced predation like that of recent times or worse. *Dinofelis* resembled the modern leopard in size and much of its morphology; it may have been phylogenetically close to the leopard's ancestry (Haile-Selassie and Howell 2009). The leopard is the greatest mammalian enemy of extant primates and *Dinofelis* may have presented a similar threat to the LCA. Archaic species of hyenas, bears, and dogs may also have been threats. Earliest hominins may have had long canine teeth (Lucas et al. 2008); however, given their arboreal capabilities, their ultimate safety was probably in the trees.

All great apes build night nests. Fruth and Hohmann (1996) held that nest building is unlikely to have evolved four different times. Therefore, implicitly using the great ape phylogenetic model, they inferred that the LCA built nests like those of other hominoids. These nests, like those of extant apes, may have been multifunctional. Physical comfort and reduction of parasite load might have reinforced the retention of this behavioral pattern. Fruth and Hohmann (1996) also viewed nests as the first results of exploratory behavior with sticks and therefore the foundation for much of great ape and hominin technology.

All great apes use tools in the wild and in captivity. This is minimal in gorillas due to their great strength, and orangutan tool use is limited because their needs are mainly arboreal. Chimpanzees probably provide the best species model for LCA tool use. Given the woodland environment of the LCA, it is likely that they used their tools for much the same purposes as living chimpanzees. The possibility that such behavior might have been prevented by less mental capability than modern apes is dispelled by the tool-using behavior of capuchin monkeys and macaques.

Taking their cue from chimpanzees, McGrew and Marchant (2005) suggested that percussive technology originated in a hominoid ancestor that smashed hard-shelled fruits against tree trunks and boughs. As terrestrial behavior increased in the LCA, they began to use anvils of stone and roots like those now used by chimpanzees and capuchin monkeys. Those primates also use stone hammers to break hard nuts and fruits. Foods that could have been obtained with sticks include social insects, honey, and shellfish.

LCA Sociality

The *Pan* species are fundamentally alike in some important aspects of social structure. Their societies are multi-male fission–fusion communities that typically contain about 25 to 75 individuals, but in some cases more than 100. This is the basis for almost every reconstruction of LCA society (King 1972, 1975; Wrangham 1987). Focus on the chimpanzee suggests that fission–fusion solved problems of male competition and use of scarce resources (Ghiglieri 1989). This is a case where the more widely distributed *Pan*, the common chimpanzee, is probably a better functional model than the bonobo. The other great apes and the gibbons have very different forms of social organization.

Isbell and Young (1996) doubted the relevance of fission–fusion to the LCA. They suggested that it was a *Pan* alternative to hominin bipedalism for access to resources; that is, they were two different ways to encompass a large enough area for subsistence. However, the fission–fusion model is a better fit with the kind of social organization that probably characterized later hominins as they became gatherer-hunters (King 1975, 1980).

There are elements of social behavior common to all African apes that can be imputed to the LCA repertoire. Wrangham (1987) presented his inferences about the LCA as an **ancestral suite**, which he derived from a phylogenetic approach. The term is intended to convey that those particular traits did not necessarily have a genetic basis (as the term homology would suggest). Instead the common social traits might arise from other shared traits, such as cognitive abilities or subtle ecological factors. According to Wrangham, LCA females typically dispersed from their natal groups while males were more likely to be philopatric. He inferred that female alliances were often weak, though females might be drawn together by a common focus such as dependent young of the same age.

Wrangham's inference about weak alliances needs modification in light of chimpanzee behavior at the Arnhem Zoo in the Netherlands. Arnhem females formed powerful alliances, indicating that the behavior is facultative in the species (de Waal 1998). Bonobo females often display similar potential in cooperating to dominate males. These patterns are consistent with Wrangham's idea that females of the African apes and the LCA are/were drawn together by a common focus. They also show that bonobo and common chimpanzee models are not always in conflict.

Ghiglieri suggested that if sexual dimorphism in the LCA was moderate as it is in extant *Pan*, males cooperated in alliances against internal and external rivals. He hypothesized that the LCA displayed other behavior patterns like those of common chimpanzees. Mating competition between males of different communities was relatively intense and in some cases males killed their rivals. Males sometimes traveled alone, which gave coalitions the opportunity to stalk and attack lone extra-group males.

This is a case where the chimpanzee and bonobo models seem to contradict each other. However, occasional tensions and aggression in bonobos suggest that some of the same proximal mechanisms for competition exist in both *Pan* species. Some may also be found in gorillas: a similar predisposition may underlie stalking of one-male units by lone males. *Pan* differs from *Gorilla*, according to Ghiglieri, in that virtually all *Pan* characteristics are traceable to female dispersal and the retention of male offspring in their natal communities.

Somewhat like Wrangham's ancestral suite, van Schaik et al. (2004) suggested a **social diagnosis** of the great apes, which we may take as hypotheses for the LCA. First, they identified a tendency toward *non-permanence of social units* (including gorillas because female membership in gorilla groups is flexible). This comparison between chimps and gorillas is supported by research indicating that gorilla social units in some populations belong to a larger flexible community, while chimpanzee subgroups may have a degree of stability in membership that makes them more similar to gorilla groups. With all three great ape genera on a continuum of non-permanence, the probability of the LCA being on this continuum is greater.

Van Schaik and colleagues see extensive flexibility in the social organization and affiliations of chimpanzees and orangutans, and possibly in the other great ape species. Relatively high subordinate leverage occurs in this fluid situation. Formal (clearly signaled) dominance between frequently associating same-sex individuals is rare in bonobos, chimpanzees (except for some adult male dyads), and Sumatran orangutans (especially females). Social tolerance is manifested in affiliative behaviors such as food sharing. A final unusual social trait among the apes is limited to the *Pan* species: intrasexual bonds among non-relatives are as common, or more so, than bonds among relatives in male chimpanzees, female bonobos, and possibly female chimpanzees. In this case attribution to the LCA depends on preference between the two *Pan* models.

Van Schaik et al. interpreted most of the features in their social diagnosis as indirect consequences of large body size, which is associated with increased potential for contest competition, especially among males for females. Because their size makes them less vulnerable to predators, both males and females can forage alone to avoid intense competition. However, this exposes them to other dangers: sexual coercion (including infanticide) for the females and assault by males of other groups for the males. Thus they will require protectors and allies at times. The situation favors facultative shifting between solitary and gregarious behavior, as well as flexibility in individual associations. This adaptation requires suppression of competitive tendencies in favor of social tolerance. Potentially dominant individuals require allies against outside dangers, and hence subordinate leverage. Cooperation with non-relatives may simply be part of the overall flexibility in social behavior, but it may arise from life history. The slow life histories of hominoids created social situations in which there were no close relatives of the preferred age, sex, and fighting ability.

Boehm (2012) used a **behavioral phylogenetic approach**, based on the principle of parsimony as used in evolutionary biology: (1) natural selection operates as simply as possible; and (2) shorter time spans between ancestor and descendant species are associated with closer resemblances. On this basis Boehm assumed that traits shared by the *Pan* species and humans (represented by recent foragers) have a common genetic basis derived from the LCA (essentially meaning that they are homologous). A number of his conclusions complement those of the researchers cited above.

Boehm assumed competition among males for food and mates, but argued that this did not usually result in conflict as long as the roles of dominants and subordinates were clear. When fights did occur, participants often reconciled with one another. In more serious disturbances, as in chimpanzees, third parties might intervene with a variety of tactics. Also as in chimpanzees, coalitions of subordinates might have threatened or attacked intolerable alphas. Boehm saw this as a possible basis for the egalitarianism that characterizes human gatherer-hunters.

A controversial concern in the study of human behavioral evolution is the antiquity of violence, including frequency and forms. Wrangham and colleagues postulated that hominins and chimpanzees share propensities for coalitional killing; that is, cooperative attacks on individuals that are usually fatal (Wrangham 1999; Wrangham and Peterson 1997). Coalitions of male chimpanzees go on patrols along their community borders or into the range of a neighboring community. They launch fatal attacks on lone individuals or small groups that they encounter. Territorial expansion results from a sufficient number of these killings, increasing food resources. It is uncertain, however, whether this is the adaptive function of the behavior or if it is the acquisition of mates from the defeated community (Mitani et al. 2010b).

Whatever the function(s), the presence of this pattern in chimpanzees and its resemblance to small-scale human warfare (Kelly 2005) suggests an origin in the LCA. Bonobos differ from chimpanzees in this regard, possibly because they have evolved a different social situation or possibly because they are rarely alone and are thus more difficult to victimize

(Aureli et al. 2006). The latter interpretation is supported by observations on chimpanzees at Tai in West Africa.

Gorillas and orangutans display only the simpler components of chimpanzee aggression, male–male hostility and aggressive competition for mates. The strongest argument for a phylogenetic origin comes from parallels between chimpanzees and spider monkeys (Aureli et al. 2006). Like chimpanzees, the male monkeys go on silent, single-file patrols and attack individuals from other communities. The timing of this behavior seems to fit the Imbalance of Power Hypothesis proposed for chimpanzees. Evidence for function is sparse, but mate acquisition seems more likely than territorial expansion (Aureli et al. 2006). More broadly, spider monkey researchers consider the similarities to be convergent responses to common features of chimpanzee and spider monkey socioecology, including fission–fusion and male bonding. This analogy suggests that coalitional killing might have existed in the LCA.

Common characteristics among African apes suggest features of LCA communication. The ape vocalizations are virtually identical in acoustic structure, consisting of variations on grunts, barks, screams, and hoots (Mitani 1996). Adult males are the most vocal age/sex category. Male chimpanzees can benefit from vocalization by attracting estrus females and by maintaining alliances with other males; similar benefits may accrue to dominant male gorillas. Females and low-ranking males usually do better to avoid attention because they are more vulnerable to food competition from within the group and to lethal aggression from outside.

Comparison among the African apes suggests that the LCA also communicated with a repertoire of about 20 gestures (Roberts et al. 2012). African apes differ from orangutans in the use of auditory gestures, such as slapping surfaces, stamping on them, drumming, or clapping hands. These constitute 5 to 20 percent of the gestural repertoire, while no such signals have been described for orangutans. This might be explained by greater freedom for the hands in a terrestrial rather than arboreal context. Chimpanzees and gorillas make greater use of gestures when they communicate about travel, including adult movement and riding by young. Perhaps this is related to frequency of movement in groups as opposed to the often-solitary travel of adult orangutans. Chimpanzees have more gestures that direct the attention of the recipient towards an object or location in the immediate environment (Roberts et al. 2012).

LCA *Life Cycle*

Great apes display a number of common features with regard to life history and we have no reason to think that the LCA differed in any profound way (Robson and Wood 2008). Brain mass is widely considered to be a good predictor of life history events. Great ape cranial capacities all cluster around 400cc, slightly larger than the fossil candidates for LCA. Body mass is another significant correlate of life history events, perhaps the most important. In this regard gorillas are far out of line with the other great apes and humans, so we might want to look at orangs and *Pan* for more precise inferences about the ancestral traits.

Females in *Pongo* and *Pan* tend to weigh about 75 pounds (~35kg). Comparative data from these genera suggest that LCA life history was slow compared to other primates, although faster than that of recent humans. Most females were probably between the ages of 16 and 18 years at the time of first birth (human foragers 19.5 years). Mothers weaned their infants when the infants were between 4 and 8 years of age (human foragers by 3 years). After a successful birth a female probably experienced an inter-birth interval of 5 to 6 years, as in chimpanzees (gorillas are shorter and orangs longer; human foragers 3.7 years). Learning to use tools was probably a significant part of the maturation process (Figure 27.4).

Figure 27.4 Young apes, like these orangutans, play with sticks. This propensity for stick play suggests a basis for the invention of stick tools in the LCA. © Rhett A. Butler/ www.mongabay.com

Chimpanzees show signs of old age at 35 years. Females are virtually infertile by 45 years and die shortly thereafter (also true of orangs). Maximum lifespan for both sexes is about 50 years (65 to 80 for humans). Few approach that point in the wild. Only 3 percent of those who survive to the age of 15 are still alive after age 45 (30 percent for human foragers). Amid all the differences between great apes and humans, the age of menopause is about the same. This indicates that what changed in the evolution from LCA to modern humans was a slowing rate of somatic aging. For females this meant a vigorous post-menopausal life rather than death soon afterwards.

LCA *Cognition*

Povinelli and Cant (1995) placed the origin of self-conception in the common ancestor of great apes and humans. By **self-conception** they meant an organism's mental representation of itself as having perceptual and causal abilities. They hypothesized that such a capability was necessary to a relatively heavy ancestor adapting to arboreal life. Survival required a positional pattern that they called **arboreal clambering**, in which the primate used a variety of non-stereotyped behaviors to negotiate a complex and fragile environment. Their single-species model for this adaptation is the orangutan (bonobos display a similar pattern). Considering their more recent work, Povinelli and colleagues might repudiate this assessment of ape mental ability today. Nevertheless, the idea is available for others to work with. One approach might be to consider how the LCA mental world might have been made even more complicated by partial bipedalism and increased terrestrialism.

Since all extant great apes make tools and nests, these are among the strongest retrodictions to the LCA. Fruth and Hohmann (1996) considered nests to have been the first results of exploratory behavior with sticks and therefore the foundation for much of great ape and LCA technology. In this connection the authors stressed the sleep function of nests. Safe and comfortable conditions facilitated REM sleep, which is important for the maturation of the fetal nervous system. Thus nests might have been the foundation for advanced cognition. Shifting to a single-species model (bonobos), Fruth and Hohmann suggested that LCA individuals located their nests in social clusters, which could have facilitated transfer of information about the quality of food patches. This might also have facilitated the spread of technological information and the development of culture.

The orangutan is probably the best single-taxon model for the technological repertoire of a primarily arboreal LCA. Sequential tool use illustrates the cognitive abilities that can be inferred. This is the use of a tool to obtain another object that will then serve as a tool for another goal (usually food). The mental requirements include foresight and planning of hierarchically organized behaviors (Hihara et al. 2003; Martin-Ordas et al. 2012). All the great apes and several monkey species are capable of spontaneously using two tools in sequence, but the *Pan* species and orangutans demonstrated in an experiment that they could use as many as five tools in sequence.

The Social Intelligence Hypothesis that relates intense social life to enhanced cognition implies a certain level of mental function in the LCA. However, it does not distinguish or explain cognitive features that set the hominoids apart from other primates. Van Schaik et al. (2004) find the key in slow life history, associated with larger body size. This permitted greater growth in the metabolically expensive brain over the course of prolonged neural development, resulting in natural selection for cognitive problem solving. Based especially on the chimpanzee model, complex social life probably presented a number of problems to the participating individuals.

Whiten (2011) used the great ape model to seek foundations for culture in the LCA. Shared patterns between *Pan* and *Pongo* suggest features of the common ancestor of great apes and humans. One of these is the existence of multiple and diverse traditions. Orang and chimp evidence also indicates that the ape–human ancestor was able to understand copying, an essential process for the transmission of culture. The apes, like human children, can learn to play games in which they must understand the nature of copying (using a do-as-I-do paradigm). Only chimpanzees display distinctive cultural profiles (combinations of traits) in different communities, but this may just reflect a greater body of research.

Based on this material, Whiten suggests that we interpret LCA culture with a degree of abstraction. Rather than project particular tool using performances, for example, we should infer the existence of a tool culture that encompassed a number of artifacts with various functions. This needs qualification. In some cases particular performance may result from stringent need and/or evolved behavioral tendencies. Great apes have a tendency to poke sticks into object and holes. Dietary needs make it likely that this will lead to fishing for insects in many different communities.

Summary

Living primates can be used to reconstruct some behavior of the human–*Pan* common ancestor (the LCA). Relevant models can be grouped into the overlapping categories of referential, conceptual, and functional models. Chimpanzees and bonobos are especially pertinent to the LCA, but they differ significantly in some of their behavior. Even where they are similar, the behavior in question might have evolved after their lineage separated from ours. A solution to this problem is to use larger phylogenetic samples that represent common ancestors earlier than the LCA.

Paleoanthropological evidence places the LCA in mosaic environments. Arboreal behavior might have persisted in the wooded components. The early fossil record indicates that bipedalism began evolving in the LCA. Living primates provide several models for this crucial development. The LCA must have eaten insects and possibly meat. It probably used a variety of tools derived from vegetation and perhaps stone.

The LCA probably lived in fission–fusion communities with social dynamics similar to those of bonobos or chimpanzees, but likely leaning more toward the latter as the earliest hominins spread through a wide variety of habitats. Gestural communication may have been elaborate. Cognition is more speculative, but great ape abilities indicate some important skills in the LCA, including culture.

Sources

Basic Bibliography

Boesch, C. (2012). *Wild cultures: a comparison between chimpanzee and human cultures*. NY: CUP.

Kinzey, W.G. (Ed.) (1987). *The evolution of human behavior: primate models*. Albany, NY: SUNY Press.

de Waal, F.B.M. (Ed.) (2001). *Tree of origin: what primate behavior can tell us about human social evolution*. Cambridge, MA: HUP.

Additional References

Aronson, J.L., et al. (2008). Hominid environments at Hadar from paleosol studies in a framework of Ethiopian climate change. *JHE*, *55*, 532–550.

Aureli, F., et al. (2006). Raiding parties of male spider monkeys: insights into human warfare? *AJPA*, *131*, 486–497.

Bobe, R. (2011). Fossil mammals and paleoenvironments in the Omo-Turkana Basin. *EA*, *20*, 254–263.

Boehm, C. (2012). Ancestral hierarchy and conflict. *Science*, *336*, 844–847.

Boesch, C., & Boesch-Achermann, H. (2000). *The chimpanzees of the Taï Forest*. NY: OUP.

Bolen, E.G. (1998). *Ecology of North America*. NY: Wiley.

Cachel, S. (2006). *Primate and human evolution*. NY: CUP.

Cameron, D., & Groves, C. (2004). *Bones, stones and molecules: out of Africa and human origins*. Burlington, MA: Academic Press.

Cerling, T.E., et al. (2011). Woody cover and hominin environments in the past 6 million years. *Nature*, *476*, 51–56.

Crompton, R.H., et al. (2008). Locomotion and posture from the common hominoid ancestor to fully modern hominins, with special reference to the common panin/hominin ancestor. *Journal of Anatomy*, *212*, 501–543.

Doran, D. (1996). Positional behavior of African apes. In L. Marchant & T. Nishida (Eds.), *Great ape societies* (pp. 210–224). NY: CUP.

Dunbar, R. (2001). Brains on two legs: group size and the evolution of social intelligence. In F. de Waal (Ed.), *Tree of Origin* (pp. 173–192). Cambridge MA: HUP.

Foley, R.A. (1987). *Another unique species*. London: Longman.

Fruth, B., & Hohmann, G. (1996). Nest building in the great apes: the great leap forward? In W.C. McGrew et al. (Eds.), *Great ape societies* (pp. 225–240). NY: CUP.

Ganas, J., et al. (2009). A preliminary study of the temporal and spatial biomass patterns of herbaceous vegetation consumed by mountain gorillas in an Afromontane rain forest. *Biotropica*, *41*, 37–46.

Ghiglieri, M.P. (1987). Sociobiology of the great apes and the hominid ancestor. *JHE*, *16*, 319–357.

Goldsmith, M.L. (2003). Comparative behavioral ecology of a lowland and highland gorilla population: where do Bwindi gorillas fit? *Gorilla biology: a multidisciplinary perspective* (pp. 358–384). NY: CUP.

Haile-Selassie, Y., & Howell, F.C. (2009). Carnivora. In Y. Haile-Selassie, & G. WoldeGabriel (Eds.), *Ardipithecus kadabba* (pp. 237–276). Berkeley: UCP.

Haile-Selassie, Y., et al. (2009). New hominid fossils from Woranso-Mille (Central Afar, Ethiopia) and taxonomy of early Australopithecus. *AJPA*, *141*, 406–417.

Hernandez, M., & Vrba, E. (2006). Plio-Pleistocene climatic change in the Turkana Basin (East Africa). *JHE*, *50*, 595–626.

Hihara, S., et al. (2003). Rapid learning of sequential tool use by macaque monkeys. *Physiol. Behavior*, *78*, 427–434.

Hunt, K. (1994). The evolution of human bipedality: ecology and functional morphology. *JHE*, *26*, 183–202.

Hunt, K.D. (1996). The postural feeding hypothesis: an ecological model for the evolution of bipedalism. *South African Journal of Science*, *92*, 77–90.

Isbell, L.A., & Young, T.P. (1996). The evolution of bipedalism in hominids and reduced group size in chimpanzees: alternative responses to decreasing resource availability. *JHE*, *30*, 389–397.

Kelly, R.C. (2005). The evolution of lethal intergroup violence. *PNAS*, *102*, 15294–15298.

King, G.E. (1972). *The use of living species in the reconstruction of hominid behavioral evolution*. University of California Berkeley: doctoral dissertation.

King, G.E. (1975). Socioterritorial units among carnivores and early hominids. *JAR*, *31*, 69–87.

King, G.E. (1980). Alternative uses of primates and carnivores in the reconstruction of early hominid behavior. *E&S*, *1*, 99–109.

Llorente, M., et al. (2012). Hunting activity among naturalistically housed chimpanzees (*Pan troglodytes*) at the Fundació Mona (Girona, Spain). *Animals*, *2*, 363–376.

Lucas, P.W., et al. (2008). Inferences regarding the diet of extinct hominins: structural and functional trends in dental and mandibular morphology within the hominin clade. *Journal of Anatomy*, *212*, 486–500.

Martin-Ordas, G., et al. (2012). The role of episodic and semantic memory in episodic foresight. *Learning and Motivation*, *43*, 209–219.

McGrew, W.C., & Marchant, L.F. (2005). Percussive technology: chimpanzee baobab smashing and the evolutionary modelling of hominid knapping. In V. Roux, & B. Bril (Eds.), *Stone knapping*. McDonald Institute Monograph, University of Cambridge.

Milton, K. (1999). A hypothesis to explain the role of meat-eating in human evolution. *EA*, *8*, 11–21.

Mitani, J.C. (1996). Comparative field studies of African ape vocal behavior. In W. McGrew, L. Marchant, and T. Nishida (Eds.), *Great ape societies* (pp. 241–254). New York: Cambridge University Press.

Mitani, J.C., et al. (2010a). Chimpanzee minds in nature. In E. Lonsdorf & 2 co-editors, *The chimpanzee mind*. Chicago: UCP.

Mitani, J.C., et al. (2010b). Lethal intergroup aggression leads to territorial expansion in wild chimpanzees. *CB*, *20*, R507–R508.

Moore, J.M. (1996). Savanna chimpanzees, referential models and the last common ancestor. In W.C. McGrew & 2 co-editors, *Great ape societies* (pp. 275–292). NY: CUP.

Potts, R. (2012). Evolution and environmental change in early human prehistory. *ARo Anthropology*, *41*, 151–167.

Povinelli, D.J., & Cant, J.G.H. (1995). Arboreal clambering and the evolution of self-conception. *Quarterly Review of Biology*, *70*, 393–421.

Reynolds, S.C., et al. (2011). Landscapes and their relation to hominin habitats: case studies from *Australopithecus* sites in eastern and southern Africa. *JHE*, *60*, 281–298.

Richmond, B.G., et al. (2001). The origin of human bipedalism: the knuckle-walking hypothesis revisited. *YPA*, *44*, 70–105.

Roberts, A.I., et al. (2012). Communicative intentions in wild chimpanzees: persistence and elaboration in gestural signalling. *AC*, *16*, 187–196.
Robson, S.L., & Wood, B.A. (2008). Hominin life history: reconstruction and evolution. *Journal of Anatomy*, *212*, 394–425.
Sarmiento, E., & Meldrum, D.J. (2011). Behavioral and phylogenetic implications of a narrow allometric study of *Ardipithecus ramidus*. *Homo*, *62*, 75–108.
Sayers, K., & Lovejoy, C.O. (2008). The chimpanzee has no clothes: a critical examination of *Pan troglodytes* in models of human evolution. *CA*, *49*, 87–114.
Spocter, M.A., & Manger, P.R. (2007). The use of cranial variables for the estimation of hominin body mass. *AJPA*, *134*, 92–105.
Stanford, C.B. (2006). Arboreal bipedalism in wild chimpanzees: implications for models of the evolution of hominid posture and locomotion. *AJPA*, *129*, 225–231.
Stanford, C.B. (2012). Chimpanzees and the biology of *Ardipithecus ramidus*. *ARo Anthropology*, *41*, 139–149.
Strier, K.B. (2001). Beyond the apes: reasons to consider the entire primate order. In F. de Waal (Ed.), *Tree of origin* (pp. 69–93). Cambridge, MA: HUP.
Strum, S.C., & Mitchell, W. (1987). Baboons: baboon models and muddles. In W.G. Kinzey (Ed.), *The evolution of human behavior: primate models* (pp. 87–104). Albany, New York: SUNY Press.
Thorpe, S.K.S., & Crompton, R.H. (2006). Orangutan positional behavior and the nature of arboreal locomotion in Hominoidea. *AJPA*, *131*, 384–401.
Thorpe, S.K.S., et al. (2007). Origin of human bipedalism as an adaptation for locomotion on flexible branches. *Science*, *316*, 1328–1331.
Tooby, J., & DeVore, I. (1987).The reconstruction of hominid behavioral evolution through strategic modeling. In W.G. Kinzey (Ed.), *The evolution of human behavior: primate models* (pp. 183–237). Albany, New York: SUNY Press.
Tutin, C.E.G. (2003). An introductory perspective: behavioral ecology of gorillas. In A.B. Taylor, & M.L. Goldsmith (Eds.), *Gorilla biology: a multidisciplinary perspective* (pp. 295–301). NY: CUP.
Tuttle, R. (1981). Evolution of hominid bipedalism and prehensile capabilities. *PTRSB*, *292*, 89–94.
van Schaik, C.P., et al. (2004). Great ape societies: structures, bonds and cognition. In A. Russon, & D. Begun (Eds.), *The evolution of thought* (pp. 192–210). NY: CUP.
de Waal, F.B.M. (1998). *Chimpanzee politics*. Johns Hopkins.
White, L.J.T., et al. (1995). Herbaceous vegetation in different forest types in the Lopé Reserve, Gabon: implications for keystone food availability. *AJE*, *33*, 124–141.
White, T.D., et al. (2009). *Ardipithecus ramidus* and the paleobiology of early hominids (authors' summaries). *Science*, *326*, 64–86.
Whiten, A. (2011). The scope of culture in chimpanzees, humans and ancestral apes. *PTRSB*, *366*, 997–1007.
Wood, B., & Lonergan, N. (2008). The hominin fossil record: taxa, grades and clades. *Journal of Anatomy*, *212*, 354–376.
Wrangham, R.W. (1987). The significance of African apes for reconstructing human social evolution. In W.G. Kinzey (Ed.) *Primate models of hominid evolution* (pp. 51–71). Albany, NY: SUNY Press.
Wrangham, R.W. (1999). The evolution of coalitionary killing. *YPA*, *42*, 1–30.
Wrangham, R.W., & Peterson, D. (1997). *Demonic males: apes and the origins of human violence*. Boston: Houghton Mifflin.
Wrangham, R.W. & Pilbeam, D.R. (2001). Apes as time machines. In B.M.F. Galdikas et al. (Eds.), *All apes great and small, volume 1*. NY: Plenum/Kluwer.
Zihlman, A. (1996). Reconstructions reconsidered: chimpanzee models and human evolution. In W. McGrew et al. (Eds.), *The great apes* (pp. 293–304). NY: CUP.
Zilhman, A. (2012). The real females of human evolution. *EA*, *21*, 270–276.
Zihlman, A., & Bolter, D.R. (2004). Mammalian and primate roots of sociality. In R. Sussman & A. Chapman (Eds.), *Origin and evolution of sociality* (pp. 23–52). NY: Aldine de Gruyter.

28 Early Hominin Evolution

The Australopiths

A dozen ape-like figures move through the woodland savanna from one copse of trees to another, walking slowly on two legs. A mother uses one of her long arms to support the infant clinging to her back. Several adult males and an adolescent carry thick branches as clubs. Some juveniles throw stones at each other and nearby animals.

Primate models can be applied to hominin evolution beyond the Last Common Ancestor. Beginning at about 4.5 million years ago, a variety of new hominin genera and species begin to appear in the fossil record. These *early hominins* occupied a geographic arc from Chad through East Africa to South Africa. They display derived human-like characteristics along with ape-like traits inherited from the LCA. Derived dental and locomotor features, including larger molar teeth and habitual bipedalism, distinguished them from their predecessors. Most of these hominins probably weighed between 30 and 45 kilograms, with significant sexual dimorphism in body size. This chapter focuses on the *australopiths*, a convenient informal term for species that are or have been assigned to the genus *Australopithecus*, famously illustrated by the skeleton called "Lucy".

Australopiths lived between 4.5 and 2 million years ago in diverse habitats that included forests, woodlands, and a variety of more open landscapes. In many cases these were two or more components of a mosaic environment. The geographic extent of the taxon ensures that these hominins engaged with variation in seasonality and topography. There were major fluctuations across time as well as space, but over the whole period of their existence australopiths probably faced a general drying trend and increase in open areas. This chapter considers primate models for the australopiths and then attempts a reconstruction of australopith behavior.

Primate Models for Australopith Behavior

Evidence for paleoenvironments has inspired two leading explanations for early hominin adaptive trends: (1) spreading grasslands (Savanna Hypothesis); and (2) continuously shifting conditions (Fluctuating Environment Hypothesis). Primate analogies can be useful in either case because various species have adapted to a wide variety of habitats and long-term studies have demonstrated successful responses to environmental fluctuation.

We use the word *savanna* here in its broadest sense to encompass various relatively hot, dry, open habitats, ranging from woodlands to treeless grasslands (Reed et al. 2013). The important point is that early hominins gradually came to exploit all of these environments, at least to some degree (Domingo-Rodriguez 2014), and persisted through changes.

Comparison between savanna populations and their conspecifics or congenerics in other habitats can suggest what happened to hominins as they expanded through those environments.

Chimpanzee Model

Chimpanzees have been observed in various kinds of savanna. The Fongoli community in Senegal is of special interest for several reasons: (1) the savanna features of their habitat are extreme; (2) they are the only savanna chimpanzees to have been habituated; and (3) their behavior differs from that of other chimpanzees in important ways that can be linked to their environment (Pruetz and Bertolani 2009).

The combination of phylogenetic closeness and ecological analogy makes chimpanzees a particularly useful model for early hominins (Zihlman 2012). They can be used as a comprehensive species model to suggest the total way of life in woodland or savanna, or as a source of limited analogies for selected features (e.g. tool use) (Jolly 2013). The latter approach resembles that of "trait models" (Mitani 2013) or attribute analysis (see previous chapter).

As discussed earlier, any model based on just one species cannot inspire a high degree of confidence. Other great apes are of little help here because they fall far short of chimpanzees in the diversity of their habitats, especially with regard to the more open kinds of savannas. Only some monkeys can provide corroboratory evidence for these ecological analogies.

Common Baboon Model

Despite intense criticism, findings in primate research and paleoanthropology tend to support common baboon models for early hominins (King 2001). One position insisted that only chimpanzees should be used because of their close relationship to hominins. However, common baboons are among the few primates that occupy the same variety of environments as chimpanzees and early hominins. As such, they can provide the corroboration needed for hypotheses generated from chimpanzee behavior alone. Baboon analogies can also complement hypotheses derived from chimpanzees and in some cases provide alternatives. For example, baboon special friendships provide a model for male–female relationships that does not occur in chimpanzees.

Early baboon analogies emphasized open savanna (Washburn and DeVore 1961), so critics portrayed baboons as forest animals (e.g. Fedigan 1982). When it emerged that early hominins probably engaged in arboreal locomotion in substantially wooded habitats, baboons were dismissed because they supposedly pertained only to open country adaptations (Susman 1987). In short, baboons were too forest for some people and too savanna for others. The fact is that common baboons resemble early hominins in occupying mosaic habitats in which the predominant environment ranges from woodland to grassland. Another common objection to the model was the "extreme" sexual dimorphism of baboons. Despite continuing controversy, the common opinion is that dimorphism in body size was substantial among the australopiths.

Other Monkeys

Papionins in general are a good source of analogies for early hominins because they are medium-sized, social, and omnivorous mammals that live in seasonal tropical environments

(Foley 1993; Jolly 2013). Different taxa provide different kinds of relevant information. For example, some macaques manipulate stones and even use them as tools (Gumert and Malaivijitnond 2013a, b). The multi-level social organization of hamadryas baboons and geladas suggests ways in which complex social structures might have arisen in early hominins (Grueter et al. 2012).

Vervets have been advanced as a hominin model because they are adapted to life on the ground in open areas (Fedigan 1982). However, baboons and early hominins are alike, and differ from vervets, in spanning diverse environments, including very open ones (Altmann and Altmann 1971) while vervets are an edge species. Baboons walk long distances, often without shade or water, in relative safety from predators (Altmann 1998). Baboons approximate the size of early hominins, weighing up to 40kg, in contrast to the vervet maximum of about 6kg. Vervet behavior can be used to test some baboon/hominin analogies, but vervets are not viable as a single-species model for hominins.

Vervets parallel chimpanzees, baboons, and early hominins in that congeners live in a variety of woodlands and forests. Thus comparison of these guenon species can be used to test hypotheses about transitions in hominin behavior (Elton 2006). Also, vervets and the more savanna-adapted patas monkeys can be used to generate hypotheses about specific behavior patterns, such as terrestrial locomotion. Finally, a few New World species offer analogies that are pertinent to hominin evolution, most notably fission–fusion organization in spider monkeys and tool use by capuchins.

Australopith Behavior

Australopith Ecology

Australopith skeletal traits suggest continuing arboreal capability. These traits include shoulder morphology, arm length, and strength of the arms and digits. Primate models indicate that such hominins still used trees to escape predators and possibly for feeding. Continuing use of the arboreal refuge, especially for sleep, suggests persistence of nest building (Figure 28.1).

Australopiths were clearly bipedal, though they may have had some limitations on long-range walking. The relevant anatomy in earlier australopiths displays mixed features (Tallman 2013) and elicits mixed opinions, but there is little doubt that ground locomotion was human-like by the end of the phase if not before (Barak et al. 2013).

Effective bipedalism does not eliminate the use of primate models. As hominins expanded through savannas, they still faced problems like those of living primates in comparable habitats and may have arrived at similar solutions. A perspective on this matter comes from ecologists who draw analogies between bison and some kangaroos, despite their drastically different modes of locomotion, because both are large mammals adapted to grassland habitats (Rana 2013).

Chimpanzees show how hominins could have coped with excessive heat and water scarcity. First and foremost, trees provide shelter from the sun. Savanna chimpanzees perform most of their daily activities in forested and wooded areas, even though these are a small fraction of the habitats. Assirik chimps spend more time in the open during the wet season when there is more cloud cover and temperatures are lower. Fongoli chimpanzees make selective use of wooded patches with closed canopies (Pruetz and Bertolani 2009). They also use caves for daytime shelter (Pruetz 2007). Baboons also use caves for thermoregulation (although for warmth during the cold winters in South Africa). Anecdotal evidence from

Figure 28.1 A plausible reconstruction of australopith behavior. Habitats included woodland components throughout almost all of early hominin evolution. © Raul Martin/MSF/ Science Photo Library

baboons suggests that early hominins could have discovered other cooling methods: adults in one desert troop took "sand baths", sitting and throwing relatively cool sand on their abdomens (Brain and Mitchell 1999).

Chimpanzees at two savanna sites obtain water during dry seasons by digging holes in sandy riverbeds (Hunt and McGrew 2002; McGrew 2007). They dig near pools of standing water, which suggests that filtration is a function and perhaps the purpose of the behavior. The well water is clear, while the nearby pools contain algae and detritus. Analysis at Assirik demonstrated that natural puddles contained pathogens that were absent from the well water obtained by the apes.

Limited availability keeps the ripe fruit proportion of the Fongoli chimpanzee diet under 60 percent, less than for most conspecifics (Pruetz 2006). Observation on common baboons suggested underground storage organs (USOs) of plants as a supplement in dry habitats. The hard work of the baboons led to speculation that a digging stick would have made a big difference to early hominins (DeVore and Washburn 1963). Baboon tool use never rises to this level, but chimpanzees use sticks to obtain USOs (Hernandez-Aguilar et al. 2007). The tools are used to penetrate the hard surface of compacted sediments, after which the apes dig with their hands much as baboons do. Some chimpanzees also obtain water from USOs. In addition to digging sticks, other organic tools like those postulated for the LCA would have been helpful to early hominins.

Australopith cranial and dental features have been interpreted as adaptations to eating hard objects, such as nuts and seeds. Comparison with living primates that consume hard objects has undermined this hypothesis (Daegling et al. 2011; Delezene et al. 2013). Hard object feeders like sooty mangabeys (*Cercocebus atys*) do not necessarily display the same craniofacial and dental features as australopiths, while some of these features can be found in the leaf-eating colobines. The australopith physical features might be related to processing tough, fibrous foods such as leaves, grasses and sedges.

Molar microwear, including pitting like that found in baboons, links australopith fossils to seasonal consumption of savanna grasses in moderate quantities. Isotopic evidence indicates that a shift to C4 resources like grasses and sedges took place more than three million years ago in several early hominin species. The alternative interpretation is that the hominins ate animals that ate those plants. Primate models can also illuminate this alternative. Since chimpanzees rarely consume C4 foods, as has been demonstrated by analysis of hairs from Fongoli chimpanzees (Sponheimer et al. 2006), baboons and perhaps other monkeys provide a more likely model for exploitation of these resources.

Eng et al. (2013) created a model to measure maximum bite forces and molar stress, which they applied to modern humans, modern primates, and hominin skulls. They concluded that australopiths had no more than ape-like bite force capabilities. They also noted that chimpanzees sometimes break hard food items with rocks, and that early hominins could have done the same thing.

With capuchins and longtailed macaques as additional trait models, the case is stronger that australopiths used stone tools to process the hardest food items. Some macaques use stones to pound open shellfish. In addition to supporting the argument for hominin stone tool use, this example suggests a particular resource that might have been readily available to hominins because they typically lived near substantial bodies of water. Further evidence for use of aquatic resources comes from baboons and orangutans that capture and eat fish.

If australopiths took relatively small land prey like that of chimpanzees, they might have made little or no use of butchering tools. Having strong fingers comparable to those of the apes, hominins might have used sharp, unmodified stone flakes only to make initial incisions or may not have used them at all. Thus, the expansion of chimpanzee-like hunting might have left few traces in the archeological record. If australopiths hunted in groups like chimpanzees pursuing colobus monkeys, they might have used this approach effectively in the simpler two-dimensional world of savanna ungulates.

Scavenging has been offered as an alternative source of meat for australopiths, but this would have entailed contacts with dangerous competitors. Lions and hyenas inhabited the savannas and leopards were dangerous in a range of environments from forest to savanna. Small mammals, presumably as now, were readily available and presented little or no danger from themselves or from scavenging carnivores. Given a primate heritage of hunting small mammals, it probably would have been safer and more efficient for hominins to expand that adaptation. Evidence from chimpanzees, baboons, and most other primates indicates that scavenging by australopiths (as well as the LCA) would have been *added* to foraging and hunting for small mammals. Archeologists have demonstrated special circumstances in which australopiths might have scavenged (e.g. Blumenschine 1989; Cavallo and Blumenschine 1989).

Whether or not carnivores attacked australopiths at scavenging sites, they probably preyed on them. Big cats occasionally take chimpanzees, which are similar in size to early hominins, and even gorillas. It seems likely that the hominins continued to use trees when they could, especially at night. As in heavier monkey species, especially baboons, hominin males may have mounted a confrontational defense on occasion.

Talon punctures in the skull of a juvenile australopith at Taung (South Africa) indicate that these hominins fell prey to large raptors. Primates killed by contemporary African eagles are marked by damage patterns that closely resemble the wounds on the Taung skull (Berger 2006; McGraw et al. 2006; Sanders et al. 2003). African crowned eagles prefer the forest environment and take their primate victims from the trees, while comparable raptors frequent more open country. This suggests that large raptors were a danger to hominins throughout their expansion from forest into savanna.

Discussions of hunting, scavenging, and predator defense lead to consideration of weapon use by australopiths. Early hominins might also have used weapons for hunting larger mammals and in aggressive competition within or between social groups, and perhaps against competitors (such as baboons). Fongoli savanna chimpanzees extract bushbabies from sleeping holes with pointed sticks and eat them (Pruetz and Bertolani 2007). Typically the tool is jabbed into an aperture multiple times with enough force to injure prey. In contrast to other chimpanzee populations, females and (mostly) young are common participants. The fieldworkers suggested that this is exploitation of a food niche not used by adult males, and explicitly suggested an analogy with early hominins.

Hunting tools like those used at Fongoli could have provided the basis for crude thrusting spears as all-purpose weapons. In an anecdotal case, a juvenile chimpanzee poked a long stick at a harassing adolescent several times (pers. obs.). Though ineffective in this instance, such behavior could have spawned the thrusting spear as a significant weapon for adult hominins. Use of clubs, as postulated for the LCA, would presumably have continued.

Increased use of such weapons may have become vital because canine teeth, the main weapon of most primates, had become small and incisiform by the middle of the australopith era. If canine reduction and use of extrasomatic weapons are connected, the direction of causality is debatable. One hypothesis is that canine size declined because the invention of more effective weapons made them superfluous. The alternative is that canines declined for other reasons, making the adoption of other weapons necessary. In one version of the latter explanation, smaller canines facilitated a more powerful bite for crushing hard objects. In another, smaller canines were related to reduced competition and aggression within social groups. Primate comparisons provide perspectives on both hypotheses.

Primate evidence suggests early use of stone tools among hominins (Figure 28.2). The first such tools would have been unmodified stones used for pounding to open hard-shelled nuts and fruits, as is done by some capuchins, macaques, and chimpanzees. The flakes that resulted might then have been used in unmodified form for simple cutting tasks, as seen in experiments with captive bonobos and an orangutan (Toth et al. 2006; Wright 1972).

Burgeoning tool use and manufacture is usually explained as adaptation to savanna environments. Savanna chimps do display unusual technological performances, such as the use of digging sticks to obtain USOs and a rudimentary spear to hunt galagos. However, some of the most elaborate tool kits are found among forest chimpanzees (Boesch and Boesch-Achermann 2000; Sanz and Morgan 2007). Capuchins in a dry forest displayed an

Figure 28.2 Compared to capuchins, chimpanzees can use heavier tools and australopiths probably did as well. © Fuse via Thinkstock

unusual variety of tools (Moura and Lee 2004), but a survey of tool using by capuchins and chimpanzees concluded that the most important factor is the degree of terrestriality (Meulman et al. 2012).

Early hominins may have protected themselves from predators by throwing stones. For a group on the move to mount an effective barrage, this would require that every individual (or at least every male) in a sizeable party carry one or two stones—which would interfere with carrying anything else of importance, such as infants or food. On the move in open country, australopiths more likely depended on other weapons, as suggested by chimpanzee behavior: clubs and spear-like sticks. However, stone-throwing could have been effective for rest periods if they had accumulated a sufficient quantity at rest locations (cf. Potts 1988, on caching butchering tools). The same strategy might have worked in dealing with frugivorous competitors at frequently used feeding sites (Osvath 2009).

The behavior of a male chimpanzee in a Swedish zoo shows how such an adaptation might have worked. In the morning, before the zoo opened, he collected rocks and cached them along the edge of the enclosure where he could readily retrieve them to throw at human visitors. A sweep of the enclosure one morning revealed five caches with three to eight stones each (Osvath 2009).

Australopith Sociality

Lee (2012) considered all African ape lineages in reconstructing australopith social organization, but relied heavily on chimpanzees. Her scenario pictured a fission–fusion society exploiting dispersed food patches. As she saw it, hominins did this with less energy expenditure because they were bipedal. If this is correct, then it is a way in which bipedalism affects the application of primate models (albeit quantitatively and subject to adjustment). However, the hypothesis that bipedalism improves energetic efficiency has been questioned.

Common baboons and chimpanzees typically live in societies with roughly the same size range, from about 15 to 150 at the extremes, so early hominin communities probably fell within this range. Hominins living in woodlands and gallery forests could have retained a fission–fusion society from the LCA, as many writers assume along with Lee (e.g. Malone et al. 2012). The behavior of savanna chimpanzees suggests that adaptation to more open and dangerous areas involved more compact communities or large parties for mutual defense. Party size for ordinary activities at Assirik is typical of chimpanzees (up to about seven individuals), but open country parties are usually larger and of mixed composition and traveling parties have a median size of about 20 (Hunt and McGrew 2002). The Fongoli community of about 35 chimpanzees cyclically joins into a single large party (Pruetz and Bertolani 2009). Common baboons tend to maintain relatively cohesive troops in most habitats, which functions well for predator defense. However, they too must balance food and danger. Common baboons in several parts of Africa will fission for at least a day during foraging (Alberts et al. 2012; Aldrich-Blake et al. 1971).

To maintain access to sufficient fruit, savanna chimpanzees inhabit large community ranges: at least $40km^2$ and probably more than $50km^2$. This is almost twice the area of the largest known ranges elsewhere. Within such ranges the apes must cross open savanna to move from one wooded patch to another. They tend to stick to small clumps of trees and travel quickly between them. Baboons may be under less constraint because of their ability to consume C4 foods. A troop can spend a day of foraging a mile or more away from trees. Hominins with increasing capability for exploiting savanna resources may have shifted from a chimp-like pattern to a more baboon-like one.

In Lee's chimpanzee-like scenario early hominin society was based on male kinship, but male competition was also important. Genetic evidence casts doubt on the frequency of kinship ties in chimpanzee communities (Mitani et al. 2000) but there is no doubt about male–male competition, which is sometimes deadly (Mitani 2009; Muller and Wrangham 2004). Lee connected competition with australopith sexual dimorphism, which averaged about 1.55 across all taxa (ranging from 1.3 to 1.9). Since chimpanzee size dimorphism is relatively moderate, hominin male–male competition may have been more intense. However, there is disagreement as to the degree of body dimorphism in australopiths, and there are theoretical questions about the degree of correlation between dimorphism and social patterns.

At some point male–female interactions shifted toward long-term relationships between particular individuals; that is, the foundation for the human family emerged. Fossil footprints of two australopiths walking close together hint that such relationships existed at about three million years ago. Size difference indicating male and female led to speculation that they were a mated pair. However, analogy with chimpanzees suggests two alternatives: (1) an adult son may maintain a close relationship with his mother and the two sometimes travel together; (2) the smaller individual may not have been a female at all; juvenile male chimpanzees are strongly attracted to adult males and may accompany them during routine activities.

A comparative test based on a broad sample of anthropoids suggested that australopiths were monogamous. The ratio of the second and fourth digits covaried with intra-sexual competition and social systems (Nelson et al. 2011). This is apparently because the digit ratio reflects levels of testosterone. The ratio displayed by one species (*A. afarensis*) is consistent with those of extant monogamous species. This argues against the competition scenario, although the degree of sexual dimorphism in body size supports it.

The baboon pattern of special friends suggests an antecedent to the human relationship. These baboon friendships can involve long-term affiliation, mating, and mutual interest in offspring. Males spend a great deal of time with nonestrus females, groom them, and huddle with them, all of which suggests affection (Smuts 1985). A narrower focus than in baboons (linking just two or three individuals) and greater intensity (especially in terms of time spent together) would have resulted in a unit much like the polygynous family that existed among human foragers and in many other human societies (King 2003). Despite occasional signs of male–female affection, no such relationship occurs in chimpanzees. Thus a baboon analogy provides a more coherent view of how this crucial hominin innovation might have begun.

This scenario leaves open the question of why the next step was taken. Based on (controversial) evidence that some male chimpanzees trade meat for sex, Gavrilets (2012a) proposed that provisioning and sexual faithfulness were the key factors. Concentration of paternity in a particular male might lead to a reliable supply of food for the female, enhancing her health and fertility (the capacity for concentrating paternity occurs in hamadryas baboons). The male could avoid competition with higher-ranking males, increase paternal certainty, and contribute to his partner's fertility.

Gavrilets noted that the food-for-sex interpretation of chimpanzee behavior has been challenged, and answered that even weak evolutionary forces can result in dramatic changes over many generations. Furthermore, meta-analysis demonstrated a correlation between male-to-female food transfers and the opportunity for female mate choice in primates (Jaeggi and van Schaik 2011).

Male Fongoli chimpanzees share plant foods and tools with females, which may be related to the savanna environment (Pruetz and Lindshield 2012). One possible factor is that these apes transport food to shaded areas during the dry season, which probably increases the value of the foods. The transfer process usually consists of male tolerance for females taking food. Females also can displace males from termiting sites. These patterns are reminiscent of some lemur behavior, which also takes place in an uncertain environment.

Based on unusual ecological differences between male and female chimpanzees, Hunt (2006) created a model that can be applied to early hominins. Access to higher quality foods is presumably based on dominance of males over females in chimpanzees. The evidence for substantial sexual dimorphism in early hominins suggests a similar situation. The implication is that male hominins concentrated on food items that were large, terrestrial, nutrient-dense, and easy to process. Females would then have been more arboreal and dependent on food items that were smaller, nutritionally inferior, and required more processing time. This would result in tool kits specialized by sex. Appropriate tool(s) for males would have been digging stick, spear, and/or club, perhaps in the shape of a single multi-purpose tool. Females would have been associated with tools that could be used arboreally for processing low-return foods. Hunt saw this system as possibly extending all the way back to the LCA.

Weston et al. (2004) used primate data to develop an unusual interpretation of mate choice in hominins with pronounced sexual dimorphism in body size and weak dimorphism in canine size. They hypothesized that dimorphism in facial breadth is inversely related to canine dimorphism, and tested this theory on 14 anthropoid species. The results confirmed that males in primate species with little canine dimorphism have shorter and broader faces relative to females. Sexual selection may favor the broader face because it arises from the relative size of the cheekbones, which in turn is based on adolescent testosterone. In other words, when canines do not signal male quality, females respond to facial breadth as an alternative.

Grueter et al. (2012) used a strategic modeling framework but they relied primarily on a single species model, the hamadryas baboon, to argue for multi-level social organization in early hominins. Their thesis is that major changes owed more to social than ecological causes. In an increasingly dry environment, foraging parties within a chimp-like community became increasingly independent. However, limited resource patches and refuges drew them together again and required a so-called "agreement" to share. These theorists postulated that a system of polygynous foraging units emerged within the larger group. Comparative primate evidence offers several reasons for this: (1) males in many primate species try to monopolize females; (2) polygyny is typical of all multi-level societies; and (3) sexual dimorphism in body size characterizes early hominins. Female acquiescence can be explained by an idea that we have encountered previously: the Bodyguard (Hired Gun) Hypothesis.

One way to explain the final transition to monogamy is the Provisioning Hypothesis favored by Gavrilets (2012a). However, Grueter and colleagues favored the Leveling Hypothesis: lethal weapons equalized male competitive abilities, making attempts to monopolize more than one female more costly. As we have seen, chimpanzee evidence suggests that clubs were available early in hominin evolution and possibly thrusting spears as well.

Carrier (2007) drew a novel inference from australopith anatomy and primate behavior. Noting that australopiths retained short legs millions of years after hominin bipedalism originated, he surmises that the trait must have been favored by natural selection. He hypothesized that short legs were advantageous in male–male aggressive competition because they provided greater stability and were more resistant to fracture. Originating the idea in research with dogs, Carrier turned to a comparison of apes and monkeys. Using sexual dimorphism as a proxy for the level of male–male aggression, he found that relative leg length was correlated with dimorphism. He concluded that leg length is also correlated with aggression in primates, including fossil hominins. Carrier and Morgan (2014) explained anatomical buttressing of the australopith face as an adaptation to fighting.

Australopith Life Cycle

Fossil evidence indicates ape-like life history in australopiths. There was little change from average ape body mass or cranial capacity, both of which correlate with life history (Robson and Wood 2008). Ape-like rates of dental development also imply similar life history characteristics. These include slow life history, late reproductive maturation (especially associated with a high level of sexual dimorphism), and long inter-birth intervals compared to Old World monkeys (Lee 2012). The equivalence was altered by infants that were moderately expensive in energetic terms due to somewhat greater brain growth than in apes. In addition the development of efficient bipedalism required additional dependency time. This implies some male investment in infants. At the least, protection against infanticide would continue. In the context of chimpanzee-like patrilocal society, there was probably no sharing of infant costs among the weakly bonded females.

Sauther (1994) suggested that ringtailed lemurs could help to identify behavioral adaptations that hominins needed to survive as they encountered increasingly seasonal habitats. Ringtails are the most terrestrial strepsirrhines and live in habitats like those of early australopiths, including forest, scrub, and arid areas. They use a wide variety of foods as they become available by season. Sauther postulated that the energy burdens of pregnancy and lactation would have induced female hominins to expand their ecological niches to obtain a more dependable food supply. However, to imagine the particular responses of these female hominins, Sauther turned to tools used by chimpanzees: digging sticks and stone hammers. The distinctive feature of Sauther's hypothesis was that pregnancy and lactation played a crucial role in technological advances made by hominin females.

According to Gavrilets (2012a), paternal care must have followed the formation of monogamous pairs because this is the general rule in mammals. However, the baboon model shows that the sequence can be different in primates and might well have been different for australopiths. In some chacma baboons (Huchard et al. 2012; Onyango et al. 2013), immature individuals associate more often with their genetic fathers than with any other males and actively manage these relationships. The young stay closer when another male is in sight and when the mother is out of sight, indicating that the genetic father offers protection. Beyond protection, the relationship increases the fitness of the offspring by enhancing nutrition: juveniles and fathers are especially close when the former are feeding. These relationships are especially strong for subordinate males, as postulated by Gavrilets.

Australopith Cognition

Perhaps we can approximate australopith capabilities in terms of the complete repertoire of all chimpanzee species, augmented by the savanna chimpanzee model for the problems and solutions of australopiths as they expanded into more open environments. One example of chimpanzee technological achievement is the Fongoli spears for killing galagos, a process that included as many as five steps, including the sharpening of the point. This example is also important because of the predominance of females in both manufacture and meat acquisition (Zihlman 2012).

Attribution of chimpanzee-like technology raises the problem of the single-species model. It is possible that chimpanzees and hominins developed more elaborate technologies separately in later stages of their evolution. Comparison with other apes does not solve the problem because their natural technologies are less elaborate than those of chimpanzees. Analogy with monkeys through attribute analysis provides more insight into the capabilities of australopiths. Of particular interest is the use of stone tools by capuchins in savanna-like environments along with bipedal transport of tools and food (Ottoni and Izar 2008). Also relevant, though anecdotal, is the use of a club by one monkey to attack a snake (Boinski 1988). If some monkeys can parallel chimpanzee achievements, it seems likely that australopiths could have done the same.

Experiments with captive animals indicate that great apes are capable of choice and foresight with regard to tools. Bonobos and orangutans tested by Mulcahy and Call (2006) had a high rate of success in selecting appropriate tools as much as 14 hours before performing a task. The researchers concluded that the foundation for future planning ability evolved as early as the common ancestor of the great apes, before 14 million years ago. If so, it would certainly have existed in our LCA.

Osvath and Osvath (2008) expanded on Mulcahy and Call (2006), using similar experimental methods. They provided evidence for two important components of future planning in chimpanzees and orangutans. First, they display self-control; that is, they can act toward a future state even when experiencing a strong conflicting motivation. Second, they can learn a function from a completely novel object and generalize it to a future use.

The behavior of the stone-throwing chimpanzee in a Swedish zoo suggests planning for a future state (Osvath and Karvonen 2012). He was calm while gathering stones and waited several hours until throwing them when in an agitated display. He stopped caching stones in winter when the zoo was closed to visitors. The researchers saw a parallel to similar forms of stone manipulation in early hominins.

Eventually the chimpanzee expanded his armaments with concrete rocks from the center of his island. He tapped on them to seek the hollow sounds denoting microscopic cracks caused by cold. With harder blows he detached discoidal pieces (sometimes breaking them into smaller fragments) and then transported them to his caches. Thus his foresight extended to manufacture as well as transport. Experiments support the existence of patience in apes. It appears that transport of stone tools, raw materials, and carcasses was based on capabilities that had evolved long before.

According to Mercader et al. (2007), the P100 chimpanzees were selective about their raw material (75 percent granitoid rocks) and chose the optimal combination of raw material, size, and weight for each particular tool using activity. Chosen items were often transported elsewhere. Archeologists are skeptical (Pelegrin 2005). Toth et al. (2006), who worked with Kanzi and Panbanisha, examined a sample of the material and perceived mostly crumbled stones that lack the markers of conchoidal fracture and other Oldowan features. However, capuchin monkeys select appropriate materials for tool use (Pouydebat et al. 2006, and others). Considering all the evidence and the doubts, it seems plausible that hominins at the cognitive level of chimpanzees might be responsible for rudimentary archeological sites.

Hopkins et al. (2011) suggested that throwing could have been involved with the development of language and speech, based on the behavior and brains of chimpanzees. The proportion of white matter to gray matter in the chimpanzee homologue to Broca's area was significantly higher in subjects that threw reliably than in those that did not. The same was true for the motor-hand area elsewhere in the brain. Both regions were larger in the hemisphere contralateral to the chimpanzee's preferred throwing hand. Finally, individuals that threw reliably displayed significantly better communication abilities than other chimpanzees. The researchers inferred that chimpanzees that have learned to throw have developed greater cortical connectivity between the primary motor cortex and the Broca's area homologue. They suggested that this system, derived from the LCA, underwent intense selection in hominin evolution in association with throwing for hunting and/or defense, and that this potentially formed the foundation for the left hemisphere specialization associated with language and speech in modern humans.

Even accepting all that has been said and inferred about chimpanzee skills in manipulating tools and learning about them, there remains the old idea that only hominins are *continuously dependent* on tools (Bartholomew and Birdsell 1953). Wynn et al. (2011) argued that technology is just as obligate for chimpanzees as it was for early hominins. In every well-studied chimpanzee population, the apes depend on technology to function adaptively. In many populations continuing use and manufacture of tools are maintained by culture. Japanese macaques also display traditions that guide a variety of stone handling techniques.

Summary

Primate models can be used to reconstruct the behavior of early hominins. Most of these models are based on ecological analogies, but capability and trait models are also pertinent. Fossil and archeological remains combined with behavioral ecology provide information that allows for the application of primate models. They have generated numerous hypotheses about the origin of bipedalism, while offering more tightly focused suggestions on how early hominins might have adapted to environmental conditions. Primates also provide insights as to the cognitive implications of these adaptations. Beyond the probability of a fission–fusion system, social inferences are more tentative.

Sources

Basic Bibliography

Begun, D., (Ed.) (2013). *A companion to paleoanthropology*. Chichester, UK: Wiley-Blackwell. [Chapters 6–10, 22, 23]

Fleagle, J.G. (Ed.) (2011). *The Turkana Basin. EA, 20*, 201–292 (Special issue).

Hublin, J-J., & Richards, M.P. (Eds.) (2009). *The evolution of hominin diets*. NY: Springer. [Chapters 1, 2, 3, 13, 17, 18]

Reed, K.E., et al. (Eds.) (2013). *The paleobiology of* Australopithecus. NY: Springer.

Reynolds, S.C., & Gallagher, A., (Eds.) (2012). *African genesis, perspectives on hominin evolution*. NY: CUP.

Sponheimer, M., et al. (Eds.) (2013). *Early hominin paleoecology*. Boulder, CO: University Press of Colorado.

Additional References

Alberts, N., et al. (2012). *Quantifying fission-fusion dynamics: a case study of wild olive baboons*. Conference Paper, Primate Society of Great Britain, Spring Meeting.

Aldrich-Blake, F.P.G., et al. (1971). Observations on baboons, *Papio anubis*, in an arid region in Ethiopia. *FP, 15*, 1–35.

Altmann, S.A. (1998). *Foraging for survival, yearling baboons in Africa*. Chicago: UCP.

Altmann, S.A., & Altmann, J. (1971). *Baboon ecology: African field research*. Chicago: UCP.

Barak, M.M, et al. (2013). Trabecular evidence for a human-like gait in *Australopithecus africanus*. *PLONE*, 8: e77687.

Bartholomew, G.A., & Birdsell, J.B. (1953). Ecology and the protohominids. AA, 55, 481–498.

Berger, L.R. (2006). Predatory bird damage to the Taung type-skull of *Australopithecus africanus* Dart 1925. *AJPA*, *131*, 166–168.

Blumenschine, R.J. (1989). A landscape taphonomic model of the scale of prehistoric scavenging opportunities. *JHE*, *18*, 345–371.

Boesch, C. (2012). *Wild cultures: a comparison between chimpanzee and human cultures*. NY: CUP.

Boesch, C., & Boesch-Achermann, H. (2000). *The chimpanzees of the Taï Forest*. NY: OUP.

Boinski, S. (1988). Use of a club by a wild white-faced capuchin (*Cebus capucinus*) to attack a venomous snake (*Bothrops asper*). *AJP*, *14*, 177–179.

Brain, C., & Mitchell, D. (1999). Body temperature changes in free-ranging baboons (*Papio hamadryas ursinus*) in the Namib Desert, Namibia. *IJP*, *20*, 585–598.

Carrier, D.R. (2007). The short legs of great apes: evidence for aggressive behavior in australopiths. *Evolution*, *61*, 596–605.

Carrier, D.R., & Morgan, M.H. (2014). Protective buttressing of the hominin face. *Biological Reviews*, *90*, 330–346.

Cavallo, J.A., & Blumenschine, R.J. (1989). Tree-stored leopard kills: expanding the hominid scavenging niche. *JHE*, *18*, 393–399.

Daegling, D.J., et al. (2011). Hard-object feeding in sooty mangabeys (*Cercocebus atys*) and interpretation of early hominin feeding ecology. *PLONE*, 6, e23095.

Delezene, L.K., et al. (2013). Premolar microwear and tooth use in *Australopithecus afarensis*. *JHE*, 65, 282–293.

DeVore, I., & Washburn, S.L. (1963). Baboon ecology in human evolution. In F.C. Howell and F. Bourlière (Eds.), *African ecology and human evolution* (pp. 335–367). Chicago: Aldine.

Dominguez-Rodrigo, M. (2014). Is the "savanna hypothesis" a dead concept for explaining the emergence of the earliest hominins? CA, 55, 59–81.

Elton, S. (2006). 40 years on and still going strong: the use of the hominin-cercopithecid comparison in human evolution. *JRAI*, *12*, 19–38.

Eng, C.M., et al. (2013). Bite force and occlusal stress production in hominin evolution. *AJPA*, *151*, 544–557.

Fedigan, L.M. (1982). *Primate paradigms: sex roles and social bonds*. Montreal: Eden Press.

Foley, R.A. (1989). The evolution of hominid social behaviour. In V. Standen & R.A. Foley (Eds.), *Comparative socioecology: the behavioural ecology of humans and other mammals* (pp. 473–494). Oxford: Blackwell.

Gaspersic, M., & Pruetz, J.D. (2008). Savanna chimpanzees (*Pan troglodytes verus*) and baobab fruits (*Adansonia digitata*): investigation of percussive technology among three chimpanzee communities in southeastern Senegal. *FP*, *79*, 332–333.

Gavrilets, S. (2012a). Human origins and the transition from promiscuity to pair bonding. *PNAS*, *109*, 9923–9928.

Gavrilets, S. (2012b). On the evolutionary origins of the egalitarian syndrome. *PNAS*, *109*, 14069–14074.

Grueter, C.C., et al. (2012). Evolution of multilevel social systems in nonhuman primates and humans. *IJP*, *33*, 1002–1037.

Gumert, M.D., & Malaivijitnond, S. (2013a). Long-tailed macaques select mass of stone tools according to food type. *PTRSB*, 368, doi:10.1098/rstb.2012.0413.

Gumert, M.D., & Malaivijitnond, S. (2013b). Marine prey processed with stone tools by Burmese long-tailed macaques (*Macaca fascicularis aurea*) in intertidal habitats. *AJPA*, *149*, 447–457.

Hernandez-Aguilar, R.A., et al. (2007). Savanna chimpanzees use tools to harvest the underground storage organs of plants. *Proceedings of the National Academy of Sciences*, *104*, 19210–19213.

Hopkins, W.D., et al. (2011). The neural and cognitive correlates of aimed throwing in chimpanzees: a magnetic resonance image and behavioural study on a unique form of social tool use. *PTRSB*, *367*, 37–47.
Huchard, E., et al. (2012). Paternal effects on access to resources in a promiscuous primate society. *BE*, doi:10.1093/beheco/ars158.
Hunt, K.D. (2006). The postural feeding hypothesis: an ecological model for the evolution of bipedalism. *SAJS*, *92*, 77–90.
Hunt, K.D., & McGrew, W.C. (2002). Chimpanzees in the dry habitats at Assirik, Senegal, and at Semliki Wildlife Reserve, Uganda. In C. Boesch & 2 co-editors, *Behavioural diversity in chimpanzees and bonobos* (pp. 35–51). NY: CUP.
Jaeggi, A.V. & van Schaik, C.P. (2011). The evolution of food sharing in primates. *BE&S*, *65*, 2125–2140.
Jolly, C.J. (2013). Analogies and models in the study of the early hominins. In M. Sponheimer et al. (Eds.), *Early hominin paleoecology* (pp. 437–455). Boulder, CO: University Press of Colorado.
King, G.E. (2001). The once and future baboon: a source of analogies for earliest hominid adaptations. *Annual meeting of the AAPA*, Kansas City, MO.
King, G.E. (2003). *Traditional cultures: a survey of nonwestern experience and achievement*. Long Grove, IL: Waveland Press.
Lee, P.C. (2012). Growth and investment in hominin life history evolution: patterns, processes, and outcomes. *IJP*, *33*, 1309–1331.
Malone, N., et al. (2012). Variation in the social systems of extant hominoids: comparative insight into the social behavior of early hominins. *IJP*, DOI 10.1007/s10764-012-9617-0.
Martin, M.K., & Voorhies, B. (1975). *Female of the species*. NY: Columbia University Press.
McGraw, W.S., et al. (2006). Primate remains from African crowned eagle (*Stephanoaetus coronatus*) nests in Ivory Coast's Tai Forest: implications for primate predation and early hominid taphonomy in South Africa. *AJPA*, *13*, 151–165.
McGrew, W.C. (2007). Savanna chimpanzees dig for food. *PNAS*, *104*, 19167–19168.
Mercader, J., et al. (2007). 4,300-year-old chimpanzee sites and the origins of percussive stone technology. *PNAS*, *104*, 3043–3048.
Meulman, E., et al. (2012). The role of terrestriality in promoting primate technology. *EA*, *21*, 58–68.
Mitani, J.C. (2009). Male chimpanzees form enduring and equitable social bonds. *AB*, *77*, 633–640.
Mitani, J.C. (2013). Chimpanzee models of human behavioral evolution. In M. Sponheimer et al. (Eds.), *Early hominin paleoecology* (pp. 397–435). Boulder, CO: University Press of Colorado.
Mitani, J.C., et al. (2000). Male affiliation, cooperation, and kinship in wild chimpanzees. *AB*, *59*, 885–893.
Moura, A.C. de A., & Lee, P.C. (2004). Capuchin stone tool use in Caatinga Dry Forest. *Science*, *306*, 1909.
Mulcahy, N.J., & Call, J. (2006). Apes save tools for future use. *Science*, *312*, 1038–1040.
Muller, M.N., & Wrangham, R.W. (2004). Dominance, aggression and testosterone in wild chimpanzees: a test of the challenge hypothesis. *AB*, *67*, 113–123.
Nelson, E.L., et al. (2011). Evidence for motor planning in monkeys: rhesus macaques select efficient grips when transporting spoons. *Developmental Science*, *14*, 822–831.
Onyango, P.O., et al. (2013). Puberty and dispersal in a wild primate population. *H&B*, *64*, 240–249.
Osvath, M. (2009). Spontaneous planning for future stone throwing by a male chimpanzee. *CB*, *19*, R190–191.
Osvath, M., & Karvonen, E. (2012). Spontaneous innovation for future deception in a male chimpanzee. *PLONE*, *7*, e36782.
Osvath, M., & Osvath, H. (2008). Chimpanzee (*Pan troglodytes*) and orangutan (*Pongo abelii*) forethought. *AC*, *11*, 661–674.
Ottoni, E.B. & Izar, P. (2008). Capuchin monkey tool use: overview and implications. *EA*, *17*, 171–178.

Pelegrin, J. (2005). Remarks about archaeological techniques and methods of knapping. In V. Roux, & B. Bril (Eds.), *Stone knapping: the necessary conditions for a uniquely human behaviour* (pp. 23–34). Cambridge: McDonald Institute for Archaeological Research.

Potts, R. (1988). *Early hominid activities at Olduvai*. NY: Aldine de Gruyter.

Pouydebat, E., et al. (2006). Substrate optimization in nut cracking by capuchin monkeys (*Cebus apella*). *AJP*, 68, 1017–1024.

Pruetz, J.D. (2006). Feeding ecology of savanna chimpanzees (*Pan troglodytes verus*) at Fongoli, Senegal. In G. Hohmann, et al. (Eds.), *Feeding ecology in apes and other primates* (pp. 161–182). NY: CUP.

Pruetz, J.D. (2007). Evidence of cave use by savanna chimpanzees (*Pan troglodytes verus*) at Fongoli, Senegal: implications for thermoregulatory behavior. *Primates*, *48*, 316–319.

Pruetz, J.D., & Bertolani, P. (2007). Savanna chimpanzees, *Pan troglodytes verus*, hunt with tools. *CB*, *17*, 412–417.

Pruetz, J.D., & Bertolani, P. (2009). Chimpanzee (*Pan troglodytes verus*) behavioral responses to stresses associated with living in a savannah-mosaic environment: implications for hominin adaptations to open habitats. *Paleoanthropology*, 2009, 252–262.

Pruetz, J.D., & Lindshield, S. (2011). Plant-food and tool transfer among savanna chimpanzees at Fongoli, Senegal. *Primates*, *53*, 133–145.

Rana, S.V.S. (2013). *Essentials of ecology and environmental science*, 5th ed. Delhi, India: PHI Learning.

Robson, S.L., & Wood, B.A. (2008). Hominin life history: reconstruction and evolution. *Journal of Anatomy*, *212*, 394–425.

Sanders, W.J., et al. (2003). Taphonomic aspects of crowned hawk-eagle predation on monkeys. *JHE*, *44*, 87–105.

Sanz, C.M., & Morgan, D.B. (2007). Chimpanzee tool technology in the Goualougo Triangle, Republic of Congo. *JHE*, *52*, 420–433.

Sauther, M.L. (1994). Wild plant use by pregnant and lactating ring-tailed lemurs (*Lemur catta*): implications for early hominid foraging. In Etkin, N.L. (Ed.), *Eating on the wild side* (pp. 240–256). Tucson: University of Arizona Press.

Schwarz, J.H. (2004). Barking up the wrong ape—australopiths and the quest for chimpanzee characters in hominid fossils. *Collections in Anthropology*, *28*, *Supplement 2*, 87–101.

Smuts, B. (1985). *Sex and friendship in baboons*. Cambridge, MA: HUP.

Sponheimer, M., et al. (2006). Do "savanna" chimpanzees consume C4 resources? *JHE*, *51*, 128–133.

Sponheimer, M., et al. (2013). Isotopic evidence of early hominin diets. *PNAS*, DOI:10.1073/pnas.1222579110.

Susman, R. (1987). Pygmy chimpanzees and common chimpanzees: models for the behavioral ecology of the earliest hominids. In W.G. Kinzey (Ed.), *The evolution of human behavior: primate models* (pp. 72–86). Albany, New York: SUNY Press.

Tallman, M. (2013). Forelimb to hindlimb shape covariance in extant hominoids and fossil hominins. *Anatomical Record*, *296*, 290–304.

Tanner, N.M. (1981). *On becoming human*. NY: CUP.

Toth, N., et al. (2006). A comparative study of the stone tool-making skills of *Pan*, *Australopithecus*, and *Homo sapiens*. In N. Toth & K. Schick (Eds.), *The Oldowan: case studies into the earliest Stone Age* (pp. 155–222). Gosport, IN: Stone Age Institute Press.

Washburn, S.L., & DeVore, I. (1961). Social behavior of baboons and early man. In S.L. Washburn (Ed.), *Social life of early man* (pp. 91–105). NY: Wenner-Gren.

Weston, E.M., et al. (2004). Wide faces or large canines? The attractive versus the aggressive primate. *PRSB*, *271*, 416–419.

Wright, R.V.S. (1972). Imitative learning of a flaked stone technology—the case of an orangutan. *Mankind*, 8, 296–306.

Wynn, J.G., et al. (2013). Diet of *Australopithecus afarensis* from the Pliocene Hadar Formation, Ethiopia. *PNAS*, doi:10.1073/pnas.1222559110.

Zilhman, A.L. (2012). The real females of human evolution. *EA*, *21*, 270–276.

29 Our Evolutionary Heritage

The Primate in Us

Two small boys in a playgroup sit near each other, each with a toy. One of them puts his toy down and pokes the other boy. They wrestle, laughing. Two little girls in the playgroup sit near each other, one with a doll. The other girl extends her hand and the first one gives her the doll.

Primate behavior patterns (or elements of them) may persist in modern humans, some homologous and some analogous. By investigating their evolutionary origins through comparative study, we can better understand ourselves. The phrase *human primates* will be used as a reminder of our connections with other primates.

Human Primate Behavior

Human Primate Ecology

Some human taste preferences seem to be homologous with other primates. Researchers tested great apes, gibbons, and humans of various ages for reactions to sweet stimuli, bitter stimuli, and water. Orofacial responses were almost identical and appeared in 27 human infants prior to their first feeding experience (Steiner and Glaser 1995). The resemblance in form across related species and in naïve human infants suggests homologous proximate mechanisms.

The Inertia Hypothesis proposed that contemporary humans (in wealthy countries) suffer from over-indulgence in the preferred foods of our phylogenetic ancestors (Hamilton 1987). When those foods were scarce or difficult to obtain, it was adaptive to maximize consumption when they were available. With the original motivational system unchanged (phylogenetic inertia), humans experienced increasing access to such foods and began to eat unhealthy quantities.

Consumption of vertebrate animals occurs among many anthropoids and reaches its peak among hominoids and by implication the LCA. The behavior of common chimpanzees and bonobos (such as begging and appropriating) indicates strong motivation. Where meat is readily available to humans it can become the source of more than half of a population's dietary energy. At this level it is implicated in colon cancer and other diseases (Hamilton and Busse 1978).

Excessive consumption of sweet foods, with effects ranging from diabetes to tooth decay, is another example. Our attraction to sweet tastes can be traced to fruit-eating ancestral primates (Hladik et al. 2002), especially a hominoid LCA that probably ate as much ripe fruit as possible, as apes do today.

A similar argument might apply to a byproduct of ripe fruit. Ethanol is the natural result of fermentation of fruit sugars by yeast and has probably been associated with angiosperm plants since the Cretaceous. The Drunken Monkey Hypothesis holds that primates are attracted to the site of ripe fruit by the scent of ethanol, which may also act as a feeding stimulant (Dominy 2004; Dudley 2004). These ancient sensory biases might underlie contemporary patterns of alcohol consumption and abuse.

Critics of this hypothesis have pointed out that ethanol concentrations in fruit are low and that wild primates prefer to eat ripe fruit rather than the over-ripe specimens containing ethanol (Levey 2004; Milton 2004). However, attraction to ethanol from over-ripe fruits may have guided primates to patches of the ripe fruits that they actually ate. Senses keener than those of modern humans might have responded to low natural concentrations. The attractive stimulus might then have expanded to include taste.

Whatever the origin, other primates parallel human responses to ethanol. For example, feral vervets spontaneously took alcohol and divided into binge drinkers, steady drinkers, social drinkers, and teetotalers (Ervin et al. 1990). Researchers have justified the damage to primate subjects in terms of seeking causes and treatments for human alcoholism. For example, an important predictor of alcohol abuse in rhesus macaques is impulsivity (Gerald and Higley 2002).

The Inertia Hypothesis might apply to other drugs. Sullivan and Hagen (2002) postulated chemical-ecological adaptations in mammals to metabolize psychotropic plant substances and defensive plant chemicals that interfere with mammalian neurotransmitters. Chimpanzees and gorillas consume plants that contain psychotropic chemicals, although it is not known if this is their motivation. If it is, the function (possible health benefits) may be distinct from the motivation (perhaps a pleasurable response).

Health benefits can explain other unusual feeding behavior. Chimpanzees, gorillas, and other primates consume various plants and soils thought to have medicinal qualities. In many cases the same plants and soils are used in similar ways by neighboring humans. Wild chimpanzees consume unusual foods twice as often as gorillas (Masi et al. 2012). Chimpanzees acquire knowledge of such foods by vertical (inter-generational) transmission, while the gorilla mode is horizontal (mainly between immatures). Consumption of unusual foods may be riskier for chimpanzees because they lack the size and gut specializations that facilitate detoxification in gorillas. Vertical transmission of food distinctions decreases the danger because the information comes from experienced individuals. It may also account for a larger body of knowledge that leads to more frequent use. Because humans are more like chimpanzees than gorillas in the relevant characteristics, self-medication in hominins may have been associated with high social tolerance and lack of gut specialization. An alternative hypothesis is that self-medication existed in the common ancestor of African apes and humans. It remained at the same level in gorillas because of changes in body size and gut specialization, but increased in chimps and hominins because they lacked these safety factors.

Other shared responses to disease may also be illuminating, including avoidance of arthropod vectors and sleeping in closed environments (Nunn and Altizer 2006). Small animals that are (in at least some cases) disease vectors stimulate disgust in humans (Davey 2011) and primates display disgust reactions similar to those of humans. Combined with adaptive significance, this suggests an evolutionary origin for the human emotion (as a motivation for avoidance) and for recognition of particular relevant stimuli.

Snakes continue to concern recent humans. The innate response to snakes appears to be attentional rather than emotional (King 1997). Young primates often watch adults mobbing snakes and presumably learn the negative response. Snakes play an important symbolic role in many human cultures, but they may be either harmful (as in the Garden of Eden) or benevolent (as is the rainbow serpent of native Australia). Whether the reaction is positive or negative, though, humans are like other primates in that they pay more attention to snakes than to most other animals (Mundkur 1976).

Psychological findings are consistent with the Attentional Hypothesis. People demonstrated preferential orientation to snakes in various experiments (Lipp and Waters 2007; Purkis and Lipp 2007). LoBue and DeLoache (2008) reported that children (aged three to five years), like adults, orient more rapidly and more often to snakes than to other stimuli. They obtained the same result by presenting pairs of animal films to infants (8 to 18 months) and interpreted the early appearance in ontogeny as support for an evolved bias in humans. They speculated that the key stimuli include slithering locomotion, the long limbless body, and coiling. Experiments have demonstrated that the fear component of snake responses is learned (Wiens et al. 2008).

Although attacks by constrictors are reported more often, Isbell (2006) maintained that this is because constrictor attacks last longer and are therefore more likely to be observed. She inferred that attacks by venomous snakes are really more common. However, the importance of constrictors is indicated by ethnographic observations of the Agta people of the Philippines when they were still hunter-gatherers. In a population of 120, 26 percent of the adult men had survived predation attempts by reticulated pythons (Headland and Greene 2011). The same population sustained six fatal attacks over the course of about 40 years. In a survey of the literature for attacks on humans and other primates, the same authors concluded that venomous snakes occasionally eat primates, but some constrictors regularly prey on them. If constrictors played the greater role in hominin evolution, then response to their greater size may be a key proximate mechanism in humans as in other primates (King 1997).

Human Primate Sociality

Touching in human life may be derived from primate grooming. Grooming probably diminished in humans with the rise of language for social communication, but we still engage in comparable behavior (Dunbar 2008). Both men and women engage in close touching, such as patting, cuddling, and embracing. Though not exactly grooming, these contacts may provide similar stimulation. Captive chimpanzees did not sharply differentiate grooming from petting in preference order or in terms of the responses they evoked (Mason et al. 1962). Humans may be more sensitive to touch than is usually appreciated. Blindfolded test subjects discerned the emotions of touchers at greater than chance levels (Hertenstein et al. 2009). Brain research with humans demonstrated neural differentiation between affective contacts and neutral tactile sensations (Gallace and Spence 2008).

With children and sometimes with adult partners, women manipulate the hair and attend to scabs and other skin blemishes. In some cultures (such as the Kalahari foragers) women form so-called *hair care cliques*, which can be compared to grooming clusters in chimpanzees and other primates. Men in many cultures interact through various kinds of touching, as do male chimpanzees. Embracing is a common greeting between men, as it is among male chimpanzees. Touching between men has been limited in the Western world by fear of being labeled homosexual. This fear is an aberration from the biological norm that may be psychologically damaging. The taboo breaks down in team sports, where touching contributes to solidarity (Carey 2010; King 1976a). Similarly, chimpanzees embrace before and during group confrontations and in other stressful situations.

Recourse to gorillas and orangutans does not tell whether embracing is homologous between humans and chimps because relations between males in those genera are largely hostile and such greetings are not to be expected. Hugging occurs in both sexes of spider monkeys, which live in fission–fusion societies comparable to those of chimpanzees. This suggests that such a greeting might evolve, either biologically or culturally, under those particular social circumstances.

Social imitation can promote social relations in humans and other primates. Psychologists have studied the "chameleon effect" in humans: people in everyday life unconsciously imitate the postures, gestures, and mannerisms of their interaction partners. This increases rapport, liking and empathy (Paulkner et al. 2009). Experiments with capuchin monkeys found that they preferred humans who imitated them to non-imitators. They looked longer at the imitators, spent more time in proximity to them, and interacted with them more frequently. Wild capuchins routinely synchronize their behavior in travel, feeding, and predator defense.

Reputation entails the ability to learn about another individual by indirect means so as to predict that individual's behavior. Experiments indicate that other primates have this faculty. Herrmann et al. (2013) allowed chimpanzees, orangutans, and human children (2.5 years old) to view scenarios in which experimenters interacted with third parties. One experimenter was generous with food and/or toys while the other withheld them. The apes and the children took into account these third-party interactions in judging the experimenters. Thus LCA individuals may have navigated their social systems with knowledge of others' reputations.

Researchers allowed chimpanzees to observe unfamiliar humans consistently giving ("generous donor") or refusing to give ("selfish donor") food to a familiar human recipient and a chimpanzee. After continued opportunities to observe the same behavioral exchanges, four of seven chimpanzees developed a preference for gesturing to the generous donor and transferred this preference to unfamiliar donor pairs. They preferred to beg from the novel generous donors on the first opportunity to do so. Then four chimpanzees observed novel selfish and generous acts directed toward other chimpanzees by human experimenters. Three of these subjects exhibited a preference for the novel generous donor on the first trial. The numbers are small, but these results suggest that chimpanzees can infer the reputation of strangers by observing third-party interactions.

Male–female bonding in primates provides a basis for the tendency of human males and females to establish long-term relationships with one another. This is widely believed to be associated with the hunting-gathering division of labor, but it may have pre-human origins. We have noted the mate-guarding and infanticide defense hypotheses. Quinlan (2008) asserted that there is little evidence to connect pair bonding in either humans or nonhuman primates with infanticide or male–male aggression. He suggested that two different types of pair bond arose under different conditions. One was a **biparental bond** that focused on child rearing. The other was a mate-defense bond arising from male–male competition and devoted to male reproductive success. The behaviors involved in the latter, directed at the female rather than another male, might range from enticement to coercion. All of these behaviors and relationships are comparable to situations among various nonhuman primates. Division of labor could have been built on these foundations, but some human behavior might still be explained on the basis of the original motivational system(s).

From a functional viewpoint the more flexible special friendship behavior of common baboons seems a more likely model for early hominins. It suggests that biparental care might have originated in either mating effort by males (common baboons) or female need for protection (chacma baboons). Mate guarding occurs in all common baboons. Baboon patterns of mating and infant care vary widely, but the same can be said about recent humans (Fernandez-Duque et al. 2009; King 2003) and presumably about our ancestors. This model suggests that the relationship was compatible with increasing terrestriality and expansion into more open habitats.

A possible evolutionary basis for human paternal care can be seen in male solicitude for young individuals among primates. In most species adult males are at least tolerant of infants or interact affiliatively with them (Fernandez-Duque et al. 2009). Such behavior in many adult male chimpanzees and gorillas suggests homology with humans. Persistent parental care by males also evolved in siamangs and numerous New World primates. In all of these species paternal or allopaternal care is associated with long-term association between a particular male and female. However, these are not necessarily examples of biparental care in that sexual relationships may be polyandrous and infant care may be cooperative among males. The most significant feature of these systems for understanding humans is probably hormonal analogies. Vasopressin in particular is implicated in the paternal behavior of humans and some of these other primates, and a decrease in paternal testosterone may be another analogous feature.

Rates of abuse and murder of stepchildren suggest that human males have at least a residue of primate infanticidal motivation (Daly and Wilson 1988). Sexually selected infanticide does not occur in all primates, but it does occur in taxa that have special significance for understanding humans: chimpanzees, gorillas, baboons, and spider monkeys. Furthermore, just as in humans, it does not occur in every situation where the circumstances seem conducive to it. It is plausible that one factor contributing to infanticide by human males is a motivational system that was adaptive in the hominin past.

Aggression in humans has received what de Waal (1992) criticized as "ideological" treatment. He meant that explanatory models are rejected as being justifications for violence; aggression is divorced from social context and peacemaking is treated as the triumph of reason over instinct. Comparative evidence from primates (and other animals) shows that aggression can be adaptive. We can find continuities between humans and other primates with regard to aggression and also peacemaking.

Primate dominance rests on the potential for aggressive behavior. A similar system can be observed in humans from an early age. Strayer (1992), for example, studied preschoolers (aged one to five years) and found that agonistic exchanges ending in the submission of one participant revealed linear dominance hierarchies at all age levels. He concluded that social dominance is developmentally the earliest stable dimension of peer group organization and that cohesive activities are increasingly coordinated with dominance rank toward the end of the preschool years.

Something like this may operate among adults, up to the highest levels of society. Two political scientists characterized the basis for politics as "playground cognition" (Fowler and Schreiber 2008). Their playground sounds very much like a primate troop: participants assess others for cooperation or avoidance, are aware of social hierarchy, join coalitions knowing that this will mean exclusion from other groups, and constantly monitor the social environment and their place in it. The political scientists note evidence for such political behavior in capuchin monkeys and chimpanzees. This cognitive interpretation can be applied to dominance competitions in a broad range of culturally framed situations, from academic departments to wine tasting clubs.

A high percentage of human assaults and homicides seem to have no obvious goal, but examination of particular cases indicated that such encounters are stimulated by real or perceived threats to social status (Daly and Wilson 1988). Modal action patterns may contribute to the problem. Humans seem to be like nonhuman primates in regarding staring (real or perceived) as a threat or challenge.

Mazur (2005) argued that several features of adult human dominance have counterparts in nonhuman primates, indicating an evolutionary basis: group members of similar rank interact with each other more often than those of dissimilar rank; high-ranking members perform services and control functions for others; low-ranking members display signs of anxiety; and dominant individuals manipulate stress on those below them. Mazur makes the important point that flexibility in human communication results in minimal use of physical aggression in connection with dominance. Although human displays often differ from those of other primates, the same motivational system may be involved: satisfaction derived from the acknowledgement by others of one's superior status.

Puts et al. (2011) explored an aspect of dominance communication that relates to the potential for physical aggression. Beginning with the fact that many primates are sexually dimorphic in vocal anatomy and production, they hypothesized that acoustic features of human male vocalizations are related to threat potential. Using samples from a foraging society (the Hadza) and an industrial society (the US), they linked each acoustic parameter studied to at least one measure of threat potential: size, strength, testosterone, and physical aggressiveness. In both samples the most dimorphic parameters were most strongly related to body size.

While formal dominance is important, there are other significant positions in the political societies of humans and chimpanzees. Based on research at Arnhem Zoo, de Waal (2007) called attention to those who have influence without obvious power. An individual who falls from high rank does not fall into oblivion; he can still affect events. One who has risen to high personal rank does not automatically have the most influence. Additional studies comparing humans and other primates emphasize that leadership is not necessarily closely connected with dominance and may in many cases be a group process (Conradt and List 2009; King et al. 2009; Stevens 2008; van Vugt 2006).

Reconciliation among preschool children in diverse cultures (American, European, Japanese) resembles postconflict behavior in other primates (Verbeek 2008). One similarity is timing, with reconciliation taking place within five minutes after a conflict ends. Another is the use of both explicit and implicit behaviors. American and Japanese subjects were like other primates in that they were more likely to stay close to an aggressor if the conflict did not involve physical contact. In humans and a few relevant studies of macaques, the peer group was an important context for learning reconciliation skills. The functional explanation most clearly supported by the human studies is the Uncertainty Reduction Hypothesis. As in many primates, children revealed changes in uncertainty through self-directed behavior. Measures of salivary cortisol also supported the hypothesis. Data from five different geographic regions showed that some peacemaking gestures are specifically cultural while others are found in all the cultures studied (Butovskaya et al. 2000). The latter include hugging, a fundamental primate behavior. This may pertain to boxers who embrace at the end of a fight (pers. obs.).

Some mediating signals in human conflict are likely to have an innate primate basis. Van Hooff and Preuschoft (2003) see the origin of human smiling in a facial expression found in the great apes and many Old World monkeys: the silent bared-teeth display, also called the grin or grimace. Comparison among primates, and with varied mammal taxa such as the horse and dog families, demonstrates that submission was the original meaning and function. In primates, however, the signal is emancipated (in the ethological sense) from fearful motivation. The expression varies, according to the Power Asymmetry Hypothesis, with the shape of the dominance hierarchy. In strong hierarchies the submission signal is formal and unambiguous. In more relaxed societies the signals are not so sharply distinguished, and dominant individuals may use the grin/grimace in a friendly way. The diversity and fluidity of these facial expressions in humans is typical of species with relaxed societies and contrasts with the sharp distinction of submissive and playful expressions in the more hierarchical chimpanzees.

Van Hooff and Preuschoft (2003) treat the play face as an adjunct to laughing. Others view it as a distinct facial expression in its own right and link it to a different kind of smiling in humans. The grimace-based smile, even though it can be friendly, is often associated with tension or aggression. The relaxed smile derived from the play face is entirely associated with positive emotions, whether in a pleasurable activity or greeting a close friend. Laughter is more often associated with this smile in a pleasurable activity than in a greeting.

The relatively long ovarian cycle shared by humans and other primates averages about 30 days. The primate version is unusual among mammals because it is not scaled to body size. The cycle may have been established in response to lunar cycling in ancient primates that were nocturnal but depended heavily on vision (Martin 2003).

There is evidence for a human estrus cycle in the sense of proceptivity and attractivity around the time of ovulation. A woman's scent and face seem to be more attractive during her fertile period and she tends to dress more attractively or provocatively (Gangestad and Thornhill 2008). Women's voices rise in pitch when they are fertile (Bryant and Haselton 2009) and are perceived as more attractive (Pipitone and Gallup 2012). Behavior patterns such as these suggest that feelings and perceptions like those associated with ovulation in other primates also occur in humans. Despite any ovulation effects, however, humans are like other anthropoids in that mating extends beyond the time of ovulation (Martin 2003), throughout the cycle and during pregnancy.

Human Primate Life Cycle

Primates typically give birth during the inactive part of the daily cycle (Jolly 1985), which means at night for anthropoids. Then they are sheltered and not obliged to keep up with group movement. Humans follow the same pattern, presumably due to the same mechanisms. Human births typically take place during the night, most often in the early morning hours (Martin 1990).

Primates have relatively long gestation periods and give birth to precocial young. Humans and great apes have the longest pregnancies, on the order of eight to nine months, so precocity is to be expected. Human neonates are sometimes characterized as altricial because they are hairless and highly dependent. However, the hairlessness is secondary and the helplessness is because the large human brain is far from having completed its growth (Martin 2003). Other than the effects of brain size, human babies are precocial like other primates. Intrauterine brain development conforms to the primate pattern; humans only differ in postnatal brain growth and its behavioral consequences.

Life history theory postulates that parental investment is governed by a trade-off between the typical number of offspring and the typical size of offspring. The generalization holds across natural-fertility human populations as long as an adjustment is made for the mother's energy availability. This suggests that natural reproductive variation in humans can be understood in terms of the basic principles of energetic allocations that apply to primates and other animals (Walker et al. 2008).

Human mothers engage in intense emotional communication with their newborns that includes gaze, facial expressions, body contact, and play. Probable homology with chimpanzees has long been acknowledged, but recent work on rhesus monkeys has revealed similar behaviors (Ferrari et al. 2009). In 14 captive mother–infant pairs, mothers gazed at their infants and engaged in intense body contact with them, including touching with the mouth that resembled kissing. In one pattern the mother held her infant and searched for its gaze, sometimes gently pulling its head toward her face. In other cases, when the two were separated, the mother moved her face towards the infant's and sometimes lowered her head. Infants responded to these overtures by reciprocating gaze. Mother–infant dyads in a semi-free-ranging context displayed similar behavior. The researchers interpreted these patterns as an emotional basis for social exchange later in life in both humans and monkeys. They suggested that the presence in an Old World monkey indicates that it is phylogenetically old. The addition of one monkey species to chimpanzees is not a strong argument for a catarrhine pattern, but it shows that such behavior is not even limited to hominoids, much less uniquely human.

Aberrant mothering may be illuminated by primate behavior (Maestripieri 1999). Large populations of three papionin species (macaques and mangabeys) were studied for about 30 years. Results showed that abusive mothers maltreated most of their infants, regardless of infant sex or birth order. Genetic transmission is suggested by two patterns. First, abuse was more likely to occur in some macaque matrilines than in others. Second, abuse in all three study species was especially likely to occur in closely related females.

Juvenile humans and nonhuman primates spontaneously separate into same-sex playgroups that display sex-differentiated play styles (Maccoby 1999). An obvious continuity between humans and other primates is the rough-and-tumble play that characterizes the juvenile period, especially for males. One of its main components, wrestling (Figure 29.1), became the most widespread sport in the human world (King 2003). Another major play

Figure 29.1 Adult captive chimpanzees wrestle and display the play face. These behavior patterns may be homologous with similar ones in humans. © Chimpanzee Sanctuary Northwest, www.ChimpsNW.org

pattern, chasing, is a component of widespread children's games such as tag, and is prominent in popular sports as well.

The first laughter of human infants occurs in play, and primate evidence supports the idea that this playful context is the evolutionary foundation for human laughter. Gervais and Wilson (2005) argued that laughter has the following characteristics of ethological ritualization in relation to comparable ape vocalizations: it has changed in structure to be more obvious, its releasing threshold has been lowered, and it has been simplified by more rhythmic repetition, greater amplitude, and greater stereotypy.

Gervais and Wilson also offered an evolutionary scenario to account for the changes, building on the hypothesis that the positive emotions of early hominins evolved in the context of short-term fluctuations in safety and the alleviation of hunger. They postulated that this situation selected for rapid communication of safe conditions to all members of a group, especially the most fearful ones. Natural selection broadened the neuroanatomical basis for laughter so that it could be triggered by a wider variety of stimuli. In addition to signaling a generally safe situation, such laughter could also convey that a particular incident (such as a fall) was not serious (Ramachandran 1998). This might explain why we are amused by slapstick humor.

Analogy with common baboons suggests that predator scares (Altmann and Altmann 1970) might have been among the greatest sources of lasting tension for our ancestors. In the absence of language, early hominin laughter might have communicated safety as well as expressing relief.

Bachorowski and Owren (2001) proposed a different and partly contradictory hypothesis, which incorporated smiling with laughter. They saw the need for cooperative behavior as the most relevant factor. Initially smiling was a reliable indicator and stimulus for positive affect, but the problem of cheaters arose. Laughter then emerged as the main vehicle for positive affect because it is more difficult to imitate on a volitional basis.

Two universal features of juvenile behavior in humans and other primates are proximity to and emulation of adults of the same sex. This is common for females throughout the anthropoids. They become averse to the increasingly rough and serious play of the males and they are attracted to the infants that they see among the adult females. Young males can have long-term role models in one-male gorilla groups, male-philopatric chimpanzee communities, and human families and traditional communities.

Human Primate Cognition

Haun and Call (2009) compared representatives of all the great apes with human children in a relational similarity test. They found that all species were alike in that they performed better when the connection between the related elements was logico-causal. Only children above the age of four years could display some ability to reason by non-causal similarity. Chimpanzees and bonobos shared this ability, but gorillas and orangutans did not. This suggests that the capability for such reasoning lies in the common ancestry of humans and the genus *Pan*.

Peeters et al. (2009) performed fMRI scans on humans, untrained monkeys, and trained monkeys as they watched videos of tool use. The stimulus elicited activity in several homologous regions of the human and monkey brains, which suggests a neural basis for tool use that goes back beyond the chimpanzee–human LCA. One section of the brain (the left inferior parietal lobule) responded only in humans, indicating a unique evolutionary development in hominins.

Another requirement of sequential tool use is **patience**, the ability to suppress immediate motivations and defer an action until appropriate preparations have been made (Martin-Ordas et al. 2012). This capability has a cognitive and emotional component: foresight and self-control. Patience is required for natural behaviors such as termiting.

Many animal species can wait only a few seconds for delayed benefits, which suggests a phylogenetic gap between animals and humans (Rosati et al. 2007). Experiments with chimpanzees, bonobos, and human adults narrow that gap considerably. First, the apes demonstrated greater patience than other animals tested so far. Second, human adults are less willing to wait for food rewards than chimpanzees. These findings suggest that the core components of future-oriented decision-making are shared derived features in *Pan* and *Homo*. Humans were more willing to wait for monetary rewards than food. This may mean that apes and humans differ in the mental mechanisms that represent biological versus abstract rewards (Rosati et al. 2007). On the other hand, nonhuman primates seem to show patience in dealing with their social situations: anthropoids, especially chimpanzees, strategically invest in social relationships in order to ascend in their dominance hierarchies (Stevens and Stephens 2007). Long-tailed macaques can also delay gratification (Pele et al. 2009).

Several studies have demonstrated a correlation between human brain anatomy and social network size, but the question of causality was left open (Miller 2011). A comparative study of captive rhesus monkeys suggests the solution. Monkeys housed in larger groups had more gray matter in several regions of the temporal cortex, as shown by MRI scans (Sallet et al. 2011). Other monkey studies have shown that one of these regions (the superior temporal sulcus or STS) is involved in responses to faces, body movements, and probably the tracking of other individuals' direction of attention. MRI studies of humans also link the STS to social network size. Kanai et al. (2011) found a relationship between the size of a person's Facebook network and gray matter density in the STS and other regions.

The study of emotions in nonhuman species invites anthropomorphism. It is important to keep in mind that such studies do not necessarily imply that the other animals have the same conscious feelings as humans (Aureli and Whiten, in Maestripieri 2003). Similar physiological correlates can establish the same emotion in humans and other primates. For example, an increase in heart rate is evidence for anxiety in humans and in a rhesus monkey approached by a dominant individual. Gentle touching in humans and grooming in other primates bring about relaxation (lowering of heart rate). As in other mammals, the human amygdala is deeply implicated in fear. With concordance of emotions supported by neurophysiology we can place increased confidence in comparison of expressive behaviors as indicators of emotion. Chimpanzees, for example, can match positive or negative video scenes with conspecific facial expressions. Anxiety in most primates is indicated by self-directed behavior such as scratching. Like the silent bared-teeth display in nonhuman primates, grimacing (not the derived smiling expression) represents pain or fear in humans.

De Waal (2008) proposed the feeling of empathy as the basis for altruistic behavior in humans and other primates. The simplest common feature of empathy across species is **emotional contagion** in which one party is affected by the emotional state of another. Continuity between apes and humans is demonstrated by similarities in emotional communication and by brain and skin temperature changes in response to emotionally charged images. Continuity is manifested in behavior, such as monkeys refusing to pull a chain to get food when it triggers a shock that causes pain for a companion. **Sympathetic concern** involves attempts to understand the cause of another's emotion. An example is consolation, the reassurance given to a defeated individual by a bystander. The most common manifestation in nonhuman primates is defense of another against aggression—in some cases against an opponent superior to both the victim and the helper. Less common but more drastic are incidents in which chimpanzees (who cannot swim) try to save conspecifics from drowning. The last level is **perspective taking**, which involves a detailed understanding of the other's situation. The mother who returns to a whimpering offspring to help it from one tree to another understands the particulars of the problem.

Experiments with chimpanzees and human children cast some doubt on the altruistic nature of the apes (Warneken and Tomasello 2009). With regard to helping, apes and children retrieved objects for experimenters without receiving any reward or praise. Providing information was another matter. Human infants as young as one year helped adults to find things by pointing at them; chimpanzees did not. Children shared food generously and chimpanzees did not. The apes were not affected by the fact that a conspecific might get food as a result of their own food-getting action, but children performed an action more often when another also received a reward.

The issue of food exemplifies some complexities in these analyses. Although the captive chimpanzees had trouble with sharing food, wild chimpanzees often do it. According to one viewpoint, this is not altruistic because it is done under social pressure (Gilby 2006). De Waal (2008) answered that top-ranking individuals are among the most generous and that sharing among captives occurs even when bars separate the participants. However, sharing in the wild is often a political tactic on the part of dominant individuals and bars do not necessarily prevent psychological intimidation based on past relationships or current displays.

Other species may contribute to understanding altruism. When marmosets and tamarins turned out to be more altruistic than chimpanzees with regard to food, some primatologists toyed with the idea that human altruism is rooted in a post-LCA evolution of cooperative breeding units (an example of attribute analysis). However, the troop-living capuchins (not cooperative breeders) reliably display prosocial behavior in experimental food sharing tasks. Subjects were more likely than chance to deliver a better marshmallow award to the receiver's chamber when the receiver was present. Capuchins delivered marshmallows to their partners even when they themselves only got celery.

Human arts may have primate roots. A fundamental element of music is the beat (a regular pulse in the auditory signal). Human children (aged 2.5 to 4.5 years) were given drums and three different external sources of tempo: a drum machine, drum sounds from a speaker, and drumming by an experimenter (Kirschner and Tomasello 2009). Like adults across cultures, the children synchronized their movements with the external beat. They achieved the highest accuracy of synchronization with the experimenter. The researchers interpreted this to mean that drumming is essentially a social behavior, based on shared understanding and intention. They considered it a special case of a *unique* human motivation to share emotions, experiences, and activities.

However, newborn infants can detect the beat in music and experiment shows that they develop an expectation for the onset of rhythmic cycles (the downbeat), which indicates that perception of the beat is innate (Winkler et al. 2009). Primate evidence supports the idea that rhythm and drumming are innate. Like the children above, some macaques actively engaged in drumming. This attracted the attention of other monkeys in a way similar to conspecific vocalizations, and influenced the way they viewed their conspecifics in a preferential looking experiment (Remedios et al. 2009). Functional imaging showed that drumming and vocalizations activated the same brain areas (caudal auditory cortex and amygdala). The researchers inferred a common origin for primate vocal and nonvocal communication systems, which they see as supporting a gestural origin for both language and music.

Wild chimpanzees also engage in drumming (Arcadi et al. 1998) and this has been compared to human music (Fitch 2006). In the drumming display a chimp (usually male, as in other displays) jumps up and pounds his hands and feet against the buttress of a large tree, making a sound that carries for long distances (Goodall 1986). Favorite trees usually trigger the behavior in several members of a traveling party. Coordinated drumming has been seen at least once in captive chimpanzees (pers. obs.). The alpha and beta males of the group sat down facing each other and each began drumming on the floor next to him on both sides while looking at the other's face. This went on for about 30 seconds and ended when both males got up and walked away in different directions.

Gorillas also perform drumming. Bimanual drumming on the chest is a prominent feature of male display, but is also done by females and (in play) by juveniles. Drumming on the body is a feature of music in numerous human cultures, native Polynesia being a familiar example (King 2003). Gorillas sometimes drum on the ground, objects, or each other. Similarly, although chimpanzees prefer tree buttresses, they adapt to other surfaces including hollow walls or overturned buckets. Drumming bouts by gorillas and chimpanzees are brief, usually about 1 second, but bonobos can maintain a steady drummed beat for up to 12 seconds. Since bimanual drumming is unique to humans and great apes, Fitch (2006) regarded it as a likely homology among them.

Although nonhuman primates do not produce melodies, they may have some elements of melody perception. Experimenters trained two rhesus monkeys to make same/different judgments and then presented them with short melodies. The monkeys correctly judged the melodies even when they were transposed by an octave (Wright et al. 2000). An infant chimpanzee displayed a clear preference for consonant music (such as minuets) over dissonant music produced by modifying the test pieces (Sugimoto et al. 2010). Snowdon and Teie (2009) suggested that affective components of human music may have evolutionary roots in the calls of other species. In this context it may be significant that the drumming of chimpanzees is accompanied by loud pant hoots.

Ape paintings have been examined as representations of mental life. Zeller (2007) analyzed 396 pictures made by chimpanzees, gorillas, orangutans, and human children. She found similarities across all species: preference for yellow as the first color, high use of diagonal lines, and preference for centered patterns. Males largely limited themselves to four colors and showed less respect for boundaries. Females used a wider range of colors and stayed inside the confines of the page nearly half the time. Direct comparison with human children in this case indicates that elementary principles of visual art are rooted in our ancestry. This is not to say that the Last Common Ancestor was a painter. As with music (McDermott and Hauser 2004), this behavior is presumably a byproduct of adaptive mental processes.

Primate communication mechanisms might have a bearing on the evolution and functions of language. Language came together as a complex system in hominins, but it is probably based on various disassociated components seen in the behavior of nonhuman primates. Hauser et al. (2002) distinguished between the *faculty of language* in a broad sense (FLB) and the *language faculty* in a narrow sense (FLN). FLN is uniquely human. FLB is composed of distinct elements that are shared with other species: sensory-motor system, conceptual-intentional system, and computational mechanisms for recursion. For example, humans share voice perception abilities with other primates that could have been the basis for speech. This suggests that speech is rooted in neural mechanisms for extracting behaviorally relevant information from conspecific vocalizations (Belin 2006). Research on humans and rhesus monkeys has identified regions in the anterior temporal part of the brain that are sensitive to species-specific vocalizations (Petkov et al. 2009).

Summary

Modern human behavior can be linked to behavior observed in other primates. Thus human behavior can be better understood through the use of the comparative method. Some similarities may be analogous and others homologous. The latter can provide insight into proximal mechanisms as well as functions. Comparison with other primates can illuminate normal human behavior, such as social imitation and rough-and-tumble play. This approach also promises to help us understand aberrant behavior, over-indulgence in consumption, and maternal abuse. Some traits that have been considered hallmarks of humanity seem to have nonhuman primate counterparts, including elements of language and the arts.

Sources

Basic Bibliography

Maestripieri, D. (Ed.) (2003). *Primate psychology: The mind and behaviour of human and nonhuman primates*. Chicago: UCP.

Additional References

Altmann, S.A., & Altmann, J. (1970). *Baboon ecology*. Chicago: UCP.

Arcadi, A., et al. (1998). Buttress drumming by wild chimpanzees. *Primates*, *39*, 503–516.

Bachorowski, J., & Owren, M.J. (2001). Not all laughs are alike: voiced but not unvoiced laughter readily elicits positive affect. *Psychological Science*, *12*, 252–257.

Belin, P. (2006). Voice processing in human and non-human primates. *PTRSB*, *361*, 2091–2107.

Bryant, G.A. & Haselton, M.G. (2009). Vocal cues of ovulation in human females. *BL*, *5*, 12–15.

Butovskaya, M., Verbeek, P., Ljungberg, T., & Lunardini, A. (2000). A multicultural view of peacemaking among young children. In F. Aureli & F.B.M. de Waal (Eds.), *Natural conflict resolution* (pp. 243–258). Berkeley: University of California Press.

Carey, B. (2010). Evidence that little touches do mean so much. *New York Times*, February 22: D5.

Conradt, L. & List, C. (2009). Group decisions in humans and animals: a survey. *PTRSB*, *364*, 719–742.

Daly, M., & Wilson, M. (1988). *Homicide*. New York: Aldine De Gruyter.

Davey, G.C.L. (2011). Disgust: the disease-avoidance emotion and its dysfunctions. *PTRSB*, *366*, 3453–3465.

de Waal, F.B.M. (1992). Aggression as a well-integrated part of primate social relationships: critical comments on the Seville Statement on Violence. In J. Silverberg & J.P. Gray (Eds.), *Aggression and peacefulness in humans and other primates* (pp. 37–56). NY: OUP.

de Waal, F.B.M. (2007). *Chimpanzee politics* (25th anniversary edition). Baltimore: Johns Hopkins University Press.

de Waal, F. (2008). Putting the altruism back into altruism: the evolution of empathy. *ARo Psychology*, *59*, 279–300.

DeLoache, J., & LoBue, V. (2009). The narrow fellow in the grass: human infants associate snakes and fear. *Developmental Science*, *12*, 201–207.

Dominy, N.J. (2004). Fruits, fingers, fermentation: the sensory cues available to foraging primates, *ICB*, *44*, 295–303.

Dudley, T.R. (2004). Ethanol, fruit ripening, and the historical origins of human alcoholism in primate frugivory. *ICB*, *44*, 315–323.

Dunbar, R. (2008). Cognitive constraints on the structure and dynamics of social networks. *Group Dynamics*, *12*, 7–16.

Ervin, F., et al. (1990). Voluntary consumption of beverage alcohol by vervet monkeys. *Pharmacology Biochemistry & Behavior*, *36*, 367–373.

Fernandez-Duque, E., et al. (2009). The biology of paternal care in human and nonhuman primates. *ARA*, *38*,115–130.

Ferrari, P.F., et al. (2009). Reciprocal face-to-face communication between rhesus macaque mothers and their infants. *CB*, *19*, 1768–1772.

Fitch, W. T. (2006). The biology and evolution of music: a comparative perspective. *Cognition*, *100*, 173–215.

Fowler, J.H., & Schreiber, D. (2008). Biology, politics, and the emerging science of human nature. *Science*, *322*, 912–914.

Gallace, A., & Spence, C. (2008). The science of interpersonal touch: an overview. *Neuroscience & Biobehavioral Reviews*, *34*, 246–259.

Gangestad, S.W., & Thornhill, R. (2008). Human oestrus. *PRSB*, *275*, 991–1000.

Gerald, M.S., & Higley, J.D. (2002). Evolutionary underpinnings of excessive alcohol consumption. *Addiction*, *97*, 415–425.

Gervais, M., & Wilson, D.S. (2005). The evolution and functions of laughter and humor: a synthetic approach. *QRB*, *80*, 395–430.

Gilby, I.C. (2006). Meat sharing among the Gombe chimpanzees: harassment and reciprocal exchange. *AB*, *71*, 953–963.

Goodall, J. (1986). *The chimpanzees of Gombe: patterns of behavior*. Cambridge MA: Harvard University Press.

Hamilton, W.J. (1987). Omnivorous primate diets and human overconsumption of meat. In M. Harris & E.B. Boss (Eds.), *Food and evolution*. Temple University Press.

Hamilton, W.J., & Busse, C.D. (1978). Primate carnivory: its significance to human diets. *Bioscience*, *28*, 761–766.

Haun, D.B.M., & Call, J. (2009). Great apes' capacities to recognize relational similarity. *Cognition*, *110*, 147–159.

Hauser, M.D., et al. (2002). The faculty of language: what is it, who has it, and how did it evolve? *Science*, *298*, 1569–1579.

Headland, T.N., & Greene, H.W. (2011). Hunter-gatherers and other primates as prey, predators, and competitors of snakes. *PNAS*, *108*, E1470–E1474.

Herrmann, E. et al. (2013). Direct and indirect reputation formation in nonhuman great apes (*Pan paniscus*, *Pan troglodytes*, *Gorilla gorilla*, *Pongo pygmaeus*) and human children (*Homo sapiens*). *JCP*, *127*, 63–75.

Hertenstein, M.J., et al. (2009). The communication of emotion via touch. *Emotion*, *9*, 566–573.

Hladik, C.M., et al. (2002). New perspectives on taste in primate evolution: the dichotomy in gustatory coding for perception of beneficent versus noxious substances as supported by correlations among human thresholds. *AJPA*, *117*, 342–348.

Huffman, M.A., et al. (2010). Leaf swallowing behavior in chimpanzees (*Pan troglodytes*): biased learning and the emergence of group-level cultural differences. AC, *13*, 871–880.

Isbell, L.A. (2006). Snakes as agents of evolutionary change in primate brains. *JHE*, *51*, 1–35.

Jolly, A. (1985). *The evolution of primate behavior*, 2nd ed. New York: Macmillan.

Kanai, R., et al. (2011). Online social network size is reflected in human brain structure. *PRSLB*, doi: 10.1098/rspb.2011.1959.

King, A.J., et al. (2009). The origins and evolution of leadership. *CB*, *19*, R911–R916.

King, G.E. (1976a). The naked jock: evolutionary perspectives on American football. Annual meeting of the Northeastern Anthropological Association, Middletown, CT.

King, G.E. (1976b). Society and territory in human evolution. *JHE*, *5*, 323–332.

King, G.E. (1997). The attentional basis for primate responses to snakes. *American Society of Primatologists annual meeting*, San Diego, CA.

King, G.E. (2003). *Traditional cultures, a survey of nonwestern experience and achievement*. Long Grove, IL: Waveland Press.

Kirschner, S., & Tomasello, M. (2009). Joint music making promotes prosocial behavior in 4-year-old children. *EHB*, *31*, 354–364.

Levey, D.L. (2004). The evolutionary ecology of ethanol production and alcoholism. *ICB*, *44*, 284–289.

Lipp, O.V., & Waters, A.M. (2007). When danger lurks in the background: attentional capture by animal fear-relevant distractors is specific and selectively enhanced by animal fear. *Emotion*, *7*, 192–200.

LoBue, V., & DeLoache, J.S. (2008). Detecting the snake in the grass: attention to fear-relevant stimuli by adults and young children. *Psychological Science*, *19*, 284–289.

Maccoby, E.E. (1999). *The two sexes, growing up apart, coming together*. Cambridge, MA: HUP.

Maestripieri, D. (1999). The biology of human parenting: insights from nonhuman primates. *Neuroscience & Biobehavioral Reviews*, *23*, 411–422.

Martin, R.D. (1990). *Primate origins and evolution, a phylogenetic reconstruction*. Princeton: Princeton University Press.

Martin, R.D. (2003). Human reproduction: a comparative background for medical hypotheses. *Journal of Reproductive Immunology*, *59*, 111–135.

Martin-Ordas, G., et al. (2012). The role of episodic and semantic memory in episodic foresight. *Learning and Motivation*, *43*, 209–219.

Masi, S., et al. (2012). Seasonal effects on great ape health: a case study of wild chimpanzees and western gorillas. *PLONE*, *7*, e49805.

Mason, W.A., et al. (1962). Differential responses of chimpanzees to social stimulation. *Journal of Comparative and Physiological Psychology*, *55*, 1105–1110.

Mazur, A. (2005). *Biosociology of dominance and deference*. Rowman & Littlefield.

McDermott, J., & Hauser, M.D. (2004). Are consonant intervals music to their ears? Spontaneous acoustic preferences in a nonhuman primate. *Cognition 94*, B11–B21.

Miller, G. (2011). The brain's social network. *Science*, *334*, 578–579.

Milton, K. (2004). Ferment in the family tree: does a frugivorous dietary heritage influence contemporary patterns of human ethanol use? *ICB*, *44*, 304–314.

Mundkur, B. (1976). *The cult of the serpent: an interdisciplinary survey of its manifestations and origins*. Albany, NY: SUNY Press.

Nunn, C., & Altizer, S. (2006). *Infectious diseases in primates, behavior, ecology and evolution*. NY: OUP.

Paulkner, A., et al. (2009). Capuchin monkeys display affiliation towards humans who imitate them. *Science*, *325*, 880–883.

Peeters, R., et al. (2009). The representation of tool use in humans and monkeys: common and uniquely human features. *Journal of Neuroscience*, *29*, 11523–11539.

Pele, M. et al. (2009). Long-tailed macaques display unexpected waiting abilities in exchange tasks. *AC*, *13*, 263–271.

Petkov, C.I., et al. (2009). Where are the human speech and voice regions, and do other animals have anything like them? *Neuroscientist*, *15*, 419–429.

Pipitone, R.N., & Gallup, G.G. (2012). The unique impact of menstruation on the female voice: implications for the evolution of menstrual cycle cues. *Ethology*, *118*, 281–291.

Purkis, H.M., & Lipp, O. V. (2007). Automatic attention does not equal automatic fear: preferential attention without implicit valence. *Emotion*, *7*, 314–323.

Puts, D.A., et al. (2011). Masculine voices signal men's threat potential in forager and industrial societies. *PRSLB*, *279*, 601–609.

Quinlan, R.J. (2008). Human pair-bonds: evolutionary functions, ecological variation, and adaptive development. *EA*, *17*, 227–238.

Ramachandran, V.S. (1998). The neurology and evolution of humor, laughter, and smiling: the false alarm theory. *Medical Hypotheses*, *51*, 351–354.
Remedios, R., et al. (2009). Monkey drumming reveals common networks for perceiving vocal and nonvocal communication sounds. *PNAS*, *106*, 18010–18015.
Rosati, A.G., et al. (2007). The evolutionary origins of human patience: temporal preferences in chimpanzees, bonobos, and human adults. *CB*, *17*, 1663–1668.
Sallet, J., et al. (2011). Social network size affects neural circuits in macaques. *Science*, *334*, 697–700.
Snowdon, C.T., & Teie, D. (2009). Affective responses in tamarins elicited by species-specific music. *BL*, 6, 30–32.
Steiner, J.E., & Glaser, D. (1995). Taste-induced facial expressions in apes and humans. *Human Evolution*, *10*, 97–105.
Stevens, J. (2008). The evolutionary biology of decision making. In C. Engel & W. Singer (Eds.), *Ernst Strüngmann Forum Report 1* (pp. 285–304). Cambridge, MA: MIT Press.
Stevens, J.R., & Stephens, D.W. (2008). Patience. *CB*, 18, R11–R12.
Strayer, F.F. (1992). The development of agonistic and affiliative structures in pre-school play groups. In J. Silverberg & J.P. Gray (Eds.), *Aggression and peacefulness in humans and other primates* (pp. 150–171). NY: OUP.
Sugimoto, T., et al. (2010). Preference for consonant music over dissonant music by an infant chimpanzee. *Primates 51*, 7–12.
Sullivan, R.J., & Hagen, E.H. (2002). Psychotropic substance-seeking: evolutionary pathology or adaptation? *Addiction*, *97*, 389–400.
van Hooff, J.A.R.A.M. & Preuschoft, S. (2003). Laughter and smiling: the intertwining of nature and culture. In F.B.M. de Waal & P.L. Tyack (Eds.), *Animal social complexity* (pp. 260–287). Cambridge MA: HUP.
van Vugt, M. (2006). The evolutionary origins of leadership and followership. *Personality and Social Psychology Review*, *10*, 354–372.
Verbeek, P. (2008). Peace ethology. *Behaviour*, *145*, 1497–1524.
Walker, R.S., et al. (2008). The trade-off between number and size of offspring in humans and other primates. *PRSLB*, *275*, 827–834.
Warneken, F., & Tomasello, M. (2006). Altruistic helping in human infants and young chimpanzees. *Science*, *31*, 1301–1303.
Wiens, S., et al. (2008). Recognizing masked threat: fear betrays, but disgust you can trust. *Emotion*, 8, 810–819.
Winkler, I., et al. (2009). Newborn infants detect the beat in music. *PNAS*, *106*, 2468–2471.
Wright, A.A., et al. (2000). Music perception and octave generalization in rhesus monkeys. *JEPG*, *129*, 291–307.
Zeller, A. (2007). "What's in a picture?" A comparison of drawings by apes and children. *Semiotica*, *2007*, 181–214.

30 Primate Conservation

Will Any Be Left?

First there was the noise, then men and machines appeared. The monkeys hid in trees and saw other trees falling, their shelter and food lying on the ground. Monkeys fell too, shot for food. The survivors moved farther away, but then they had to fight strangers of their own species for food. Eventually there was nowhere to go—most of the forest was gone.

Nearly half of extant primate taxa (species, subspecies, populations) face extinction (Mittermeier et al. 2012). These mainly arboreal animals are highly susceptible to the rapid deforestation occurring in many parts of the world. Extinction begins with a decline in numbers that makes a population more vulnerable to random demographic and environmental fluctuations. Reduced genetic diversity leads to inbreeding depression and reduced adaptability. Declining numbers can also disrupt social organization, an effect of great importance to most primates. Mammal studies show that some common primate traits are associated with high vulnerability: (1) unusually large or small body size; (2) long generation time; and (3) high reproductive skew. Relatively slow reproductive rates mean lower probability of recovery from population decline. This chapter begins with general problems in primate conservation and then considers issues that vary by taxa. Finally, it samples some of the solutions so far.

Worldwide Problems

Habitat Disturbance

There are degrees to which habitats are degraded. **Habitat loss** means complete elimination of a population's environment, resulting in extinction or (if the option exists) emigration. There is a time lag between habitat loss and extinction that can hide the problem until it is too late for remedies. Many primate taxa may be in that phase right now. In **habitat fragmentation** an extensive environmental area is reduced to small, separated remnants that may be too small to sustain populations. Survival is more likely in larger and less isolated fragments that maintain ecological features needed by the particular taxon. **Habitat modification** is the least severe change. Continuing food availability seems to be the key factor for primate survival in these situations. Survival is more likely when these changes mimic the natural dynamics of a habitat, such as forest succession.

Natural forces can devastate habitats. For instance, storms can destroy large swaths of trees, and severe droughts affect the food supply. However, habitat disturbance is mostly **anthropogenic**, i.e. caused by humans. Commercial activities have the greatest impact:

logging, road building, and agricultural expansion. Development of industry and infrastructure has also been implicated. Anthropogenic causes of habitat disturbance often reinforce one another. For instance, roads created by loggers may be used to transport agricultural products, which encourages forest clearance for farmland. Global climate change is now a serious consideration. Computer simulation indicates that this may result in an 85 percent loss of Amazonian rainforest by 2150 (Jones et al. 2009).

Conservation issues require detailed understanding of ecology and behavior. For example, many trees taken by loggers are mature, fruit-producing trees. The impact of such human choices is clearer with recognition that most primates are inclined to frugivory. On the other side of the coin, primates strongly adapted to leaf eating are more likely to survive. However, folivores and other primates may be severely disturbed by adjacent logging activities. In extreme cases they move away from the disturbance into the ranges of conspecifics, potentially causing high levels of aggression. Some primates seem especially susceptible to logging: the populations remain diminished several decades after logging has stopped (Paciulli 2009). The forest itself may not recover as quickly as anticipated, as at Kibale.

Local economic activities can also be detrimental, e.g. gathering firewood and clearing land for farming or grazing. A study in India (Srivastava et al. 2001) demonstrated the effect of local economy on a viable forest fragment. This fragment of about 5km^2 harbored five primate species with canopy cover of 20 to 50 percent over two-thirds of the forest. After three years less than one-third of the forest had more than 20 percent cover and all primate populations had drastically declined. Small-scale harvesting of forest products with hand tools accounted for the losses.

Hunting and Harvesting

People hunt primates to get food for personal use or for local markets. Intrusive workers, such as loggers and road-builders, also hunt for food. Farmers kill primates to defend their crops. Omnivorous primates find some human foods congenial, but often they are driven to crop raiding by habitat disturbance. Finally, people hunt primates for pelts and body parts that are used as decoration or in the practice of traditional medicine. More than 100 primate species are used in traditional medicines, rituals, and other folk practices (Alves et al. 2010) and two-thirds of these are classified as Endangered. Primates taken for these purposes are also used in other ways, which makes it difficult to quantify the impact of ritual use. Pelts and body parts often find their way into national and international markets, which encourages a larger scale of depredation.

Hunting may simply reduce population numbers, but research on other mammals shows that it can alter the behavior of survivors. Changes in sex ratio might change patterns of dispersal, and diminished dispersal could result in heightened aggression and inbreeding. Hunting could affect vigilance and alarm behavior, resulting in shifts of activity patterns.

Harvesting is the capture and removal of live animals from the wild. Primates are harvested for biomedical research and other experimentation, as well as for zoos, circuses, and for use as pets. Harvesting young animals often involves killing adults. Even if no animals are killed, the population effects of harvesting and hunting are similar. They can reduce population size to a point below viability. However, differences in motivation between hunters and harvesters might in some cases affect the approaches taken by conservationists.

Hunters and harvesters typically live near or among their primate prey, so they can exert almost constant pressure on local animal populations. Their choices often lessen the chances of population recovery. Hunters either make no distinctions among their prey or seek the largest and most robust individuals, rather than taking those less likely to reproduce. They often kill mothers in order to get their infants to use or sell as pets. This is the most devastating form of direct exploitation because the number of reproductive females in a population regulates the number of births.

Disease

Disease is a special problem in primate conservation because physiological similarities increase the likelihood of transmission from humans. The many potential human sources include veterinarians, park personnel, local people, and tourists (Wallis and Lee 1999). This applies to primatologists as well, especially in their typical attempts to gain close proximity to their subjects.

Some diseases have dramatic effects, such as ebola, but there are many mundane illnesses that are deadly sooner or later. This includes parasites that many people might not think of as disease organisms. These might not be transmitted directly from humans, but anthropogenic changes can cause them to become more prevalent or force primates into places or activities that make them more susceptible.

Taxa in Trouble

Various primate taxa have problems of their own in addition to the worldwide threats. These can arise from their intrinsic biological and behavioral features, from the nature of their habitats, or from culturally varied human activities. Some examples follow.

Strepsirrhines

Compared to anthropoids, strepsirrhines get less attention from the public because most of them are relatively small and occupy limited geographic ranges. Most are inconspicuous because they are nocturnal, which puts them at greater risk during logging or forest clearance because they are asleep when it happens. If their trees are felled for timber, strepsirrhines are easily collected for animal markets to become pets or food. If trees are burned to clear agricultural land, the sleeping primates are likely to be burned alive.

Despite devastating loss of habitat, Asian lorisoids face even greater immediate threats in the form of their popularity as pets and uses in traditional medicines. Lorises are desirable prey for profit because they can be captured easily with no special equipment, they are easily transported, and they are rarely confiscated by authorities. Slow lorises are particularly vulnerable because of their limited speed. After a forest is clear-cut people pick through the trees to find lorises, which cling to branches rather than fleeing.

Trade in lorises varies with cultural perceptions (Nekaris et al. 2010). Sri Lankans consider slender lorises to be ugly and bringers of bad luck. Cambodians have similar ideas about slow lorises, but vigorous trade is promoted in that country by traditional healers who claim that loris parts can cure one hundred diseases. Lorises usually become pets in Indonesia, a practice deeply rooted in Indonesian culture.

Trade in slow lorises exceeds trade in slender lorises by an order of magnitude, largely due to the international market. The United States, Japan, and the European Union are the main importers. The pet appeal of the Javan slow loris is high because it has a panda-like face, a striking combination of colors, and it seems cuddly. Expert Anna Nekaris provided important perspectives in an interview (Hance 2009). A supposedly cuddling loris holds still in a person's hands because stillness and silence are defense mechanisms. The frightened animal does not bite because its front teeth have been removed or cut out, usually without anesthetic, by traders who fear its toxic bite. Most of these animals die of abscesses or pneumonia and survivors cannot eat their preferred food, gum, or perform social grooming with the toothcomb. This means that they are unlikely to survive if they are confiscated by authorities and released in the wild. Most pet lorises die from diabetes, obesity, or kidney failure, all due to inappropriate diet (we know little about what an appropriate captive diet would be). Widely circulated videos stimulate the market and provide false information, especially the notion that available lorises have been bred in captivity. Even the best scientific breeding facilities have great difficulty with loris reproduction and survival.

Hunting of lemuroids (Butler 2005; Handwerk 2009) is a two-pronged problem. Many Malagasy people hunt and trap because they are poverty-stricken. At the same time there is a demand from the affluent for primate meat as a luxury food (Barrett and Ratsimbazafy 2009). Obviously two different cultural solutions are needed.

Commercial habitat destruction is especially severe on the island of Madagascar. For instance, an estimated 100 million dollars' worth of Madagascar hardwood was cut down and sold in one year, mostly to China for furniture manufacture (Smith 2009). The effects of forest degradation are not always obvious. In some ringtail populations this meant heavier use of tamarind fruit as a fallback food, and the physical properties of fruit resulted in a high incidence of severe tooth wear and loss (Sauther and Cuozzo 2009).

Conservation problems were exacerbated by a coup that ousted an eco-friendly president and created anarchic conditions. Armed loggers and poachers swarmed into protected areas. Other countries suspended aid to Madagascar, doing severe harm to the 45 national parks that received 90 percent of their funds from foreign aid. Ecotourism, which generated almost 400 million dollars per year, dropped by 60 percent (Smith 2009).

The most distinctive problem for lemuroids is illegal rum production, which is an important part of Malagasy culture (Irwin and Ravelomanantsoa 2004). This industry contributes to forest degradation in two ways: the conversion of land for cultivation of sugarcane and the use of large amounts of firewood in the distillation process. Large-scale rum production threatens forests that are otherwise protected or relatively undisturbed.

Anthropoids

The diurnal anthropoids are more visible than most strepsirrhines. Relatively large size makes many of them attractive as sources of meat and good targets for weapons. Size also enhances their appeal for decorative products (especially furs) and conspicuous novelty items (such as shrunken monkey heads passed off as humans). Large numbers of anthropoids have been harvested for biomedical research because of their anatomical and physiological similarities to humans. This is particularly true of catarrhines, to the special detriment of chimpanzees and several monkey species.

Marmosets and Tamarins

Intrinsic characteristics are in some cases an advantage with regard to survival. The status of marmosets is relatively favorable compared to most other platyrrhines (Ferrari 2009). As the smallest primates, they are of little interest to hunters. Small size, combined with ecological flexibility, also tends to give them resistance to habitat fragmentation. In fact, fragmentation may actually benefit marmosets because their preferred habitats are disturbed, edge, and secondary forest. In addition, fragmentation may cause local extinction of competing species. Though they may be aided by habitat modification, marmosets are as vulnerable to habitat loss as other primates, if not more so. Because most marmosets have limited geographic distributions, relatively localized destruction of habitat can eliminate an entire species.

Habitat loss has been much more severe in the Atlantic Forest of Brazil than in Amazonia. Only about 7 percent of the former original forest remains. Primates struggling in the Atlantic Forest include lion tamarins. Large numbers were exported throughout the twentieth century for the pet trade, exotic animal collectors, zoo exhibits, and biomedical research. Brazil had outlawed all trade by 1975, but trapping continues in response to international demand. One bright spot is that the Brazilian environmental agency recovered enough illegally exported tamarins to begin a genetically diverse breeding colony that has supplied breeding programs in other countries.

In the past some lion tamarins were helped by cocoa plantations that maintained old growth forest because they used *cabruca*, a system that replaces middle and understory trees with cacao trees but leaves some old growth trees for shade (Oliveira et al. 2009). Cabruca provided greater quantities of fruit, flowers, animal prey, and sleeping sites than other available forest types. However, international prices fell after 1986 and a fungal disease attacked the crops. Owners had to abandon cacao, log the old growth trees for money, switch to alternative crops grown with different systems, and clear land for grazing. In addition, failure of the plantations sent displaced workers into the forests to clear land for subsistence farming. The case of the cocoa plantations illustrates several points: (1) the possibility of coexistence between humans and other primates; (2) the impact of global patterns; and (3) the impact of unexpected events.

Old World Monkeys

A recent modeling study elicited several traits that identified extinction risk for African colobine species: greater female body mass, greater mature leaf consumption, fewer habitat types occupied, and maximum latitude occupied (Kamilar and Paciulli 2008). The threat posed by body mass is probably related to food needs, which is mitigated for species more able to use mature leaves as a fallback food. Survival is more likely for species that are widely distributed and adaptable, and less likely for those with narrow environmental requirements. A less intuitive conclusion from this study is that species living closer to the equator are in greater danger. The explanation may be cultural: a factor that is roughly connected with latitude is that hunting of colobines is especially prevalent in Central and West Africa, where the meat feeds both local people and migrant workers.

The gray langur occupies a special position in India, which is the major part of its range of distribution. Also called the Hanuman langur, it is associated with an important and popular figure in Hindu mythology. As a result, gray langurs have enjoyed an unusual level of tolerance, even though they raid crops and houses. In addition to custom, Indian law protects gray langurs. However, some forms of exploitation are consistent with Hindu beliefs. The langurs are often kept as pets and some Hindu priests keep them for religious purposes. Furthermore, the protection provided by Hinduism is declining as India becomes more secular. People are less likely to tolerate the monkeys or object to their exploitation, and many do not know that the animals are legally protected. Of course, indulgence of langur behavior never protected them from habitat disturbance. One population declined to about 38 percent of its former size in about 50 years (Kumara et al. 2010).

Rhesus macaques, the most widespread and numerous species of the genus, are unthreatened even though original habitats have been increasingly lost to development (IUCN Red List). Tolerance is most common in India (rhesus as well as gray langurs are associated with the monkey god), while hunting pressure is high in some other countries. However, increasing cohabitation has been associated with waning levels of human tolerance in India.

A different kind of problem arises when successful primates have a negative impact on others. Rhesus monkeys have expanded as a natural process in some regions and as a result of introduction by humans in others. The process poses a serious threat to declining populations of bonnet macaques, which are restricted to peninsular India (Kumar et al. 2011). In Vietnam rhesus are threatened by the release of confiscated longtailed macaques.

Japanese macaques present some of the most complex issues in primate conservation (Mito and Sprague 2012). Japan is a highly urbanized country in which contact with the monkeys is in many places unavoidable. The macaques have coexisted with dense populations and Japanese civilization for thousands of years, so conservationists must explain to the people how the monkeys can be endangered. Changes in land use have introduced new threats. Logged forests are replaced, but often with commercially useful conifers. These provide inferior food for the monkeys, which presumably encourages crop raiding. Crop raiding may have been facilitated by the decline in farming populations, which has removed a barrier of human activity separating the monkeys from fields and orchards. The same law that protects the monkeys from other kinds of hunting permits the killing of crop raiders.

Gibbons

Almost all gibbon taxa are seriously threatened. Gibbon populations have been declining for 30 to 40 years (Cheyne 2009; Whittaker and Lappan 2009) and reintroduction has generally been unsuccessful (Harding 1998). Several distinctive gibbon traits have an impact on conservation. Brachiation requires a continuous canopy, so habitat disturbance has a greater effect on gibbons than on many other primates (Kakati 2006). Being strictly arboreal, they depend on stretches of forest to move between food patches; therefore, fragmentation and even lesser forms of habitat degradation seriously inhibit natural behaviors. Even secondary forest is a challenge.

The social unit contributes to gibbon vulnerability. After an adult pair have been shot (often for meat), several young may be easily collected (usually for the pet trade). By destroying entire family groups, hunters can significantly affect the overall population in a short time (Leighton 1987). At least some gibbon species are relatively easy to hunt (Geissmann 2007; Yimkao and Srikosamatara 2006), so a small number of hunters can seriously damage a population. Indigenous people have probably hunted gibbons for a long time, but the loss of forest cover seems to have encouraged hunting (Eudey 1994).

Inbreeding may have negative effects on the survival of small populations of gibbons (Leighton 1987). Because of their natural dispersal patterns, neighboring groups have some likelihood of being related. Where suitable habitat is limited because of habitat degradation, young males and females are likely to range close to their natal territories. In some species, when a male or female is widowed, the offspring of the opposite sex inherits the territory and mating may occur between parent and offspring.

China seems to have become particularly inhospitable for gibbons. Of six taxa that have lived there in recent times, one is assumed to be extinct, two more are at the brink, and one has not been confirmed since the 1980s (Grueter et al. 2009). This is a special loss because the apes are part of a cultural as well as a natural heritage. They have been prominent in the Chinese arts for more than 2000 years (Whittaker and Lappan 2009). Gibbons appeared in Chinese literature by about 1000 BC and in visual art by about 1000 AD, serving as Buddhist and Daoist symbols (Geissmann 2008). The genre spread from China to Japan and Korea, where wild gibbons did not exist. Gibbon paintings displayed a dramatic increase in the twentieth century, but by the end of the century the animals were approaching extinction.

The pet trade has perhaps had a surprising impact on white-handed gibbons, *Hylobates lar* (Geissmann 2007). Though large compared to "cuddly" strepsirrhines, these gibbons can be quite tame in captivity. In Thailand, for example, they are often kept as pets in both rural and urban areas and are sometimes found in animal markets for sale (Eudey 1994).

Great Apes

Great apes are the primates most closely related to humans, but not all humans consider this kinship important. Great apes share several intrinsic characteristics that affect survivability in the modern setting (Miles et al. 2005). These traits distinguish them from other primates, at least quantitatively. Their size marks them as targets for hunting because they are more visible than many other primates and their bodies provide more meat. Replacement of population losses is especially slow in great apes because of their low reproductive rates and slow development. It takes a long time to replace adult casualties because great apes have the longest inter-birth intervals and life history stages among the primates (Morrogh-Bernard et al. 2003).

Because great apes all build nests in trees and require large ranges, forest clearance affects them as much as most other primates and more than some. Timber that came primarily from great ape habitats fed a global market for tropical wood in the late twentieth century. When the supply from Southeast Asia was depleted, the industry turned to Central and West Africa.

As our closest relatives, great apes are at least as likely as other primates to contract human diseases. Researchers and tourists have been in contact with a wide range of pathogens (presumably developing immunity to many of them) while the apes occur in small, previously isolated populations. Diseases known to affect great apes include paralytic poliomyelitis, pneumonia, measles, hepatitis, cholera, malaria, HIV, and tuberculosis (Carne et al. 2014; Woodford et al. 2002).

Governments in great ape countries have supported conservation for the most part and the law protects the apes in all of these countries. However, enforcement is difficult when resources are limited. Of the 23 great ape countries, 21 are among the poorest in the world (Miles et al. 2005). Armed conflicts have caused heavy casualties among the apes. Rebellions or wars have taken place in at least 13 of the 23 great ape countries since 1990. Meanwhile, trade in live great apes seems to have increased across many countries in recent years.

Orangutans

Orangutans differ from other great apes in several characteristics related to conservation: lower population density, smaller population numbers, lower reproductive rates, a longer period of development, the longest inter-birth interval, and a strong dependence on arboreality. These characteristics, combined with severe external threats, make orangutans extremely vulnerable to extinction in the near future (Morrogh-Bernard et al. 2003). For example, the spread of disease can be disastrous in such small and fragmented populations (Woodford et al. 2002).

Because orangutans are the most arboreal great apes, destruction of forests devastates every aspect of their lives (Felton et al. 2003). Due to the very patchy distribution of fruiting trees, the loss of any important species significantly reduces foraging efficiency. Females with limited access to high quality foods have lower reproductive output (Knott 1998). Orangutans are poor dispersers where habitats are severely disturbed or fragmented, because they are very reluctant to travel across open areas (Delgado and van Schaik 2000). Habitat loss is largely due to unsustainable logging practices in Indonesia and Malaysia, stimulated by the global market for tropical wood (Miles et al. 2005). During the later twentieth century, more than half of the Borneo forest disappeared, taking with it more than 80 percent of the orangutan habitat (Rijksen 2001).

Logging in peat swamp forests is of special concern because they support the highest observed population densities of orangutans, along with commercially valuable timber (Felton et al. 2003; Morrogh-Bernard et al. 2003; Russon et al. 2001). As trees are extracted, peat swamp forests lose their value to orangutans and these pockets of dense orangutan populations decline or are eliminated. Displaced orangutans do not generally survive, succumbing to disease, hunting, or capture for trade (Delgado and van Schaik 2000; van Schaik et al. 2001). Oil palm plantations are an increasingly serious threat to orangutan habitat as land is completely converted from recovering logged forest to monoculture plantations (Robertson and van Schaik 2001).

Other dangers are also of special significance for orangutans. Their deliberate and slow locomotion makes them easy targets, even from a distance. Though indigenous people in much of this area are Muslim and therefore have religious taboos against eating primates, the ape population is dwindling because even the slightest hunting pressure on areas of precariously low population densities has significant effects. As elsewhere, killing adults is further encouraged by the value of the young for the pet market. The effect of the international market is illustrated by the boom that apparently resulted from a popular television show in Taiwan that featured an orangutan as an ideal pet and companion (Miles et al. 2005).

African Apes

In addition to being hunted, African apes suffer from snares set for other animals (Plumptre et al. 2003). The resultant wounds can lead to loss of limbs or painful and lingering deaths by infection. Snares can disfigure chimpanzees to the point that they can no longer obtain and eat food. In Kibale snares injure chimpanzees at a rate of 3.7 percent per year. Even with vigilant patrolling, snare removal is a constant challenge where apes are sympatric with desirable game animals. Though chimpanzees have learned to recognize snares, they are often caught when their attention is distracted (Hill et al. 2001).

Profits from agricultural products traditionally harvested in semi-rural areas and sold in markets in urban centers plummeted due to deterioration of the economy, roads, and river transport systems associated with political instability (Dupain et al. 2000). While this may have temporarily mitigated habitat destruction due to agriculture, the resulting trend was a re-immigration of people into the forests where they practiced small-scale cultivation and subsistence hunting as well as collection of wood for their own fuel and shelter (Reinartz and Bila Isia 2001). Many of these people settled in and exploited supposedly protected areas, such as national parks (Coxe et al. 2000).

Gorilla priorities may be somewhat different than for other great apes. On the positive side, they can coexist with logging because they tend to favor the areas of secondary vegetation that grow after an area has been clear-cut (Plumptre et al. 2003; White and Tutin 2001). However, hunting for meat and capture for collections greatly affect gorilla populations across Africa. Humans who share their range often perceive gorillas as dangerous and occasional attacks occur on land adjacent to parks, eliciting retaliation (Butynski 2001). Crop raiding also contributes to a negative image. The social structure of western gorillas makes them especially susceptible to hunting, because a few hunters can wipe out entire groups (Tutin and Vedder 2001). Confronting and displaying against the threat, the silverback can be killed quickly. The rest of the group can then be dispatched.

Common chimpanzees are among the most threatened primates in Africa for many reasons (Goodall 2001). Some intrinsic factors, shaped by human-induced environmental changes, threaten chimpanzee populations. High juvenile mortality and sex differences in mortality threaten already slow population growth rates (Hill et al. 2001). Because of the long lifespan and reproductive characteristics, a female chimpanzee is expected to produce only 0.8 daughters, on average, in her entire lifetime. This is well below the population replacement rate, so slight changes in population composition, caused by diseases, habitat fragmentation, or poaching could negatively affect growth rate, causing extirpation.

More than 70 percent of chimpanzee habitat has been affected by infrastructure. If current human population growth rates are maintained, it has been estimated that by the year 2030 less than 10 percent of chimpanzee habitat will remain unaffected by development (Nellemen and Newton 2002). Even in Gabon and Congo, widely considered to be stronghold countries for chimpanzees, populations were declining at a rate of at least 4.7 percent per year (Walsh et al. 2003).

Though trade in chimpanzees is restricted, the unstable political situation and corruption in many countries translates into little enforcement (Ammann 2001). Inbreeding necessitated by habitat fragmentation could also pose a threat. Habitat fragmentation could isolate female chimpanzees and either force them to mate with community members that may be related to them or inhibit successful dispersal from their natal communities (Gagneux et al. 1999). The potential problems associated with inbreeding are particularly devastating in small communities.

Conservation Initiatives

Priorities

Solutions require time, effort, and money, which means that priorities must be established. The **World Conservation Union**, often designated by its French acronym **IUCN**, is an international organization devoted to the protection of threatened species. Its **Red List** assesses the status of threatened species, based on the total number of individuals in a species, the size and distribution of its populations, the rate at which occurrence has declined, and the causes of decline. The **Primate Specialist Group** of the IUCN applies these criteria to primates. About 300 primate taxa are considered threatened, and more than 200 are classified as *Endangered* or *Critically Endangered*, which means that they face the greatest risks of extinction.

Red List criteria are applied equally to all species, but extinctions do not all have the same significance. The loss of a species that is the lone survivor of a lineage has greater implications than the loss of a species with numerous close living relatives. **Conservation units** can be identified and ranked by various criteria and weighting systems, including taxon richness, character richness, endemism, degree of threat, and viability.

Conservation goals are affected by the tendency in taxonomy to raise populations to full species level, which implies equal concern for each of these entities. An extinction that was considered the loss of one species in the past is now likely to be seen as the loss of several different species. A broad strategy is to identify a set of areas that complement one another so as to maximize the biological diversity of the whole network.

Evaluation of taxa has to be relatively specific. For example, a serious error can occur if healthy diurnal primate populations lead to the conclusion that the less visible nocturnal species in the same habitat are in good condition. Various surveys demonstrate an inverse relationship in the presence of strepsirrhines and anthropoids in particular habitats. Aside from the effects of human intrusion, the substrate and sleeping requirements of nocturnal primates are significantly different from those of anthropoids.

Protection

Creation of protected areas has been a major accomplishment in many parts of the world. Some protected areas are mainly for primates and some for other animals, but all are likely to benefit. However, benefit depends on the quality of the areas chosen. The tolerant macaques of Sulawesi are in a general decline (Riley and Priston 2010). About 80 percent of Sulawesi's forests have been destroyed or altered and only a third of the remainder are in good condition. Protected areas constitute the main conservation strategy, but there are questions about efficacy: much of the protected environment consists of montane forest that may not be optimal for the macaques; not all the species are well represented in protected areas; and not all the protected areas are protected well.

Continuing success depends on sufficient finances for enforcement and provision of economic alternatives for local people who might otherwise need the resources in a protected area. The expense of enforcement should be curtailed by boundaries designed to minimize patrol needs and poacher access. Another problem is the risk of isolation. One solution is to maximize the size of protected areas. Another is to establish **habitat corridors** between them. These are pathways that facilitate dispersal and genetic interchange.

Animals may roam beyond the boundaries of protected areas, especially in search of food. This is especially true of **forest fragments**, the small remnants of natural forests. Consideration of the primate's perspective is based on population characteristics such as ecological requirements and patterns of dispersal from natal communities. Fragmentation is a large-scale process, so research must go beyond individual fragments to the more heterogeneous landscape (Arroyo-Rodriguez and Mandujano 2009). Activity beyond the boundaries may take the form of **landscape supplementation** (Asensio et al. 2009). This means that the animals augment their diet by accessing sources such as other fragments or isolated trees. With human toleration the animals can help themselves. Locally, fences and roads can provide pathways rather than barriers. The larger scale can be much more complex, as exemplified by monkey use of the mosaic environment created by the agricultural pattern on the island of Bali.

Most humans will not tolerate crop raiding, so conservation requires ways to diminish this activity without killing the animals. This entails understanding the animals' preferences. At one African site primates raided cultivated bananas only when fruit supplies in the forest were low. On the other hand, maize was a constant target regardless of forest conditions (Naughton-Treves et al. 1998).

Sustainable utilization is an alternative to full protection. It means allowing human exploitation of animals and habitats at levels that do not result in irreversible damage. This is a poor solution for most primates because of their long generations and small litter sizes.

Permissible levels of exploitation would in most cases be unsatisfactory to the humans involved.

Ecotourism, guided animal watching, is a kind of utilization that minimizes removal of animals from the population. It can provide finances for park protection and economic opportunities for local people as guards, guides, and provisioners. Ecotourism requires careful management to avoid harm to the animals, especially the transmission of diseases from humans.

Assisted Recovery

There are ways to help severely depleted populations that seem unable to recover on their own. One is **reintroduction**, the release of new individuals or groups into the wild. In some cases these are young animals confiscated from poachers or their customers. After rehabilitation (physical and behavioral) they are released where they can make contact with their own species. Since primates are so dependent on learning, it is often necessary to teach them survival skills. The same is true for many primates from captive breeding programs. These are also problematic because of cost and space limitations. Few reintroduction candidates can be provided for the larger-bodied species. The small numbers make inbreeding a serious problem, though this can be alleviated when animals are traded among participating facilities.

A less costly and somewhat more natural alternative is **translocation**, the transport of some animals from a viable population to a threatened one. Trauma can be reduced if related or otherwise acquainted individuals are transferred together. Reintroduction and translocation have some problems in common besides the obvious transportation issues. One is to minimize the risk of spreading diseases and parasites. Another is to arrange releases so as to minimize trauma to released individuals and to the populations they are projected to join. Habitat corridors can alleviate the latter difficulty, by allowing the animals to make contact at their own rate. Disease transmission is still a concern but less so than with animals that have been outside their natural environments.

Ethnoprimatology

Eliciting cooperation from humans in primate countries and conservation areas is a complex undertaking. **Ethnoprimatology** studies the interface between primate and human behavior (Sponsel 1997). For example, it is important to distinguish between subsistence efforts by ordinary local people and depredations by commercial poachers who are outsiders. Their effects on the animals may be similar, but conservation solutions will necessarily be different. The larger cultural context for conservation is varying sociopolitical systems and their ideas about property. These include private, communal, and state ownership. Each presents different costs and benefits for conservation action.

Ethnoprimatology postulates that the effects of humans on primates should be viewed as potential causes of primate behavioral patterns rather than just disruptions of a natural state. Anthropogenic processes can influence the most basic components of our socioecological models (such as competition and predator pressure).

Nekaris et al. (2010) exemplified ethnoprimatology in their assessment of human–loris relations in Asia. They studied historical records, assessed international trade patterns from the database of a conservation organization, and interviewed a variety of relevant people (e.g. traders and hunters) in various countries. Like cultural anthropologists they used local languages, built rapport with their subjects, and spent long periods at each locality that they visited.

Most humans perceive a special similarity between humans and other primates (Hill and Webber 2010; Lee 2010). This can aid conservation. Primates are often judged by human moral standards, but may also benefit from human empathy. For example, some farmers tolerate crop-raiding primates if they believe that the animals are only taking what they need. In some cases farmers speak of the primates as criminal, but still provision them out of pity when natural foods are scarce. Awareness of such complex attitudes is essential to the formulation of effective conservation strategies. Straddling the human–animal boundary gives primates special significance in many cultures that is usually interpreted as magical, religious, or mythological. A negative cultural view is that the intermediate position violates an important boundary.

Education can potentially get people in primate habitat countries involved with conservation on a systematic and long-term basis. Wallis and Lonsdorf (2010) recommended that conservation education should begin with attention to the needs of the local population. People in need are less inclined to curtail harmful activities or to heed outsiders' reasons for doing so. Other recommendations were more specifically about education: (1) teach basic ecological principles that are relevant to the audience, such as the dependence of forests on seed dispersal, which is a function of many frugivorous primates; (2) train local people to be educators and try to integrate conservation education with local school curriculums; and (3) use the expertise of those knowledgeable about teaching non-university students.

A project in Colombia illustrates consideration of the people in a primate area. Local collection of firewood was damaging the understory of the forest, which diminished the supply of insects for tamarins. The ultimate result could have been habitat loss: if wood is the only cooking fuel, a family of five may consume 15 logs per day. Working with local communities, the conservation organization Proyecto Tití determined that modification of a traditional clay oven would allow it to use cornhusks, coconut shells, and other vegetative waste products as efficient fuel sources. These low-cost ovens were distributed to community members and used with success.

Conclusions

Lee (2010) sees disturbing trends in the globalization of conservation. Conflict paradigms have become prevalent in political circles and scientific literature, emphasizing primate behaviors such as crop raiding and attacks on humans. The worst manifestations appear in the language of the press, such as "monkeys wreak havoc" and "monkeys have turned violent".

Probably even worse, primate conservation has become enmeshed in polarized political philosophies. Igoe and Brockington (2007) argued that conservation will benefit from a market-driven neoliberal approach. Lee (2010) asked a series of cogent questions. How can markets drive conservation if material costs outweigh benefits? Who has access to these markets? What are the local effects? When benefits are at issue, self-interest typically wins at all levels. Lee concluded that big business and big conservation have combined to reach a position where neither can evaluate its own actions in relation to sustaining wildlife.

Many people who learn about nonhuman primates come to view them as fascinating and beautiful animals. Others may develop a philosophical and ethical appreciation of the fact that they are our closest relatives. Even the most hardheaded pragmatist must recognize the scientific value of primates in our effort to understand ourselves. I hope you will be moved by any or all of these considerations to support the conservation of nonhuman primates.

At a minimum you can speak out and write against exploitation, especially to elected officials who formulate relevant laws and policies. Beyond that you can join and support any of the numerous organizations that work to preserve primates in general or various primate species. Most of them can be found online. Finally, you can be active in the field as a volunteer or a scientist. More and more research projects are getting intensely involved in conservation efforts, usually in cooperation with local people. People from non-primatological and non-scientific fields, such as education, may play greater roles in the future.

Summary

Many primates are among the most endangered animals in the world today. They have been devastated by habitat disturbance, hunting, harvesting, disease, war, and politics. Different taxa have varying problems arising from the combination of their intrinsic characteristics, the attitudes of the humans in contact with them, and the environmental circumstances imposed on them. Effective conservation requires the establishment of priorities. Protection includes the establishment of habitat corridors and support for landscape supplementation. Reintroduction and translocation give populations a chance to recover. Ethnoprimatology studies the interaction of humans and primates to achieve a better understanding of the humans as well as the animals.

Sources

Basic Bibliography

C.A. Campbell, et al. (Eds.) (2011), *Primates in perspective*, 2nd ed. NY: OUP. [Chapters 45 and 46; Part Two for particular taxa]

Cowlishaw, G., & Dunbar, R. (2000). *Primate conservation biology*, 2nd ed. Chicago: UCP.

Fuentes, A., & Hockings, K.J. (2010). The ethnoprimatological approach in primatology. *AJP*, *72*, 841–847.

Marsh, L.K., & Chapman, C.A. (Eds.) (2013). *Primates in fragments, complexity and resilience*. NY: Springer.

Mittermeier, R.A., et al. (Eds.) (2012). *Primates in peril: the world's 25 most endangered primates 2010–2012*. Arlington, VA: IUCN/SSC Primate Specialist Group, International Primatological Society, and Conservation International.

Additional References

Alves, R.R.N., et al. (2010). Primates in traditional folk medicine: a world overview. *Mammal Review*, *40*, 155–180.

Ammann, K. (2001). Bushmeat hunting and the great apes. In B.B. Beck et al. (Eds.), *Great apes and humans: the ethics of coexistence* (pp. 71–85). Washington DC: Smithsonian Institution Press.

Ancrenaz, M., et al. (2014). Of Pongo, palms and perceptions: a multidisciplinary assessment of Bornean orang-utans *Pongo pygmaeus* in an oil palm context. *Oryx*, doi:10.1017/S0030605313001270.

Arroyo-Rodriguez, V., & Mandujano, S. (2009). Conceptualization and measurement of habitat fragmentation from the primates' perspective. *IJP*, *30*, 497–514.

Asensio, N., et al. (2009). Landscape supplementation provided by neighbouring small patches: a case study for howler monkeys in Los Tuxtlas, Mexico. *Biotropica*, *41*, 768–773.

Barrett, M.A., & Ratsimbazafy, J. (2009). Luxury bushmeat trade threatens lemur conservation. *Nature*, *461*, 470 (letter).

Bitanyi, S. et al. (2012). Awareness and perceptions of local people about wildlife hunting in western Serengeti communities. *Tropical Conservation Science*, *5*, 208–224.

Blanco, V., & Waltert, M. (2013). Does the tropical agricultural matrix bear potential for primate conservation? A baseline study from Western Uganda. *Journal for Nature Conservation*, *21*, 383–393.

Bonnell, T.R., et al. (2011). Post-logging recovery time is longer than expected in an East African tropical forest. *Forest Ecology and Management*, *261*, 855–864.

Butler, R. (2005). Lemur hunting persists in Madagascar, rare primates fall victim to hunger. http://news.mongabay.com/2005/07/lemur-hunting-persists-in-madagascar-rare-primates-fall-victim-to-hunger/ (last accessed 8.3.15).

Butynski, T.M. (2001). Africa's great apes. In B.B. Beck et al. (Eds.), *Great apes and humans: the ethics of coexistence* (pp. 3–56). Washington DC: Smithsonian Institution Press.

Carne, C., et al. (2014). The risk of disease to great apes: simulating disease spread in orang-utan (*Pongo pygmaeus wurmbii*) and chimpanzee (*Pan troglodytes schweinfurthii*) association networks. *PLoS One*, *9*, e95039.

Chapman, C.A., et al. (2010). Understanding long-term primate community dynamics: implications of forest change. *Ecological Applications*, *20*, 179–191.

Cheyne, S.M. (2009). The role of reintroduction in gibbon conservation: opportunities and challenges. In S. Lappan & D.J. Whittaker (Eds.), *The gibbons* (pp. 477–496). New York: Springer.

Cowlishaw, G., et al. (2009). High variability in patterns of population decline: the importance of local processes in species extinctions. *PRSB*, *276*, doi: 10.1098/rspb.2008.0767.

Coxe, S., et al. (Eds.) (2000). *Bonobo conservation assessment workship final report*. IUCN/SSC Conservation Breeding Specialist Group.

Delgado, R.A., & van Schaik, C.P. (2000). The behavioral ecology and conservation of the orangutan (*Pongo pygmaeus*): a tale of two islands. *Evolutionary Anthropology*, 9, 201–218.

Dupain, J., et al. (2000). Current status of the bonobo (*Pan paniscus*) in the proposed Lomako Reserve (Democratic Republic of Congo). *Biological Conservation*, *94*, 265–272.

Eudey, A. (1994). Temple and pet primates in Thailand. *Rev. Em/. (Terre Vie)*, *49*, 273–280.

Felton, A.M, Engström, L.M., Felton, A., & Knott, C.D. (2003). Orangutan *Pongo p. pygmaeus* population density, forest structure and fruit availability in hand-logged and unlogged peat swamp forests in West Kalimantan, Indonesia. *Biological Conservation*, *114*, 91–101.

Ferrari, S.R. (2009). Conservation of the marmosets and callimicos. In S.M. Ford et al. (Eds.), *The smallest anthropoids* (pp. 465–477). NY: Springer.

Gagneux, P., et al. (1999). Female reproductive strategies, paternity and community structure in wild West African chimpanzees. *AB*, *57*, 19–32.

Geissmann, T. (2007). Status reassessment of the gibbons: results of the Asian primate red list workshop 2006. *Gibbon Journal*, *3*, 5–15.

Geissmann, T. (2008). Gibbon paintings in China, Japan, and Korea: historical distribution, production rate and context. *Gibbon Journal*, *4*, 1–38.

Goodall, J. (2001). Problems faced by wild and captive chimpanzees: finding solutions. In B.B. Beck et al. (Eds.), *Great apes and humans: the ethics of coexistence* (pp. xii–xxiv). Washington DC: Smithsonian Institution Press.

Grueter, C.C., et al. (2009). Are *Hylobates lar* extirpated from China? *IJP*, *30*, 553–567.

Hance, J. (2009). YouTube videos may be imperiling cuddly primate. http://news.mongabay.com/2009/02/youtube-videos-may-be-imperiling-cuddly-primate/ (last accessed 8.3.15)

Handwerk, B. (2009). Lemurs hunted, eaten amid civil unrest, group says. *National Geographic News*, http://news.nationalgeographic.com/news/2009/08/090821-lemurs-killing-madagascar.html (last accessed 8.3.15).

Harball, E. (2014). Agribusiness drives most illegal deforestation. *Scientific American* (reprinted from *Climatewire*), http://news.nationalgeographic.com/news/2009/08/090821-lemurs-killing-madagascar.html (last accessed 8.3.15).

Harding, J.L. (1998). The first stage of rehabilitation of a family of white handed gibbons *Hylobates lar*. *Australasian Primatology*, *12*, 2–8.

Hill, C.M., & Webber, A.D. (2010). Perceptions of nonhuman primates in human–wildlife conflict scenarios. *AJP*, *72*, 919–924.

Hill, K., et al. (2001). Mortality rates among wild chimpanzees. *JHE*, *40*, 437–450.

Hoffman, T.S., & O'Riain, M.J. (2012). Landscape requirements of a primate population in a human-dominated environment. *Frontiers in Zoology*, *9*, http://www.frontiersinzoology.com/content/9/1/1 (last accessed 8.3.15).

Igoe, J., & Brockington, D. (2007). Neoliberal conservation: a brief introduction. *Conservation and Society*, *5*, 432–449.

Irwin, M.T., & Ravelomanantsoa, H.V. (2004). Illegal rum production threatens health of lemur populations at Tsinjoarivo, eastern central Madagascar: brief report and request for information. *Lemur News*, *9*, 16–17.

Jenkins, C.N., et al. (2013). Global patterns of terrestrial vertebrate diversity and conservation. *PNAS*, *110*, E2602–E2610. www.pnas.org/cgi/doi/10.1073/pnas.1302251110 (last accessed 8.3.15).

Jones, C., et al. (2009). Committed terrestrial ecosystem changes due to climate change. *Nature Geoscience*, *2*, 484–487.

Jones-Engel, L., & Engel, G.A. (2006). Disease risk analysis: a paradigm for using health-based data to inform primate conservation and public health. *AJP*, *68*, 851–854.

Kakati, K. (2006). Fragment-living: a study of hoolock gibbons in Assam, India. *Gibbon Voice*, *8*, 1–7.

Kamilar, J.M., & Paciulli, L.M. (2008). Examining the extinction risk of specialized folivores: a comparative study of colobine monkeys. *AJP*, *70*, 1–12.

Knott, C.D. (1998). Changes in orangutan caloric intake, energy balance and ketones in response to fluctuating fruit availability. *International Journal of Primatology*, *19*, 1061–1079.

Kumar, R., et al. (2011). Of least concern? Range extension by rhesus macaques (*Macaca mulatta*) threatens long-term survival of bonnet macaques (*M. radiata*) in peninsular India. *IJP*, DOI:10.1007/s10764-011-9514-y.

Kumara, H.H., et al. (2010). Of how much concern are the 'least concern' species? Distribution and conservation status of bonnet macaques, rhesus macaques and Hanuman langurs in Karnataka, India. *Primates*, *51*, 37–42.

Lee, P.C. (2010). Sharing space: can ethnoprimatology contribute to the survival of nonhuman primates in human-dominated globalized landscapes? *AJP*, *71*, 925–931.

Lehman, J., et al. (2010). Apes in a changing world—the effects of global warming on the behaviour and distribution of African apes. *Journal of Biogeography*, *37*, 2217–2231.

Leighton, D.R. (1987). Gibbons: territoriality and monogamy. In B.B. Smuts et al. (Eds.), *Primate societies* (pp. 135–145). Chicago: UCP.

Linder, J.M., & Oates, J.F. (2011). Differential impact of bushmeat hunting on monkey species and implications for primate conservation in Korup National Park, Cameroon. *Biological Conservation*, *144*, 738–745.

Matthews, L.J. et al. (2011). Primate extinction risk and historical patterns of speciation and extinction in relation to body mass. *PRSB*, *278*, 1256–1263.

Merrode, E., et al. (2007). The impact of armed conflict on protected-area efficacy in Central Africa. *BL*, *3*, 299–301.

Miles, L., et al. (2005). Challenges to great ape survival. In J. Caldecott & L. Miles (Eds.), *World atlas of great apes and their conservation* (pp. 217–241). Berkeley: University of California Press.

Mito, Y., & Sprague, D. (2012). The Japanese and Japanese monkeys: dissonant neighbors seeking accommodation in a shared habitat. In S. Radhakrishna et al. (Eds.), *The macaque connection* (pp. 33–51). NY: Springer.

Morrogh-Bernard, H., et al. (2003). Population status of the Bornean orang-utan (*Pongo pygmaeus*) in the Sebangau peat swamp forest, Central Kalimantan, Indonesia. *Biological Conservation, 110*, 141–152.

Naughton-Treves, L., et al. (1998). Temporal patterns of crop-raiding by primates: linking food availability in croplands and adjacent forest. *Journal of Applied Ecology*, *35*, 596–606.

Nekaris, K.A.I., et al. (2010). Exploring cultural drivers for wildlife trade via an ethnoprimatological approach: a case study of slender and slow lorises. *AJP*, *72*, 877–886.

Nekaris, K.A.I., et al. (2013). An ethnoprimatological approach to assessing levels of tolerance between human and commensal non-human primates in Sri Lanka. *Journal of Anthropological Sciences*, *91*, 1–14.

Nellemen, C., & Newton, A. (Eds.) (2002). The great apes—the road ahead. Nairobi, Kenya: UNEP report. Available as pdf: http://www.globio.info/downloads/249/Great+Apes+-+The+Road+Ahead (last accessed 8.4.15).

Nijman, V., & Nekaris, K.A.I. (2014). Trade in wildlife in Bali, Indonesia, for medicinal and decorative purposes. *TRAFFIC Bulletin*, *26*, 31–36.

Nijman, V., et al. (2011). Primate conservation: measuring and mitigating trade in primates. *Endangered Species Research*, *13*, 159–161.

Oliveira, L.C., et al. (2009). Key tree species for the golden-headed lion tamarin and implications for shade-cocoa management in southern Bahia, Brazil. *Animal Conservation*, *13*, 60–70.

Paciulli, L.M. (2009). The relationship between nonhuman primate densities and vegetation on the Pagai, Mentawai Islands, Indonesia. In S. Gursky-Doyen & J. Supriatna (Eds.), *Indonesian primates* (pp. 199–215). New York: Springer.

Plumptre, A.J., et al. (2003). The current status of gorillas and threats to their existence at the beginning of a new millennium. In A.B. Taylor & M.L. Goldsmith (Eds.), *Gorilla biology: a multidisciplinary perspective* (pp. 414–431). New York: CUP.

Pozo-Montuy, G., et al. (2011). Influence of the landscape matrix on the abundance of arboreal primates in fragmented landscapes. *Primates*, *52*, 139–147.

Rijksen, H.D. (2001). The orangutan and the conservation battle in Indonesia. In B.B. Beck et al. (Eds.), *Great apes and humans: the ethics of coexistence* (pp. 57–70). Washington DC: Smithsonian Institution Press.

Riley, E.P., & Fuentes, A. (2010). Conserving social–ecological systems in Indonesia: Human–Nonhuman primate interconnections in Bali and Sulawesi. *AJP*, *71*, 1–13.

Riley, E.P., & Priston, N.E.C. (2010). Macaques in farms and folklore: exploring the human–nonhuman primate interface in Sulawesi, Indonesia. *AJP*, *72*, 848–854.

Robertson, J.M.Y., & van Schaik, C.P. (2001). Causal factors underlying the dramatic decline of the Sumatran orang-utan. *Oryx*, *35*, 26–38.

Russon, A.E., et al. (2001). The population and distribution of orangutans (*Pongo pygmaeus pygmaeus)* in and around the Danau Sentarum Wildlife Reserve, W. Kalimantan, Indonesia. *Biological Conservation*, *97*, 21–28.

Sauther, M.L., & Cuozzo, F.P. (2009). The impact of fallback foods on wild ring-tailed lemur biology: a comparison of intact and anthropogenically disturbed habitat. *AJPA*, *140*, 671–686.

Schulte-Herbruggen, B., et al. (2013). The importance of bushmeat in the livelihoods of West African cash-crop farmers living in a faunally-depleted landscape. *PLONE*, *8*, e72807.

Schwitzer, C., et al. (2011). Responses of animals to habitat alteration: an overview focussing on primates. *Endangered Species Research*, *14*, 31–38.

Smith, D. (2009). Madagascar's lemurs in danger from political turmoil and "timber mafia". *The Guardian*, http://www.theguardian.com/environment/2009/nov/17/madagascar-lemurs-conservation-forests-extinction (last accessed 8.10.15).

Srivastava, A., et al. (2001). Primate population decline due to habitat loss: Borajan Reserve Forest of Assam, India. *Primates*, *42*, 401–406.

Strier, K.B., & Ives. A.R. (2012). Unexpected demography in the recovery of an endangered primate population. *PLONE*, *7*, e44407.

Travis, D.A., et al. (2006). Disease risk analysis: a tool for primate conservation planning and decision making. *AJP*, *68*, 855–867.

Trayford, H.R., & Farmer, K.H. (2012). An assessment of the use of telemetry for primate reintroductions. *Journal for Nature Conservation*, *20*, 311–325.

Tutin, C.E.G., & Vedder, A. (2001). Gorilla conservation and research in central Africa. In W. Weber et al. (Eds.), *African rain forest ecology and conservation* (pp. 429–448). New Haven, CT: Yale University Press.

van Schaik, C.P., et al. (2001). Dramatic decline in orang-utan numbers in the Leuser Ecosystem, northern Sumatra. *Oryx*, *35*, 14–25.

Wallis, J., & Lee, D.R. (1999). Primate conservation: the prevention of disease transmission. *International Journal of Primatology*, *20*, 803–826.

Wallis, J. and Lonsdorf, E.V. 2010. A summary of recommendations for primate conservation education. *AJP*, *72*, 441–444.

Walsh, P.D., et al. (2003). Catastrophic ape decline in western equatorial Africa. *Nature*, *442*, 611–614.

White, L.J.T., & Tutin, C.E.G. (2001). Why chimpanzees and gorillas respond differently to logging. In W. Weber et al. (Eds.), *African rain forest ecology and conservation* (pp. 449–462). New Haven, CT: Yale University Press.

Whitehead, H. (2010). Conserving and managing animals that learn socially and share cultures. *Learning & Behavior*, *38*, 329–336.

Whittaker, D.J., & Lappan, S. (2009). The diversity of small apes and the importance of population-level studies. In S. Lappan & D.J. Whittaker (Eds.), *The gibbons* (pp. 3–10). New York: Springer.

Wich, S.A., et al. (2008). Distribution and conservation status of the orang-utan (*Pongo spp.*) on Borneo and Sumatra: how many remain? *Oryx*, *42*, 329–339.

Wiederholt, R., & Post, E. (2009). Tropical warming and the dynamics of endangered primates. *BL*, doi:10.1098/rsbl.2009.0710.

Woodford, M.H. et al. (2002). Habituating the great apes: the disease risks. *Oryx*, *36*, 153–160.

Yimkao, P., & Srikosamatara, S. (2006). Ecology and site-based conservation of the white-handed gibbon (*Hylobates lar* L.) in human-use forests in Mae Hong Son province, northern Thailand. *Natural History Bulletin of the Siam Society*, *54*, 109–138.

Young, H., et al. (2013). Does habitat disturbance increase infectious disease risk for primates? *Ecology Letters*, *16*, 656–663.

Glossary

abstract conceptual representation use of general criteria to categorize objects
adaptation controversial term that has been defined as a trait that facilitates survival and/or reproduction, or as the evolutionary process that produces such a trait
additional references sources in this work that serve to augment the Basic Bibliographies (see entry), represent divergent opinions on controversial issues, or provide interesting sidelights
ad libitum sampling recording behavior without a predetermined scheme
agonistic involving aggressive behavior and/or submissive or fearful behavior
alarm calls vocalizations that warn conspecifics about predators and/or notify predators that they have been seen
allocare infant care by any individual other than the mother
Amazonia the basins of the Amazon and Orinoco rivers in South America
analogical model heuristic comparison of two taxa based on phylogenetic closeness or ecological similarity
ancestral suite a set of ancestral behaviors that might have been linked across taxa by homology or by other shared traits, such as cognitive abilities or ecological factors
ancestral traits shared biological traits in a clade that have undergone little evolutionary change (also called **primitive**, used as a technical term)
anecdote description of a unique or unusual event
anonymous mimetism system of social interaction in which an individual imitates the actions of a larger group, regardless of affiliations
anthropogenic caused by humans
anthropomorphism interpretation of animal behavior as if the subjects were human
arboreal active primarily in trees
arboreal clambering primate locomotion using a variety of non-stereotyped behaviors to negotiate a complex and fragile environment
attentional mechanisms proximate mechanisms that make certain stimuli more likely to be noticed than others
attractivity female signals that stimulate sexual interest from the male
attribute analysis explanation of an ancestral trait by analogy with an extant taxon; i.e. use of known functional explanations for an extant trait provide a model
baboon a term often applied to any of several papionin genera that are not closely related; more specifically, monkeys of the genus *Papio*
baculum penis bone
bais marshy clearings in West Africa that facilitate observation of animals

basal metabolic rate (BMR) an adult's minimal rate of energy expenditure when resting and not directly affected by factors such as digestion or stress
basic bibliographies references in this volume that are general works providing fundamental information and concepts agreed upon by most experts
behavior reading taking cues from the behavior of others without understanding the underlying mental states
behavioral phylogenetic approach reconstruction of ancestral behavior based on parsimonious assumptions about natural selection and about resemblances between ancestor and descendant species
bimanual role differentiation simultaneous performance of different but complementary actions with hands
binocular vision fixing both eyes on one thing simultaneously (making depth perception possible)
biparental bond relationship between a male and a female focused on child rearing
bipedalism regular use of upright two-legged locomotion
brow raise threat display that exposes eyelids of a lighter color than the rest of the integument
cache and carry parking an infant at several different sites during a night so the mother can be near
callitrich informal and taxonomically noncommittal term for the taxon that contains marmosets and tamarins
canopy continuous layer of leaves in a rainforest
capital breeders females that rely on previously stored fat during reproduction (cf. income breeders)
captive study research on animals that are to some degree restrained by humans
carnivory consumption of animals with the connotation of meat (vertebrate muscle tissue)
cathemerality the ability to be active for significant periods at any time of the 24-hour cycle
cheek pouches flexible areas of tissue in the lining of cercopithecine cheeks that allow food to be accumulated
chest beating unique gorilla display in which cupped hands are slapped against the chest in rapid alternation
circum-oral gland gland peripheral to the mouth in tarsiers
clade a monophyletic ancestor–descendant group
clan social unit composed of affiliated one-male units in hamadryas baboons
clasping various intimate body contacts such as grasping and hugging that are relatively frequent in tolerant macaques
classification ordering living things into groups on the basis of relatedness
coda male gibbon vocalization in reply to a female's great call
cognitive style distinctive ways in which different species may approach the same problems, according to their natural adaptations
colobus informal term for several genera of African colobines, which were at one time all placed in the genus *Colobus*
common baboons numerous and geographically extensive *Papio* subtaxa that typically live in troops
community a society with multiple adults (usually applied to fission–fusion societies)
comparative method mapping trait variation among living species onto their known phylogenetic relationships to reconstruct evolution of the trait
competing hypotheses alternative explanations for the same phenomenon

conceptual model theoretical representation of a phenomenon consisting of analytically specified relationships within sets of concepts or variables (see strategic modeling)
conservation units taxa that have been ranked according to degree of threat and implications of extinction
contest competition direct conflict over use of a resource
copulatory plug congealed sperm left in the vaginal tract of a mate to impede fertile mating by other males
core knowledge hypothesized set of mental systems dedicated to learning various adaptive behaviors (cf. domain and module)
correlational model statistical description of the relationship between two variables across a taxon by means of a regression equation
CP_3 complex catarrhine dental arrangement in which long upper canines are honed against the lower premolars behind them
crèches infant play groups in patas monkeys that are supervised by a single adult or juvenile female while the mothers are away
crepuscular active at dawn and/or dusk
cross fostering giving an experimental subject a foster mother of a nonprimate species to achieve learning deprivation
crypsis ability to avoid notice by other organisms
cultural primatology the study of primate traditions, i.e. learned behavioral patterns that are common in a society and transmitted across generations
dental formula numerical representation of the types of teeth on one side of the mouth, upper and lower
deprivation experiment research in which the opportunity to learn a behavior is taken away from the subjects
dermatoglyphics friction surfaces on the skin of appendages, or the study of them
descriptive statistics representation of behavioral patterns in data, usually as frequencies
despotic societal pattern of emphasis on hostile and competitive behaviors
development ontogeny before maturity
dichromatic vision limited color vision, comparable to red–green color blindness in humans
dispersal emigration from the natal group
dispersed social system society in which members usually move separately but maintain contact by vocal and olfactory communication
dispersion distribution of individuals in space
distributed cognition mental processes associated with interaction among conspecific individuals
distributed leadership decision-making shared by a whole group or by all members of a subgroup (also called democratic or egalitarian)
domain hypothesized set of mental processes with a limited area of functioning (cf. module)
dominance relationship in which an individual habitually defers to another
dormancy a state in which body temperature is close to ambient temperature and metabolic rate can be extremely low
ecotourism guided animal watching that attempts to minimize damage to the ecosystem while obtaining support for conservation
embodied cognition mental processes associated with integration of brain, body, and environment
emergents widely-spaced trees rising above others in a rainforest
emotional contagion process by which one party is affected by the emotional state of another

enhancement learning learning what can be used to solve a problem without learning how to use it
episodic memory recollection of a personal event that combines where, what, and when information
estrus the time during an ovarian cycle when a female is especially receptive
estrus cycles ovarian cycles in which sexual behavior is tightly regulated by hormonal changes
ethnoprimatology studies of the interface between primate and human behavior, usually in service to conservation
evolution heritable changes in biological features of a group of organisms across generations, resulting from shifts in the gene pool
explicit reconciliation reconciliation effected by conspicuous behaviors that rarely occur in other contexts
exploratory behavior investigation of novel objects and places
exudates substances such as sap, secreted by trees and some other plants
fallback foods foods eaten when preferred foods are scarce
faunivory consumption of any kind of animal
female dilemma need of females in multi-male societies to simultaneously assure and confuse various males about the probability of paternity
female dominance social pattern in which females are typically dominant over males
flexion bending of limbs at the joints
floaters males who have no permanent home ranges and may travel widely
focal animal sampling recording all activities of a single individual over a predetermined period of time
folivorous specializing in leaf consumption to some degree
follower a relatively young male who accompanies a one-male unit as a strategy for obtaining one or more mates
forced mating coercive mating instigated by a male
forest floor gleaners description applied to mandrills because they get so much food from the ground
forest fragments small remnants of natural forests
fossa *Cryptoprocta ferox*, the top native predator of Madagascar; a viverrid (related to civets and genets)
fossils physical remains of living things preserved from the past by some transformation, e.g. mineralization
fovea specialized section of the retina with closely packed light-sensitive cells, found in tarsioids and anthropoids
free riders individuals benefitting from others' behavior though they never reciprocate
function how a trait works and what the results are
gaze following the act of following another individual's direction of attention
gene pool totality of genes circulating among members of a population from one generation to the next
general study gathering data on a variety of observable behaviors
genomics molecular approach that studies the complete DNA sequence of an organism
geophagy consumption of soils and/or clays
grade a notional intermediate stage between two taxa; not a valid taxon in itself
great call female gibbon vocalization composed of a highly rhythmic series of long notes with increasing tempo and/or peak frequency

grooming claws elongated and curved nails standing at an angle to the dorsal surfaces of second toes in strepsirrhines
grunting deep rhythmic sound that conveys various positive meanings in *Papio* species
gums exudates favored by primates, found immediately under tree bark
habitat corridors pathways between habitat fragments that facilitate dispersal and genetic interchange
habitat fragmentation reduction of an extensive environmental area to small, separated remnants
habitat loss complete elimination of a population's environment
habitat modification anthropogenic changes in the environment that may or may not mimic natural dynamics
habituation process in which animals become sufficiently accustomed to researcher presence to allow close observation
hallux the first toe of any primate (the "big toe" in humans)
harvesting capture and removal of live animals from the wild
heterothermy the ability to make substantial adjustments in body temperature
hibernation a long period of dormancy, up to seven months in some strepsirrhines
hindlimb dominance weight distribution in which the load on the rear limbs is equal to or greater than the load on the front limbs
hominins clade that produced the human species, including all fossil forms more closely related to us than to any of the apes
horde unique kind of foraging group found among some mandrills that is relatively stable and cohesive despite containing many hundreds of individuals
hypothesis testing use of various procedures to decide whether a postulated relationship or explanation is probable beyond the level of chance
implicit reconciliation reconciliation effected by affiliative behaviors that regularly occur in other contexts
inbreeding avoidance inhibition of reproduction with close genetic relatives
income breeders females that use the maximum amount of available resources during reproduction rather than relying on previously stored fat (cf. capital breeders)
infancy the period between birth and weaning
inferential statistics mathematical tools used to determine the probability that a described relationship is not due to chance
innate under a significant degree of genetic influence
ischial callosities sitting pads attached to bones in the pelvic girdle of Old World monkeys
instantaneous scan survey and recording of all sample subjects' behavior at regular intervals
inter-birth interval (IBI) amount of time between successive births from a given female; sum of infant care period and following pregnancy
isosexual behavior genital stimulation between individuals of the same sex
Jarman/Bell principle metabolic rate increases in proportion to body mass, but more slowly
juvenile individual independent of the mother but infertile and short of full growth
Kay's Threshold constraint of body size by insectivory to about 500g
kin bias intensity degrees of preferential treatment for relatives
kin recognition ability to distinguish genetic relatives from others
kin selection process of natural selection that favors positive behavior or tolerance toward genetic relatives
knuckle walkers misnomer for unique locomotion of African apes in which weight on the forearms is supported by the middle phalanges of the fingers

landscape supplementation augmentation of diet by animals living in forest fragments by accessing separate sources such as other fragments or isolated trees
Last Common Ancestor (LCA) most recent ancestor shared with any nonhuman primate, i.e. the common ancestor of *Pan* and *Homo*
locomotion any movement from one place to another
lunar philia more active during full moons
male greeting ritual complex *Papio* interaction that involves exchange of various signals of affiliation or appeasement
mast fruitings irregular Southeast Asian cycle in which the majority of trees in a forest bear fruit simultaneously, usually followed by a period of scarcity
mate choice behavior that increases the likelihood of copulating with certain individuals rather than others
mate guarding trying to prevent other males from copulating with a female mate
matrilines groups of individuals related through females
matrilocal cluster association of females marked by spatial proximity and friendly or tolerant behavior
maturation developmental process that makes some behavior patterns fully operative
mental time travel the ability to anticipate future events
metabolism an organism's extraction of energy from the environment and application to life processes
metacognition a strong form of representation in which mental images can be called to mind in the absence of sensorimotor cues
mind reading synonym for theory of mind
mobbing approaching a potential predator while emitting frequent loud calls and engaging in vigorous physical activities
modal action pattern behavior that has a relatively stereotypical form in a species
module hypothesized set of mental processes with a limited area of functioning (cf. domain)
molecular primatology the comparative study of biomolecules, such as DNA, in primates
monophyly criterion for a taxon in which all members derive from one common ancestor
Morgan's Canon principle that "higher mental processes" should not be attributed to nonhuman animals without a strong foundation
multi-level society a society organized into two or more tiers based on the aggregation of one-male units
mutually exclusive relationship between hypotheses such that acceptance of one precludes the other
mutualism actions that benefit participants without active cooperation
natural selection evolutionary process in which genes become more or less common because the traits they influence affect reproductive success
neonates newborn infants
niches different ways of relating to a given environment
nonpursuit hunting quick capture after waiting for prey to approach
noyau community in which one male occupies and defends a territory that encompasses the ranges of several females
nulliparous not having had first birth
ontogeny growth and change of an individual from conception to death
opportunistic omnivores animals that eat most of the digestible foods in their environment
opposable thumb thumb with the flexibility to touch its pulp to the pulp of another finger
orbits bony sockets that hold the eyes

order a group of animals with a common ancestor dated to tens of millions of years
ovarian cycles female physiological cycles in which fertile eggs become available
pair long-term relationship between a particular adult male and female
pair-group social unit clearly centered on a single pair that may contain one or more peripheral adults
parietal cortex part of the brain that processes sensory information such as taste, temperature, and touch
parking leaving an infant in a specific location while foraging
patchy distribution irregular location of food clumps
paternity concentration female behavior that gives a male mate a high level of paternal certainty
patience ability to suppress immediate motivations and defer an action
pendular brachiation swinging from hand to hand while maintaining continuous contact with the superstrate
pentadactyly five digits on each hand and foot
percussive foraging tapping a structure to locate hidden food items
peri-ovulatory period the time around ovulation
perspective taking matching the mental processes of another individual; detailed understanding of another's situation
phenotypic gambit assumption that evidence for immediate function implies long-term effect on fitness
philopatry tendency of one sex to remain in the natal group or area
phylogenetic signals similarities between species that are hypothesized to arise from their common ancestry
phylogeny ancestor–descendant relationships; or the pattern of such relationships in a particular clade
play behavior with no immediate practical result that seems to be pleasurable
postorbital septum bone that completes the cup behind the eyeball
postpartum estrus fertile estrus that occurs while a female is still nursing an infant
precision handling gripping an object with the thumb opposing the fingertips
predation rate frequency of successful attacks on a group or population
predation risk rate of unsuccessful attacks or other encounters between a group or population and its predators
preferred foods foods eaten when there is a choice
prehensile tail tail with a tip that can grasp somewhat like a hand
Primate Specialist Group subgroup of the IUCN that assesses threats to primates
primiparous having given birth once
proceptivity female behavior that invites sexual approaches
prognathous protruding (part of face)
provisioning regular offering of food to animals to bring them closer to researchers
proximate mechanisms immediate causal antecedents of particular behaviors in individuals
quadratic interaction exchange of behaviors involving four individuals
quadrupedal walking on four limbs on a relatively level surface
quadrumanous scrambling bonobo arboreal pattern of using all four appendages opportunistically without any particular gait
queue jumping deviations from the Priority-of-Access Model for reproductive competition
rainforest complex ecological community with vegetational layers
receptivity female readiness (physical and behavioral) to accept male sexual attempts

reconciliation affiliative behavior between opponents after a conflict
Red List IUCN assessment of the status of threatened species
referential model representation that links one real phenomenon to another that is more difficult to study
reintroduction release of new individuals or groups into a wild habitat
reverse contingency task cognitive test in which the subject chooses between two quantities of food and receives the array that was *not* chosen
rhinarium hairless, moist area around the nostrils that takes scent molecules from the air
ricochetal brachiation swinging from hand to hand with aerial phases between contacts with the superstrate
roaming departure from the usual home range to travel widely in search of sexual partners
routine trichromacy ubiquitous occurrence across a taxon of full color vision equally in both sexes
sampling observing a number of individuals who accurately represent a total study population
scramble competition obtaining as much of a limited resource as possible without engaging in direct conflict
scream-embrace special display in reunions among male capuchin monkeys
scrounging snatching bits of food from the mother's hand or mouth by an infant
secondary compounds chemicals secreted by plants to protect themselves from animals
secondary representations mental images reflecting external situations (representations) that can be separated from reality in order to model hypothetical situations
seed predator animal that digests seeds for nutrition rather than passing them through the alimentary tract intact
self-conception an animal's mental representation of itself as having perceptual and causal abilities
self-directed behavior behavior in which the performer is also the object, such as scratching
sensitive period time span during a life when a particular behavior is readily learned
serial processing mental activity that supports manipulation of the physical environment
sexual conflict conflicting reproductive strategies of male and female
sexual dichromatism differentiation of the sexes in a species by color(s) and/or color pattern
sexual pair exclusive copulatory relationship between two adults within a group that may or may not include other adults
shared derived characters common traits that distinguish a clade because they have changed from the condition of earlier ancestors
silverback an adult male gorilla at a stage in life where the hair on his back is silvery gray
single prolonged intromission (SPI) continuous genital contact with a duration from four minutes to four hours
social brain theory of relationship between brain structure/function and various aspects of social life
social diagnosis comparative generalization about social behavior across a taxon (cf. ancestral suite)
social organization characteristics of a society, such as sexual composition and cohesion
social pair association between two non-kin adults characterized by spatial proximity and affiliative behaviors and/or energetic investment in offspring or territorial defense
society a set of conspecific animals that regularly interact more with each other than with other conspecifics
solitary species animals in which adults are usually alone; applies to the nightly foraging period of many strepsirrhines

special friendship an adult male–female relationship signified by elevated rates of proximity and grooming, and in some cases by male attention to the female's infant

specific intelligence cognitive mechanisms directed toward a distinct and limited set of functions

sperm competition differential probability of conception for two males who copulate with the same female within a short time

stimulus generalization association of objects based on perceptual features or cues

strata vegetational layers of a rainforest

strategic modeling method for producing conceptual model based on the gene as the unit of selection, and depiction of animals as strategists promoting inclusive fitness

strategy shorthand way of referring to functions of a particular kind of behavior

structurally defended resources food items with hard or dangerous casings

sustainable utilization permitted human exploitation of animals and habitats at levels that do not result in irreversible damage

sympathetic concern attempting to understand the cause of another's emotion

sympatric living in the same natural habitat

systems model functional representation of behavioral ecology in a group smaller than a species, constructed by means of multivariate regression equations

tannins common kind of secondary compound that interferes with protein uptake

tapetum lucidum layer of tissue at the rear of the strepsirrhine retina that reflects light back toward the pupil, a recycling effect that enhances night vision

targeted aggression repeated severe attacks on a particular individual by one or more others without provocation

taxonomy the study and practice of classification; creation of formal groups called **taxa** (singular **taxon**)

telemetry automatic measurement and transmission of data from remote sources to receiving stations

theory of mind an individual's ability to understand the perceptions and cognition of another individual

tolerance societal pattern of relatively relaxed social relationships, especially between subordinates and dominant individuals (also called **egalitarian**)

toothcomb dental structure in the lower jaw of a strepsirrhine, composed of protruding incisors and canines

torpor temporary slowing of metabolism to save energy

translocation transport of some animals from one habitat to another to save the transported ones or to refresh a threatened population

trichromacy ability to see colors in the red–green range as well as the blue–yellow range due to three kinds of cones in the retina

troop a society with multiple adults

ultimate causes conditions thought to account for the origins of particular behaviors

understory lowest stratum of a rainforest

urine marking communication by depositing urine at selected places in the environment

urine washing urination by an individual on its own hands and feet, sometimes followed by rubbing; seems to be associated with aggressive or stressful situations

vertical clinging and leaping (VCL) positional pattern in which an animal rests in a vertical position on a vertical substrate and jumps from one such substrate to another by pushing off and landing on the hindlimbs

viable offspring offspring that can reproduce

vibrissae "whiskers" at the sides of a face that provide sensory information
vigilance special visual attention monitoring for predators and/or competitors
vomeronasal organ auxiliary olfactory apparatus at the base of the nasal cavity
vulnerability likelihood that a particular individual or kind of individual will be taken by a predator
woodlands areas with varying proportions of tree and grass cover
World Conservation Union international organization devoted to protection of threatened species, often designated by its French acronym **IUCN**
yawn threat ritualized signal in the "baboon" genera that displays the large male canines

Index

Page numbers in italic indicate figures and those followed by a letter g indicate glossary items.

For Product Safety Concerns and Information please contact our EU representative GPSR@taylorandfrancis.com Taylor & Francis Verlag GmbH, Kaufingerstraße 24, 80331 München, Germany

Batch number: 10726434

Printed by Printforce, the Netherlands